MOTORS

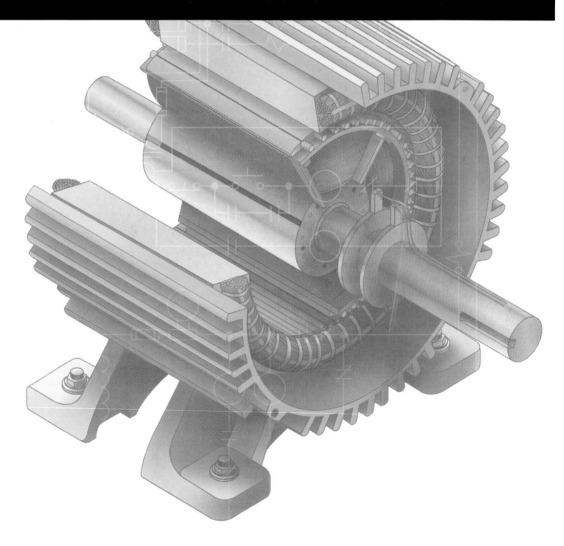

AMERICAN TECHNICAL PUBLISHERS, INC.
HOMEWOOD, ILLINOIS 60430-4600

IN PARTNERSHIP WITH NJATC

Motors contains procedures commonly practiced in industry and the trade. Specific procedures vary with each task and must be performed by a qualified person. For maximum safety, always refer to specific manufacturer recommendations, insurance regulations, specific job site and plant procedures, applicable federal, state, and local regulations, and any authority having jurisdiction. The material contained herein is intended to be an educational resource for the user. Neither American Technical Publishers, nor the National Joint Apprenticeship & Training Committee for the Electrical Industry is liable for any claims, losses, or damages, including property damage or personal injury incurred by reliance on this information.

American Technical Publishers, Inc., Editorial Staff

Editor in Chief:
    Jonathan F. Gosse
Vice President—Production:
    Peter A. Zurlis
Art Manager:
    James M. Clarke
Technical Editor:
    Eric F. Borreson
Copy Editors:
    Valerie A. Deisinger
    Diane J. Weidner
Cover Design:
    Mark S. Maxwell
Illustration/Layout:
    Mark S. Maxwell
    Thomas E. Zabinski
    Samuel T. Tucker
    William J. Sinclair
    Nicole S. Polak
Multimedia Coordinator:
    Carl R. Hansen
CD-ROM Development:
    Robert E. Stickley
    Gretje Dahl

Adobe, Acrobat, and Reader are either registered trademarks or trademarks of Adobe Systems Incorporated in the United States and/or other countries. Quick Quiz and Quick Quizzes are registered trademarks of American Technical Publishers, Inc. Intel and Pentium are registered trademarks of Intel Corporation or its subsidiaries in the United States and other countries. Microsoft, Windows Vista, Windows XP, Windows 2000, Windows NT, and Internet Explorer are either registered trademarks or trademarks of Microsoft Corporation in the United States and/or other countries. Netscape is a registered trademark of Netscape Communications Corporation in the United States and other countries. Boston Gear is a registered trademark of Boston Gear. Glastic is a registered trademark of The Glastic Corporation. Megger is a registered trademark of Megger Limited. National Electric Code and NEC are registered trademarks of the National Fire Protection Association, Inc, Quincy, MA 02269.

© 2008 by the National Joint Apprenticeship & Training Committee for the Electrical Industry and American Technical Publishers, Inc.
All rights reserved

1 2 3 4 5 6 7 8 9 – 08 – 9 8 7 6 5 4 3 2 1

Printed in the United States of America

    ISBN 978-0-8269-1975-5

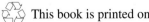 This book is printed on 10% recycled paper.

Acknowledgments

In loving memory of

Stan Klein

Technical information and assistance was provided by the following companies, organizations, and individuals:

ABB Inc., Drives and Power Electronics
Alemite Corp.
ASI Robicon
Baldor Electric Co.
Boston Gear Works Corporation
Cone Drive Operations Inc./Subsidiary of Textron Inc.
Electrical Apparatus Service Association, Inc.
Emerson Power Transmission
Englehardt Gear Co.
Furnas Electric Co.
GE Motors and Industrial Systems
Heidelberg Harris, Inc.
Kebco Power Transmission
Leeson Electric Co.
NTN Bearing Corporation of America
Pacific Bearing Company
Rockwell Automation, Allen-Bradley Company, Inc.
Rockwell Automation/Reliance Electric
Ruud Lighting
The Timken Company

NJATC Acknowledgments
Technical Editor
    William R. Ball, NJATC Staff

Technical Reviewers
    Jim Dewig; Evansville Electrical JATC
    Curtis Duncan; North Alabama Electrical JATC
    Mike Flynn; Hampden Engineering Corp.
    Bob Kosky, Ann Arbor Electrical JATC
    Chris MacCreey; Battle Creek Electrical JATC
    Ron Michaelis; South Bend Electrical JATC
    Tony Naylor; Wichita Electrical JATC
    James Orban Jr.; Flushing, Joint Industry Board for the Electrical Industry (LU 3)
    Jim Overmyer; South Bend Electrical JATC
    Gary Strouz; Houston JATC
    Brent Tyroff; South Texas Electrical JATC

Table of Contents

1 MAGNETISM AND INDUCTION — 2

Magnetism • Electromagnetism • Current Flow • Induction and Inductance
- Application–Solenoids and Motor Starters 16
- Summary, Glossary, and Review 18

2 MOTOR NAMEPLATES — 20

Motor Nameplate Data • Electrical Ratings • Operating Ratings • Mechanical-Design Codes
- Application–Applying Motor Nameplate Specifications 46
- Application–Motor Insulation Life 48
- Summary, Glossary, and Review 49

3 MOTOR PROTECTION — 52

Motor Protection Requirements • Short-Circuit and Ground Fault Protection • Motor Overload Protection
- Application–Troubleshooting Fuses 65
- Application–Troubleshooting Circuit Breakers 66
- Supplemental Topic–Overload Ambient Temperature Compensation 73
- Supplemental Topic–Motor Control Circuit Protection 74
- Application–Sizing Overload Protection 76
- Summary, Glossary, and Review 77

4 THREE-PHASE MOTORS — 80

Three-Phase Motor Construction • Operating Principles • Motor Load and Torque • Motor Power • Motor Efficiency
- Application–Motor Line and Motor Wiring Diagram 107
- Application–Voltage and Current Unbalance 108
- Summary, Glossary, and Review 109

5 INDUCTION MOTORS — 112

Induction Motor Construction • Operating Principles
- Application–Troubleshooting 3-Phase Motors 120
- Application–Troubleshooting Motor Control Circuits 121
- Summary, Glossary, and Review 122

6 WOUND-ROTOR MOTORS 124

Wound-Rotor Motor Construction • Operating Principles
- Supplemental Topic–Reactance and Phase Angle 133
- Application–Crane and Elevator Operation 134
- Application–Current-Control Resistance Switching.................................... 135
- Summary, Glossary, and Review.. 136

7 SYNCHRONOUS MOTORS 138

Synchronous Motor Construction • Operating Principles • Power Factor
- Supplemental Topic–Hunting... 159
- Application–Synchronous Motor Power Factor... 163
- Summary, Glossary, and Review.. 164

8 SINGLE-PHASE MOTORS 166

Single-Phase Motors • Shaded-Pole Motors • Split-Phase Motors • Capacitor Motors
- Supplemental Topic–Repulsion Motors 182
- Application–Centrifugal Switches... 183
- Application–Resistors... 183
- Application–Single-Phase AC Motor Terminal Designations...................... 184
- Application–Motor Winding Lead Terminations 185
- Summary, Glossary, and Review.. 186

9 AC ALTERNATORS 188

Alternator Construction • Operating Principles • Alternator Ratings • Paralleling Alternators
- Application–Selecting Portable Generators.. 202
- Summary, Glossary, and Review.. 204

10 DC MOTORS AND GENERATORS 206

DC Motors and Generators • DC Motor Types • DC Generators • DC Generator Types
- Supplemental Topic–DC Brushless Motors 229
- Supplemental Topic–DC Coreless Motors.. 230
- Application–DC Motor Reduced-Voltage Starting 231
- Application–Reversing DC Series Motors 232
- Summary, Glossary, and Review.. 233

11 STARTING 236

Motor Starting • Full-Voltage Starting • Reduced-Voltage Starting • Starting-Method Comparison
- Supplemental Topic–NEMA and IEC Ratings 257
- Supplemental Topic–Solid-State Switches 258
- Application–Wye-Delta Starting Overload Protection 259
- Summary, Glossary, and Review.. 260

Table of Contents

12 BRAKING — 262

Braking • Friction Braking • Plugging • Electric Braking • Dynamic Braking • Braking Comparison

- Supplemental Topic—Determining Brake Torque 275
- Supplemental Topic—AC Motor Dynamic Brake Sizing 276
- Application—Braking Solenoid Connections 277
- Summary, Glossary, and Review 278

13 MULTISPEED MOTORS — 280

Multispeed Motor Types • Consequent-Pole Motor Circuits

- Supplemental Topic—Schrage Motor Control Voltages 289
- Application—Two-Speed Separate-Winding Motors 290
- Application—Two-Speed Consequent-Pole Motors 291
- Summary, Glossary, and Review 292

14 ADJUSTABLE-SPEED DRIVES — 294

Adjustable-Speed Drive Components • AC Motor Speed Control • DC Motor Speed Control

- Supplemental Topic—Shunt Motor Drive with Compound-Wound Motor 314
- Supplemental Topic—DC Drive Classification 316
- Supplemental Topic—Soft Start and Soft Stop 317
- Application—Volts-per-Hertz Ratio 318
- Application—Motor Lead Lengths 319
- Summary, Glossary, and Review 320

15 BEARINGS — 322

Bearings and Loads • Bearing Installation • Bearing Operation • Bearing Removal

- Supplemental Topic—Bearing Currents 342
- Summary, Glossary, and Review 344

16 DRIVE SYSTEMS AND CLUTCHES — 348

Flexible Drives • Mechanical Drives • Clutches

Supplemental Topic—Gear Reducers 363
Supplemental Topic—Calculating V-Belt Length 364
Application—Belt Tensioning 365
Application—Selecting Motor Couplings 366
Summary, Glossary, and Review 367

17 MOTOR ALIGNMENT — 370

Shaft Alignment • Motor Placement • Dial Indicators • Alignment Methods

Application—Rim-and-Face Alignment 396
Summary, Glossary, and Review 399

18 TROUBLESHOOTING MOTORS — 402

Motor Failure • AC Motors • DC Motors • Motor Controls • Motor Lead Identification • Test Tools

Application—Electric Motor Drive Troubleshooting 425
Application—Troubleshooting Guide 443
Summary, Glossary, and Review 448

19 SPECIAL-APPLICATION MOTORS — 450

Motion Control Motors • Universal Motors • Linear Induction Motors • Rotary Phase Converters

Supplemental Topic—Stepper Motor Resolution 460
Supplemental Topic—Servomotor Encoder Resolution 462
Application—Stepper Motor Shaft Position 464
Summary, Glossary, and Review 464

CD-ROM CONTENTS

Using this CD-ROM • Quick Quizzes® • Illustrated Glossary • Flash Cards • Motor Animations • ATP eResources

ANSWER KEY 466
APPENDIX 471
GLOSSARY 483
INDEX 493

About the Authors

Mr. Otto Taylor has over 40 years of experience as an electrician, business executive, and JATC instructor. After completing his formal education at Central Vocational Tech, he started working in the service side of the electrical industry. He rose to the position of superintendent of a large company before he moved on to a position with a small motor repair facility. Under his guidance, the company rose to prominence serving the large steel mills in the Chicago area. As an executive vice president, he attended many educational seminars offered by major manufacturers in the motor and magnet industry. Serving as a JATC instructor offered a chance to share the knowledge gained over many years in this field. He continues as a motors lab instructor at the South Bend, Indiana, Electrical JATC and the NTI in Knoxville.

Mr. Jim Overmyer has over 40 years of experience as an electrician and JATC instructor. He, too, began work in the service side of the electrical industry. He completed his apprenticeship at Indiana Vocational Tech and later earned a degree in electronics from the Radio, Electronics, and Television Schools (RETS). Upon completion of his electronics degree, he started teaching for the local IBEW motor winders apprenticeship committee. His employment by a large motor repair facility gave him the opportunity to work directly on the equipment and the opportunity to attend the many seminars offered by leading companies in the field. Serving as a JATC instructor has offered him an opportunity to give back to the industry that has been so good to him. He continues teaching fourth-year curriculum at the South Bend, Indiana, Electrical JATC and at the NTI in Knoxville.

Mr. Ron Michaelis has over 35 years of experience as an electrician, electrical inspector, and JATC instructor and training director. He completed his apprenticeship at Indiana Vocational Tech and enrolled at Indiana University South Bend to continue his education. He has worked in locations ranging from the Alaskan pipeline to the oil fields in Saudi Arabia. He is now the South Bend, Indiana, Electrical JATC training director. He also serves as the electrical inspector for a large township in Michigan. His affiliation with Indiana University continues today, as students at the South Bend JATC attend classes staffed by the university.

Mr. Roger Mutti has over 35 years of experience in the construction industry as a project engineer, superintendent, university instructor, and business owner. His field experience as an IBEW member has been primarily in the high-voltage service department of an electrical contractor. He completed his apprenticeship with the South Bend, Indiana, Electrical JATC and has taught fourth-year and journeyman classes there. He holds a bachelor's degree in agriculture and a master's degree in construction management from Purdue University.

Introduction

Motors provides a comprehensive overview of motor operation, maintenance, installation, and troubleshooting. This textbook is designed to develop basic competencies in electrical apprentices and beginning learners.

Motors begins with a thorough discussion of magnets, magnetism, and electromagnetism, and how they apply to motor operation. Subsequent chapters include the latest information on the operation of many types of motors, reading and understanding motor nameplates, starting and braking, adjustable-speed drives, bearings, drive systems and clutches, and motor alignment. Installation, maintenance, and troubleshooting of motors are discussed in detail. The text presents correct safety procedures in compliance with the National Electric Code® (NEC®) and National Fire Protection Association (NFPA 70E).

Motors contains 19 chapters. At the end of each chapter, a chapter Summary, chapter Glossary, and Review help learners review key concepts and reinforce common operation, maintenance, and troubleshooting aspects of typical motor installations. Answers to odd-numbered questions are included at the end of the book. Key terms are italicized and defined in the text for additional clarity. Tech Facts, Applications, and Supplemental Topics throughout the text provide information that augments text content. An extensive Glossary and Appendix provide useful, easy-to-find information. A comprehensive Table of Contents and Index simplify navigation and make finding desired information easy.

The *Motors* CD-ROM located at the back of the book is designed as a study aid to complement information presented in the book, and includes Quick Quizzes®, an Illustrated Glossary, Flash Cards, and Motor Animations. The Quick Quizzes® provide an interactive review of topics in a chapter. The Illustrated Glossary provides a helpful reference to terms commonly used in industry. The Motor Animations are a collection of animated graphics that illustrate fundamental motor operating concepts. Clicking on the American Tech web site button (www.go2atp.com) or the American Tech logo accesses information on related electrical training products.

Features

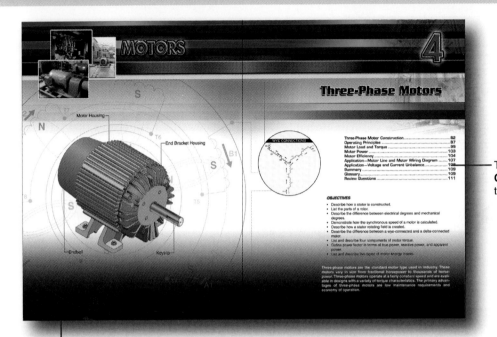

The chapter opener provides an overview of key content found in the chapter.

The chapter **Table of Contents** makes it easy to find relevant information.

Definitions are emphasized throughout to ensure understanding of important concepts.

Tech Fact

Motor windings consist of coils of insulated copper wire, insulated aluminum wire, or heavy, rigid insulated conductors. Since windings are real conductors, all windings have electrical losses while conducting power. These losses create heat that must be removed.

Tech Facts are used to provide supplemental background information of interest to electricians.

Detailed drawings illustrate the fine points of motor construction.

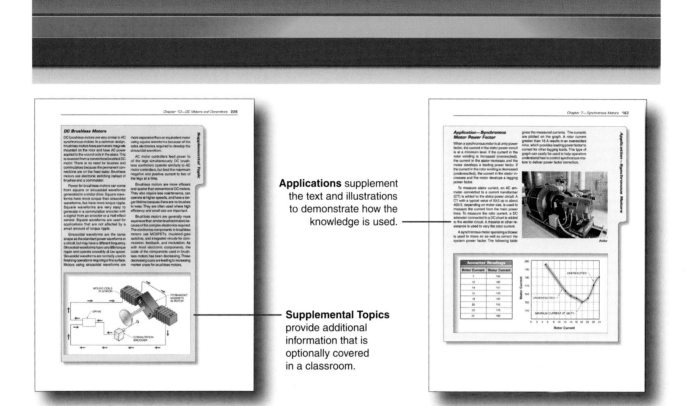

Applications supplement the text and illustrations to demonstrate how the knowledge is used.

Supplemental Topics provide additional information that is optionally covered in a classroom.

Using This CD-ROM provides information about components included on the CD-ROM.

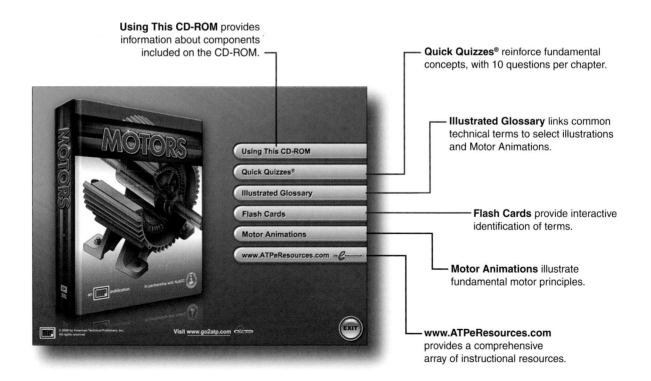

Quick Quizzes® reinforce fundamental concepts, with 10 questions per chapter.

Illustrated Glossary links common technical terms to select illustrations and Motor Animations.

Flash Cards provide interactive identification of terms.

Motor Animations illustrate fundamental motor principles.

www.ATPeResources.com provides a comprehensive array of instructional resources.

MOTORS

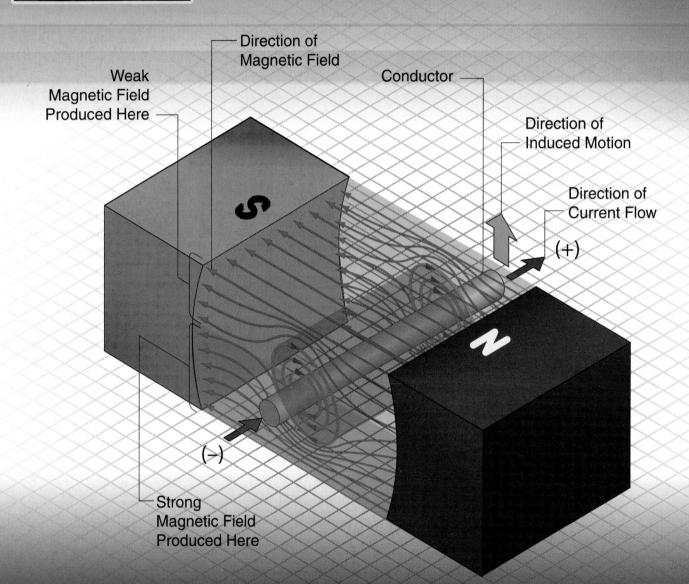

Magnetism and Induction

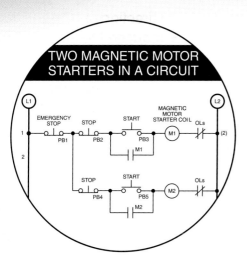

TWO MAGNETIC MOTOR STARTERS IN A CIRCUIT

Magnetism 4
Electromagnetism 7
Current Flow 9
Induction and Inductance 11
Application—Solenoids and Motor Starters 16
Summary 18
Glossary 18
Review 19

OBJECTIVES

- Describe the difference between a permanent magnet and an electromagnet.
- Describe the difference between magnetic flux and magnetic flux density.
- Describe the difference between electron current flow and conventional current flow.
- Explain how self-induction in an inductor produces inductive reactance.
- Explain how mutual induction creates current and a magnetic field in a nearby circuit.
- Explain how a loop of conductor in a magnetic field creates thrust and rotary motion.

The operation of motors is based on electromagnetism and induction. In order to understand how motors operate, it is important to understand how electricity and magnetism work together to create the forces that cause a motor to turn. Conventional current flow is used to describe the directions of current flow, magnetic fields, and thrust.

MAGNETISM

Magnetism is a force caused by a magnet that acts at a distance on other magnets. Magnetism is used to produce most of the electricity consumed, develop rotary motion in motors, and develop linear motion in solenoids.

Generators and electric motors operate using the same basic principles of magnetism. A generator uses magnetism to produce electrical energy. An electric motor uses the electrical energy to develop the electromagnetism inside the motor. Electromagnetism is used to develop a rotating mechanical force on the shaft of the motor. The strength of the magnetic forces determines the amount of force the motor's shaft can deliver.

Magnets

A *magnet* is a substance that produces a magnetic field. A *magnetic field* is a force produced by a magnet that exerts a force on moving electric charges or on other magnets. Most magnets are constructed of alloys of various magnetic materials. Three common naturally occurring magnetic metals are iron, nickel, and cobalt. Less common magnetic materials include neodymium, samarium, and other rare earth elements. Magnets are commonly used in electric motors and generators, speakers and microphones, and many industrial products.

A *permanent magnet* is a magnet that can hold its magnetism for a long period. Permanent magnets include natural magnets (magnetite) and manufactured magnets. Permanent magnets are used in electrical applications such as permanent-magnet DC motors and reed switches. Common permanent magnets are horseshoe magnets, compasses, and bar magnets. **See Figure 1-1.**

A *temporary magnet* is a magnet that retains only trace amounts of magnetism after a magnetizing force has been removed. *Retentivity* is a measure of the ability of a magnet to retain magnetism after the magnetizing force has been removed. Temporary magnets are used in most electrical applications such as motors, transformers, and solenoids. The most common temporary magnets are coils of conductor used as electromagnets.

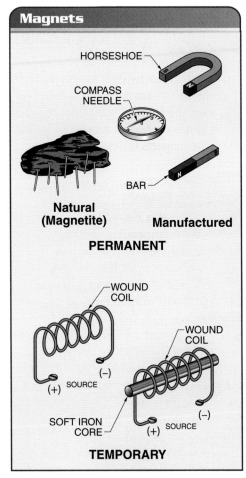

Figure 1-1. Permanent magnets include natural magnets and manufactured magnets. Temporary magnets include wound coils of wire with a source of electricity.

Magnetic Flux

Michael Faraday proposed using magnetic flux to visualize a magnetic field. *Magnetic flux,* or *field flux,* is the invisible lines of force that make up the total quantity of a magnetic field. **See Figure 1-2.** *Flux density* is the amount of concentration of magnetic flux through a specific area. A more concentrated, or denser, magnetic flux makes

---

Definition

Magnetism is a force caused by a magnet that acts at a distance on other magnets.

A magnet is a substance that produces a magnetic field.

A magnetic field is a force produced by a magnet that exerts a force on moving electric charges or on other magnets.

A permanent magnet is a magnet that can hold its magnetism for a long period.

A temporary magnet is a magnet that retains only trace amounts of magnetism after a magnetizing force has been removed.

Retentivity is a measure of the ability of a magnet to retain magnetism after the magnetizing force has been removed.

Magnetic flux, or field flux, is the invisible lines of force that make up the total quantity of a magnetic field.

Flux density is the amount of concentration of magnetic flux through a specific area.

a stronger magnetic force. Magnetic flux is densest at the ends of a magnet. For this reason, the magnetic force is strongest at the ends of a magnet.

All magnets have a north (N) pole and a south (S) pole. The magnetic flux leaves the north pole and enters the south pole of a magnet. The basic law of magnetism states that unlike magnetic poles (N and S) attract each other and like magnetic poles (N and N or S and S) repel each other. The force of attraction between two magnets increases as the distance between the magnets decreases. Likewise, the force of attraction between two magnets decreases as the distance between the magnets increases.

Magnetic Polarity

The polarity of a bar magnet can be determined by suspending it from a point overhead and allowing it to turn freely. The end of the bar magnet that points north is the north pole of the magnet. Since unlike magnetic poles attract, Earth's magnetic field interacts with the bar magnet's magnetic field so that the bar magnet's north pole points toward Earth's south magnetic pole. From experience, we all know that the north pole of a magnet points toward Earth's geographic north pole. This shows that the Earth's geographic north pole is actually Earth's south magnetic pole.

A bar magnet is surrounded by its own magnetic field that exits the north pole and enters the south pole. This can be confirmed by placing a small compass at different positions around the magnet. The needle of the compass marked N aligns with the field to point to the opposite pole. **See Figure 1-3.**

Tech Fact

The sheet steel used in modern motor cores contains a sufficient amount of silicon to prevent the steel from aging. Aging is the characteristic of a magnetic circuit that decreases the permeability as the motor continues to operate over a period of time.

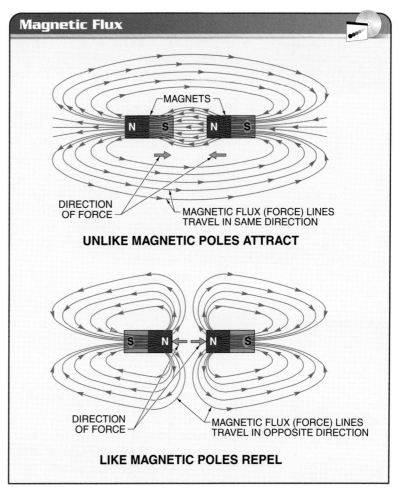

Figure 1-2. With two magnets, unlike magnetic poles attract and like magnetic poles repel.

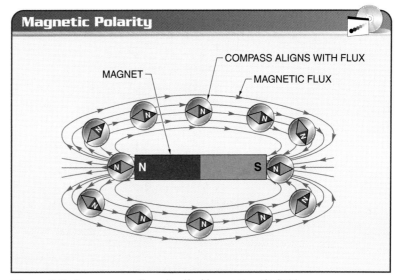

Figure 1-3. A compass aligns itself with a magnetic field.

Definition

Permeability *is a measure of the ability of a material to conduct magnetic flux.*

Molecular Theory of Magnetism

The molecular theory of magnetism states that ferromagnetic materials are made up of a very large number of molecular domains acting like molecular magnets that can be arranged in either an organized or disorganized manner. **See Figure 1-4.** A material is magnetic if it has organized molecular domains so that the individual fields add together. A material is nonmagnetic if it has disorganized molecular domains so that the individual fields cancel each other. The individual magnetic domains consist of many atoms or molecules that are individually aligned so their electrons all spin in the same direction. The moving charges of the electrons create a magnetic field.

The molecular theory of magnetism explains how certain materials used in control devices react to magnetic fields. For example, it explains why hard steel is used for permanent magnets, while soft iron is used for the temporary magnets found in control devices. Permanent magnets can be manufactured by using another magnetic field to align the magnetic domains during manufacture. Hard steel is difficult to magnetize and demagnetize, making it a good permanent magnet. The dense molecular structure of hard steel does not easily disorganize once a magnetizing force has been removed. Hard steel has high retentivity. However, any permanent magnet can be demagnetized by a sharp blow or by heat that causes the molecular arrangement to become disorganized. Vibrations over an extended time may also demagnetize a permanent magnet.

Magnetic domains in temporary magnets are aligned by a magnetic field created by electric current flowing through a coil. Soft iron is ideal for use as a temporary magnet in control devices because it does not retain residual magnetism very easily. By not retaining residual magnetism, a temporary magnet can be turned off while de-energized, such as a coil in a magnetic starter.

Permeability

Permeability is a measure of the ability of a material to conduct magnetic flux. Permeability can be compared to specific conductivity in an electric circuit. A material or device with a high specific conductivity can conduct more current than a material or device with a low conductivity. Similarly, a material with a high permeability can conduct more magnetic flux than a material with a low permeability.

Permeability is usually described relative to the permeability of a perfect vacuum. For example, if the permeability of a nickel alloy is 500 times the permeability of vacuum, the relative permeability of the alloy is 500. **See Figure 1-5.** Materials are often described as being ferromagnetic, paramagnetic, or diamagnetic materials.

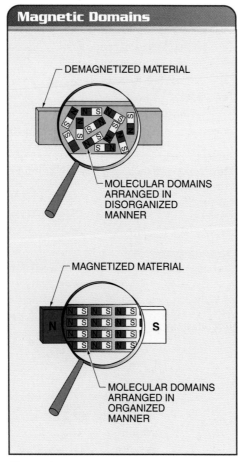

Figure 1-4. The molecular theory of magnetism states that ferromagnetic materials are made up of a very large number of molecular domains that can be arranged in either an organized or disorganized manner.

| Typical Permeability Values ||
Material	Relative Permeability
Copper	0.999991
Vacuum	1
Air	1.0000004
Aluminum	1.00002
Nickel	400 to 1000
Ferrite	2300 to 5000
Silicon steel	5000 to 10,000
Pure iron	6000 to 8000
Supermalloy	800,000

Figure 1-5. Permeability values refer to the relative ability of a material to conduct magnetic flux.

Ferromagnetic Materials. A *ferromagnetic material* is a material that is easily magnetized and has high permeability. All materials have their relative permeability near 1 except ferromagnetic materials. Iron is relatively inexpensive and has a much higher permeability than air or copper. In addition, *ferrite* is a ferromagnetic nonconducting ceramic alloy with a high permeability. Ferromagnetic materials are often used as the cores of coils in motors and transformers to concentrate the magnetic fields.

Paramagnetic Materials. A *paramagnetic material* is a material that can be weakly magnetized in the same direction as the applied magnetic field. Paramagnetic materials have a permeability slightly more than 1 and do not retain any magnetism in the absence of a magnetic field. This means that paramagnetic materials are weakly affected by a magnetic field, but cannot become magnets. A common paramagnetic material is aluminum. When aluminum is used as the core of an inductor, it has a very weak ability to concentrate the magnetic field.

Diamagnetic Materials. A *diamagnetic material* is a material that can be very weakly magnetized in the opposite direction as the applied magnetic field. Paramagnetic materials have a permeability slightly less than 1 and do not retain any magnetism in the absence of a magnetic field. This means that diamagnetic materials are weakly affected by a magnetic field, but cannot become magnets. A common diamagnetic material is copper. Diamagnetic materials used as the cores of coils have a very weak ability to weaken a magnetic field. When copper is used as the core of an inductor, it has a very weak ability to disperse the magnetic field.

ELECTROMAGNETISM

A magnetic field is produced any time electricity passes through a conductor (wire). An *electromagnet* is a temporary magnet produced when electricity passes through a conductor, such as a coil, that concentrates the magnetic field. *Electromagnetism* is the temporary magnetic field produced when electricity passes through a conductor. Electromagnetism is a temporary magnetic force because the magnetic field is present only as long as current flows. The magnetic field is reduced to zero when the current flow stops.

The magnetic field around a straight conductor is relatively weak and must be concentrated for use in an electric motor. **See Figure 1-6.** The strength of the magnetic field is increased by wrapping the conductor into a coil, increasing the amount of current flowing through the conductor, or wrapping the conductor around an iron core. A strong, concentrated magnetic field is developed when a conductor is wrapped into a coil. The strength of a magnetic field is directly proportional to the number of turns in the coil and the amount of current flowing through the conductor. An iron core increases the strength of the magnetic field by concentrating the field.

> **Tech Fact**
>
> Reluctance is the ability of a magnetic circuit to block magnetic flux, similar to how resistance blocks current. Permeability refers to the ability to conduct magnetic flux. Pure iron and silicon steel are often used for the core material of motors because of their high permeability.

> **Definition**
>
> A *ferromagnetic material* is a material that is easily magnetized and has high permeability.
>
> *Ferrite* is a ferromagnetic nonconducting ceramic alloy with a high permeability.
>
> A *paramagnetic material* is a material that can be weakly magnetized in the same direction as the applied magnetic field.
>
> A *diamagnetic material* is a material that can be very weakly magnetized in the opposite direction as the applied magnetic field.
>
> An *electromagnet* is a temporary magnet produced when electricity passes through a conductor, such as a coil, that concentrates the magnetic field.
>
> *Electromagnetism* is the temporary magnetic field produced when electricity passes through a conductor.

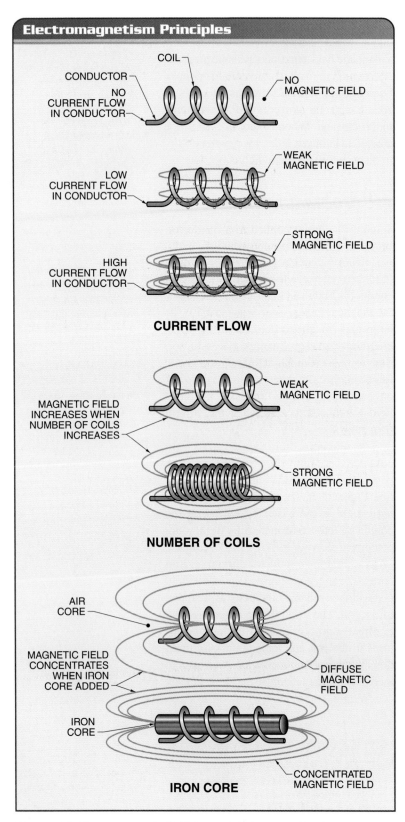

Figure 1-6. Electromagnetic field strength is increased by increasing the current flow or by increasing the number of coils, and concentrated by adding an iron core.

Electromagnets

An electromagnet is a magnet whose magnetic energy is produced by the flow of electric current. Some electromagnets are so large and powerful that they can be used with a crane to lift tons of scrap metal at one time. Other electromagnets are very small and are used in electrical and electronic circuits, such as those found in solenoids and relays. An electromagnet typically consists of an iron core inserted into a coil. Because of its high permeability, an iron core concentrates the magnetic flux produced by the coil. With the core in place and the coil energized, the polarity of the magnet can be determined by the left-hand rule for coils.

The advantages of electromagnets are that they can be made stronger than permanent magnets and that magnetic strength can be easily controlled by regulating the electric current. The main characteristics of an electromagnet include the following:

- When electricity flows through a conductor, an electromagnetic field is created around that conductor.
- An electromagnetic field is stronger close to the conductor and weaker farther away.
- The strength of the electromagnetic field and the current are directly related: more current, the stronger the electromagnetic field; less current, the weaker the electromagnetic field.
- The direction of the electromagnetic field is determined by the direction of the current flowing though the conductor.
- The more permeable the core, the greater the concentration of magnetic flux.

Saturation

Saturation is the condition where a magnetic core has substantially all the magnetic domains aligned with the field and any increases in current no longer result in a stronger electromagnet. As the current flow in a coil is increased, the strength of the magnetic field (field intensity) increases. As the field intensity increases, the magnetic flux density increases. In other words, at low current, only a relatively small number of magnetic domains in the core are aligned with the magnetic field. At higher and higher currents, more and more of the magnetic domains align with the magnetic field.

CURRENT FLOW

All electric circuits depend on current flowing through the circuit. Current can be direct current (DC), where the current flows in one direction only, or alternating current (AC), where the current flow periodically reverses direction. In addition, the direction of current flow, relative to the polarity of the voltage source, depends on the convention used to describe the current flow.

Direct Current

Direct current applied to a conductor starts at zero and goes to its maximum value almost instantly when a switch is closed. **See Figure 1-7.** The magnetic field around the conductor also starts at zero and goes to its maximum strength almost instantly. The current and the strength of the magnetic field remain at their maximum value as long as the load resistance does not change. The current and the strength of the magnetic field increase if the resistance of the circuit decreases. The current and the magnetic field drop to zero when the switch is opened.

Alternating Current

Alternating current applied to a conductor causes the current to continuously vary in direction and magnitude and the magnetic field to continuously vary in strength. **See Figure 1-8.** Current flow and magnetic field strength are at their maximum value at the positive and negative peaks of the AC sine wave. The current is zero and no magnetic field is produced at the zero points of the AC current sine wave through the coil.

Definition

Saturation is the condition where a magnetic core has substantially all the magnetic domains aligned with the field and any increases in current no longer result in a stronger electromagnet.

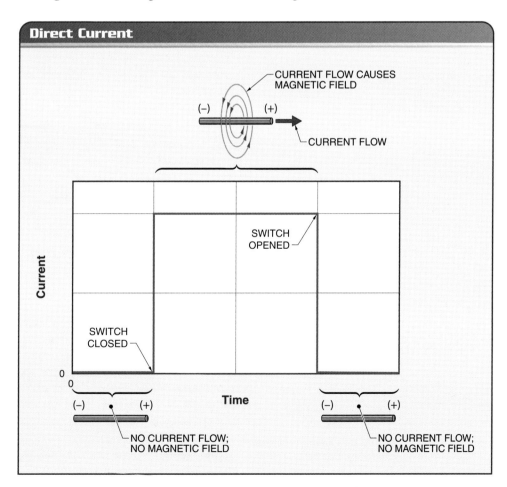

Figure 1-7. Direct current applied to a conductor causes the current and magnetic field to both increase instantly to the maximum magnitude and remain at a constant value as long as the current flows.

Definition

Electron current flow is a description of current flow as the flow of electrons from the negative terminal to the positive terminal of a power source.

The direction of current flow and polarity of the magnetic field change every time the current passes the zero point of the AC sine wave. On standard 60 Hz (cycle) power frequencies, the current passes the zero point 120 times per second.

Current Flow Direction

Early physicists established the polarity of voltage sources, with positive cathode and negative anode terminals. Because a voltage source has polarity, current flow has a direction. These physicists believed that current flowed from the positive to the negative terminals. Later, when atomic structure was studied, electron flow from the negative to the positive terminals was introduced. The two different theories are called electron current flow and conventional current flow. **See Figure 1-9.**

Electron Current Flow. *Electron current flow* is a description of current flow as the flow of electrons from the negative terminal to the positive terminal of a power source. In a battery or other power source, electrons move to the negative terminal, creating a separation of charge, with the negative terminal having a negative charge and the positive terminal having a positive charge.

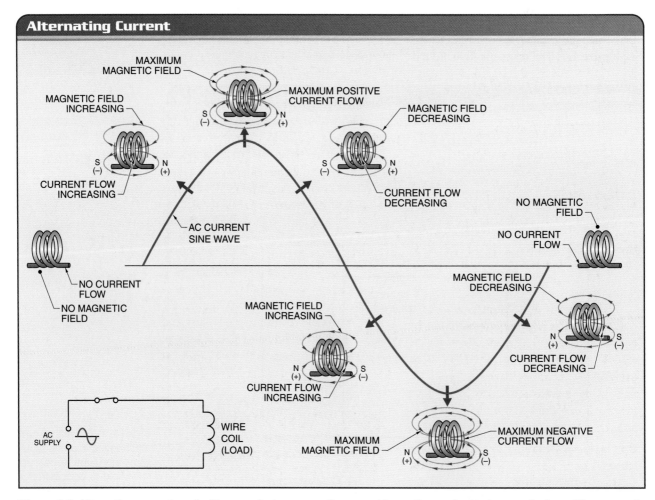

Figure 1-8. Alternating current applied to a conductor causes the current to continuously vary in magnitude and the magnetic field to continuously vary in strength. Current flow and magnetic field strength are at their maximum value at the positive and negative peaks of the AC sine wave.

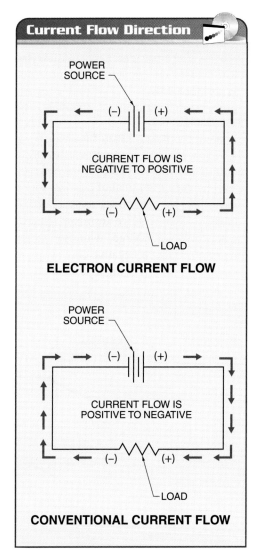

Figure 1-9. Conventional current flow is current flow from positive to negative. Electron current flow is current flow from negative to positive.

This is how a power source produces a potential difference. The potential difference is used to move electrons through the external circuit and through the load to do useful work. Electron current flow is used more in the electronic semiconductor field to assist in explaining the operation of solid-state electronic components.

Conventional Current Flow. *Conventional current flow* is a description of current flow as the flow of positive charges from the positive terminal to the negative terminal of a power source. In a battery or other power source, conventional current flow states that positive charges move to the positive terminal, creating a separation of charge and a potential difference. The potential difference is used to move positive charges through the external circuit and through the load to do useful work.

Conventional current flow is used more in the electrical field and by electrical engineers to aid in explaining electrical circuit properties. A positive potential is considered to be higher than a negative potential, and thus a description of positive charges flowing down from a higher potential to a lower potential works well to explain circuit flow theory. While either method can be used to describe current flow, this book uses electron current flow.

INDUCTION AND INDUCTANCE

Induction has two meanings depending on whether it refers to a single circuit or to two or more circuits. Self-induction refers to the electromagnetic properties of a single circuit. Mutual induction refers to the electromagnetic properties of two or more circuits. *Inductance* is the property of a device or circuit that causes it to store energy in an electromagnetic field.

The three requirements for induction are a conductor, a magnetic field, and relative motion between the conductor and the magnetic field. In a motor, the conductor is the wire making up the winding (coil). The AC power flowing through a conductor generates an expanding and collapsing magnetic field. The expanding and collapsing magnetic field flows through the winding and provides the relative motion between a second conductor and the magnetic field. In a generator, the conductor is moved through the fixed magnetic field and current is induced in the conductor.

Tech Fact

In 1819, it was discovered that when an electric current flows in a copper wire (conductor), a magnetic field exists in the space around the conductor. A fixed relationship exists between the direction of the current in a conductor and the magnitude and direction of the resulting magnetic field.

Definition

Conventional current flow is a description of current flow as the flow of positive charges from the positive terminal to the negative terminal of a power source.

Inductance is the property of a device or circuit that causes it to store energy in an electromagnetic field.

12 MOTORS

Definition

Self-induction is the ability of an inductor in a circuit to generate inductive reactance, which opposes change in the current.

Self-Induction

Self-induction is the ability of an inductor in a circuit to generate inductive reactance, which opposes change in the current. The current flowing in a coil produces a magnetic field that expands out of and surrounds the coil. Energy is stored in the magnetic field. When an AC source voltage rises and the magnetic flux expands around the conductors, an opposing voltage, or counter-electromotive force (CEMF), is induced in the circuit. **See Figure 1-10.** Lenz's law states that the polarity of the induced voltage is such that it produces a current whose magnetic field opposes the change that produced it.

As the AC source voltage falls back to zero and the field collapses back into the circuit, the CEMF acts to prevent the current from falling. This shows that the first 90° of a cycle is spent charging the inductor. The electrical energy is converted into magnetic energy in the inductor. When the voltage peaks, the current is at the minimum, the field stops expanding, and the magnetic field is fully charged. When the source voltage starts to drop from peak, the magnetic field starts to collapse back into the inductor and aids the current provided by the source.

An inductor or coil in a circuit adds inductive reactance. In a typical situation, the inductive reactance of a coil is much greater than the resistance. Therefore, a coil is a reactive circuit. The CEMF is 180° out of phase from the source and the current lags the source by 90°.

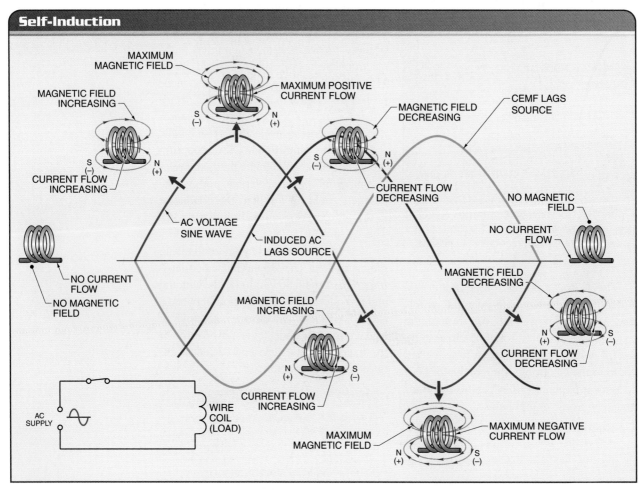

Figure 1-10. When an AC source voltage rises and the magnetic flux expands around the conductors, an opposing voltage, or counter-electromotive force (CEMF), is induced in the circuit.

Mutual Induction

Mutual induction is the ability of an inductor (coil) in one circuit to induce a voltage in another circuit or conductor. **See Figure 1-11.** The magnetic field surrounding the coil interacts with another wire or coil. The interaction between the magnetic field around the first coil and the second coil induces a current in the second coil. The induced current in the second coil creates its own magnetic field that opposes the initial magnetic field.

For an AC motor, current flows through the stator windings and generates a moving magnetic field. The moving magnetic field in the stator interacts with the stationary conductors in the rotor windings and induces a current in the rotor windings. The current in the rotor windings generates a magnetic field that follows the moving magnetic field in the stator. This creates torque that starts rotating the rotor.

Magnetic Field Direction

Just as voltage has a polarity and current has a direction of flow, magnetic fields have a direction. In a magnet, the magnetic field enters into the south pole and exits from the north pole. In a conductor, coil, or in the windings (coils) of generators and motors, the magnetic field also has a direction. These rules for field direction use electron current flow and apply only to DC, as the changing polarities of AC reverse the direction of the magnetic flux twice every cycle.

Left-Hand Conductor Rule. The *left-hand conductor rule* is an explanation of the direction of a magnetic field around a conductor relative to the direction of the current in the conductor, where a left hand is used to illustrate the relationship. When a conductor is wrapped with the left hand with the thumb in the direction of the current flow, the fingers point in the direction of the magnetic field. The direction of current is from negative to positive and can be determined by the polarity of the source. **See Figure 1-12.**

Left-Hand Coil Rule. The *left-hand coil rule* is an explanation of the direction of a magnetic field around a coil relative to the direction of the current in the coil, where a left hand is used to illustrate the relationship.

Definition

Mutual induction is the ability of an inductor (coil) in one circuit to induce a voltage in another circuit or conductor.

The *left-hand conductor rule* is an explanation of the direction of a magnetic field around a conductor relative to the direction of the current in the conductor, where a left hand is used to illustrate the relationship.

The *left-hand coil rule* is an explanation of the direction of a magnetic field around a coil relative to the direction of the current in the coil, where a left hand is used to illustrate the relationship.

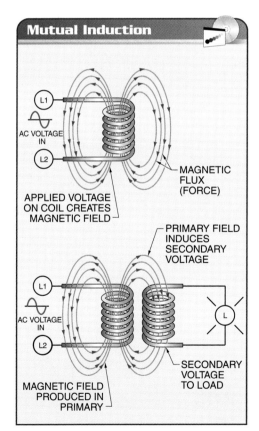

Figure 1-11. Mutual induction is the process where one coil induces current flow in another coil.

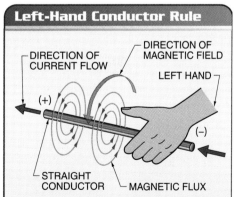

Figure 1-12. When a conductor is wrapped with the left hand with the thumb in the direction of the current flow, the fingers point in the direction of the magnetic field.

A coil has poles just like a bar magnet. When a coil is wrapped with the left hand with the fingers in the direction of the current flow, the thumb points in the direction of the north pole. The coil acts like an electromagnet where the magnetic field enters into the south pole and exits from the north pole. **See Figure 1-13.**

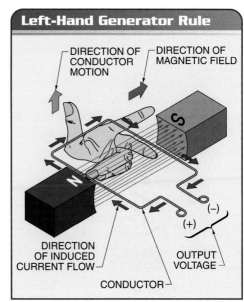

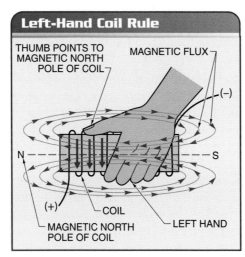

Figure 1-13. When a coil is wrapped with the right hand with the fingers in the direction of the current flow, the thumb points in the direction of the north pole.

Left-Hand Generator Rule. In a generator, an applied force moves a conductor through a magnetic field and a current is induced. The *left-hand generator rule* is an explanation of the relationship between the direction of motion of the conductor within a magnetic field in a generator, the direction of the magnetic field existing around the conductor, and the direction of induced current in a conductor. The left-hand generator rule states that with the thumb, index finger, and middle finger of the left hand set at right angles to each other, the index finger points in the direction of the magnetic field, the thumb points in the direction of the motion of the conductor, and the middle finger points in the direction of the induced current. When using the left-hand generator rule, it is assumed that the magnetic field is stationary and that the conductor is moving through the field. **See Figure 1-14.**

Figure 1-14. The left-hand generator rule explains the relationship between the direction of the magnetic field, the direction of the motion of the conductor, and the direction of the induced current.

Right-Hand Motor Rule. In a motor, an electric current flows through a conductor in a magnetic field and a movement is induced. The *right-hand motor rule* is an explanation of the relationship between the direction of the applied current in a conductor, the direction of the magnetic field around the conductor, and the direction of the induced motion of the conductor within a motor. The right-hand motor rule states that with the thumb, index finger, and middle finger of the right hand set at right angles to each other, the index finger points in the direction of the magnetic field, the thumb points in the direction of the induced motion of the conductor, and the middle finger points in the direction of the current. When using the right-hand motor rule, it is assumed that the magnetic field is moving and the conductor is initially stationary. **See Figure 1-15.**

Definition

*The **left-hand generator rule** is an explanation of the relationship between the direction of motion of the conductor within a magnetic field in a generator, the direction of the magnetic field existing around the conductor, and the direction of induced current in a conductor.*

*The **right-hand motor rule** is an explanation of the relationship between the direction of the applied current in a conductor, the direction of the magnetic field around the conductor, and the direction of the induced motion of the conductor within a motor.*

Tech Fact

Electromagnetic interference (EMI) is unwanted electrical noise on a communications line caused by induced currents from the magnetic field around a nearby line.

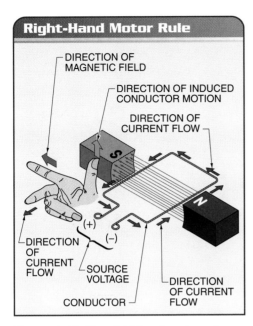

Figure 1-15. The right-hand motor rule explains the relationship between the direction of the magnetic field, the direction of the induced motion of the conductor, and the direction of the current.

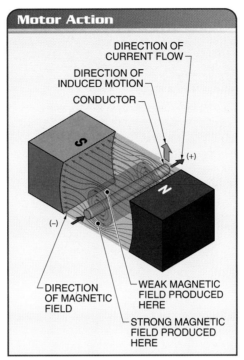

Figure 1-16. When a current-carrying wire is placed in a magnetic field, the wire is pushed out of the field.

Motor Action

The magnetic field direction can be used to explain how the magnetic fields in a motor develop into motion. There are many different ways to design a motor and the implementation details of many types of motors are explained in later chapters. However, the basic principles of all motor designs are similar.

Current in a conductor creates a magnetic field around the conductor. The left-hand conductor rule can be used to determine the direction of the magnetic field. When the conductor is placed within a magnetic field between poles, the two fields interact. **See Figure 1-16.** The field around the conductor distorts the magnetic field between the poles. On one side of the conductor, the fields combine and become concentrated, creating a strong magnetic field. On the other side of the conductor, the fields oppose and cancel, creating a weak magnetic field. The distortion creates a thrust that pushes the conductor out of the magnetic field. The right-hand motor rule shows the direction of the thrust.

When a loop of conductor is placed within a magnetic field, two lengths of conductor are exposed to the same magnetic field. When a source voltage is applied to the loop, current flows from the negative terminal to the positive terminal.

The field around the conductors distorts the field and causes a twisting of the loop. This moves the loop out of the magnetic field and toward the neutral plane, where there is no interaction of the magnetic fields. Momentum carries the loop past the neutral plane. However, the direction of the thrust prevents the loop from entering the magnetic field again.

In order to create the continuous motion needed in a motor, the direction of the current must be reversed. With a DC motor, a commutator is used to reverse the direction of the current. With an AC motor, the current automatically reverses direction. In an actual motor, multiple loops and poles are used to provide much more power than is possible with a single loop and a single pair of poles.

Application—Solenoids And Motor Starters

A solenoid is an electromagnetic device that converts electrical energy into a linear mechanical force. In a solenoid, an electric coil uses current to create a magnetic field in order to produce the power required to move a plunger. The movement of the plunger produces a linear mechanical force that is used to produce work in many industrial applications.

The current drawn by a solenoid coil depends on the applied voltage and size of the coil. Manufacturers list coil specifications to assist in installation and sizing of components. Because magnetic coils are encapsulated and cannot be repaired, they must be replaced when they fail.

For example, the coil used in a size 2 motor starter may have two, three, or four poles and draw 0.14 A sealed current when connected to a 208 V power source. This current value is used when selecting circuit fuse, wire, and transformer sizes. The manufacturer-listed coil specifications are required when designing and troubleshooting a circuit. For example, when troubleshooting blown motor control circuit fuses, both the inrush current and the sealed current (operating current) ratings of the coil must be considered.

In a circuit that uses two magnetic motor starters (M1 and M2), it is possible for the two motor starters to be energized separately or at the same time. This situation (using two size 2 starters to control a three-phase motor with a 240 V control circuit) could cause the following current conditions:

- If M1 starts alone, there would be 1.04 A inrush current and 0.14 A sealed current.
- If M2 starts alone, there would be 1.04 A inrush current and 0.14 A sealed current.
- If M1 is ON and M2 is started, there would be 1.17 A as M2 starts and 0.28 A when both M1 and M2 are ON.
- If M2 is ON and M1 is started, there would be 1.17 A as M1 starts and 0.28 A when both M1 and M2 are ON.
- If M1 and M2 are started at the same time, there would be 2.08 A inrush current and 0.28 A sealed current.

Although starting both M1 and M2 simultaneously is unlikely, this situation must be considered as it could cause a 2 A (or less) fuse to blow. Understanding the circuit operation and the circuit component ratings is required during design and troubleshooting

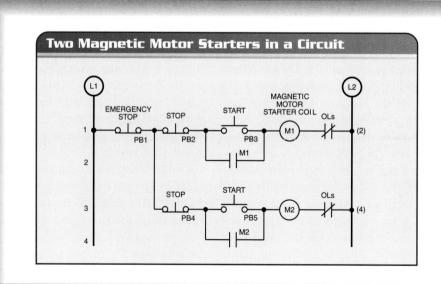

Motor Starter Coil Specifications

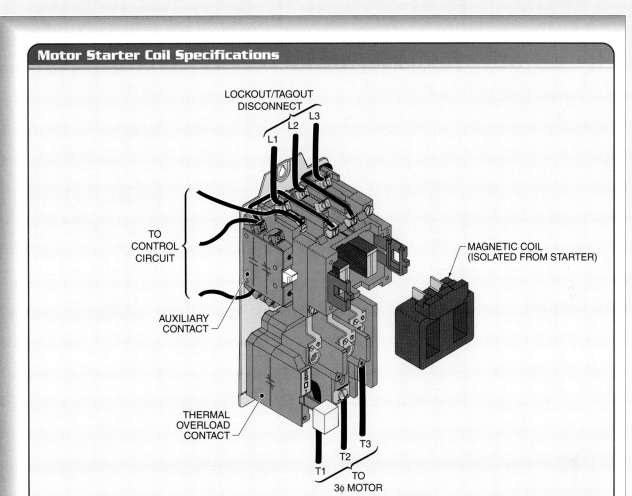

Size	Number of poles	Inrush current* 60 Hz					Sealed current* 60 Hz					Approximate operating time†	
		120 V	208 V	240 V	480 V	600 V	120 V	208 V	240 V	480 V	600 V	Pick-up	Drop-out
00	1-2-3	0.50	0.29	0.25	0.12	0.07	0.12	0.07	0.06	0.03	0.02	28	13
0	1-2-3-4	0.88	0.50	0.44	0.22	0.17	0.14	0.08	0.07	0.04	0.03	29	14
1	1-2-3-4	1.54	0.89	0.77	0.39	0.31	0.18	0.10	0.09	0.04	0.04	26	17
2	2-3-4	1.80	1.04	0.90	0.45	0.36	0.25	0.14	0.13	0.06	0.05	32	14
3	2-3	4.82	2.78	2.41	1.21	0.97	0.36	0.21	0.18	0.09	0.07	35	18
	4	5.34	3.08	2.67	1.33	1.07	0.39	0.23	0.20	0.10	0.08	35	18
4	2-3	8.30	4.80	4.15	2.08	1.66	0.54	0.31	0.27	0.14	0.11	41	18
	4	9.90	5.71	4.95	2.47	1.98	0.61	0.35	0.31	0.15	0.12	41	18
5	2-3	16.23	9.36	8.11	4.06	3.25	0.81	0.47	0.41	0.20	0.16	43	18

* in A
† in ms

Summary

- A magnet is a substance that produces a magnetic field. An electromagnet can be energized or de-energized by applying or removing an electric current. Electromagnets are used in many electrical devices.

- Magnetic flux is the total amount of a magnetic field. Flux is concentrated to increase the flux density and increase the power available to a motor.

- Electron current flow is a description of current flow as the flow of electrons from the negative terminal to the positive terminal of a power source. Conventional current flow is a description of current flow as the flow of positive charges from the positive terminal to the negative terminal of a power source.

- When an AC source voltage rises and the magnetic flux expands around an inductor, an opposing voltage, or counter-electromotive force (CEMF), is induced in the circuit. Lenz's law states that the polarity of the induced voltage is such that it produces a current whose magnetic field opposes the change that produced it.

- The interaction between a magnetic field around one coil and a second coil induces a current in the second coil. The induced current in the second coil creates its own magnetic field that opposes the initial magnetic field.

- Current in a conductor creates a magnetic field around the conductor. When the conductor is placed within a magnetic field between poles, the two fields interact. On one side of the conductor, the fields combine and become concentrated, creating a strong magnetic field. On the other side of the conductor, the fields oppose and cancel, creating a weak magnetic field. The distortion creates a thrust that pushes the conductor out of the magnetic field

Glossary...

Magnetism is a force caused by a magnet that acts at a distance on other magnets.

A **magnet** is a substance that produces a magnetic field.

A **magnetic field** is a force produced by a magnet that exerts a force on moving electric charges or on other magnets.

A **permanent magnet** is a magnet that can hold its magnetism for a long period.

A **temporary magnet** is a magnet that retains only trace amounts of magnetism after a magnetizing force has been removed.

Retentivity is a measure of the ability of a magnet to retain magnetism after the magnetizing force has been removed.

Magnetic flux, or **field flux,** is the invisible lines of force that make up the total quantity of a magnetic field.

Flux density is the amount of concentration of magnetic flux through a specific area.

Permeability is a measure of the ability of a material to conduct magnetic flux.

A **ferromagnetic material** is a material that is easily magnetized and has high permeability.

Ferrite is a ferromagnetic nonconducting ceramic alloy with a high permeability.

A **paramagnetic material** is a material that can be weakly magnetized in the same direction as the applied magnetic field.

A **diamagnetic material** is a material that can be very weakly magnetized in the opposite direction as the applied magnetic field.

...Glossary

An **electromagnet** is a temporary magnet produced when electricity passes through a conductor, such as a coil, that concentrates the magnetic field.

Electromagnetism is the temporary magnetic field produced when electricity passes through a conductor.

Saturation is the condition where a magnetic core has substantially all the magnetic domains aligned with the field and any increases in current no longer result in a stronger electromagnet.

Electron current flow is a description of current flow as the flow of electrons from the negative terminal to the positive terminal of a power source.

Conventional current flow is a description of current flow as the flow of positive charges from the positive terminal to the negative terminal of a power source.

Inductance is the property of a device or circuit that causes it to store energy in an electromagnetic field.

Self-induction is the ability of an inductor in a circuit to generate inductive reactance, which opposes change in the current.

Mutual induction is the ability of an inductor (coil) in one circuit to induce a voltage in another circuit or conductor.

The **left-hand conductor rule** is an explanation of the direction of a magnetic field around a conductor relative to the direction of the current in the conductor, where a left hand is used to illustrate the relationship.

The **left-hand coil rule** is an explanation of the direction of a magnetic field around a coil relative to the direction of the current in the coil, where a left hand is used to illustrate the relationship.

The **left-hand generator rule** is an explanation of the relationship between the direction of motion of the conductor within a magnetic field in a generator, the direction of the magnetic field existing around the conductor, and the direction of induced current in a conductor.

The **right-hand motor rule** is an explanation of the relationship between the direction of the applied current in a conductor, the direction of the magnetic field around the conductor, and the direction of the induced motion of the conductor within a motor.

Review

1. What is the difference between a permanent magnet and an electromagnet? Why is this difference important?

2. What is the difference between magnetic flux and magnetic flux density?

3. What the difference between electron current flow and conventional current flow?

4. How does self-induction in an inductor produce inductive reactance?

5. How does mutual induction create current and a magnetic field in a nearby circuit?

6. How does a loop of conductor in a magnetic field create thrust and rotary motion?

Refer to the CD-ROM for Quick Quiz® questions related to chapter content.

MOTORS

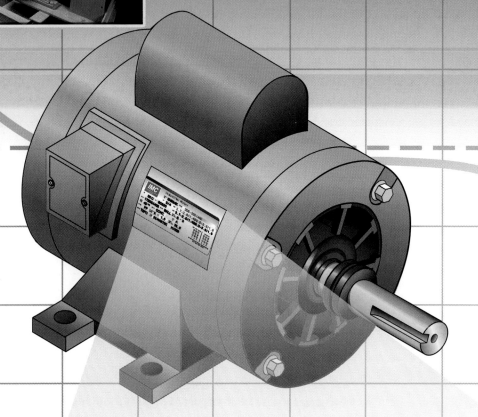

Industrial Motor

IMC
INTERNATIONAL MOTOR COMPANY

MOD NO.	2N924K	VOLTS	208-220/440
HP	3/4	AMP	2.7-2.8/1.4 & 3.0-3.6/1.9
FR	56 C	HZ 60&50	SFA 2.7-2.8/1.4 & 3.0-3.6/1.9
RPM	1725&1425	BRGS BALL	NEMA DESIGN B PH 3
MAX AMB	40 °C	DUTY CONT.	INS CL A
TYPE	PF	LR KVA CODE K SF 1.0	NEMA NOM EFF 78.5
ENCL	TEFC	MTR REF RO65876K K1028	

208 - 220 VOLTS
④─⑤─⑥
⑦ ⑧ ⑨
① ② ③
L1 L2 L3

440 VOLTS
④ ⑤ ⑥
⑦ ⑧ ⑨
① ② ③
L1 L2 L3

TO REVERSE ROTATION INTERCHANGE ANY TWO LINE LEADS

Motor Nameplates

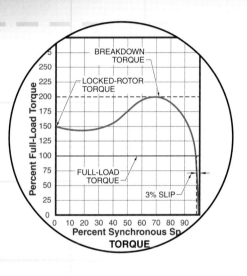

Motor Nameplate Data 22
Electrical Ratings 22
Operating Ratings 27
Mechanical-Design Codes 35
Application—Applying Motor Nameplate Specifications 46
Application—Motor Insulation Life 48
Summary 49
Glossary 49
Review 51

OBJECTIVES

- Describe the purpose of electrical ratings.
- List and describe the common electrical ratings on the motor nameplate.
- Describe the purpose of operating ratings.
- List and describe the common operating ratings on the motor nameplate.
- Describe the purpose of mechanical-design code information.
- List and describe the common mechanical-design codes on the motor nameplate.
- Describe the difference between a general-purpose motor and a definite-purpose motor.
- Explain why an energy-efficient motor may be more economical than a standard motor, even though the energy-efficient motor may cost more to purchase.
- Describe the different types of motor enclosures.

The information needed to service a motor can be found on the nameplate. Understanding the motor nameplate data is the first requirement when servicing or replacing a motor. The nameplate includes information about the electrical ratings, the operating ratings, and the mechanical design of the motor. This information is used when selecting a motor for a particular application and when maintaining a motor.

22 MOTORS

> **Definition**
>
> A *nameplate* is a metal tag permanently attached to an electric motor frame that gives the required electrical ratings, operating ratings, and mechanical-design codes of the motor.
>
> The *voltage rating* is the voltage level that a motor can use.

MOTOR NAMEPLATE DATA

A *nameplate* is a metal tag permanently attached to an electric motor frame that gives the required electrical ratings, operating ratings, and mechanical-design codes of the motor. **See Figure 2-1.** Electric motors are used to produce work. In order for a motor to safely produce work as required for the expected life of the motor, the motor's electrical and operating ratings and mechanical-design codes must be considered.

Since the nameplate has limited space to convey the ratings, most information is abbreviated or coded to save space. In addition to the written information, most motor nameplates also include the motor's wiring diagram. Understanding the abbreviated and coded information provided on a motor nameplate is required when selecting, installing, and troubleshooting electric motors. For motors that are already in service, the information provided on the motor nameplate is often the only information available.

Manufacturer. The name of the manufacturer of the motor is usually prominently displayed on a nameplate. This section of the nameplate may also include an address or other contact information. The nameplate usually contains an abbreviation for the model number (MOD), type (TP), or catalog (CAT) number. A serial number is often included. A serial number is very important if the manufacturer is to be contacted for information about the motor.

ELECTRICAL RATINGS

Electrical ratings describe the electrical requirements for the operation of a motor. Electrical ratings included on motor nameplates include a voltage rating, current rating, frequency rating, and phase.

Voltage Rating

The *voltage rating* is the voltage level that a motor can use. The nameplate voltage rating of a motor is abbreviated V or VOLTS. **See Figure 2-2.** All motors are designed for optimum performance at a specific voltage level. When selecting a motor, a motor that is rated close to the actual supply voltage should be used. Likewise, when troubleshooting a motor, the nameplate rated voltage should be checked against the actual measured supply voltage.

A motor's nameplate voltage rating is the optimal voltage that should be connected to the motor for best operating performance.

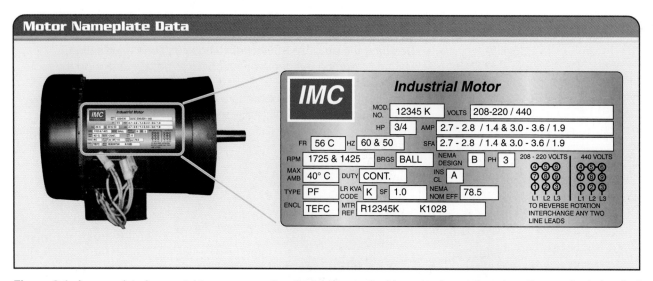

Figure 2-1. A nameplate is a metal tag permanently attached to an electric motor frame that gives the required electrical ratings, operating ratings, and mechanical-design codes of the motor.

The actual voltage applied to the motor usually varies somewhat from the nameplate rated voltage. Any variation in applied voltage from the nameplate rated voltage will change the operating characteristics of a motor and must be considered when installing and troubleshooting the motor. The available (applied) voltage to a motor can be easily measured using a standard voltmeter or a DMM set to measure voltage. Voltage measurements should always be taken prior to installing, servicing, or troubleshooting a motor.

Voltage Unbalance. The actual voltage applied to a motor should be within 5% of the motor's nameplate rated voltage. However, with 3-phase motors, the applied voltage must also be balanced by having the same applied voltage on each of the three power lines. For every 1% voltage unbalance, there can be up to a 6% current unbalance. The greater the voltage or current unbalance, the higher the operating temperature of the motor.

The power lines can be checked for voltage unbalance by using a voltmeter or DMM to measure the voltage between L1-L2, L2-L3, and L1-L3. The voltage unbalance should be measured with the motor OFF. To offset a voltage unbalance of up to 5%, the motor can be derated or a larger motor can be used. **See Figure 2-3.**

AC Motors. The actual supply voltage will vary depending upon the source of voltage. For example, a 120/208 V, 3-phase, 4-wire wye service can supply a single-phase motor of either 120 V or 208 V. A 115/230 V, 3-phase, 4-wire delta service can supply a single-phase motor of either 115 V or 230 V. Because of the differences in supply voltages, motor manufacturers offer motor models with different voltage ratings. Typical motor voltage ratings are required for specific applications as follows:
- 115 V (single-phase motor rating, typically 2 HP or less)
- 115/230 V (single-phase motor rating, typically 2 HP or less)
- 115/208–230 V (single-phase motor rating, typically 2 HP or less)

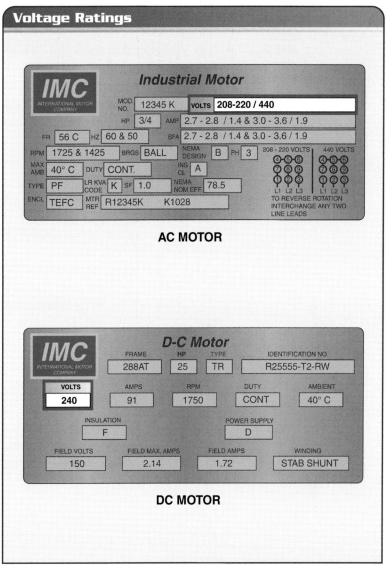

Figure 2-2. The voltage rating is the voltage level that a motor can use. All motors are designed for optimum performance at a specific voltage level.

Voltage Unbalance	
Unbalance*	Motor Derating*
2	5
3	10
4	18
5	25
over 6	Do not operate motor

* in %

Figure 2-3. When voltage unbalance is present, the motor may need to be derated.

Definition

*The **current rating** is the amount of current a motor draws when delivering full rated power output.*

- 110/220 V (single-phase motor rating, typically 2 HP or less)
- 230 V (single-phase or 3-phase motor rating, typically up to 7.5 HP)
- 208–230 V (single-phase or 3-phase motor rating, typically up to 7.5 HP)
- 208–230/460 V (3-phase motor rating, typically up to 75 HP)
- 230/460 V (3-phase motor rating, typically up to 75 HP)
- 460 V (3-phase motor rating, typically up to 250 HP)
- 460/796 V (3-phase motor rating, typically up to 250 HP)
- 575 V (3-phase motor rating, typically 250 HP motors)
- 2300/4160 V (3-phase motor rating, typically 250 HP to 500 HP motors)

DC Motors. DC motors will typically use voltage levels of 24 V, 48 V, 90 V, or 180 V. However, other voltage levels may be seen on the nameplate. The maximum voltage to be applied to the armature of a DC motor is abbreviated on the nameplate as ARM. The voltage to be applied to the field of a DC motor is abbreviated on the nameplate as FLD.

Current Rating

The *current rating* is the amount of current a motor draws when delivering full rated power output. The nameplate current rating of a motor is abbreviated A or AMP. The total current drawn per horsepower output varies with the power supply and motor design. **See Figure 2-4.** For multiple-speed motors, the current rating should be given for each speed. However, for single-phase shaded-pole and split-phase motors, the current is only required for the maximum speed.

The current rating is used to size the motor and overload heaters. If possible, a 3-phase motor should be used instead of a single-phase motor to reduce the amount of required current draw by the motor. Likewise, a motor should be connected to the higher nameplate voltage rating to reduce the amount of current draw by the motor. Since the load on most motors varies, the actual current draw of a motor will vary from the nameplate rated current as the motor operating conditions change and the motor ages. Motors that are not fully loaded draw less than their rated nameplate current. Motors that are overloaded draw more than their rated nameplate current.

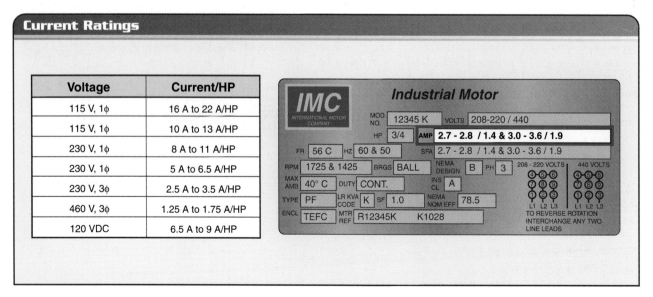

Figure 2-4. The current rating is the amount of current a motor draws when delivering full rated power output.

The amount of current a motor draws can easily be measured with a clamp-on ammeter. Current measurements should be taken during different motor operating conditions to get a better understanding of the operating condition of the motor. A meter with a recording mode can be used to record current measurements over time.

Locked-Rotor Indicating Code Letters. The *locked-rotor indicating code letter* is a designation for the range of locked-rotor current draw per motor horsepower. *Locked-rotor current (LRC)*, or *starting current*, or *inrush current*, is the amount of current a motor draws on startup or when the rotor is locked. The locked-rotor indicating code letter or the locked-rotor current must be listed on the nameplate for all motors over ½ HP. The code letter is omitted on polyphase wound-rotor motors. The locked-rotor indicating code letter is abbreviated on the nameplate as LR KVA CODE.

The list of motor nameplate code letters begins with the letter "A" and ends with "V." **See Figure 2-5.** The closer the nameplate listed code letter is to "A," the lower the motor's starting current at a given power output. Likewise, the closer the nameplate listed code letter is to "V," the higher the motor starting current at a given power output. For example, a motor with a listed code letter of "G" will have a lower starting current than a motor with a listed code letter of "H", as long as the power rating is the same. Most motors have a code letter in the H to N range.

Frequency Rating. The *frequency rating* is the power line frequency at which a motor is designed to operate. The frequency rating of a motor is abbreviated on the nameplate as HZ. A motor's nameplate frequency is determined by the manufacturer and will be either 50 Hz or 60 Hz. **See Figure 2-6.** When the nameplate of a motor shows only one frequency rating (50 Hz or 60 Hz), the listed current ratings apply to that frequency. When the nameplate of a motor lists a dual-frequency rating (50/60 Hz), the first set of current ratings apply to the first listed frequency rating and the second set of current ratings apply to the second listed frequency rating.

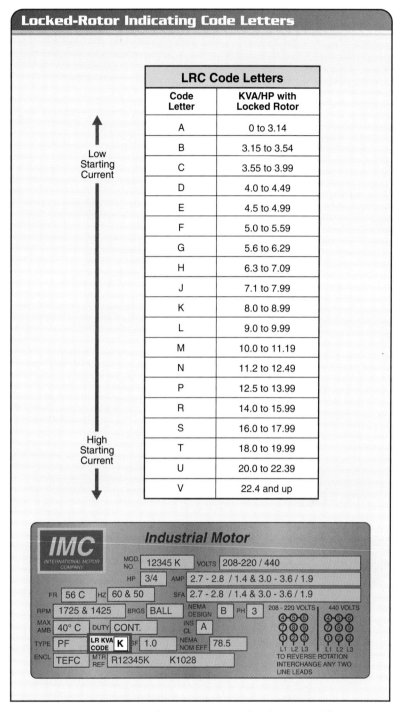

Figure 2-5. A locked-rotor indicating code letter is a designation for the range of locked-rotor current draw per motor horsepower.

Tech Fact

The National Electrical Manufacturers Association Standard MG1 sets the basic requirements for the information to be marked on motor nameplates.

Definition

*The **frequency rating** is the power line frequency at which a motor is designed to operate.*

26 MOTORS

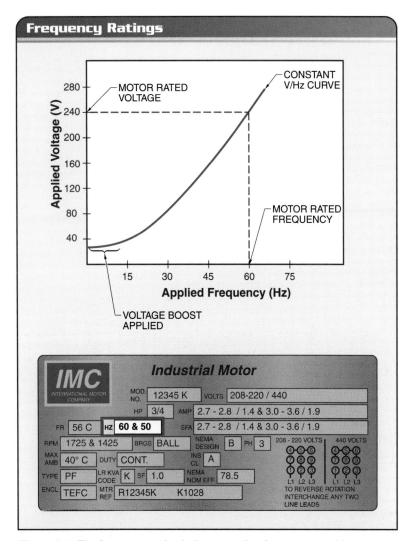

Figure 2-6. The frequency rating is the power line frequency at which a motor is designed to operate. For variable-frequency drives, the V/Hz ratio can be manipulated to modify torque characteristics.

Definition

*The **locked-rotor indicating code letter** is a designation for the range of locked-rotor current draw per motor horsepower.*

***Locked-rotor current (LRC)**, or **starting current**, or **inrush current**, is the amount of current a motor draws on startup or when the rotor is locked.*

Motors typically tolerate a ±5% frequency variation without affecting the motor's load operation. A higher frequency will increase motor speed but will reduce motor torque. A lower frequency will decrease motor speed and increase motor torque.

Directly applying any frequency to a motor other than the nameplate rated frequency will change the motor's operating characteristics (speed and output torque) and can damage the motor. When the speed of a motor needs to be varied during operation, a variable-speed drive is used. Variable-speed drives are designed to maintain the motor's nameplate volts-per-hertz ratio (V/Hz). As the drive reduces the frequency applied to the motor, the drive also reduces the voltage applied to the motor by the same ratio.

The V/Hz ratio for 230 V 3-phase motors is 3.8 V/Hz (230 ÷ 60 = 3.8) and the V/Hz ratio for 460 V 3-phase motors is 7.7 V/Hz (460 ÷ 60 = 7.7). The V/Hz ratio can be manipulated to modify the torque characteristics. Below about 15 Hz, the voltage applied to the motor stator may be boosted to compensate for the large power loss AC motors typically have at low speed.

Phase

The *phase* is the power phase (1ɸ, 3ɸ, or DC) that a motor requires for operation. The phase rating of a motor is abbreviated on the nameplate as PH. Electric motors typically require either direct current (DC), single-phase alternating current (1ɸ AC), or three-phase alternating current (3ɸ AC). The phase rating of a motor is listed as 1, 3, or DC. **See Figure 2-7.**

Before the development of motor drives, the type of current (DC, 1-phase, or 3-phase) and voltage level (120 V, 240 V, 480 V, etc.) supplied to a motor starter had to match the nameplate rating of a motor. For example, a 3-phase, 240 V rated motor controlled by a magnetic motor starter needed to be connected to a 3-phase, 240 V power supply.

When electric motor drives are used to control motors, the type and level of the input supply voltage does not have to match the rated type and voltage level of the motor. Some electric motor drives allow the input power to the drive to be of a different type and at a different voltage level than the power required by the motor. For example, an electric motor drive supplied with single-phase, 120 V power can be used to control a 3-phase, 240 V motor.

Single-phase motors are used primarily in applications requiring power of less than 1 HP and where 3-phase power is not available. Electric motor drives that can be powered by a single-phase supply and deliver a 3-phase output allow 3-phase motors of less than 5 HP to be used in applications in which only single-phase power is available. Thus, 3-phase motors are standard for HVAC systems in which only single-phase power is available but energy efficiency is important.

OPERATING RATINGS

Operating ratings describe how a motor is designed and how that design relates to where the motor is to be used. Operating ratings include the power rating, usage rating, service factor rating, speed rating, operating time rating (duty cycle), motor efficiency rating, ambient temperature rating, and insulation class. These ratings are displayed on the nameplate and are critical in performing installation and troubleshooting procedures.

Power Rating

The *power rating* is the amount of power a motor can deliver to a load. For motors of ⅛ HP or more, the horsepower rating must be placed on the motor nameplate. The nameplate power rating of a motor is abbreviated HP or KW. **See Figure 2-8.** A motor's nameplate power rating is determined by the manufacturer and cannot be changed.

Definition

*The **phase** is the power phase (1φ, 3φ, or DC) that a motor requires for operation. The phase rating of a motor is abbreviated on the nameplate as PH.*

*The **power rating** is the amount of power a motor can deliver to a load.*

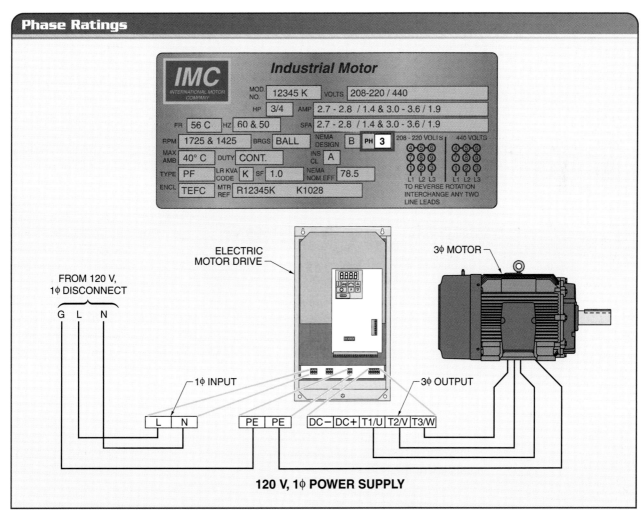

Figure 2-7. The phase rating is the power phase (1φ, 3φ, or DC) that a motor requires. Some electric motor drives allow the input power to the drive to be of a different type and at a different voltage level than the power required by the motor.

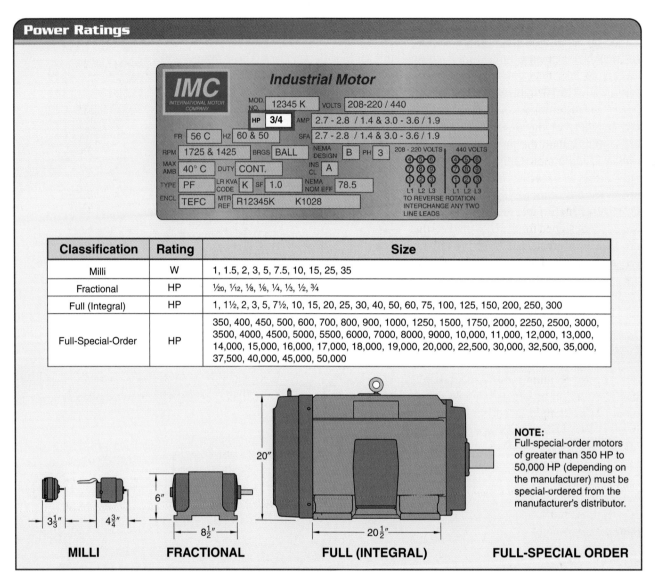

Figure 2-8. The power rating is the amount of power a motor can deliver to a load. Motors designed for the U.S. market that are 1/20 HP or greater are typically rated in HP, and motors that are less than 1/20 HP are typically rated in watts (W).

Motors convert electrical energy into rotating mechanical energy. The amount of rotating mechanical energy produced by a motor determines how much work the motor can perform. The amount of power a motor produces is not easily measured with standard test instruments. A power meter, such as a power quality meter, is required to measure the amount of power a motor is producing or using.

A motor that is not connected to a load will produce less than the nameplate rated power. A motor operating the maximum load the motor was designed to safely handle will produce the nameplate rated power (HP or kW). An overloaded motor will operate the load by trying to produce more power than the motor is rated for. The more power produced by a motor, the higher the current draw and motor temperature. Because higher operating temperatures reduce the life of a motor, motors should not be operated with loads that are greater than what they are designed for. Since the operational life of a motor is important, selecting the next size higher-rated motor for an application will ensure longer motor life.

The power rating of a motor is used when selecting or servicing an electric motor drive. Typical electric motor drive power ratings start at fractional values under 1 HP up to 500 HP or more. When ordering an electric motor drive, the minimum power rating of the drive must be equal to or greater than the power rating of the motor. Electric motor drive manufacturers build drives to sizes that follow the standard power ratings of motors.

Motors manufactured in the United States (or designed for the U.S. market) that are 1/20 HP or greater are typically rated in HP, and motors that are less than 1/20 HP are typically rated in watts (W). Motors manufactured in Europe are usually rated in kilowatts (kW), regardless of size.

For conversion purposes, 0.746 kW equals 1 HP. If a motor is rated in kW, a general comparison to horsepower can be made by dividing the kW rating of the motor by 0.746 to get the equivalent HP rating. For example, a 3 kW rated motor is equal to a 4.02 HP (about 4 HP) motor. Likewise, a general comparison of kilowatts to horsepower can be made by multiplying the HP rating of the motor by 0.746 to get the equivalent kW rating. For example, a 10 HP rated motor is equal to a 7.46 kW (about 7.5 kW) motor.

Usage Rating

The *usage rating* is a description of the expected or allowed application of a motor. Motors are rated for general-purpose usage or definite-purpose usage. Motors that are rated for general-purpose usage are used in a wide range of applications and for mechanical loads such as conveyors, machine tools, and belt-driven equipment. **See Figure 2-9.** They are also used with reciprocating pumps and for moving-air applications. Motors rated for general-purpose usage are designed to operate under the following conditions:
- ambient temperature not over 40°C
- altitude not over 3300 ft
- rigid mounting surface
- free airflow

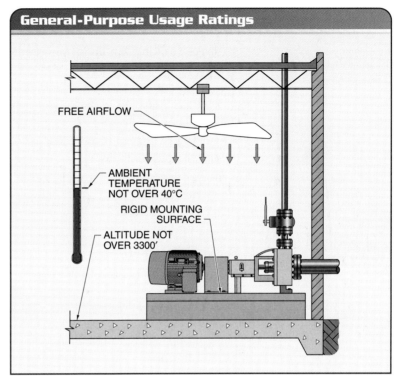

Figure 2-9. The usage rating is a description of the expected or allowed application of a motor.

Definite-purpose motors are designed for usage in a specific application, such as the following:
- washdown-rated motors (in food, beverage, and chemical plants)
- submersible pumps (sump pumps, drainage pumps, water wells, and septic systems)
- other pumps (for wastewater treatment, water treatment, and water recirculation systems)
- hazardous-location, severe-duty motors (in dry-cleaning plants, paint factories, and grain elevators)
- instantly reversible motors (for hoists, gates, cranes, and mechanical doors)
- extra-high-torque motors (for hard-starting loads, etc.)
- farm-duty or agricultural-rated motors (having added protection against dust, dirt, and chemicals)
- irrigation (motors that resist corrosion from high-moisture environments and/or environments where chemicals are present)

Definition

*The **usage rating** is a description of the expected or allowed application of a motor.*

Definition

The service factor rating is a multiplier that represents the amount of load, beyond the rated load, that can be placed on a motor without causing damage.

The speed rating of a motor is the approximate speed at which the rotor of a motor rotates when delivering rated power to a load.

Slip is the difference between the synchronous speed and rated speed of a motor.

- auger drive (for augers and drilling systems)
- HVAC (heating/ventilating/air conditioning)
- inverter-duty-rated (motors designed to be controlled by variable-frequency drives)
- pool (for swimming pools, water parks, and whirlpool hot tubs)
- AC/DC vacuum (for commercial vacuum systems, carwash, and sprayer or fogger systems)

Service Factor Rating

The *service factor rating* is a multiplier that represents the amount of load, beyond the rated load, that can be placed on a motor without causing damage. The service factor rating is abbreviated on the nameplate as SF. The amount of current at the increased load is abbreviated on the nameplate as SFA (service factor amps). **See Figure 2-10.**

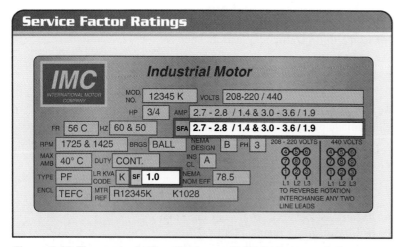

Figure 2-10. The service factor rating is a multiplier that represents the amount of extra load that can be placed on a motor without causing damage.

A motor will attempt to drive a load, even if the load exceeds the motor's power rating. A nameplate service rating of 1.00 (or no listed rating) indicates the motor is not designed to safely handle an overloaded condition above the motor's rated power. A nameplate service rating higher than 1.00 indicates the motor was designed to develop more than its nameplate rated power without causing damage to the motor's insulation. Thus, the service factor rating is the motor's margin of overload rating. For example, a 10 HP rated motor with a service factor of 1.15 can be operated as an 11.5 HP motor (10 × 1.15 = 11.5) for a short period. Drawbacks to running a motor above the rated horsepower are reduced motor speed and efficiency and increased operating temperature.

Allowable service factor ratings are 1.00, 1.15, 1.25, and 1.35. The service factor rating is based on the motor operating within all other motor specifications (frequency rating, ambient temperature rating, balanced voltage supply, etc.). When the nameplate of a motor does not list a service factor rating, a service factor rating of 1.00 is assumed and there is no built-in safety margin for the motor. Current and temperature rise when motors are operated at more than rated power (above a service factor of 1.00).

Speed Rating

The *speed rating* of a motor is the approximate speed at which the rotor of a motor rotates when delivering rated power to a load. The speed rating is abbreviated on the nameplate as RPM. **See Figure 2-11.** *Slip* is the difference between the synchronous speed and rated speed of a motor. Nonsynchronous motors typically slip approximately 3% to 5%.

The synchronous speed of an AC motor is based on the number of stator poles and the applied frequency. The operating speed is the actual nameplate listed speed at which the motor develops rated horsepower at rated voltage and frequency. For DC motors, the speed is determined by the supply voltage and/or the amount of field current.

The top speed of a motor depends upon the voltage limits of the motor and the motor's mechanical balancing. Most motor manufacturers balance their motors (rotors) for speeds up to 25% over nameplate rated speeds. Thus, it is recommended to not operate a motor at more than 20% over its nameplate speed rating. Operating a motor below its rated speed usually does not produce a dangerous situation in which the motor produces an accident, but can still damage the motor. Motor damage is usually a result of the increased current and reduced cooling (with fan-cooled motors) at lower operating speeds. Reduced cooling problems are solved by using a motor rated for operation with electric motor drives. The cooling fan operates at full speed regardless of actual motor shaft speed.

Operating Time Rating (Duty Cycle)

The *operating time rating,* or *duty cycle,* is the amount of time a motor can be operated without being turned OFF to allow for cooling. The operating time rating (duty cycle) is abbreviated on the nameplate as DUTY, DUTY CYCLE, or TIME RATING. Motors designed to operate for unlimited periods are marked CONT (for "continuous") on the motor nameplate, or have no designation. Motors designed to operate for intermittent duty are marked INTER on the nameplate, or will have a time rating listed. Intermittent-duty motor time ratings are listed as 5, 15, 30, or 60 minutes. **See Figure 2-12.**

Most motors can be operated for any length of time. However, some motors are designed to operate only for short periods. Continuous-duty motors are designed to operate at higher temperatures. Intermittent-duty motors are not designed for high-temperature operation and require time to cool off between operating periods. For example, a compressor motor may need to shut off for one hour after running one hour, to allow time for it to cool.

Intermittent-rated motors are used in applications such as compressors, waste-disposal systems (garbage disposals), electric hoists, gate-opening motors, and other applications in which the motor need only be turned on for short time periods to meet the application requirements. Using an intermittent-rated motor instead of a continuous-rated motor in applications in which the motor is to be ON for short periods is more cost-effective because intermittent-rated motors are less expensive than continuous-rated motors with the same power rating.

Motor Efficiency Rating

Motor efficiency is a measure of the effectiveness with which a motor converts electrical energy to mechanical energy. The motor efficiency rating is abbreviated on the nameplate as EFF or NEMA NOM EFF. **See Figure 2-13.** Efficiency is the ratio of motor power output to supply power input. All motors take more power to operate than they can produce, because of power losses within the motor. Power losses occur because of losses from friction and heat within the motor.

> **Definition**
>
> The ***operating time rating***, or ***duty cycle***, is the amount of time a motor can be operated without being turned OFF to allow for cooling.
>
> ***Motor efficiency*** is a measure of the effectiveness with which a motor converts electrical energy to mechanical energy.

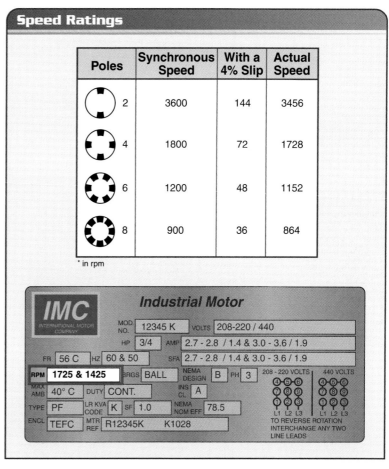

Figure 2-11. The speed rating of a motor is the approximate speed at which a motor rotates when delivering rated power to a load. Slip is the difference between the synchronous speed and rated speed of a motor.

32 MOTORS

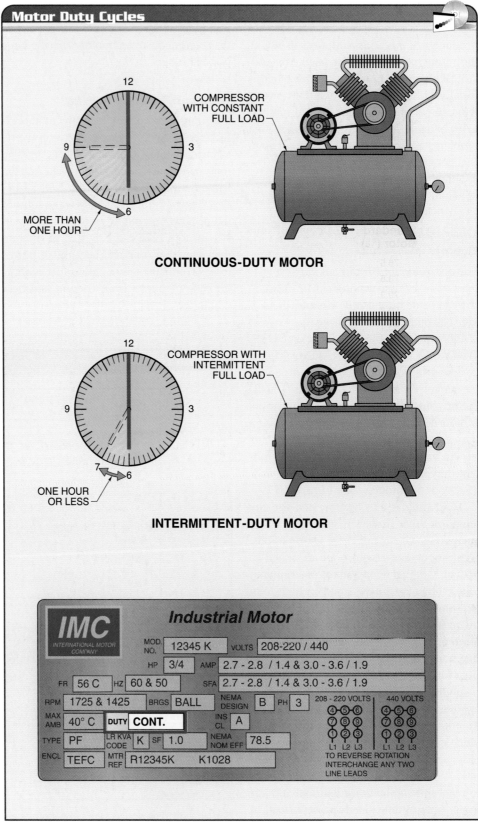

Figure 2-12. The duty cycle is the amount of time a motor can be operated without being turned OFF to allow for cooling.

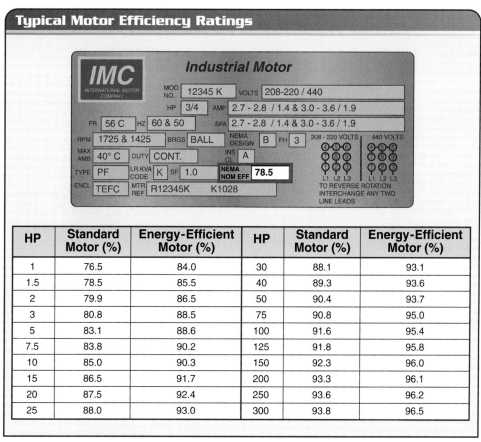

Figure 2-13. Motor efficiency is a measure of the effectiveness with which a motor converts electrical energy to mechanical energy. Energy-efficient motors are more expensive than standard motors but cost less to operate.

Typically, efficiency ratings listed on motor nameplates are nominal efficiencies at full load. A nominal efficiency value represents a value not greater than the average efficiency of a large population of motors tested of the same type and design.

Motor manufacturers produce both standard and energy-efficient motors. Energy-efficient motors are more expensive than standard motors but cost less to operate. Energy-efficient motors are built more ruggedly than standard motors and include larger rotor bars and increased laminations. This reduces losses and increases torque. Since this type of motor is designed to run cooler, the life of the insulation and bearings is improved. Energy-efficient motors can endure excessive voltage and current, overload, and unbalance conditions better than standard motors. Likewise, energy-efficient motors operate better on nonsinusoidal waveforms and are better suited for use with variable-frequency drives, which distort sine waves.

Because motors consume so much energy, a higher efficiency rating results in lower operating cost to the user and requires less energy produced by the utility company. For this reason, the *Energy Policy Act of 1992* (EPACT 92) was created and took effect in 1997. EPACT 92 sets minimum energy-efficiency levels for most general-purpose, 3-phase motors from 1 HP to 200 HP sold in the United States. A motor's energy efficiency rating is listed on the motor nameplate.

Tech Fact

Motor efficiency is usually stated numerically. On some motors, efficiency may be stated as a motor efficiency index letter.

Definition

*The **ambient temperature rating** is the maximum allowable temperature of the air surrounding an object.*

*The **temperature rise** is the difference between the winding temperature of a running motor and the ambient temperature.*

*The **permissible temperature rise** is the difference between the ambient temperature and the nameplate ambient rating of a motor.*

*The **insulation class** is a code letter signifying the maximum operating temperature of the insulation used in a motor.*

*The **frame size** is a number designating standard dimensions of a motor housing, shaft, and mounting holes.*

Ambient Temperature Rating

The *ambient temperature rating* is the maximum allowable temperature of the air surrounding an object. The temperature of a motor rises as work is performed. Ambient temperature rating is abbreviated on the nameplate as AMB, MAX AMB, or DEG. The *temperature rise* is the difference between the winding temperature of a running motor and the ambient temperature. The *permissible temperature rise* is the difference between the ambient temperature and the nameplate ambient rating of a motor.

A typical ambient temperature rating is 104°F (40°C). **See Figure 2-14.** A motor without a rating is assumed to be rated at 40°C. Motors designed to operate in higher-ambient-temperature areas should have additional cooling or have a higher rating (such as a 55°C nameplate rating). Although a motor nameplate only lists the maximum ambient temperature at which the motor was designed to operate, a motor also has a lower temperature rating that can be found in the motor specifications. This lower limit is typically –25°C, unless stated otherwise.

Higher temperatures caused either by an increase in the ambient air temperature or by overloading the motor will damage the motor's insulation and break down bearing lubricants. Typically, for every 10°C above the rated temperature limit of a motor, the motor's life will be cut in half. Heat destroys insulation, and the higher the heat, the greater and faster the damage occurs.

Motor installations with an ambient temperature above the rated ambient temperature of the motor require the use of a temperature correction factor. The correction factor derates motor specifications to prevent damage caused by environmental conditions other than what is stated by the manufacturer on the motor nameplate.

Ambient temperature correction charts provide temperature correction factors to derate motor specifications. Correction charts provide correction factors for temperatures above and below ambient temperature, but derating is typically applied when ambient temperature around the motor is above the listed ambient temperature rating of the motor. **See Figure 2-15.** For example, when the ambient temperature is 50°C (122°F), a motor must be derated to about 95% of full load.

Insulation Class

The *insulation class* is a code letter signifying the maximum operating temperature of the insulation used in a motor. Insulation class is abbreviated on the nameplate as INS CL. **See Figure 2-16.** Motor insulation prevents motor coils (windings) from shorting to each other or to ground (frame of motor). Insulation breakdown is the most common cause of motor failure.

Insulation is graded by its resistance to thermal aging and failure. The insulation class represents the maximum temperature at which the insulation can be operated to yield an average life of 20,000 hours. Each insulation class step—from A to B, B to F, and F to H—represents a 45°F (25°C) jump in the maximum operating temperature. If there is no temperature rating listed on the nameplate, the insulation is typically Class B. Class A is the least common insulation in use. Class F is the most commonly used motor insulation. Class H is the best rated and should be used in any application in which a motor drive is used to operate the motor.

The class of winding insulation affects the service factor rating. As the quality of the insulation rises, the service factor rating also improves. If the motor has been taken out of service and rewound, the insulation class of the winding is usually higher than the class listed on the nameplate. Most motor repair technicians use a higher class of winding conductor insulation during the rewinding process of a used or damaged motor. The higher class of insulation ensures that the motor's new service factor rating is equal to or greater than the motor's original service factor rating.

All insulation deteriorates over time due to the effects of thermal stress (heat), high-voltage spikes (transient voltages), contaminants, and mechanical stress. Heat buildup in a motor is caused by several factors such as the following:

- incorrect motor type or size for an application
- improper cooling, often from dirt build-up
- excessive load, often from improper use
- excessive friction, often from misalignment or vibration
- electrical problems, such as voltage unbalance, phase loss, or surge voltages
- harmonics on the supply lines, especially negative sequence harmonics (5th, 11th, 17th, etc.)
- nonsinusoidal waveforms produced by electric motor drives
- frequent starting and stopping

MECHANICAL-DESIGN CODES

Mechanical-design codes describe the different types of motor design features. Mechanical-design codes on a motor nameplate include frame size, NEMA design letter, enclosure type, and type of motor bearings. Following the guidelines on mechanical-design codes when installing or servicing a motor results in improved efficiency and lower operating cost.

Frame Size

The *frame size* is a number designating standard dimensions of a motor housing, shaft, and mounting holes. Frame size ratings are abbreviated on the nameplate as FR. All motors have a frame to protect the working parts of the motor and provide a means of mounting. Standardized dimensions allow for interchangeability among different motor manufacturers.

Frame size follows standards established by either the National Electrical Manufacturers Association (NEMA) or the International Electrotechnical Commission (IEC). Dimensionally, NEMA standards are expressed in English units and IEC standards are expressed in metric units. For both NEMA and IEC standards, the larger the frame size number, the larger the motor.

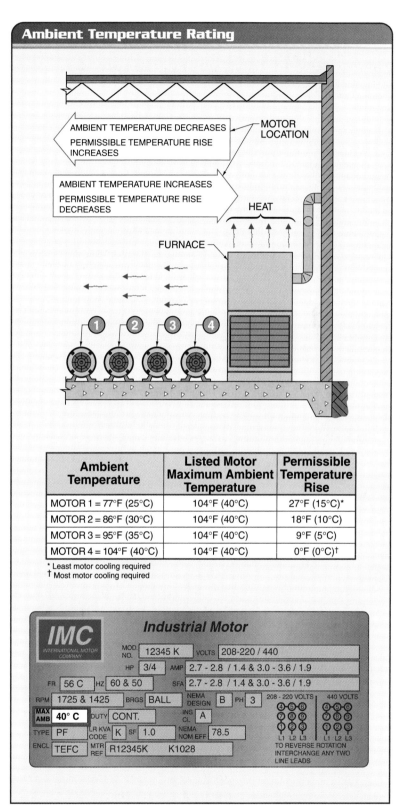

Figure 2-14. The ambient temperature rating is the maximum allowable temperature of the air surrounding an object. The permissible temperature rise is the difference between the ambient temperature and the nameplate ambient temperature rating of a motor.

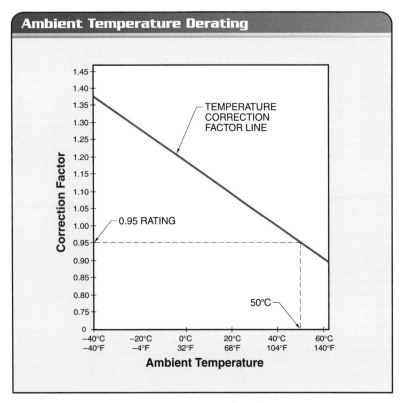

Figure 2-15. Ambient temperature correction charts provide temperature correction factors to derate motor specifications.

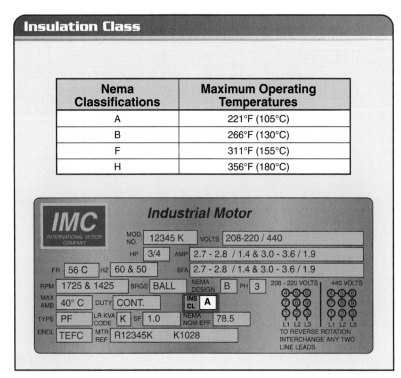

Figure 2-16. The insulation class is a code letter signifying the maximum operating temperature of the insulation used in a motor.

Motors manufactured before 1952 were made with standard frame sizes and are designated as "original." In 1952, new frame assignments were made to reflect improvements in motor design and manufacture. These motor frame sizes are designated as "U" frames. Some of these older motors are still operating in industry. In 1964, the modern "T" frame sizes were introduced. This designation is still in use. The actual dimensions are readily available in tables. **See Figure 2-17.** In some cases, a letter is added after the frame number to designate special-purpose motors, such as an S for short shaft for direct connections.

Fractional-Horsepower Frames. Fractional-HP motors all have two-digit frame numbers. Frame numbers for most NEMA motors with a dimension of 3½″ or less from the bottom of the base to the center of the shaft have two digits. For example, common NEMA frame sizes for small frame motors include 42, 48, and 56. Typically, all two-digit frame numbers are for motors with a power rating of less than 1 HP. However, there are a few motors with a power rating of more than 1 HP that fit into these frames, so it is somewhat misleading to refer to these as fractional-HP frames.

For fractional-HP frame sizes, the frame number represents the distance from the bottom of the base to the center of the shaft in 16ths of an inch. For example, a size 56 frame has a dimension of 3½″ (56 ÷ 16 = 3.5) from the bottom of the base to the center of the shaft. **See Figure 2-18.**

Integral-Horsepower Frames. Integral-HP motors have three- or four-digit frame numbers. For integral-HP frame sizes, the first two digits of the frame number represent the distance from the bottom of the base to the center of the shaft in 4ths of an inch. For example, a size 505 frame has a dimension of 12½″ (50 ÷ 4 = 12.5) from the bottom of the base to the center of the shaft.

Tech Fact

For motors larger than 200 HP, motor frames are not generally standardized because these motors need to be special-ordered from the manufacturer.

Motor Frame Dimensions

Frame No.	Shaft U	Shaft V	Key W	Key T	Key L	Dimensions — Inches A	B	D	E	F	BA
48	½	1½*	flat	3/64	—	5⅝*	3½*	3	2⅛	1⅜	2½
56	⅝	1⅞*	3/16	3/16	1⅜	6½*	4¼*	3½	2 7/16	1½	2¾
143T	⅞	2	3/16	3/16	1⅜	7	6	3½	2¾	2	2¼
145T	⅞	2	3/16	3/16	1⅜	7	6	3½	2¾	2½	2¼
182	⅞	2	3/16	3/16	1⅜	9	6½	4½	3¾	2¼	2¾
182T	1⅛	2½	¼	¼	1¾	9	6½	4½	3¾	2¼	2¾
184	⅞	2	3/16	3/16	1⅜	9	7½	4½	3¾	2¾	2¾
184T	1⅛	2½	¼	¼	1¾	9	7½	4½	3¾	2¾	2¾
203	¾	2	3/16	3/16	1⅜	10	7½	5	4	2¾	3⅛
204	¾	2	3/16	3/16	1⅜	10	8½	5	4	3¼	3⅛
213	1⅛	2¾	¼	¼	2	10½	7½	5¼	4¼	2¾	3½
213T	1⅜	3½	5/16	5/16	2⅜	10½	7½	5¼	4¼	2¾	3½
215	1⅛	2¾	¼	¼	2	10½	9	5¼	4¼	3½	3½
215T	1⅜	3½	5/16	5/16	2⅜	10½	9	5¼	4¼	3½	3½

* not NEMA standard dimensions

Designation

Letter	Designation
G	Gasoline pump motor
K	Sump pump motor
M and N	Oil burner motor
S	Standard short shaft for direct connection
T	Standard dimensions established
U	Previously used as frame designation for which standard dimensions are established
Y	Special mounting dimensions required from manufacturer
Z	Standard mounting dimensions except shaft extension

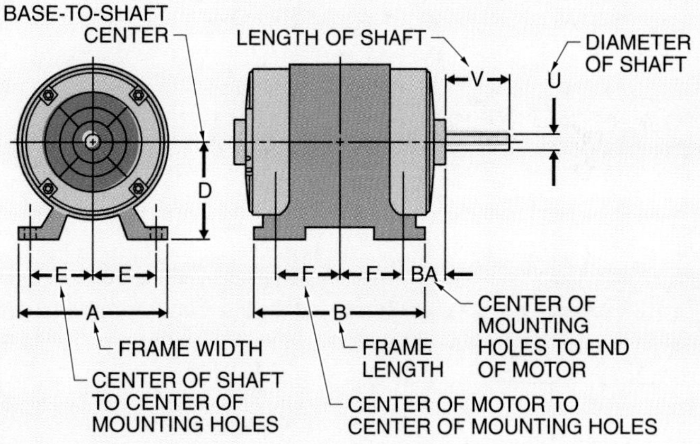
BASE-TO-SHAFT CENTER
LENGTH OF SHAFT
DIAMETER OF SHAFT
V
U
D
E
E
F
F
BA
A
B
FRAME WIDTH
CENTER OF SHAFT TO CENTER OF MOUNTING HOLES
FRAME LENGTH
CENTER OF MOUNTING HOLES TO END OF MOTOR
CENTER OF MOTOR TO CENTER OF MOUNTING HOLES

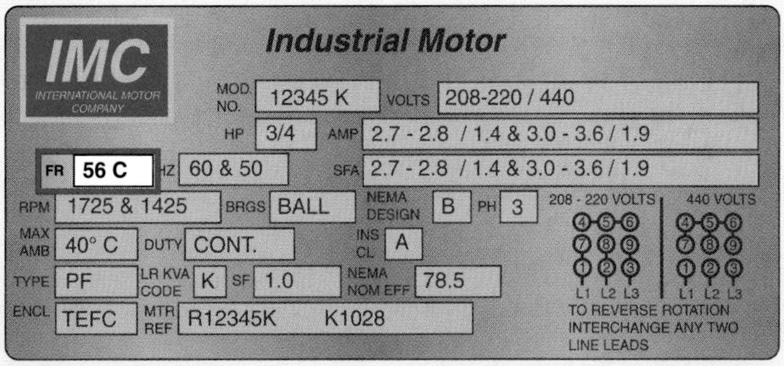
IMC
INTERNATIONAL MOTOR COMPANY
Industrial Motor
MOD. NO. 12345 K
VOLTS 208-220 / 440
HP 3/4
AMP 2.7 - 2.8 / 1.4 & 3.0 - 3.6 / 1.9
FR 56 C
HZ 60 & 50
SFA 2.7 - 2.8 / 1.4 & 3.0 - 3.6 / 1.9
RPM 1725 & 1425
BRGS BALL
NEMA DESIGN B
PH 3
208 - 220 VOLTS
440 VOLTS
MAX AMB 40° C
DUTY CONT.
INS CL A
TYPE PF
LR KVA CODE K
SF 1.0
NEMA NOM EFF 78.5
ENCL TEFC
MTR REF R12345K K1028
L1 L2 L3
L1 L2 L3
TO REVERSE ROTATION INTERCHANGE ANY TWO LINE LEADS

Figure 2-17. The frame size is a number designating standard dimensions of a motor housing, shaft, and mounting holes. Standardized dimensions allow for interchangeability among different motor manufacturers.

Fractional-HP Frames

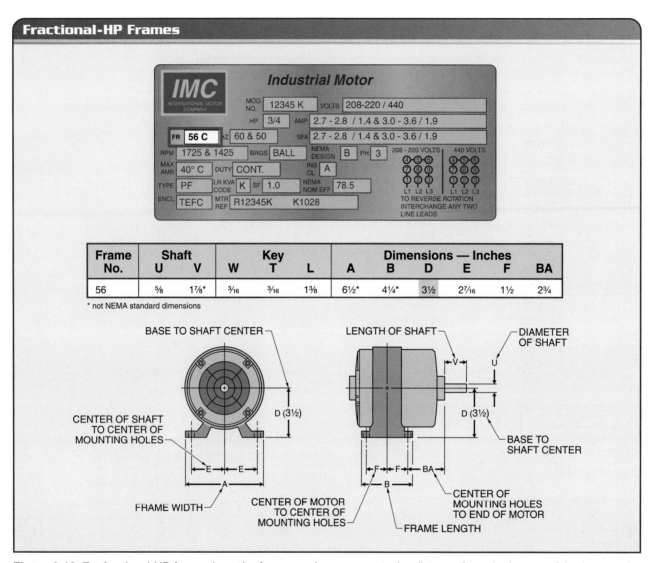

Figure 2-18. For fractional-HP frame sizes, the frame number represents the distance from the bottom of the base to the center of the shaft in 16ths of an inch.

> **Definition**
>
> The *NEMA design letter* is a code letter representing a National Electrical Manufacturers Association (NEMA) motor classification for the torque and current curves of a motor

The third and fourth (if required) numbers in the frame number are assigned by doubling dimension F of the motor and applying the Motor Frame Table. **See Figure 2-19.** A 505 frame has a value of 18.00 in the table at the intersection of row 500 and column 5. This means that dimension F of the motor is half that, or 9″. In other words, a 505 frame motor has a dimension of 12½″ from the base to the center of the shaft and 9″ from the center of the motor to the center of the mounting holes.

NEMA Design Letter

The *NEMA design letter* is a code letter representing a National Electrical Manufacturers Association (NEMA) motor classification for the torque and current curves of a motor. NEMA design letters are shown on the nameplate as NEMA DESIGN. **See Figure 2-20.** Motors with the same horsepower rating may need different torque curves and starting current. Motor designs may be listed as designs A through D. The design code letter indicated different rotor designs, with different resistance and reactance characteristics that affect the available torque and current.

The design rating listed on motor nameplates is used mostly to indicate the amount of starting torque a motor has in comparison to its running torque. Motors are suited for specific applications because each motor type has specific power, torque, and speed characteristics. The basic characteristics of each motor are determined by the design of the motor. Motors are designed for high efficiency, high starting torque, or high power factor, but generally cannot be optimized for all three characteristics.

Design code letters indicate motor speed and torque characteristics. Design code letters should not be confused with the locked-rotor indicating code letters. Design code letters indicate motor design characteristics while the locked-rotor indicating code letters indicate the current draw while the rotor is locked.

Integral-HP Frames

Frame No. Series	Third/Fourth Digit of Frame No.							
	D	1	2	3	4	5	6	7
140	3.50	3.00	3.50	4.00	4.50	5.00	5.50	6.25
160	4.00	3.50	4.00	4.50	5.00	5.50	6.25	7.00
180	4.50	4.00	4.50	5.00	5.50	6.25	7.00	8.00
200	5.00	4.50	5.00	5.50	6.50	7.00	8.00	9.00
210	5.25	4.50	5.00	5.50	6.25	7.00	8.00	9.00
220	5.50	5.00	5.50	6.25	6.75	7.50	9.00	10.00
250	6.25	5.50	6.25	7.00	8.25	9.00	10.00	11.00
280	7.00	6.25	7.00	8.00	9.50	10.00	11.00	12.50
320	8.00	7.00	8.00	9.00	10.50	11.00	12.00	14.00
360	9.00	8.00	9.00	10.00	11.25	12.25	14.00	16.00
400	10.00	9.00	10.00	11.00	12.25	13.75	16.00	18.00
440	11.00	10.00	11.00	12.50	14.50	16.50	18.00	20.00
500	12.50	11.00	12.50	14.00	16.00	18.00	20.00	22.00
580	14.50	12.50	14.00	16.00	18.00	20.00	22.00	25.00
680	17.00	16.00	18.00	20.00	22.00	25.00	28.00	32.00

Frame No. Series	Third/Fourth Digit of Frame No.							
	D	8	9	10	11	12	13	14	15
140	3.50	7.00	8.00	9.00	10.00	11.00	12.50	14.00	16.00
160	4.00	8.00	9.00	10.00	11.00	12.50	14.00	16.00	18.00
180	4.50	9.00	10.00	11.00	12.50	14.00	16.00	18.00	20.00
200	5.00	10.00	11.00	—	—	—	—	—	—
210	5.25	10.00	11.00	12.50	14.00	16.00	18.00	20.00	22.00
220	5.50	11.00	12.50	—	—	—	—	—	—
250	6.25	12.50	14.00	16.00	18.00	20.00	22.00	25.00	28.00
280	7.00	14.00	16.00	18.00	20.00	22.00	25.00	28.00	32.00
320	8.00	16.00	18.00	20.00	22.00	25.00	28.00	32.00	36.00
360	9.00	18.00	20.00	22.00	25.00	28.00	32.00	36.00	40.00
400	10.00	20.00	22.00	25.00	28.00	32.00	36.00	40.00	45.00
440	11.00	22.00	25.00	28.00	32.00	36.00	40.00	45.00	50.00
500	12.50	25.00	28.00	32.00	36.00	40.00	45.00	50.00	56.00
580	14.50	28.00	32.00	36.00	40.00	45.00	50.00	56.00	63.00
680	17.00	36.00	40.00	45.00	50.00	56.00	63.00	71.00	80.00

Figure 2-19. For integral-HP frame sizes, the first two digits of the frame number represent the distance from the bottom of the base to the center of the shaft in 4ths of an inch. The third and fourth (if required) numbers in the frame number are assigned by doubling dimension F of the motor and applying the Motor Frame Table.

NEMA Design Letters

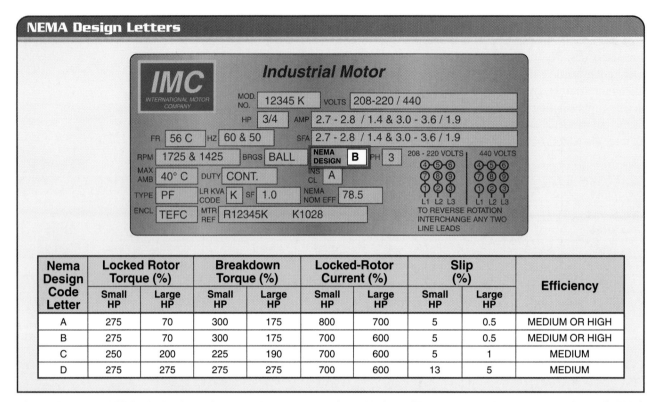

Figure 2-20. The NEMA design letter is a code letter representing a National Electrical Manufacturers Association (NEMA) motor classification for the torque and current curves of a motor.

Definition

A design A motor is a seldom-used integral-horsepower 3-phase induction motor design made for full-voltage starting, with normal values of breakdown torque and locked-rotor torque.

A design B motor is an integral-horsepower 3-phase induction motor design made for full-voltage starting, with normal values of breakdown torque and locked-rotor torque not exceeding a specified value.

A design C motor is an integral-horsepower 3-phase induction motor design made for full-voltage starting and high locked-rotor torque, with locked-rotor current not exceeding a specified value.

NEMA Design A Motors. A *design A motor* is a seldom-used integral-horsepower 3-phase induction motor design made for full-voltage starting, with normal values of breakdown torque and locked-rotor torque. The locked-rotor current is higher than for design B, C, and D motors. Design A motors have maximum 5% slip.

NEMA Design B Motors. A *design B motor* is an integral-horsepower 3-phase induction motor design made for full-voltage starting, with normal values of breakdown torque and locked-rotor torque not exceeding a specified value. A design B motor has maximum 5% slip. **See Figure 2-21.**

Design B motors have a locked-rotor torque of about 150% of full-load torque. The breakdown torque is about 200% of full-load torque. The starting current is up to about 600% to 650% of running full-load current.

Design B motors are the most common type of industrial motors in use today. Design B motors are designed for a broad variety of applications. They are often seen in HVAC applications providing power for fans and blowers and in industrial applications providing power for pumps.

NEMA Design C Motors. A *design C motor* is an integral-horsepower 3-phase induction motor design made for full-voltage starting and high locked-rotor torque, with locked-rotor current not exceeding a specified value. A design C motor has maximum 5% slip. Design C motors are suitable for equipment with high-inertia starts, such as positive-displacement pumps and heavily loaded conveyors. **See Figure 2-22.**

NEMA Design D Motors. A *design D motor* is an integral-horsepower 3-phase induction motor design made for full-voltage starting and locked-rotor torque of at least 275% of full-load torque, with locked-rotor current not exceeding a specified value. A design D motor has minimum 5% slip. Design D motors are suitable for equipment with very-high-inertia starts, such as cranes and hoists. **See Figure 2-23.**

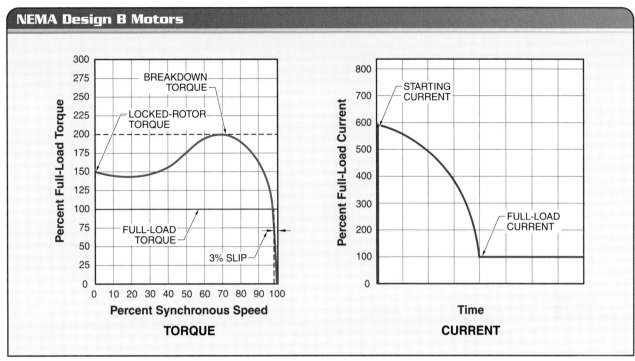

Figure 2-21. A design B motor has normal locked-rotor torque, low slip, and fairly high starting current.

Single-Phase Motor Designs. A *design L motor* is an integral-horsepower single-phase motor design made for full-voltage starting and locked-rotor torque higher than for design M, N, and O motors. A *design M motor* is an integral-horsepower single-phase motor design made for full-voltage starting and locked-rotor current not exceeding specified values. A *design N motor* is a fractional-horsepower single-phase motor design made for full-voltage starting and locked-rotor current not exceeding specified values. A *design O motor* is a fractional-horsepower single-phase motor designed for full-voltage starting and locked-rotor current not exceeding specified values, which are higher than for design N motors.

Enclosure Type

Enclosure type is the type of protection given to a motor to shield the motor from the outside environment as well as to protect individuals from the electrical and rotating parts of the motor. Enclosure type is abbreviated on the nameplate as ENCL.

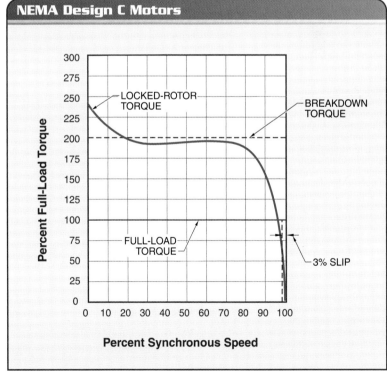

Figure 2-22. A design C motor has higher locked-rotor torque than a design B motor, with low slip.

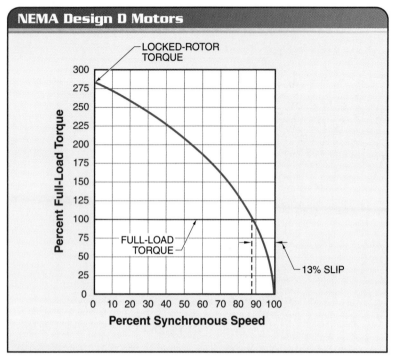

Figure 2-23. A design D motor has the highest locked-rotor torque, with higher slip than design B and design C motors.

Definition

A *design D motor* is an integral-horsepower 3-phase induction motor design made for full-voltage starting and locked-rotor torque of at least 275% of full-load torque, with locked-rotor current not exceeding a specified value.

Enclosure type is the type of protection given to a motor to shield the motor from the outside environment as well as to protect individuals from the electrical and rotating parts of the motor.

An *open motor enclosure* is a motor enclosure with openings to allow passage of air to cool the windings.

A *totally enclosed motor enclosure* is a motor enclosure that prevents air from entering the motor.

A wide variety of motor enclosures is available, with the type of enclosure affecting the cost of the motor. The different enclosure types provide simple and economical enclosure designs for standard operating conditions, while allowing for more-expensive enclosure designs for motors that require more protection from the environment. Typical operating environments for totally enclosed motor enclosures include the following:

- Wet locations (rain, snow, sleet, washdown, etc.)
- Dirty locations (dirt, noncombustible dust, etc.)
- Oily locations (lubricants, cutting oils, coolants, etc.)
- Corrosive locations (salt, chlorine, fertilizers, chemicals, etc.)
- Extremely low- and high-temperature locations (ranging from well below zero to above boiling water temperatures)
- Gas and other hazardous locations (liquid and solid)
- Combustible dust locations (grain and chemical)

Because of the many types of operating environments, two general classifications for motor enclosures have been developed. The two classifications are open motor enclosures and totally enclosed motor enclosures.

Open Motor Enclosures. An *open motor enclosure* is a motor enclosure with openings to allow passage of air to cool the windings. Open motor enclosures include general, drip-proof, splashproof, guarded, semiguarded, and drip-proof fully guarded types. **See Figure 2-24.**

Totally Enclosed Motor Enclosures. A *totally enclosed motor enclosure* is a motor enclosure that prevents air from entering the motor. Protecting individuals from the electrical and rotating parts of a motor can be accomplished using a basic enclosure. However, protecting a motor from all the outside environments motors must operate in requires much more than a basic enclosure. Totally enclosed motor enclosures include fan-cooled, nonventilated, pipe-ventilated, water-cooled, explosionproof, dust-ignition-proof, and waterproof types. **See Figure 2-25.**

Common Enclosure Abbreviations. The descriptions of the many types of open and totally enclosed motor enclosures are too long to include on the nameplate. Therefore, abbreviations are used. Typical motor enclosure abbreviations include the following:

- ODP (open drip-proof) with a fan and air openings for use in clean, dry, nonhazardous locations
- TENV (totally enclosed nonventilated) without a fan for use in moist, dirty, nonhazardous locations where a fan would clog
- TEFC (totally enclosed fan-cooled) with a small fan at the rear shaft of the motor for use in common industrial and commercial applications
- TEAO (totally enclosed air over) without a fan for use installed in the airstream of a driven blower or fan
- TEBC (totally enclosed blower-cooled) with a constant-speed fan for use with variable-speed drives

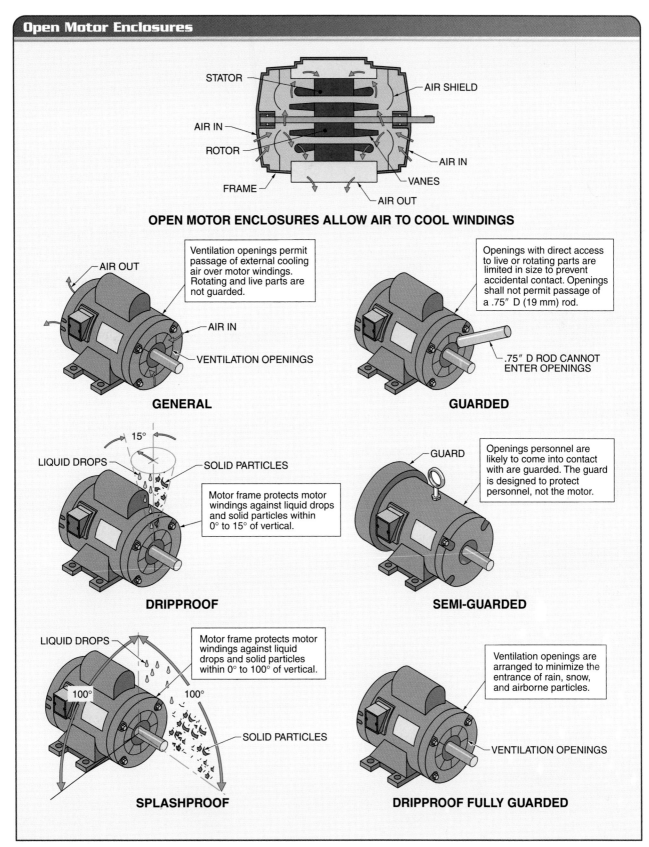

Figure 2-24. An open motor enclosure is a motor enclosure with openings to allow passage of air to cool the windings.

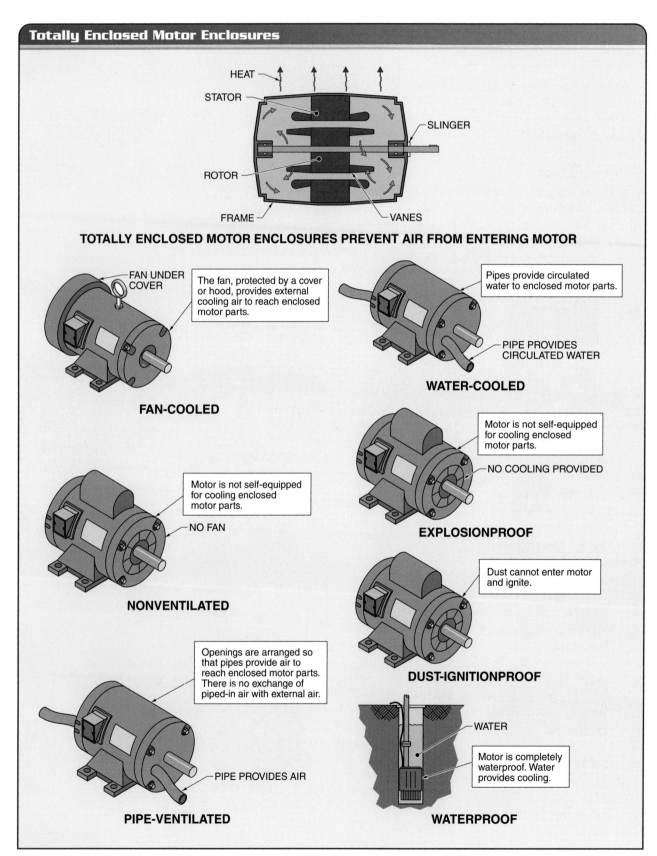

Figure 2-25. A totally enclosed motor enclosure is a motor enclosure that prevents air from entering the motor.

- TEWC (totally enclosed water-cooled) with water cooling for use in enclosed areas or for high power at low speeds

Hazardous-location motors include a hazardous-location listing specifying the hazardous classification and group rating (such as a Class 1, Group D rating).

Motor Bearings

A *motor bearing* is a machine component used to reduce friction and maintain clearance between stationary and moving parts. Bearing type is abbreviated on the nameplate as BRGS. **See Figure 2-26.** The type of bearing used on a motor is listed on the motor nameplate as "BALL" or "SLEEVE." Bearings may be either sleeve or ball designs. Both ball bearings and sleeve bearings are used with different-size motors ranging from fractional-HP to motors with hundreds of horsepower. Sleeve bearings are used where a low noise level is important, such as fan and blower motors. Ball bearings are used where higher load capacity is required or periodic lubrication is not practical.

Most motors include light-duty-rated or medium-duty-rated bearings. Heavy-duty-rated bearings are also used in some application-rated motors in which the load driven by the motor produces forces on the motor shaft. In some application-rated motors, the drive-end bearing has a higher load rating for longer motor life. Some ball and sleeve bearings include a method of lubricating the bearings (slot, fitting, etc.). Some bearings, such as those from mostly lower-power-rated motors, cannot be lubricated.

Motors are often lubricated at the factory to provide long operation without lubrication under normal service conditions. Frequent or excessive lubrication may damage a motor. The length of time between lubrications depends on the motor's service conditions, ambient temperature, and environment. The proper lubrication instructions are typically found on the motor nameplate or terminal box cover. Sleeve bearings should be lubricated according to the instructions provided with the motor. Ball bearings are designed for many years of operation without lubrication. Typical ball bearing lubrication schedules vary from 1 yr to 10 yr. While the vast majority of bearing failures are due to mechanical and thermal problems, a significant number of failures are due to bearing currents.

Tech Fact

Lubricant containers should be disposed of in an environmentally safe manner and in accordance with local, state, and federal regulations for petroleum distillates.

Definition

*A **motor bearing** is a machine component used to reduce friction and maintain clearance between stationary and moving parts. Bearing type is abbreviated on the nameplate as BRGS.*

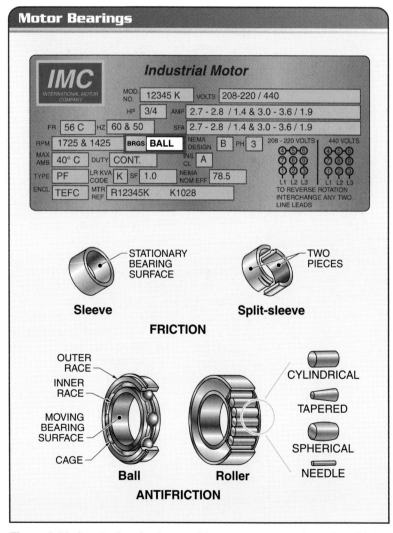

Figure 2-26. A motor bearing is a machine component used to reduce friction and maintain clearance between stationary and moving parts.

Application—Applying Motor Nameplate Specifications

Motors are purchased from motor manufacturers, electrical distributors, and/or electrical supply houses. A specific motor model is selected either because it meets a certain design established by an original equipment manufacturer (OEM) installing the motor as part of an original installation, or because it is needed to replace a damaged motor already in service.

If a motor is being selected as part of an original design, the information provided in the motor catalog or specification sheet is used to select the best model that meets the design requirements. If a motor is being selected to replace a damaged motor, the damaged motor's nameplate is used to select the best model to replace the damaged motor. For example, a replacement motor is required for a specific application. The requirements are for 5 HP, 230 V, 3-phase, 1800 rpm, 182T frame, and insulation class F. From the Motor Selection Table, the motor specified as stock number C-6 should work properly in this application.

When selecting a motor model to be used in either an original equipment and application design or as a replacement for a damaged motor, several factors must be considered. Considerations include the following:

- The motor cost. For a manufacturer of equipment that requires a motor, the purchase cost of the motor is important because it must be factored into the total cost of the equipment's list price. For customers who require the supplier to provide a written bid on equipment that they purchase, a few dollars' difference may mean the difference between a certain supplier getting the order or not. For the customer paying to operate the motor, the operating cost of the motor is taken into consideration.

- The amount of current the motor will draw. For any given size motor (in HP or kW), there is a choice of voltages that the motor can be connected to. A motor that can be connected to 460 V instead of 230 V will draw about half the amount of line current and still deliver the same amount of power. Since it is the current that determines conductor (wire) size, fuse size, and motor control size, the lower the current, the less the installation cost. See Conductor Ratings.

- The motor control method. If a motor is to be controlled with mechanical controls (magnetic motor starters, etc.), the mechanical controls will not add any additional stresses on the motor. However, if the motor is to be controlled by a variable-frequency drive, the drive will produce additional stress on the motor (voltage spikes, harmonics, etc.) that needs to be considered. Such problems are compensated for by using a better class of motor insulation (Class H instead of F, F instead of B, etc.), or by using a motor one size larger.

- The available voltage. Any change in the voltage applied to a motor will change the amount of current the motor draws. For example, a motor can be rated to operate on 208–230/460 volts. The reason for the 208 V rating is that it is the typical voltage delivered by a 120/208 V wye service, which is common in schools and other commercial locations. The reason for the 230 V rating is that it is the typical voltage delivered by a 120/240 V delta service, which is common in industrial locations. A ¾ HP 208–230/460 V rated motor will typically have a 2.7–2.6/1.3 A rating. Note that, although the motor can be connected to either 208 V or 230 V, the motor will actually draw more current at 208 V than at 230 V.

- The location in which the motor will be placed. A new motor will operate in almost any environment the motor is placed in. How long and how safely the motor will operate depends upon factors such as the motor's enclosure rating, insulation rating, service factor rating, and design code.

Motor Selection Table

NEMA RATED MOTORS

HP	Nameplate RPM	NEMA Frame	Enclosure	Volts 60 HZ	Full-load Amps	Service Factor	Nominal Efficiency	Insulation Class	Stock Number	Shpg. Wt.
¼ 1φ	3450	48	TEFL	115/230	6.4/3.2	1.25	85.5		1	17.0
¼ 1φ	1725	48	TEFL	115/230	6.2/3.1	1.25	82.5		2	18.5
½ 1φ	1725	48	TEFL	115/230	8.9/4.5	1.25	77.0		3	18.0
½ 1φ	1725	56	TEFL	100–120/200–240	7.3–6.6/4.1–3.8	1.25	82.5		4	22.5
1 1φ	1725	56	TEFL	115/208–230	12.3/5.2–5.4	1.35	82.5		5	26.5
1 1φ	1140	56	TEFL	115/230	11.0/5.5	1.35	85.5	F	6	30.2
¼ 3φ	3450	48	TEFL	208–230/460	1.4–1.2/0.7	1.25	77.0		1	14.0
¼ 3φ	1725	48	TEFL	208–230/460	1.2–1.1/0.65	1.25	81.5		2	16.0
½ 3φ	1725	56	TEFL	208–230/460	2.5–2.3/1.2	1.15	81.5	F	3	18.2
½ 3φ	1725	56	TEFL	230/460	2.0/1.0	1.25	84.0		4	19.2
1 3φ	1725	56	TEFL	208/230/460	3.4–3.2/1.6	1.25	82.5	F	5	21.0
1 3φ	1140	56	TEFL	208/230/460	3.2–3.1/1.5	1.25	84.0	F	6	22.0
2 3φ	3450	145T	ODP	230/460	5.5/2.8	1.15	84.0		1	33.0
2 3φ	1725	145T	ODP	230/460	6.0/3.0	1.15	84.0		2	35.5
3 3φ	3450	145T	ODP	230/460	9.2/4.6	1.15	86.5	F	3	41.0
3 3φ	1725	145T	ODP	230/460	9.0/4.5	1.15	86.5	F	4	45.0
5 3φ	3510	182T	ODP	230/460	14.4/6.7	1.15	87.5	F	5	75.0
5 3φ	1755	182T	TEFL	230/460	12.2/6.1	1.15	89.5	F	6	75.0
15 3φ	1770	215T	ODP	208–230/460	40.3–36.4/18.2	1.15	89.5	F	D 1	100
20 3φ	1765	254T	ODP	208–230/460	54–47.2/23.6	1.0	90.2	F	D 2	181
25 3φ	1775	256T	ODP	230/460	59/29.5	1.15	91.0	F	D 3	235
30 3φ	1775	286T	ODP	230/460	71.4/35.7	1.15	93.6	F	D 4	334
40 3φ	1780	324T	ODP	230/460	92.0/46.0	1.15	94.1	F	D 5	390
50 3φ	1775	326T	ODP	460	58.9	1.15	94.5	F	D 6	642

IEC RATED MOTORS

HP	kW	RPM at 60 Hz	IEC Frame	Enclosure IP-55 (Washdown)	Nameplate Volts at 60 HZ*	Full-load Amps	Service Factor	Nominal Efficiency	Insulation Class	Stock Number	Shpg. Wt.
¼ 3φ	0.18	3600	71A	IP-55	255–275/440–480	1.2/0.7	1.15	71.6	F	L-1	10.0
¼ 3φ	0.18	1800	71A	IP-55	255–275/440–480	1.5/0.8	1.15	68.3	F	L-2	12.4
½ 3φ	0.37	1800	71A	IP-55	255–275/440–480	2.0/1.2	1.25	72.8	F	L-3	13.1
½ 3φ	0.37	1800	80A	IP-55	255–275/440–480	1.7/0.9	1.15	74.7	F	L-4	13.6
1 3φ	0.75	1800	80A	IP-55	255–275/440–480	2.7/1.6	1.15	81.0	F	L-5	19.0
1 3φ	0.75	1200	80B	IP-55	255–275/440–480	3.0/1.8	1.15	78.2	F	L-6	18.2
2 3φ	1.5	3600	90L	IP-55	255–275/440–480	5.0/2.9	1.15	86.8	F	M-1	35.0
2 3φ	1.5	1800	90L	IP-55	255–275/440–480	5.2/3.1	1.15	85.6	F	M-2	33.5
3 3φ	2.2	3600	100LB	IP-55	255–275/440–480	7.1/4.1	1.25	87.8	F	M-3	43.0
3 3φ	2.2	1800	100LC	IP-55	255–275/440–480	7.4/4.6	1.25	86.5	F	M-4	42.0
5 3φ	4.0	3600	112M	IP-55	255–275/440–480	10.2/5.9	1.15	90.2	F	M-5	81.0
5 3φ	4.0	1800	112M	IP-55	255–275/440–480	11.4/6.6	1.15	87.8	F	M-6	79.0

* operable at 230/460

Application—Motor Insulation Life

Insulation breakdown is the main cause of motor failure. Different types of insulation have different tolerances for high operating temperatures. Although several factors, such as chemical corrosion, physical damage, heat, or moisture, can cause motor insulation to fail, overheating causes most insulation failure. Motors overheat when they are overloaded, required to operate in a hotter environment than the insulation is rated for, are supplied with any voltage other than the voltage specified, have blocked ventilation passageways, or are not properly mounted and balanced.

If motor is properly sized and mounted and the motor insulation is not damaged, a typical motor usually will have a life expectancy of 20,000 hr to 60,000 hr. Motor manufacturers use charts to show the expected insulation life based on the type of insulation and the maximum temperature the insulation can withstand.

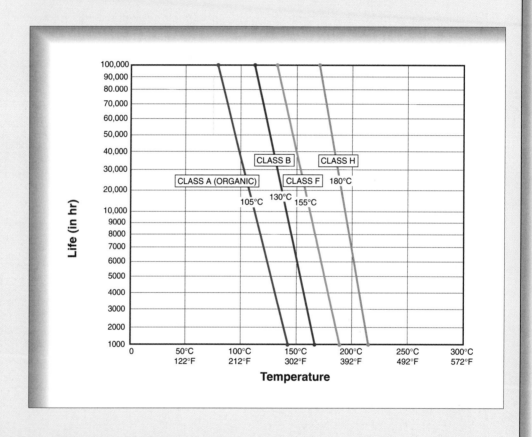

Summary

- Electrical ratings describe the electrical requirements for the operation of a motor.

- Electrical ratings included on motor nameplates include a voltage rating, current rating, frequency rating, and phase rating.

- Operating ratings describe how a motor is designed and how that design relates to where the motor is to be used.

- Operating ratings include the power rating, usage rating, service factor rating, speed rating, operating time rating (duty cycle), motor efficiency rating, ambient temperature rating, and insulation class.

- Mechanical-design codes describe the different types of motor design features.

- Mechanical-design codes on a motor nameplate include frame size, NEMA design letter, enclosure type, and type of motor bearings.

- Motors are rated for general-purpose usage or definite-purpose usage. Motors that are rated for general-purpose usage are used in a wide range of applications and for mechanical loads such as conveyors, machine tools, and belt-driven equipment. Definite-purpose motors are designed for applications in which the motor's usage is for a specific application.

- Energy-efficient motors are more expensive than standard motors but cost less to operate. Because motors consume so much energy, a higher efficiency rating results in lower operating cost to the user and requires less energy produced by the utility company.

- A wide variety of motor enclosures are available, with the type of enclosure affecting the cost of the motor. The different enclosure types provide simple and economical enclosure designs for simple operating conditions, while allowing for more-expensive enclosure designs for motors that require more protection from a harsher environment.

Glossary...

A *nameplate* is a metal tag permanently attached to an electric motor that gives the required electrical ratings, operating ratings, and mechanical-design codes of the motor.

The *voltage rating* is the voltage level that a motor can use.

The *current rating* is the amount of current a motor draws when delivering full rated power output.

The *locked-rotor indicating code letter* is a designation for the range of locked-rotor current draw per motor horsepower.

Locked-rotor current (LRC), or *starting current*, or *inrush current*, is the amount of current a motor draws on startup or when the rotor is locked.

The *frequency rating* is the power line frequency at which a motor is designed to operate.

The *phase* is the power phase (1φ, 3φ, or DC) that a motor requires for operation.

The *power rating* is the amount of power a motor can deliver to a load.

...Glossary

The ***usage rating*** is a description of the expected or allowed application of a motor.

The ***service factor rating*** is a multiplier that represents the amount of load, beyond the rated load, that can be placed on a motor without causing damage.

The ***speed rating*** of a motor is the approximate speed at which the rotor of a motor rotates when delivering rated power to a load.

Slip is the difference between the synchronous speed and rated speed of a motor.

The ***operating time rating***, or ***duty cycle***, is the amount of time a motor can be operated without being turned OFF to allow for cooling.

Motor efficiency is a measure of the effectiveness with which a motor converts electrical energy to mechanical energy.

The ***ambient temperature rating*** is the maximum allowable temperature of the air surrounding an object.

The ***temperature rise*** is the difference between the winding temperature of a running motor and the ambient temperature.

The ***permissible temperature rise*** is the difference between the ambient temperature and the nameplate ambient temperature rating of a motor.

The ***insulation class*** is a code letter signifying the maximum operating temperature of the insulation used in a motor.

The ***frame size*** is a number designating standard dimensions of a motor housing, shaft, and mounting holes.

The ***NEMA design letter*** is a code letter representing a National Electrical Manufacturers Association (NEMA) motor classification for the torque and current curves of a motor.

A ***design A motor*** is a seldom-used integral-horsepower 3-phase induction motor design made for full-voltage starting, with normal values of breakdown torque and locked-rotor torque.

A ***design B motor*** is an integral-horsepower 3-phase induction motor design made for full-voltage starting, with normal values of breakdown torque and locked-rotor torque not exceeding a specified value.

A ***design C motor*** is an integral-horsepower 3-phase induction motor design made for full-voltage starting and high locked-rotor torque, with locked-rotor current not exceeding a specified value.

A ***design D motor*** is an integral-horsepower 3-phase induction motor design made for full-voltage starting and locked-rotor torque of at least 275% of full-load torque, with locked-rotor current not exceeding a specified value.

A ***design L motor*** is an integral-horsepower single-phase motor design made for full-voltage starting and locked-rotor torque higher than for design M, N, and O motors.

A ***design M motor*** is an integral-horsepower single-phase motor design made for full-voltage starting and locked-rotor current not exceeding specified values.

A ***design N motor*** is a fractional-horsepower single-phase motor design made for full-voltage starting and locked-rotor current not exceeding specified values.

A ***design O motor*** is a fractional-horsepower single-phase motor designed for full-voltage starting and locked-rotor current not exceeding specified values, which are higher than for design N motors.

Enclosure type is the type of protection given to a motor to shield the motor from the outside environment as well as to protect individuals from the electrical and rotating parts of the motor.

An ***open motor enclosure*** is a motor enclosure with openings to allow passage of air to cool the windings.

A ***totally enclosed motor enclosure*** is a motor enclosure that prevents air from entering the motor.

A ***motor bearing*** is a machine component used to reduce friction and maintain clearance between stationary and moving parts.

Review

1. Why are electrical ratings provided on a motor nameplate?

2. Explain how the voltage and current ratings are used when purchasing or installing a motor.

3. Why are operating ratings provided on a motor nameplate?

4. Explain how high operating temperatures shorten the operating life of a motor. Explain why intermittent-duty motors require time to cool off.

5. Why are mechanical-design codes provided on a motor nameplate?

6. Explain the difference between a NEMA design B motor and a NEMA design C motor. Why would one type be selected over another?

7. List at least three applications where a definite-purpose motor must be used instead of a general-purpose motor.

8 Describe some of the design components that improve the energy efficiency of motors.

9. Explain the difference between open motor enclosures and totally enclosed motor enclosures. List several situations where a totally enclosed motor enclosure should be used instead of an open motor enclosure.

Refer to the CD-ROM for Quick Quiz® questions related to chapter content.

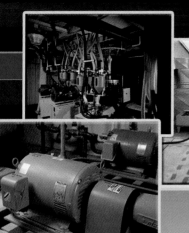

MOTORS

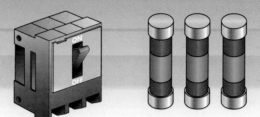
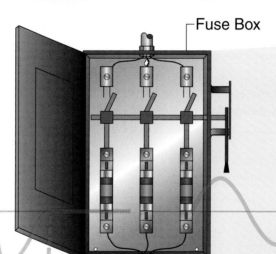

Fuse Box

Circuit Breaker

Motor Starter

Motor

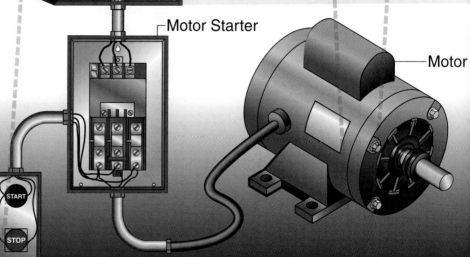

3

Motor Protection

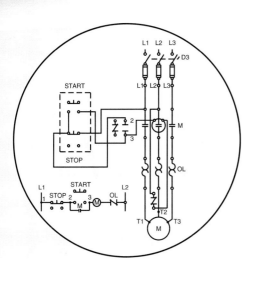

Motor Protection Requirements	54
Short Circuit and Ground Fault Protection	56
Application—Troubleshooting Fuses	65
Application—Troubleshooting Circuit Breakers	66
Motor Overload Protection	67
Supplemental Topic—Overload Ambient Temperature Compensation	73
Supplemental Topic—Motor Control Circuit Protection	74
Application—Sizing Overload Protection	76
Summary	77
Glossary	78
Review	79

OBJECTIVES

- Explain the difference between an overcurrent, a short circuit, a ground fault, and an overload.
- Describe why an overcurrent protection device is used.
- List and describe the different types of plug fuses and cartridge fuses.
- Describe the different ratings for overcurrent protection devices.
- Describe motor-starting protection and motor-running protection.
- Demonstrate how to troubleshoot fuses and circuit breakers.
- List and describe the different types of circuit breakers.
- Describe the operation of thermal overload relays, electromagnetic overload relays, and electronic overload relays.

Motors and circuits must be protected from overcurrents and overloads. A motor draws a much higher current during startup than when operating under load. The protection requirements for starting a motor are very different than the protection requirements to operate a motor under load. A combination of fuses, circuit breakers, and overload relays can be used to provide motor and circuit protection.

Definition

An *overcurrent* is any current over the normal current level.

A *short circuit* is an excessive current that leaves the normal current-carrying path by going around the load and back to the power source.

MOTOR PROTECTION REQUIREMENTS

Motors must be protected from short circuits, ground faults, and overloads. Fuses, circuit breakers (CBs), or overload relays are typically used to protect the motor. Fuses and CBs are used for ground fault and short-circuit protection. Overload relays are used for overload protection. **See Figure 3-1.** When properly selected, these devices provide a safety link in the motor circuit. They disconnect the circuit when there is a problem and are the first place to start when troubleshooting a motor or motor circuit.

Overcurrent

An *overcurrent* is any current over the normal current level. Current flows through and is confined to the conductive paths provided by conductors and other components when a load is turned ON. Under normal circumstances, a load draws the amount of current for which the load was designed. The NEC® requires overcurrent protection of motor installations.

Some overcurrents are to be expected and are normal. Others are very dangerous. One normal overcurrent occurs whenever a motor is started. A motor draws many times its running (normal) current when starting. **See Figure 3-2.** An overcurrent may be caused by a short circuit, a ground fault, or an overload.

Short Circuits. A *short circuit* is an excessive current that leaves the normal current-carrying path by going around the load and back to the power source. **See Figure 3-3.** Overloads occur in most circuits at modest levels. However, a short circuit causes the current to rise hundreds of times higher than normal at a very fast rate. A typical short circuit occurs when the insulation between two wires is broken. The current bypasses the load and returns to the source through the damaged insulation and not through the load.

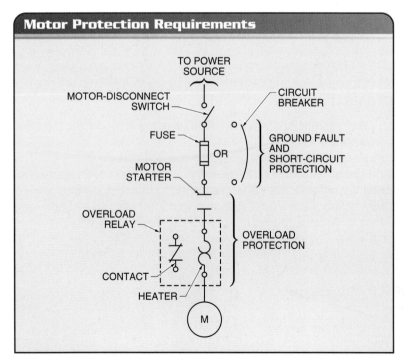

Figure 3-1. Fuses, circuit breakers, and relays are used to protect a motor from short circuits, ground faults, and overloads.

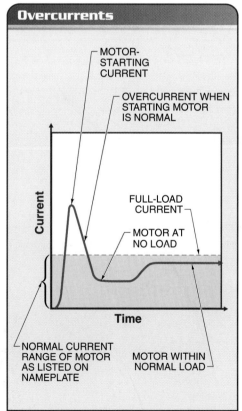

Figure 3-2. An overcurrent is any current over the normal current level.

Ground Faults. A *ground fault* is an unintentional connection between an ungrounded conductor and any grounded raceway, box, enclosure, or fitting. Like short-circuit currents, ground-fault currents follow an abnormal path. In this case however, the path of current is from an ungrounded conductor to ground. Ground faults can occur due to insulation failures or more commonly at conductor terminations.

Ground-fault currents are generally of a lower magnitude than short-circuit currents. According to the NEC®, care shall be taken when installing electrical conductors to ensure that insulation is not damaged, as this could contribute to a ground fault.

Overloads. An *overload* is an excessive current that is confined to the normal current-carrying conductors and is caused by a load that exceeds the full-load torque rating of the motor. An overload occurs whenever the current in a circuit rises above the normal current level for which the circuit has been designed to operate. Overloads are usually between one and six times the normal current level of the circuit.

A temporary overload can occur when a motor starts up and draws much more current than the full-load current (FLC). This type of overload is very common and typically is not damaging. A continuous overload occurs when a motor is driving a load greater than the design load. This may occur when a motor is used in the wrong application or when there is motor damage, such as worn bearings. When overloaded, the motor still attempts to drive the load. To produce the higher torque, the motor draws a higher current. **See Figure 3-4.**

Tech Fact

Manufacturers often provide manuals and troubleshooting charts with their products to assist in locating the cause of a problem. These manuals should be consulted when a problem occurs because they provide detailed information for finding a permanent solution to a specific problem. These charts and manuals may also be found on the Internet.

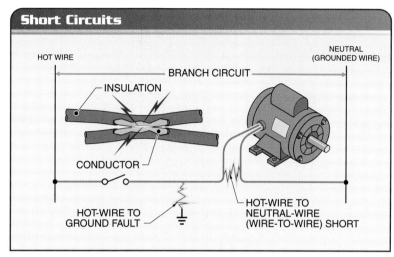

Figure 3-3. A typical short circuit occurs when the insulation between two wires is broken. The current bypasses the load and returns to the source through the bare wire and not through the load.

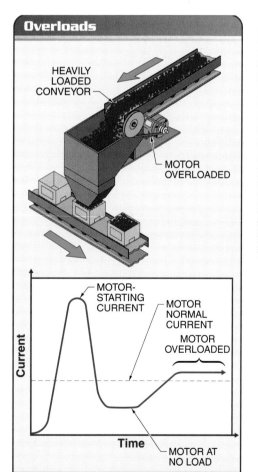

Figure 3-4. An overload is usually caused by the temporary surge current that occurs whenever motors are started. An overload can also occur if a motor is overloaded while running.

Definition

*A **ground fault** is an unintentional connection between an ungrounded conductor and any grounded raceway, box, enclosure, or fitting.*

*An **overload** is an excessive current that is confined to the normal current-carrying conductors and is caused by a load that exceeds the full-load torque rating of the motor.*

Definition

An overcurrent protection device (OCPD) is a set of fuses or circuit breakers added to provide overcurrent protection of a switched circuit.

SHORT-CIRCUIT AND GROUND FAULT PROTECTION

An *overcurrent protection device (OCPD)* is a set of fuses or circuit breakers added to provide overcurrent protection of a switched circuit. An OCPD must be used to provide protection from short circuits and ground faults. Fuses and CBs are OCPDs designed to automatically stop the flow of current in a circuit that has an overcurrent.

Overcurrent protection devices are classified by how quickly they activate. **See Figure 3-5.** A non-current-limiting device operates slowly, allowing damaging short-circuit currents to build up to full values before opening. If non-current-limiting devices are used, short-circuit currents may reach 10,000 A or more in the first half cycle of a fault. The high current also causes very strong magnetic fields that can cause severe damage to a circuit or equipment.

A current-limiting device opens the circuit in less than one-quarter cycle of short-circuit current, before the current reaches its highest value and limiting the amount of destructive energy allowed into the circuit. Most modern OCPDs are current-limiting devices. Current-limiting devices restrict fault currents to levels low enough to minimize damage.

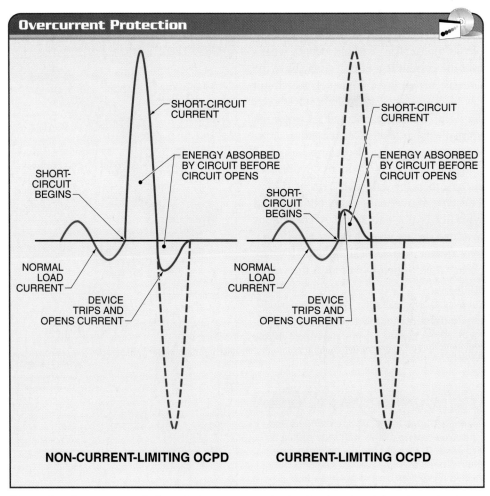

Figure 3-5. A non-current-limiting device operates slowly, allowing damaging short-circuit currents to build up to full values before opening. A current-limiting device opens the circuit in less than one-quarter cycle of short-circuit current, before the current reaches its highest value and limiting the amount of destructive energy allowed into the circuit.

Motor Circuit Protection

The function of an OCPD is to protect branch-circuit conductors, control equipment, and motors from damage caused by an overcurrent, such as a short circuit or ground fault. The OCPD must clear this type of fault, but must not open the circuit as a result of the normal momentary inrush current of a motor starting. OCPDs are installed in the combination starter, safety switch, or fuse panel. The OCPD is located before the motor starter. The OCPD will automatically activate if there is a problem when the motor is started. This will turn the motor OFF. The motor can be restarted by fixing the problem that caused the overcurrent and by changing the fuses or resetting the CBs.

Because power conductors must be protected against overcurrents, a fuse or CB must be installed in every ungrounded power conductor. **See Figure 3-6.** One OCPD is required for low-voltage, single-phase circuits (120 V or less) and all DC circuits. The neutral line (in AC circuits) or the negative line (in DC circuits) does not include an OCPD. Two OCPDs are required for single-phase, high-voltage circuits (208 V, 230 V, or 240 V). Both ungrounded power lines include an OCPD. Three OCPDs are required for all 3-phase circuits (any voltage). All three ungrounded power lines include an OCPD.

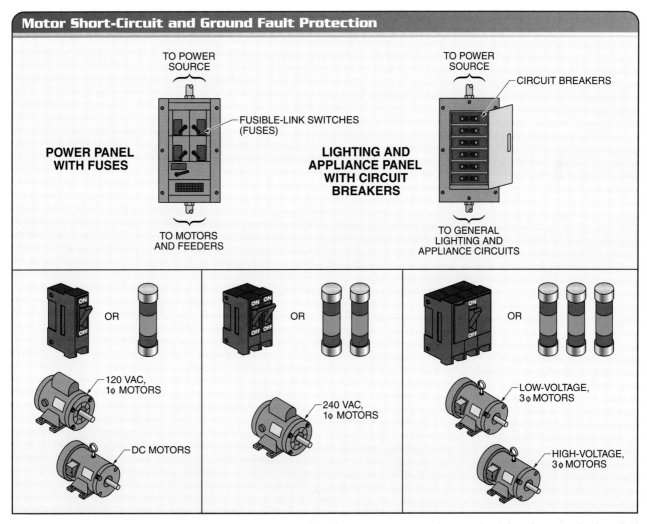

Figure 3-6. Because power conductors must be protected against overcurrents, a fuse or circuit breaker must be installed in every ungrounded power conductor.

Definition

A fuse is an overcurrent protection device with a fusible link that melts and opens the circuit when an overcurrent occurs.

A plug fuse is a screw-in fuse that uses a metallic strip that melts when a predetermined amount of current flows through it.

Fuses

A *fuse* is an overcurrent protection device with a fusible link that melts and opens the circuit when an overcurrent occurs. Fuses are very low-resistance devices connected in series with the circuit's conductors. When the circuit current exceeds the rating of the fuse, the fuse opens and prevents current from flowing in that part of the circuit.

Fuse Operation. The basic component of the fuse is the fusible link. During an overcurrent condition, the high current causes resistive heating. Because of the resistive heating, the fusible link overheats and melts, creating an open circuit. A fuse may have multiple links, depending on the fuse rating. The link is contained entirely within a tube or cartridge. If an overcurrent occurs, the link melts and opens the circuit. Fuses themselves do not need to be maintained, while the terminations, disconnects, and clips should be inspected regularly. Fuses are commonly available as plug fuses and cartridge fuses.

Plug Fuses. A *plug fuse* is a screw-in fuse that uses a metallic strip that melts when a predetermined amount of current flows through it. **See Figure 3-7.** Plug fuses are inserted in series with ungrounded conductors to protect the conductors against overcurrents that could damage the conductor or the connected equipment. At one time, plug fuses were the overcurrent protection device most commonly used in residential and light commercial electrical installations.

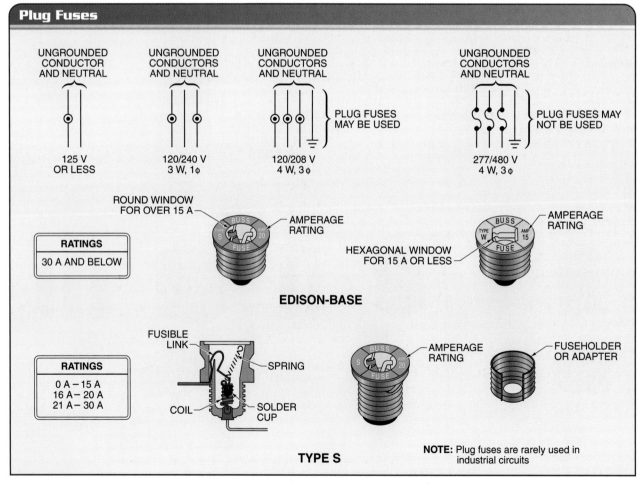

Figure 3-7. Plug fuses are inserted in series with ungrounded conductors to protect the conductors against overcurrents that could damage the conductor or the connected equipment.

An *Edison-base fuse* is a plug fuse that incorporates a screw configuration that is interchangeable with fuses of other amperage ratings. Edison-base fuses are the older of the plug-fuse designs. Edison-base fuses are classified as not over 125 V and are designed to be used in circuits with amperage ratings of 30 A and below. Because the Edison-base fuse is not designed to be noninterchangeable with fuses of other amperage ratings, they are permitted to be installed only as replacements for existing installations where there is no evidence of overfusing.

A *type S fuse* is a plug fuse that incorporates a screw-and-adapter configuration that is not interchangeable with fuses of another amperage rating. Like Edison-base plug fuses, type S fuses are designed for circuits not exceeding 125 V. Type S fuses have three amperage rating classes: 0–15 A, 16–20 A, and 21–30 A. The fuses are noninterchangeable with amperage ratings of a lower rating, which protects the circuit from the possibility of overfusing. Additionally, Type S fuses are designed with fuseholders and adapters so that the adapters fit Edison-base fuseholders and are nonremovable. Once a type S adapter is installed, a type S fuse shall be used.

Cartridge Fuses. A *cartridge fuse* is a snap-in cylindrical fuse constructed of a fusible metallic link or links that is designed to open at predetermined current levels to protect circuit conductors and equipment. **See Figure 3-8.** Cartridge fuses are classified according to their performance, operation, and construction characteristics.

Cartridge fuses are available in many more types and in more amperage ratings than are plug fuses. Amperage ratings for cartridge fuses are available from 0 A to 6000 A. Cartridge fuses are available in either the one-time type or the renewable type. A one-time cartridge fuse must be replaced after it operates, whereas a renewable cartridge fuse contains a fusible link that can be replaced after it operates.

Cartridge fuses can be constructed with a ferrule or knife-blade configuration. Both of the types are inserted into fuseholders that make the connection from the line to the load. A cartridge fuse can also be classified as either a non-time-delay fuse (NTDF) or a time-delay fuse (TDF).

Three-phase circuits require one fuse for each phase.

Definition

*An **Edison-base fuse** is a plug fuse that incorporates a screw configuration that is interchangeable with fuses of other amperage ratings.*

*A **type S fuse** is a plug fuse that incorporates a screw-and-adapter configuration that is not interchangeable with fuses of another amperage rating.*

*A **cartridge fuse** is a snap-in cylindrical fuse constructed of a fusible metallic link or links that is designed to open at predetermined current levels to protect circuit conductors and equipment.*

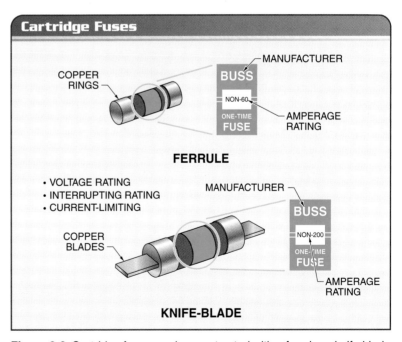

Figure 3-8. Cartridge fuses can be constructed with a ferrule or knife-blade configuration. Both of the types are inserted into fuseholders that make the connection from the line to the load.

Definition

A *non-time-delay fuse (NTDF)* is a single-element fuse that can detect an overcurrent and open the circuit almost instantly.

A *time-delay fuse (TDF)* is a dual-element fuse that can detect and remove a short circuit almost instantly, but allow small overloads to exist for a short period.

A *non-time-delay fuse (NTDF)* is a single-element fuse that can detect an overcurrent and open the circuit almost instantly. NTDFs contain a fusible link with a notched section that melts and opens the circuit at a set overcurrent. **See Figure 3-9.** The fusible link is heated by the current. The higher the current, the higher the temperature. Under normal operation and when the circuit is operating at or below its amperage rating, the fusible link simply functions as a conductor. However, if an overcurrent occurs in the circuit, the temperature of the fusible link reaches the melting point at the notched section. The notched section melts and burns back, causing an open in the fuse. If the overcurrent is caused from an overload, only one notched link opens. If the overcurrent is caused from a short circuit, several notched links open. NTDFs are also known as single-element fuses. Non-time-delay fuses are fast acting and provide excellent overcurrent protection.

A *time-delay fuse (TDF)* is a dual-element fuse that can detect and remove a short circuit almost instantly, but allow small overloads to exist for a short period. Time-delay fuses allow overcurrents to exist for short periods to avoid nuisance tripping due to equipment start-up characteristics.

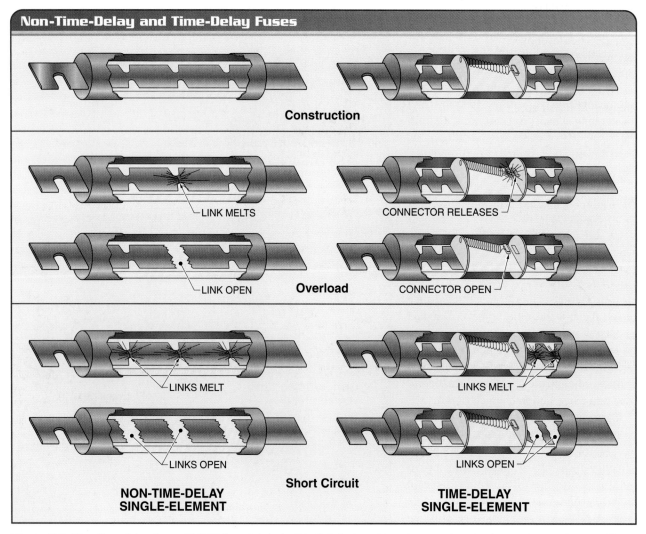

Figure 3-9. Non-time-delay fuses (NTDFs) contain a fusible link that melts and opens the circuit at a set overcurrent. Time-delay fuses (TDFs) are fuses that may detect and remove a short circuit almost instantly, but allow small overloads to exist for a short period.

Time-delay fuses include two elements. One element removes overloads from the circuit, and the other element removes short circuits from the circuit. The short-circuit element is the same type of fusible-link element used in a non-time-delay fuse. The overload element is a spring-loaded device that opens the circuit when solder holding the spring in position melts. The solder melts after an overload exists for a short period (normally several seconds). The overload element protects against temporary overloads up to approximately 800% of rated current. TDFs are also known as dual-element fuses.

TDFs are used in circuits that have temporary, low-level overloads. Motor circuits are the most common type of circuit that includes low-level overloads. Motors draw an overload current when starting. As the motor accelerates, the current is reduced to a normal operating level. Therefore, TDFs are used in motor applications.

Circuit Breakers

A *circuit breaker (CB)* is an overcurrent protection device with a mechanical mechanism that automatically opens a circuit when an overload condition or short-circuit occurs. **See Figure 3-10.** CBs are designed to perform the same basic functions as fuses. CBs offer the added benefit of not damaging themselves and they are resettable after an overload is cleared. CBs that clear short circuits and ground faults can be subjected to extremely high-current levels. CBs are available in single-, two-, and three-pole configurations.

CBs are designed to be trip free. Essentially, this design permits the internal operation of the CB to occur regardless of whether the handle is locked in the closed position or not. CBs are mechanical devices and require periodic maintenance. Many manufacturers include testing requirements in their instructions for CBs that have cleared short circuits or ground faults. Whenever the CB contacts are opened, an arc is drawn across the movable and stationary contacts. A short circuit produces a much hotter arc than an overload. Therefore, a CB that has tripped numerous times from short circuits may be damaged to the point of requiring replacement.

Circuit Breaker Operation. CBs can be designed to be operated either manually, automatically, or both manually and automatically. Although the design characteristics may be different, all CBs contain elements that sense a fault or overload condition and open or interrupt the flow of current. The design elements determine the operational characteristics of the CB.

Definition

*A **circuit breaker (CB)** is an overcurrent protection device with a mechanical mechanism that automatically opens a circuit when an overload condition or short-circuit occurs.*

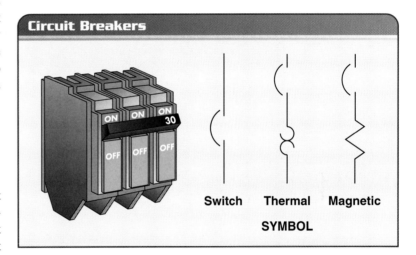

Figure 3-10. CBs are designed to perform the same basic functions as fuses.

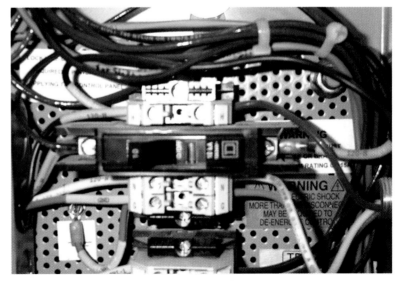

Single-phase circuit breakers are used with single-phase motors.

Thermal CBs contain a spring-loaded electrical contact that opens the circuit. The spring is used to open and close the contacts with a fast snap action. A handle is added to the contact assembly so the contacts may be manually opened and closed. The contacts are automatically opened on an overcurrent by a bimetal strip and/or an electromagnetic tripping device. **See Figure 3-11.** The contacts have one stationary contact and one movable contact. The movable contact is attached to a spring that provides a fast snap action when tripped.

The bimetal strip is made of two dissimilar metals that expand at different rates when heated. The bimetal strip is connected in series with the circuit and is heated by the current flowing through it. The strip bends when heated and opens the contacts. The higher the circuit current, the hotter the bimetal strip becomes. Likewise, the higher the current, the shorter the time required to trip the CB.

Like the bimetal strip, the electromagnetic CB is connected in series with the circuit. As current passes through the coil, a magnetic field is produced. The higher the circuit current, the stronger the magnetic field. When the magnetic field becomes strong enough, the magnetic field pulls in the armature and opens the contacts.

Electromagnetic CBs use the magnetic effect created when current flows in a conductor. One of the most common CB designs combines a magnetic element with a thermal element. These thermal-magnetic CBs use both the thermal and magnetic effects associated with current flow to sense current and operate the CB. Generally, the thermal element is used for overload protection while the magnetic element provides faster protection for ground faults and short circuits.

Recent developments in CB design have led to electronically operated "smart" CBs. Smart CBs incorporate electronic components to accomplish the monitoring, sensing, and tripping mechanisms of the CB. Smart CBs are true-rms (root-mean-square) symmetrical current-sensing devices that use a microprocessor to gain closer control over the performance of their operations. Typical settings and adjustments for these CBs include continuous amperes, long time delay, short time pickup, short time delay, instantaneous pickup, and ground-fault pickup. Common types of CBs include inverse-time CBs, adjustable-trip CBs, non-adjustable-trip CBs, and instantaneous-trip CBs.

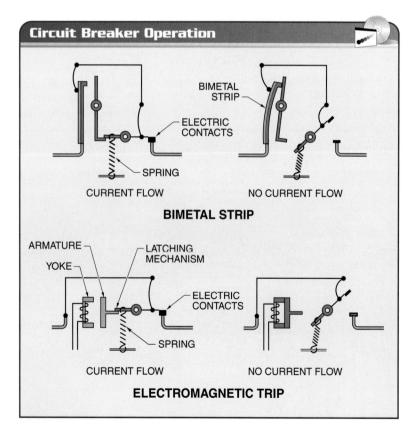

Figure 3-11. CBs contain a spring-loaded electrical contact that opens the circuit.

Tech Fact

When more than one live conductor is present in a branch circuit, each conductor must be protected by a breaker. All live conductors in a circuit must be interrupted whenever one conductor is overloaded. Therefore, a common-trip breaker must be used. For small breakers, the operating handles are typically tied together. For larger breakers, there are two or three tripping mechanisms tied together within the base. For 120/240 V systems, two-pole common-trip breakers are used for 240 V loads on the circuit.

Inverse-Time Circuit Breaker. An *inverse-time circuit breaker (ITCB)* is a circuit breaker (CB) with an intentional delay between the time when the fault or overload is sensed and the time when the CB operates. **See Figure 3-12.** The relationship between the trip action and the delay is such that the smaller the overload or fault, the longer the delay. Conversely, the larger the overload or fault, the faster the device operates.

Circuit Breaker Types	
Type	**Characteristics**
ITCB	Fixed current setpoint with time delay inversely proportional to overload
ATCB	Adjustable current and/or trip-time setpoints
NATCB	Factory-present setpoints with no provision for adjustments
ITCB	Instant trip, for use with listed motor controllers only

Figure 3-12. Common types of CBs include inverse-time CBs (ITCB), adjustable-trip CBs (ATCB), non-adjustable-trip CBs (NATCB), and instantaneous-trip CBs (ITCB).

ITCBs are used extensively in electrical distribution systems and for many different types of applications. ITCBs provide protection from normal overloads and against damaging currents from short circuits and ground faults. This is accomplished in a thermal-magnetic CB by using the thermal element for overload protection and the magnetic element for short-circuit and ground-fault protection.

Adjustable-Trip Circuit Breaker. An *adjustable-trip circuit breaker (ATCB)* is a circuit breaker (CB) whose trip setting can be changed by adjusting the current setpoint, trip-time characteristics, or both, within a particular range. These CBs are used in industry where equipment may be frequently changed and different overcurrent protection characteristics are required.

For example, an ATCB with an 800 A frame size may be installed to protect a feeder for a downstream subpanel. The initial setpoint of the CB is 600 A. If the load on the subpanel increases and the feeder conductors are sized for the additional load, the setpoint of the ATCB can be increased to the new value. This provides great flexibility to the electrical installation at a considerable cost savings.

Non-Adjustable-Trip Circuit Breaker. A *non-adjustable-trip circuit breaker (NATCB)* is a circuit breaker (CB) whose settings for the amperage-trip setpoint or the time-trip setpoint cannot be changed. These CBs are used when equipment with known and constant operational characteristics requires protection. For example, if a piece of equipment with a full-load amperage rating of 60 A is installed, an NATCB can be installed because the operational characteristics of the equipment are constant and there is no need to increase the size or rating of the circuit supplying the equipment.

Instantaneous-Trip CB. An *instantaneous-trip circuit breaker (ITB)* is a circuit breaker (CB) with no delay between the fault- or overload-sensing element and the tripping action of the device. Because many types of equipment, such as motors, are subject to surge or start-up currents much higher than normal operating levels, ITBs are sized well above the amperage rating for normal operation. For example, motors with ITBs for the branch-circuit, short-circuit, and ground-fault protection may have the CB sized at up to 1300% of the FLC when starting current is a problem. ITBs may only be used in combination with a listed motor controller.

> **Definition**
>
> An *inverse-time circuit breaker (ITCB)* is a circuit breaker (CB) with an intentional delay between the time when the fault or overload is sensed and the time when the CB operates.
>
> An *adjustable-trip circuit breaker (ATCB)* is a circuit breaker (CB) whose trip setting can be changed by adjusting the current setpoint, trip-time characteristics, or both, within a particular range.
>
> A *non-adjustable-trip circuit breaker (NATCB)* is a circuit breaker (CB) whose settings for the amperage-trip setpoint or the time-trip setpoint cannot be changed.
>
> An *instantaneous-trip circuit breaker (ITB)* is a circuit breaker (CB) with no delay between the fault- or overload-sensing element and the tripping action of the device.

> **Tech Fact**
>
> Circuit breakers used on high-voltage power transmission lines are almost always solenoid operated because of the difficulty of measuring current at high voltages. A current transformer is used to reduce the current in the transmission line to a level that can be measured with standard equipment.

Definition

*The **current rating** is the continuous amount of current that can be safely carried by an overcurrent protection device without blowing or tripping.*

*The **voltage rating** is the maximum amount of voltage that can safely be applied to an overcurrent protection device and determines the ability of a fuse to suppress the internal arcing after the fusible link melts.*

*The **interrupting rating** is the maximum amount of current that can be safely applied to an OCPD while still maintaining its physical integrity when reacting to fault currents.*

Overcurrent Protection Device Ratings

Like other electrical devices and machines, an OCPD must be rated for the conditions to which it will be operated. For example, fuses and CBs have current ratings, voltage ratings, and interrupting ratings.

Current Ratings. The *current rating* is the continuous amount of current that can be safely carried by an overcurrent protection device without blowing or tripping. The current rating of the OCPD is determined by the size and type of conductors, control devices used, and loads connected to the circuit. The current rating should not normally exceed the circuit capacity. Consideration must be given to the type of load and code requirements. There are some specific circumstances given in the NEC® where the current rating is permitted to exceed the circuit capacity. For electric motors, fuses must be sized to survive the inrush current inherent in motor starting.

Fuses and ITCBs are available in standard ratings. **See Figure 3-13.** The NEC® also permits the use of fuses and inverse-time CBs with nonstandard amperage ratings. In addition to the values in the table, standard amperage ratings for fuses can include 1, 3, 6, 10, and 601. For ATCBs, the rating is the maximum possible setting.

Fuses and Inverse-Time Circuit Breakers (ITCBs)

Increase	Standard Amperage Ratings
5	15, 20, 25, 30, 35, 40, 45, 50
10	50, 60, 70, 80, 90, 100, 110
25	125, 150, 175, 200, 225, 250
50	250, 300, 350, 400, 450, 500
100	500, 600, 700, 800
200	1000, 1200
400	1600, 2000
500	2500, 3000
1000	3000, 4000, 5000, 6000

Figure 3-13. Fuses and ITCBs are available in standard ratings.

Voltage Ratings. The *voltage rating* is the maximum amount of voltage that can safely be applied to an overcurrent protection device and determines the ability of a fuse to suppress the internal arcing after the fusible link melts. Article 240 of the National Electric Code® (NEC®) contains a listing of other articles that apply to the use of OCPDs. All fuses must have voltage ratings suitable for the electrical circuits in which they are used. Cartridge fuses rated at 600 V are commonly used in 480 V motor circuits. However, an allowed maximum voltage of 300 V line-to-neutral is allowed where a single-phase circuit is supplied from a three-phase, 4-wire, solidly grounded neutral system. Additionally, cartridge fuses shall not be used in circuits greater than that for which they are rated. Typical voltage ratings of cartridge fuses are 250 V, 300 V, and 600 V.

Plug fuses have a voltage rating of 125 V between conductors in most applications. The maximum voltage rating, however, for circuits supplied from a system using a grounded neutral is 150 V to ground. Thus, for a standard 120/240 V, single-phase, 3-wire system, plug fuses can be used for the protection of both 120 V and 240 V circuits.

All plug fuses shall be marked to indicate their amperage rating. By design, the plug fuse amperage rating is identified by the shape of the window on the fuse. Plug fuses with amperage ratings of 15 A or less have a hexagonal window, while those with ratings greater than 15 A have a round window.

Interrupting Rating. The *interrupting rating* is the maximum amount of current that can be safely applied to an OCPD while still maintaining its physical integrity when reacting to fault currents. An OCPD must be able to withstand the damaging effects of an overcurrent without rupturing and causing additional damage. The NEC® requires that equipment intended to break current at fault levels have an interrupting rating sufficient for the current that must be interrupted.

Application—Troubleshooting Fuses

Fuses are connected in series with a circuit to protect a circuit from overcurrents or shorts. Fuses can be checked using a continuity tester placed across a fuse that has been removed from a circuit. Fuses can also be checked using a DMM or voltmeter. All electrical work done on potentially live circuits must be done while wearing the proper PPE for the job. To troubleshoot fuses, apply the procedure:

1. Turn the handle of the safety switch or combination starter to the OFF position.
2. Open the door of the safety switch or combination starter. The operating handle must be capable of opening the switch. If it is not, replace the switch.
3. Check the enclosure and interior parts for deformation, displacement of parts, and burning. Such damage may indicate a short, fire, or lightning strike. Deformation requires replacement of the part or complete device. Any indication of arcing damage or overheating, such as discoloration or melting of insulation, requires replacement of the damaged part(s).
4. Check the voltage between each pair of power leads. Incoming voltage should be within 10% of the voltage rating of the motor. A secondary problem exists if voltage is not within 10%. This secondary problem may be the reason the fuses have blown.
5. Test the enclosure for grounding if voltage is present and at the correct level. To test for grounding, connect one side of a voltmeter to an unpainted metal part of the enclosure and touch the other side to each of the incoming power leads. If the enclosure is grounded, a voltage difference will be present. The line-to-ground voltage probably does not equal the line-to-line voltage reading taken in Step 4.
6. Check the fuses. Remove the fuses with a fuse puller. Use the DMM set to resistance to measure the resistance of each fuse. If the fuse is blown, the DMM reading will show OL. A high resistance also indicates that the fuse has been damaged and should be replaced. The resistance of a good fuse should be very low, typically less than a few ohms. In addition, verify that all three fuses have very similar resistances.
7. Replace any bad fuses. Replace all bad fuses with the correct type and size replacement. It is very important to match the interrupting rating. Close the door on the safety switch or combination starter and turn the circuit ON.

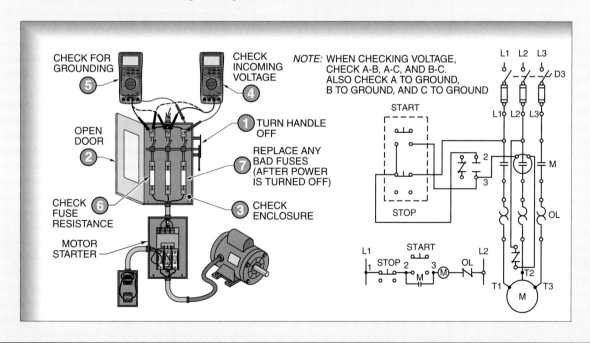

Application—Troubleshooting Circuit Breakers

CBs are connected in series with a circuit to protect a circuit from overcurrents or shorts. CBs are thermally or magnetically operated and are reset after a short-circuit or ground fault has cleared. A multimeter or voltage meter is used to test CBs. CBs perform the same function as fuses and are tested the same way. All electrical work done on potentially live circuits must be done while wearing the proper PPE for the job. To troubleshoot CBs, apply the procedure:

1. Turn the handle of the safety switch or combination starter to the OFF position.
2. Open the door of the safety switch or combination starter. The operating handle must be capable of opening the switch. If it is not, replace the switch.
3. Check the enclosure and interior parts for deformation, displacement of parts, and burning. Such damage may indicate a short, fire, or lightning strike. Deformation requires replacement of the part or complete device. Any indication of arcing damage or overheating, such as discoloration or melting of insulation, requires replacement of the damaged part(s).
4. Check the voltage between each pair of power leads. Incoming voltage should be within 10% of the voltage rating of the motor. A secondary problem exists if voltage is not within 10%. This secondary problem may be the reason the CBs have tripped.
5. Test the enclosure for grounding if voltage is present and at the correct level. To test for grounding, connect one side of a voltmeter to an unpainted metal part of the enclosure and touch the other side to each of the incoming power leads. If the enclosure is grounded, a voltage difference will be present. The line-to-ground voltage probably does not equal the line-to-line voltage reading taken in Step 4.
6. Examine the CB. It will be in one of three positions, ON, TRIPPED, or OFF.
7. If no evidence of damage is present, reset the CB by moving the handle to OFF and then to ON. CBs must be cooled before they are reset. CBs are designed so they cannot be held in the ON position if a short is present. If resetting the CB does not restore power, use the multimeter to check the voltage of the reset CB. Never try to service a faulty CB.

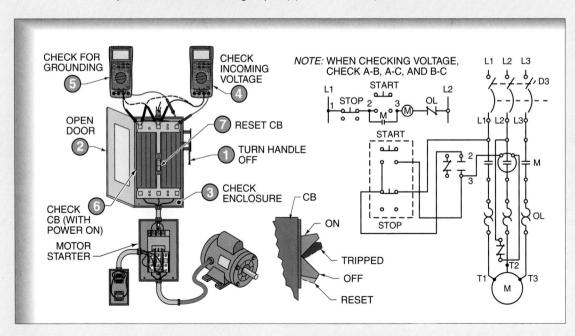

MOTOR OVERLOAD PROTECTION

All overloads shorten motor life by deteriorating motor insulation. Small overloads of short duration cause little damage. However, a sustained overload, even though small, will damage the motor. An overload protection device is designed to carry harmless overloads, but quickly remove the motor from the supply voltage when an overload has remained too long. **See Figure 3-14.** The NEC® requires overload protection of motor installations. Common overload protection devices include thermal and magnetic overload relays.

Motor-Running Protection

After a motor has started and has accelerated to its rated speed, the motor draws enough current from the power lines to remain running. The amount of current a motor draws from the power lines while running is primarily dependent upon the connected load. *Full-load current (FLC)* is the amount of current drawn when the motor is connected to the maximum load the motor is designed to drive. The FLC is listed on the motor nameplate. The load a motor can turn is based on the motor design. If a motor is connected to an easy-to-turn load, the motor draws less than FLC. If a motor is connected to a hard-to-turn load, the motor draws more than FLC. If a motor is not connected to a load, the motor draws only the amount of current required to keep the shaft rotating at the rated rpm.

The overload relays in a motor starter are time-delay devices that allow temporary overloads without disconnecting the motor. If an overload is present for longer than the preset time, the overload relays trip and disconnect the motor from the circuit. Overload relays on manual starters can be reset after tripping by pressing the stop button and then the start button, or by pressing the reset button on units that have one. On magnetic starters set for manual reset, the overloads are reset by pressing the reset button next to the overload contact. On magnetic starters set for automatic reset, the overloads reset automatically after the unit has cooled.

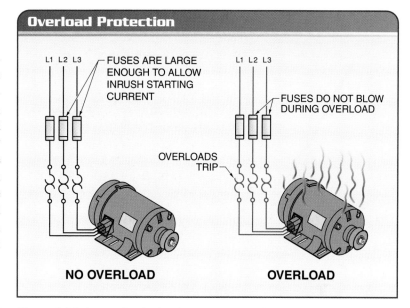

Figure 3-14. An overload relay does not open a circuit while a motor is starting, but opens the circuit if the motor becomes overloaded and the fuses do not open.

Tech Fact

When installing (or replacing) fuses, install so that the written information on the fuse label is displayed facing out and can be clearly read.

Definition

Full-load current (FLC) is the amount of current drawn when the motor is connected to the maximum load the motor is designed to drive.

ASI Robicon
Overload protection must be provided to ensure that a motor is stopped if a load jams and stops moving.

Definition

*The **overload class rating** is an indicator of the maximum length of time it takes for the overload relay to trip at 600% overload.*

*A **thermal overload relay** is an overload relay that uses the resistive heating to open a set of contacts*

*A **melting-alloy overload relay**, or **heater**, is an overload relay that uses a heater coil to produce the heat to melt a eutectic alloy.*

*A **eutectic alloy** is an alloy that has a constant melting-point temperature due to the combination of the given components.*

Overload Class Ratings

The *overload class rating* is an indicator of the maximum length of time it takes for the overload relay to trip at 600% overload. **See Figure 3-15.** For example, a Class 10 overload relay trips in 10 seconds or less and a Class 20 overload relay trips in 20 seconds or less at a 600% overload. The time it takes to trip is more at a smaller overload and less at a larger overload.

The motor application must be considered when selecting an overload relay class. Most typical applications should use a Class 20 overload relay. If a longer trip time is required, such as for a motor driving a high-inertia load, a Class 30 overload relay should be selected. If a short trip time is required, such as for a motor with a limited capacity to disperse heat, a Class 10 overload relay should be selected.

Thermal Overload Relays

A *thermal overload relay* is an overload relay that uses the resistive heating to open a set of contacts. The normal resistive heating in a conductor increases the temperature. As the level of current passing through the relay reaches an overload condition, the increased temperature opens a set of contacts. The increased temperature opens the contacts by melting an alloy or by bending a bimetallic strip to activate a mechanism that operates the contacts.

Melting-Alloy Overload Relays. A *melting-alloy overload relay*, or *heater*, is an overload relay that uses a heater coil to produce the heat to melt a eutectic alloy. **See Figure 3-16.** A *eutectic alloy* is an alloy that has a constant melting-point temperature due to the combination of the given components. This temperature never changes and is not affected by repeated melting and resetting.

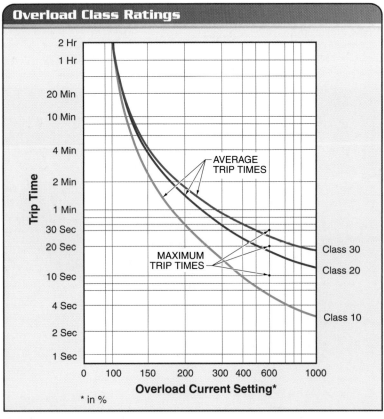

Figure 3-15. The overload class rating is a measure of the maximum length of time it takes for the overload relay to trip at 600% overload.

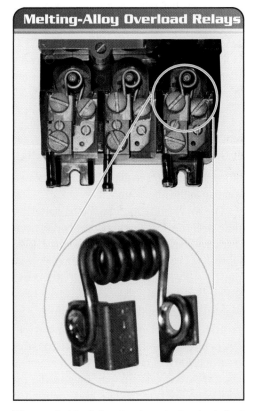

Figure 3-16. A heater has to be installed into a motor starter, with one heater for each ungrounded power line.

Many different types of heater coils are available. The operating principle of each type of coil is the same. A heater coil converts the current drawn by a motor into heat. Current in excess of the rating generates enough heat to melt the eutectic alloy. The time to melt the alloy also depends on the ambient temperature. The eutectic alloy is used in conjunction with a mechanical mechanism to activate a tripping device when an overload occurs.

The eutectic alloy tube consists of an outer tube and an inner shaft connected to a ratchet wheel. The ratchet wheel is held firmly in the tube by the solid eutectic alloy. The inner shaft and ratchet wheel are locked into position by a pawl (locking mechanism) so that the wheel cannot turn when the alloy is cool. **See Figure 3-17.**

The main device in an overload relay is the eutectic alloy tube. The compressed spring tries to push the normally closed overload contacts open when motor current conditions are normal. The pawl is caught in the ratchet wheel and does not let the spring push up to open the contacts. During an overload condition, the heat melts the alloy, which allows the ratchet wheel to turn. The spring pushes the reset button up, which opens the contacts to the voltage coil of the contactor.

The contactor opens the circuit to the motor, which stops the current flow through the heater coil. The heater coil cools, which solidifies the eutectic alloy tube. Only the normally closed overload contacts open during an overload condition. The normally closed overload contacts can be manually reset to the closed position. The actual heaters installed in the motor starter do not open during an overload. The heaters are only used to produce heat. The higher the current draw of the motor, the more heat produced.

Tech Fact

An overload current produces dangerous excess heat in the conductors. The heat causes the eutectic alloy to melt, opening the contacts and stopping the motor. After the eutectic alloy cools and solidifies, the heater can be reset and the motor restarted.

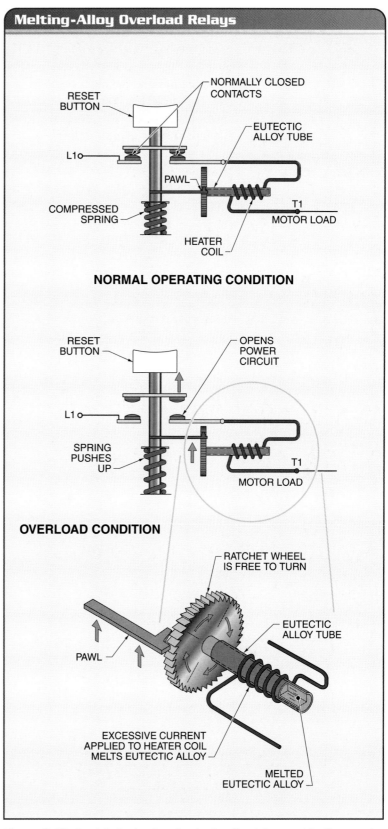

Figure 3-17. A ratchet wheel and eutectic alloy tube combination can be used to activate a trip mechanism when an overload occurs.

Definition

A *bimetallic-strip overload relay* is a relay consisting of two joined pieces of dissimilar metals with different expansion rates constructed into a strip in such a way that the strip bends when heated and opens a set of contacts.

A heater must be installed in a motor starter. One heater is required for each ungrounded power line. **See Figure 3-18.** For all single-phase, 115 V, AC motors, one line is the hot line and the other is the neutral line. The neutral line is connected to ground at the service panel. Therefore, one heater is required in the hot power line.

For single-phase, 230 V, AC motors, both power lines are hot. Therefore, one heater in each hot power line is required when controlling a single-phase, 230 V, AC motor. For all three-phase motors, one heater is required in each power line. For all DC motors, the positive line is not grounded and the negative line is grounded. One heater in the positive power line is required when controlling a DC motor.

Bimetallic-Strip Overload Relays. A *bimetallic-strip overload relay* is a relay consisting of two joined pieces of dissimilar metals with different expansion rates constructed into a strip in such a way that the strip bends when heated and opens a set of contacts. **See Figure 3-19.** A bimetallic-strip overload relay can be designed to reset automatically. In certain applications, such as walk-in meat coolers, remote pumping stations, and some chemical-processing equipment, overload relays that reset automatically to keep the unit operating up to the last possible moment may be required.

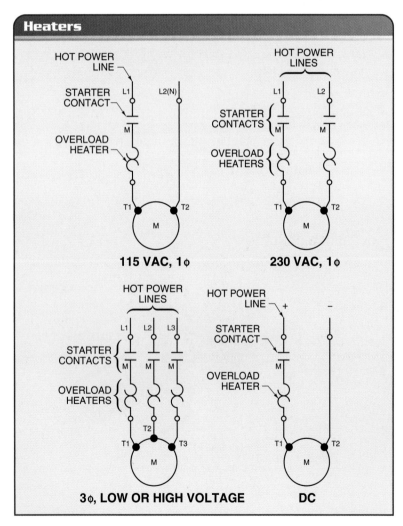

Figure 3-18. A heater coil is used to produce the heat generated by excessive current. The heat created through ambient temperature rise also helps melt the eutectic alloy.

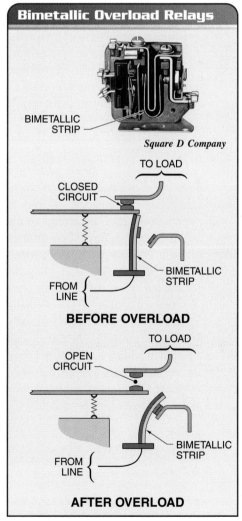

Figure 3-19. The bending effect of a bimetallic strip is used to open the contacts.

Internal Motor Temperature Protection

Internal motor temperature protectors are typically assembled as an integral part of a motor. These protectors are used to provide protection from overheating caused by an overload or from failure to start. Depending on the design, these protectors may automatically reset or may require a manual reset. The types of internal motor protection are current-sensitive protection, which responds to the current draw, and temperature-sensitive protection, which responds to the actual motor temperature.

Current-responsive protectors prevent overheating from common causes of overload where the current increases above the expected line current. However, they do not prevent overheating caused by hot ambient conditions or blocked ventilation. Current-responsive protectors are often current relays that open the circuit when the current exceeds a specified maximum.

For high-slip induction motors, the hottest spot in a motor is often in the rotor. Since the rotor is the main moving part of a motor, a temperature measurement in the rotor also requires slip rings to transmit the measurement out of the motor. This is expensive and unreliable, so current-responsive protectors are usually used in this situation.

Temperature-responsive devices protect against overloads that produce an increase in winding temperature. Thermistors or RTDs are often used as the temperature-sensing device and can be placed almost anywhere in the motor. Special motors may be designed for continuous locked-rotor current, but can still burn out from blocked ventilation. Temperature-responsive devices are usually used in this situation.

Thermostats are sometimes mounted to the motor frame. The thermostats are usually connected in series with the motor starter holding coil. If the motor overheats, the thermostat opens the circuit and stops the motor. The motor cannot be restarted until the thermostat cools down. The thermostat leads are usually labeled P or P1 and P2.

When automatically reset, a motor restarts even when the overload has not been cleared, and trips and resets itself again at given intervals. Care must be exercised in the selection of a bimetallic overload relay because repeated cycling can eventually cause the motor to burn out.

Electromagnetic Overload Relays

An *electromagnetic overload relay* is a relay that operates on the principle that as the level of current in a circuit increases so does the strength of the magnetic field produced by that current flow. When the level of current through a current coil in the circuit reaches the preset value, the increased magnetic field acts as a solenoid and opens a set of contacts. **See Figure 3-20.** A current-sensing core is pulled in when the current in the coil reaches the overload level. When the core is pulled in, it presses against a trip pin that opens the contacts.

> **Definition**
>
> An **electromagnetic overload relay** is a relay that operates on the principle that as the level of current in a circuit increases so does the strength of the magnetic field produced by that current flow.

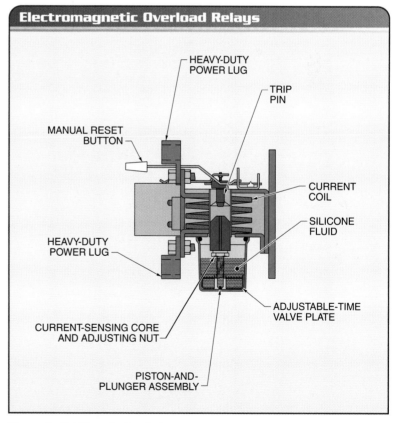

Figure 3-20. When the level of current in the circuit reaches the preset value, the increased magnetic field around the coil pulls in the armature and opens a set of contacts.

72 MOTORS

Definition

An electronic overload relay is a device that has built-in circuitry to sense changes in current and temperature.

Electromagnetic overload relays are often used in heavy-duty applications or special applications where normal time/current curves of thermal overload relays do not provide satisfactory operation. This flexibility is possible because an electromagnetic overload relay may be set for either instantaneous or inverse-time characteristics. Some models include a viscous silicone fluid and an adjustable-time valve plate that are used to adjust the time before the relay trips.

Electronic Overload Relays

An *electronic overload relay* is a device that has built-in circuitry to sense changes in current and temperature. New manual starters normally include an electronic overload relay instead of heaters. An electronic overload relay monitors the current in the motor directly by measuring the current in the power lines leading to the load. The electronic overload is built directly into the motor starter. **See Figure 3-21.**

An electronic overload measures the strength of the magnetic field around a wire instead of converting the current into heat. The higher the current in the wire leading to a motor, the stronger the magnetic field produced. An electronic circuit is used to activate a disconnecting device that opens the starter power contacts. Electronic overloads have an adjustable range. The setting is based on the nameplate current listed on the motor.

Large-HP motors have currents that exceed the values of standard overload relays. To make the overload relays larger would greatly increase their physical size, which would create a space problem in relation to the magnetic motor starter. To avoid such a conflict, current transformers are used to reduce the current by a fixed ratio. **See Figure 3-22.** A current transformer is not used to change the amount of current flowing to a motor but reduces the secondary current to a lower value to allow a smaller overload relay to be used. For example, if 50 A were flowing to a motor, only 5 A would flow to the overload relay through the use of the current transformer. Standard current transformers are normally rated for a ratio of primary to secondary current, such as 50:5 or 100:5.

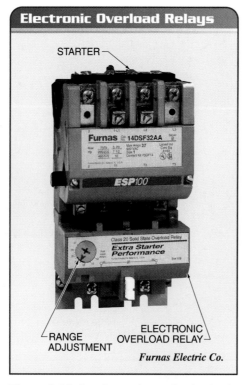

Figure 3-21. An electronic overload relay is built directly into a motor starter.

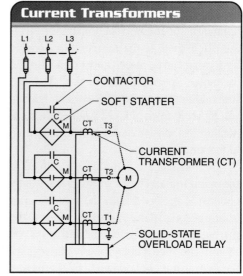

Figure 3-22. A current transformer reduces the current to a lower value to allow a smaller overload relay.

Because the ratio between the current in the power circuit and the current to the relay is always the same, an increase in the current to a motor also increases the current to the overload relay. If the correct current transformer and overload relay combination is selected, the same overload protection can be provided to a motor as if the overload relay were in the load circuit. The overload relay contacts open and the coil to the magnetic motor starter is de-energized when excessive current is sensed, shutting the motor OFF. Several different current transformer ratios are available to provide this type of overload protection.

Overload Ambient Temperature Compensation

Thermal overloads are heat-sensing devices that provide a means of monitoring the current drawn by a motor. Thermal overloads trip when the heat generated by motor windings approaches a harmful level. Because thermal overloads are temperature dependent, ambient temperature must be accounted for in applications where the ambient temperature varies above or below the standard rating temperature of 104°F. If the ambient temperature of the motor is different than that of the overloads, the overloads can cause nuisance tripping or motor burnout.

For example, if non-temperature-compensated overloads are used for a well pump motor that is at a different temperature than the surface-mounted overloads, the overloads are adjusted for the difference in temperature. Graphs show temperature adjustments for different ambient temperatures.

Once the corrected overload trip current is found, it is used instead of the nameplate current when sizing overcurrent protection. To find the overload trip current using an ambient temperature correction, apply the procedure:

1. Determine ambient temperature.
2. Find the rated current (%) multiplier for the ambient temperature on the graph.
3. Multiply the FLC on the motor nameplate by the rated current multiplier to obtain the corrected overload trip current to be used.

For example, a motor has an FLC of 25 A. The ambient temperature is 130°F. The overload trip current is found as follows:

1. Determine ambient temperature.

   The ambient temperature is 130°F.

2. Find the rated current (%) for the ambient temperature on the graph.

   From the graph, the rated current multiplier is 0.9 (90%).

3. Multiply the FLC on the motor nameplate by the rated current multiplier to obtain the corrected overload trip current to be used.

   Overload trip current = 25 × 0.9

   Overload trip current = 22.5 A

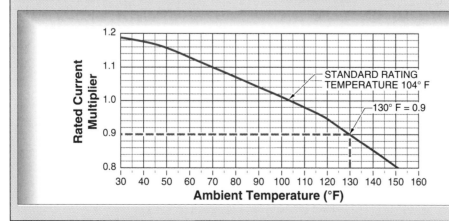

Supplemental Topic

Motor Control Circuit Protection

A motor control circuit is the circuit of a motor control apparatus or system that carries electric signals directing the performance of the controller, but does not carry the main power current. The motor control circuit is the circuit that controls the operation of the magnetic coil within the controller itself. Within this circuit are pilot devices and indicating devices. A pilot device is a sensing device that controls the motor controller. An indicating device is a pilot light, buzzer, horn, or other type of alarm. Often, the wiring of a motor control circuit is complex and requires 10 times the amount of wiring of the motor power circuit.

Control-Circuit Overcurrent Protection

The motor control circuit includes all of the pilot devices, indicating equipment, and magnetic coils within the circuit. The main concern of NEC® Article 430 regarding the control circuit is the protection of these items, the conductors, and the control-circuit transformer.

The NEC® requires that the control circuits be provided with protection in accordance with their ampacity. It does not require that the control circuit conductors have an ampacity large enough to carry the current required of a motor circuit. In some cases, the ampacity of the overcurrent device may be higher than the ampacity of the conductor.

A motor control circuit may originate in one of two ways. First, the motor control circuit may be tapped from the load side of a motor branch-circuit fuse or CB. A control circuit tapped from the load side is not considered a branch circuit and is considered protected by either a supplementary or branch-circuit overcurrent device. Second, the motor control circuit may be derived from a panelboard or a control transformer. In this case, the control conductors shall be protected. The NEC® gives requirements for specific overcurrent protection applications.

For control circuit conductors supplied from a 2-wire, single-voltage transformer secondary, the NEC® allows the circuit to have overcurrent protection provided by the transformer's primary overcurrent protection device.

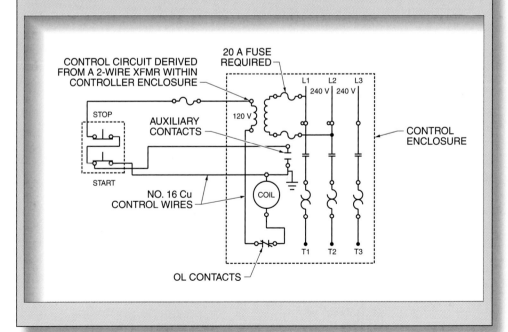

Supplemental Topic

The amperage rating is calculated from Table 430.72(B). For example, a control circuit derived from a 2-wire transformer within a controller enclosure with #16 AWG wire and a transformer that reduces the voltage from 240 V to 120 V. From the Table, the maximum current for a #16 AWG copper wire is 40 A for conductors within an enclosure. This current is multiplied by the ratio of the secondary voltage and the primary voltage. In this case, the maximum-size fuse required for the primary of the control transformer is 20 A (40 × 120/240 = 20).

The NEC® permits five methods of providing overcurrent protection for motor control-circuit transformers. The first method is described in NEC® Article 725 for Class 1 power-limited circuits or Class 2 or Class 3 remote-control circuits. The second method is described in NEC® Section 450.3 for transformer overcurrent protection.

The third method allows a control transformer to be protected by the primary overcurrent devices, provided that the transformer is 50 VA or smaller and is part of the motor controller and located within the controller enclosure. The fourth method permits a control transformer with a rated primary current of 2 A or less to be considered protected at up to 500% of the rated primary current, provided the overcurrent device is in each of the ungrounded conductors of the supply circuit. The fifth method for motor control circuit transformer protection is the use of any protection scheme that is approved.

Application—Sizing Overload Protection

Usng the nameplate FLC to size running overload protection protects a motor up to the maximum current allowed. The running overload protection device disconnects the motor only when the current has reached its maximum permissible level. If a motor is connected to a load that does not require full motor power, the running overload protection can be rated less than FLC, but greater than the current required to turn the load.

For a separate overload device, the level at which the running overload protecting device is set to open is determined by one of three possible circumstances, as long as motor starting is not a problem. First, overloads for motors with a marked service factor of 1.15 or higher can be set as high as 125% of FLC. Second, overloads for motors with a marked temperature rise of 40°C or less can be set as high as 125% of FLC. Third, overloads for all other motors not meeting these special restrictions can be set as high as 115% of FLC. For motors that are difficult to start, the respective percentages are increased from 125% to 140% and from 115% to 130%. The requirements for thermally protected motors are given in the NEC®.

A properly sized running overload protection device removes the motor from the circuit if an overload is present for a long enough period of time to do damage to the motor. For example, overloads may occur when the motor bearings begin to lock or the load increases. When this happens, excessive heat builds up in the windings, damaging the insulation.

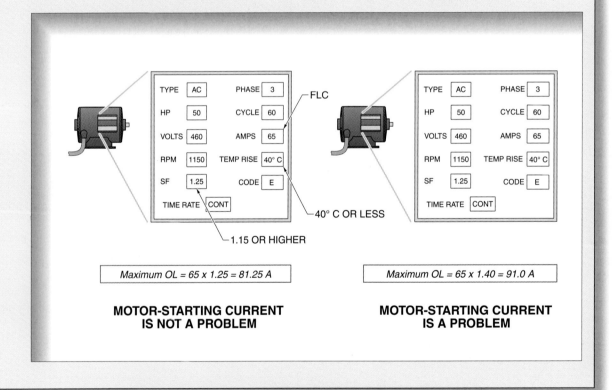

Maximum OL = 65 × 1.25 = 81.25 A

MOTOR-STARTING CURRENT IS NOT A PROBLEM

Maximum OL = 65 × 1.40 = 91.0 A

MOTOR-STARTING CURRENT IS A PROBLEM

Summary

- An overcurrent is any current over the normal current level.

- A short circuit is an excessive current that leaves the normal current-carrying path by going around the load and back to the power source.

- A ground fault is a type of short circuit consisting of an unintentional connection between an ungrounded conductor and any grounded raceway, box, enclosure, or fitting.

- An overload is an excessive current that is confined to the normal current-carrying conductors and is caused by a load that exceeds the full-load torque rating of the motor.

- An overcurrent protection device (OCPD) is used to provide overcurrent protection of a switched circuit.

- A plug fuse is a screw-in fuse. The two common types of plug fuses are Edison-base fuses and type S fuses.

- A cartridge fuse is a snap-in cylindrical fuse constructed of a fusible metallic link or links. Cartridge fuses may be time-delay and non-time-delay types.

- OCPDs are commonly rated by the current, voltage, and interrupting ratings. The current rating is the maximum amount of current that can be carried without opening the device. The voltage rating is the maximum amount of voltage that can safely be applied. The interrupting rating is the maximum amount of current that the device can safely interrupt.

- Motors typically draw a large inrush current during startup. An OCPD must clear short circuits and ground faults, while allowing for the inrush current. Fuses and CBs are sized for the inrush current.

- Motors typically draw up to their normal full-load current (FLC) while operating. An OCPD must open the circuit if excessive operating current is drawn. Overload relays are sized for the operating current.

- Fuses and CBs can be tested during troubleshooting motor problems as part of an overall evaluation of the power distribution and circuit.

- Common types of CBs include inverse-time CBs, adjustable-trip CBs, non-adjustable-trip CBs, and instantaneous-trip CBs. An inverse-time CB (ITCB) has an intentional delay before tripping. An adjustable-trip CB (ATCB) is a CB whose trip setting can be changed. A non-adjustable-trip CB (NATCB) is a CB with settings that cannot be changed. An instantaneous-trip CB (ITB) is a CB with no delay before tripping.

- The NEC® requires protection of all pilot devices, indicating equipment, magnetic coils, conductors, and the control circuit transformer according to their ampacity.

Glossary...

An **overcurrent** is any current over the normal current level.

A **short circuit** is an excessive current that leaves the normal current-carrying path by going around the load and back to the power source.

A **ground fault** is an unintentional connection between an ungrounded conductor and any grounded raceway, box, enclosure, or fitting.

An **overload** is an excessive current that is confined to the normal current-carrying conductors and is caused by a load that exceeds the full-load torque rating of the motor.

An **overcurrent protection device (OCPD)** is a set of fuses or circuit breakers (CBs) added to provide overcurrent protection of a switched circuit.

A **fuse** is an overcurrent protection device with a fusible link that melts and opens the circuit when an overcurrent occurs.

A **plug fuse** is a screw-in fuse that uses a metallic strip that melts when a predetermined amount of current flows through it.

An **Edison-base fuse** is a plug fuse that incorporates a screw configuration that is interchangeable with fuses of other amperage ratings.

A **type S fuse** is a plug fuse that incorporates a screw-and-adapter configuration that is not interchangeable with fuses of another amperage rating.

A **cartridge fuse** is a snap-in cylindrical fuse constructed of a fusible metallic link or links that is designed to open at predetermined current levels to protect circuit conductors and equipment.

A **non-time-delay fuse (NTDF)** is a single-element fuse that can detect an overcurrent and open the circuit almost instantly.

A **time-delay fuse (TDF)** is a dual-element fuse that may detect and remove a short circuit almost instantly, but allow small overloads to exist for a short period.

A **circuit breaker (CB)** is an overcurrent protection device with a mechanical mechanism that automatically opens a circuit when an overload condition or short-circuit occurs.

An **inverse-time circuit breaker (ITCB)** is a circuit breaker (CB) with an intentional delay between the time when the fault or overload is sensed and the time when the CB operates.

An **adjustable-trip circuit breaker (ATCB)** is a circuit breaker (CB) whose trip setting can be changed by adjusting the current setpoint, trip time characteristics, or both, within a particular range.

A **non-adjustable-trip circuit breaker (NATCB)** is a circuit breaker (CB) whose settings for the ampere trip setpoint or the time-trip setpoint cannot be changed.

An **instantaneous-trip circuit breaker (ITB)** is a circuit breaker (CB) with no delay between the fault- or overload-sensing element and the tripping action of the device.

The **current rating** is the continuous amount of current that can be safely carried by an overcurrent protection device without blowing or tripping.

The **voltage rating** is the continuous amount of voltage that can safely be applied to an overcurrent protection device and determines the ability of a fuse to suppress the internal arcing after the fusible link melts.

The **interrupting rating** is the maximum amount of current that can be safely applied to an overcurrent protection device while still maintaining its physical integrity when reacting to fault currents.

Full-load current (FLC) is the amount of current drawn when the motor is connected to the maximum load the motor is designed to drive.

The **overload class rating** is an indicator of the maximum length of time it takes for the overload relay to trip at 600% overload.

A **thermal overload relay** is an overload relay that uses the resistive heating to open a set of contacts.

...Glossary

A ***melting-alloy overload relay***, or ***heater***, is an overload relay that uses a heater coil to produce the heat to melt a eutectic alloy.

A ***eutectic alloy*** is an alloy that has a constant melting-point temperature due to the combination of the given components.

A ***bimetallic-strip overload relay*** is a relay consisting of two joined pieces of dissimilar metals with different expansion rates constructed into a strip in such a way that the strip bends when heated and opens a set of contacts.

An ***electromagnetic overload relay*** is a relay that operates on the principle that as the level of current in a circuit increases so does the strength of the magnetic field produced by that current flow.

An ***electronic overload relay*** is a device that has built-in circuitry to sense changes in current and temperature.

Review

1. Describe the difference between a short circuit, a ground fault, and an overload.

2. Use an example to demonstrate why overcurrent protection is needed in a circuit.

3. Describe the difference between an Edison-base fuse and a type S fuse. Why are type S fuses preferred?

4. Describe the difference between a non-time-delay fuse and a time-delay fuse. Explain which type would typically be used in a motor circuit.

5. List the most common types of CBs.

6. Describe the different ratings used with OCPDs.

7. Describe the similarities and differences between motor-starting protection and motor-running protection.

8. Describe the operation of thermal overload relays.

9. Describe the operation of electromagnetic overload relays and electronic overload relays.

Refer to the CD-ROM for Quick Quiz® questions related to chapter content.

MOTORS

- Motor Housing
- End Bracket Housing
- Endbell
- Keyslip

4

Three-Phase Motors

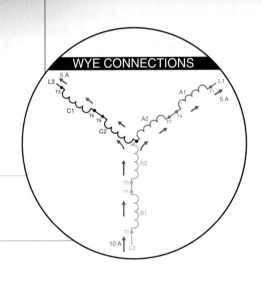

WYE CONNECTIONS

Three-Phase Motor Construction	82
Operating Principles	87
Motor Load and Torque	99
Motor Power	103
Motor Efficiency	104
Application—Motor Line and Motor Wiring Diagrams	107
Application—Voltage and Current Unbalance	108
Summary	109
Glossary	109
Review	111

OBJECTIVES

- Describe how a stator is constructed.
- List the parts of a rotor.
- Describe the difference between electrical degrees and mechanical degrees.
- Demonstrate how the synchronous speed of a motor is calculated.
- Describe how a stator rotating field is created.
- Describe the difference between a wye-connected and a delta-connected motor.
- List and describe four components of motor torque.
- Define power factor in terms of true power, reactive power, and apparent power.
- List and describe five types of motor energy losses.

Three-phase motors are the standard motor type used in industry. These motors vary in size from fractional horsepower to thousands of horsepower. Three-phase motors operate at a fairly constant speed and are available in designs with a variety of torque characteristics. The primary advantages of three-phase motors are low maintenance requirements and economy of operation.

Definition

A stator is the fixed, unmoving part of a motor, consisting of a core and windings, or coils, that converts electrical energy to the energy of a magnetic field.

THREE-PHASE MOTOR CONSTRUCTION

A 3-phase motor transfers energy from electric power to the load through the use of a rotating magnetic field. The rotating magnetic field is generated as current flows through the stator coils. The extensive use of 3-phase motors for almost all industrial applications is due to the following:

- Three-phase power is the standard power supplied by electrical power companies to almost all commercial and industrial locations. A 3-phase motor can be connected to the supplied power with little modification or control required.
- Three-phase motors are simple in construction, rugged, and require little maintenance. It is not uncommon to find 3-phase motors that have run in an application for ten or more years without a failure.
- Three-phase motors are less expensive than other motor types of the same horsepower rating. They are also available in a larger selection of sizes, speeds, and frames than other motors.
- Three-phase motors cost less to operate per horsepower than single-phase motors or DC motors.
- Three-phase motors have a fairly constant speed characteristic and are available in a wide variety of torque characteristics.
- Three-phase motors are self-starting. Under many circumstances, no special starting method is required.

Stator Construction

A *stator* is the fixed, unmoving part of a motor, consisting of a core and windings, or coils, that converts electrical energy to the energy of a magnetic field. The stator of a three-phase motor is enclosed within a housing made of cast iron, rolled steel, or cast aluminum. **See Figure 4-1.** The windings are coils of wire wrapped through slots in the stator core.

Stator Cores. The stator core consists of many thin iron sheets laminated together, pressed into a frame, and secured in place. These sheets are round and resemble a large, flat doughnut, with a large hole in the middle. The iron sheets are electrically separated from each other by an insulating coating. The separation reduces the cross-sectional area of the core and shortens the conduction path for damaging eddy currents.

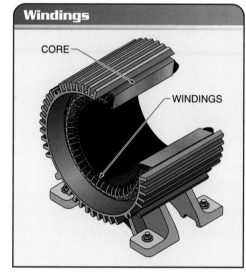

Figure 4-1. The stator consists of a core and windings and is enclosed within a housing.

The laminated sheets of the core have notches punched around their inner diameter and are stacked in the housing. When the sheets are stacked and pressed into the housing, these notches become the slots for the windings. **See Figure 4-2.**

Stator Windings. The stator windings consist of coils of copper wire placed in slots 120 electrical degrees apart. The windings are well insulated and can tolerate high voltage and high current. For smaller motors, the stator coils are assembled into one component. During maintenance, the entire motor must be removed and replaced.

For larger stators, the coils are individually constructed and connected. These coils are wound into the desired form with rectangular wire. The end of each coil is made so that it can be bolted to the next coil. This facilitates in-field replacements by eliminating the need to remove the stator and completely rewind it.

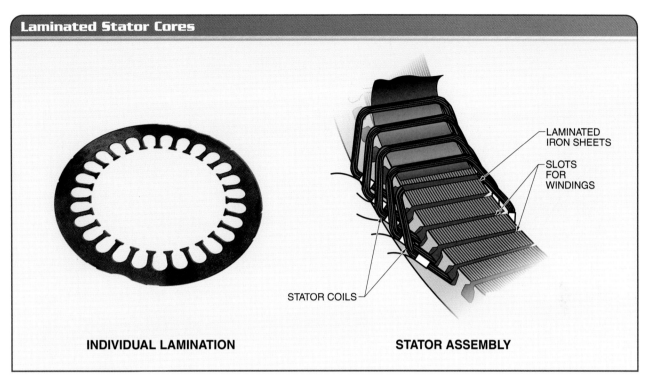

Figure 4-2. Many individual laminated sheets are pressed together into the housing, becoming the stator frame.

A 3-phase stator is wound with coils that are connected to produce the three separate phases, A, B, and C. **See Figure 4-3.** Each phase must have the same number of coils, so the total number of coils must be divisible by the number 3. For example, if the total number of coils wound in the stator is 36, then each phase contains 12 coils equally divided among the poles.

During motor manufacture, an insulating material called slot paper is first laid in the slot to provide protection and electrical insulation. When the windings are laid in the slots, a holder made of Glastic® or fiberglass, called a topstick, is driven into the slot over each winding to hold it in place. **See Figure 4-4.** In some cases, a midstick is also used to hold the windings in place.

The coil connections are made around the outer perimeter and the coils are tied together to prevent movement. The stator is lowered into a tank containing insulating varnish, where it is submerged long enough to allow the varnish to saturate the windings. The stator is then suspended over the tank long enough to allow the excess varnish to drip off. Next the stator is placed in an oven for curing. The cured varnish not only insulates the windings, but also holds the windings together to prevent movement.

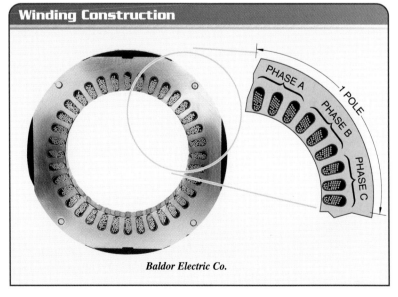

Figure 4-3. A 3-phase stator is wound with coils that are connected to produce the three separate phases, A, B, and C.

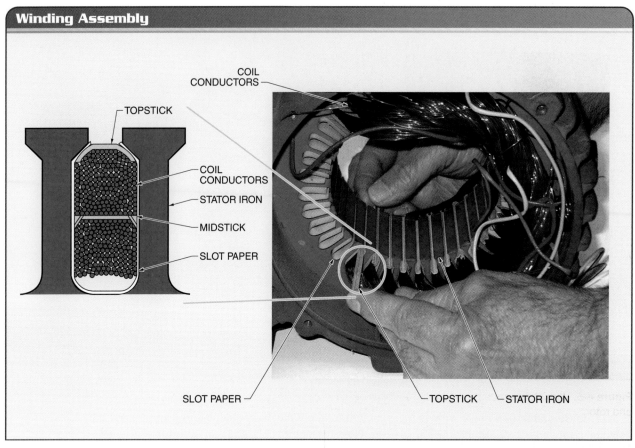

Figure 4-4. During motor manufacture, an insulating material called slot paper is first laid in the slot to provide protection and electrical insulation.

If the coils are allowed to move, the wires can rub through the insulation and short between different turns of the wire or to ground. Turn-to-turn shorts lower the flux, reducing induction to the rotor. This also reduces the inductive reactance in the stator, raising the line current to potentially dangerous levels. Heat is produced at the shorted connection and the stator eventually fails.

Operation of the overload device might not occur with only a few shorted turns because a small load may draw much lower current than required for the overloads to trip. A short to ground produces large currents and affects the operation of the short-circuited devices. This is why proper grounding of the motor is so important. Unless the motor is grounded, the frame is common to the phase that shorted, and a dangerous voltage is present on the housing.

When the stator is removed from the oven and cooled, the varnish is sanded from the bore of the core to allow for a minimal air gap between the stator and rotor. **See Figure 4-5.** The size of the air gap must be very small to prevent loss of magnetic flux between the stator and the rotor. The flux created in the stator must cross the air gap to induce current in the rotor windings. Any loss of flux affects torque, increases slip, and decreases the motor's efficiency.

Tech Fact

In a 3-phase motor, there are three stator windings for each pole. The windings are combined to give the required number of poles. A common 4-pole stator design has 36 slots, so each pole has nine windings. The nine windings are placed in three groups of three, with each group connected in series for each phase.

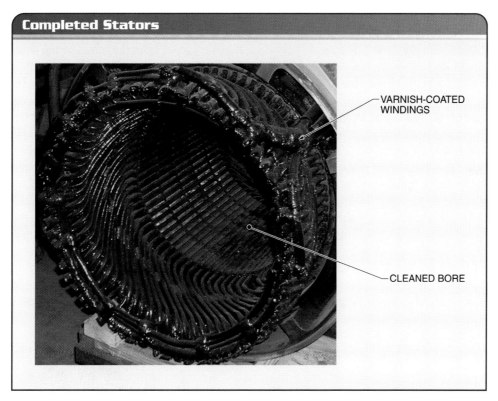

Figure 4-5. Varnish is sanded from the bore to allow for a minimal air gap between the stator and rotor.

Definition

*A **rotor** is the rotating, moving part of a motor, consisting of a core and windings, which convert the rotating magnetic field of the stator into the torque that rotates the shaft.*

Housing. Feet are attached to the housing to provide a method of mounting the motor to a base. **See Figure 4-6.** The core is pressed into the housing. The end of the housing is machined to allow precise fit of the end bell as it is attached to the housing. Close tolerances of the endbell to the housing are critical, as the rotor must be centered in the core to ensure an even air gap. The air gap is the space between the rotor and stator. The endbells contain the bearings to support the rotor, and in some types of motors, close the ends of the housing as well. Larger motors also may have cooling fins to help remove heat and a lifting hook to facilitate installation.

Rotor Construction

A *rotor* is the rotating, moving part of a motor, consisting of a core and windings, which convert the rotating magnetic field of the stator into the torque that rotates the shaft. The rotor is mounted on a shaft that is used to transfer the power to the load.

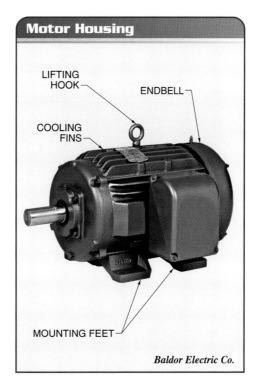

Figure 4-6. Feet are attached to the housing to provide a method of mounting the motor to a base.

> **Definition**
>
> A *salient pole*, or *projecting pole*, is a pole that extends away from the core toward the stator or extends away from the stator toward the rotor.
>
> The *motor shaft* is a cylindrical bar used to carry the revolving rotor and to transfer power from the motor to the load.

Rotor Cores. Like the stator, the rotor core consists of many thin iron sheets laminated together. **See Figure 4-7.** The rotor sheets are usually thicker than the stator sheets. Since the rotor frequency is lower than line frequency, other than at locked rotor, the induced eddy currents are much smaller than in the stator. These sheets are slightly smaller in outer diameter than the inner bore of the sheets of the stator to allow a small space for the air gap. In the most common design, the notches are punched on the outer diameter of the sheet.

In an alternate design, the iron sheets are manufactured with a salient pole. A *salient pole*, or *projecting pole*, is a pole that extends away from the core toward the stator or extends away from the stator toward the rotor. Stacked together, the laminated sheets resemble a drum, with the notches aligned to become slots that will hold the rotor conductors or with salient poles for the windings.

Shafts. The *motor shaft* is a cylindrical bar used to carry the revolving rotor and to transfer power from the motor to the load. The end of the shaft is machined with a keyway to contain a bar-type key or with a circular slot that contains a half-moon key. **See Figure 4-8.** This key prevents the coupling device from slipping on the shaft. Some smaller motors have a flat spot milled on the shaft against which a setscrew on the coupling device can be set.

The shaft is machined on its ends to allow the bearings to be pressed on to support the shaft in the center of the stator. This shaft also holds the fan and connects the rotor to the load. The shaft is pressed into a hole in the center of the rotor. When the motor is assembled, the rotor is set inside the stator with as small an air gap as possible.

Figure 4-7. The rotor core consists of many thin iron sheets laminated together.

Rotor Windings. While almost all 3-phase motors have very similar stator designs, the main difference between different types of 3-phase motors is the design of the rotors and windings. The three most common types of rotors are used in squirrel-cage induction motors, wound-rotor motors, and synchronous motors.

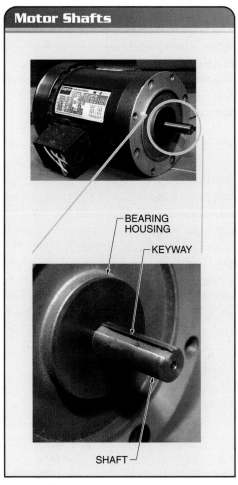

Figure 4-8. The end of the shaft is machined with a keyway to contain a bar-type key.

OPERATING PRINCIPLES

An electric motor converts electrical energy into torque. An electric power source, usually an external source of supply from the electric utility, supplies rated voltage to the motor terminals. The electric energy induces a voltage within the rotor, causing the rotor to rotate. A mechanical load coupled to the motor shaft receives a torque that enables mechanical work to be performed.

Electrical Degrees

The flow of electricity is represented by sine waves, or curves. For 3-phase power, three sine waves placed 120 electrical degrees apart are used. Electrical degrees and mechanical degrees differ. There are 360° (mechanical) in one revolution of the rotor. There are 360 electrical degrees in one cycle of a sine wave. Each cycle of a sine wave in the stator coils results in a north pole and a south pole. Therefore, there are two poles in 360 electrical degrees, or 180 electrical degrees per pole.

During one rotor revolution in a 2-pole motor, the rotor passes from one north pole to a south pole and back to the original north pole, completing 360 mechanical degrees for each 360 electrical degrees. During one revolution in a 4-pole motor, a rotor passes from one north pole, through a south pole, through a north pole, through a south pole, and back to the original north pole. Since there are 180 electrical degrees per pole, it takes 720 electrical degrees, or two electrical cycles, to complete one revolution in a 4-pole motor. **See Figure 4-9.**

Synchronous Speed

The *synchronous speed* is the theoretical speed of a motor based on line frequency and the number of poles of the motor. The synchronous speed of a motor is calculated as follows:

$$\omega_S = \frac{120 \times f}{P}$$

where
ω_S = synchronous speed, in rpm
f = source voltage frequency, in Hz
P = number of poles

Electrical Degrees

Poles per Phase	Electrical Degrees per Revolution
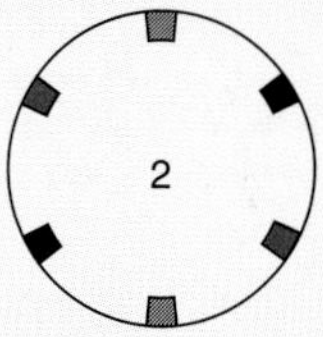 2	2 × 180 = 360
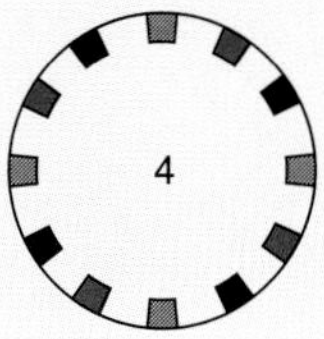 4	4 × 180 = 720

Figure 4-9. It takes 720 electrical degrees, or two electrical cycles, to complete one revolution in a 4-pole motor.

Different motors are designed with different numbers of poles. For example, a 2-pole motor has two A-phase, two B-phase, and two C-phase coils in the stator frame. The synchronous speed is calculated as follows:

$$\omega_S = \frac{120 \times f}{P}$$

Since 60-Hz power is used almost exclusively in the United States, the equation can be simplified as follows:

$$\omega_S = \frac{7200}{P}$$

For a 2-pole motor, the synchronous speed is calculated as follows:

$$\omega_S = \frac{7200}{P}$$

$$\omega_S = \frac{7200}{2}$$

$$\omega_S = \mathbf{3600\ rpm}$$

> **Definition**
>
> The *synchronous speed* is the theoretical speed of a motor based on line frequency and the number of poles of the motor.

> **Tech Fact**
>
> There are 360 electrical degrees in one sine wave. Since each cycle of a sine wave results in a north pole and a south pole, there are two poles in 360 electrical degrees.

Definition

*The **rotor frequency** is the rate at which the stator magnetic field passes the poles in the rotor.*

Very large slow motors often have many poles to reduce the speed. For example, a large synchronous motor could be designed with 8 poles. The synchronous speed is calculated as follows:

$$\omega_s = \frac{7200}{P}$$

$$\omega_s = \frac{7200}{8}$$

$$\omega_s = \mathbf{900 \text{ rpm}}$$

Rotor Frequency

The *rotor frequency* is the rate at which the stator magnetic field passes the poles in the rotor. The line frequency is 60 Hz, or 60 cycles per second. Therefore, the magnetic field in the stator of a 2-pole motor is also rotating at 60 revolutions per second, or 3600 revolutions per minute (rpm). At startup, the stator magnetic field passes the rotor poles at 3600 rpm, since the rotor is not yet turning. Therefore, the rotor poles are exposed to a magnetic field rotating at 3600 rpm (60 Hz).

At startup when a motor is in locked rotor, the winding resistance and reactance are the only limits to current flow. Inductive reactance increases with increasing frequency and decreases with decreasing frequency. The rotor conductors have very low resistance, but have much higher reactance because of the relatively high frequency. Since the circuit is at least 10 times more reactive than resistive, it is a reactive circuit and the current lags the voltage induced in the rotor by 90 electrical degrees. **See Figure 4-10.**

As the rotor begins to move, the frequency of the induced current in the rotor decreases, until the rotor approaches synchronous speed, where the rotor frequency approaches 0. As the rotor frequency decreases, the inductive reactance also decreases and the total impedance decreases. The circuit becomes more resistive and the angle at which the current lags the voltage decreases. At very low frequencies, the poles of the rotor are moving at almost the same speed as the poles of the stator magnetic field. At synchronous speed, there would be no induced current because there is no longer any relative motion.

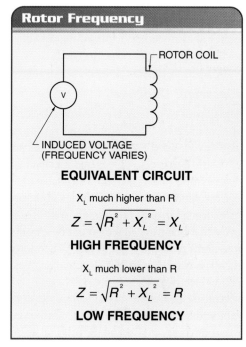

Figure 4-10. Inductive reactance increases with increasing frequency and decreases with decreasing frequency.

Stator Rotating Field

Looking at the 3-phase sine curves, an arbitrary point in time is chosen at the beginning and 0 electrical degrees is assigned to the alternating current at that point. **See Figure 4-11.** At this beginning point, the sine waves show –5 A for phase A, +10 A for phase B, and –5 A for phase C. The + and – signs indicate the direction of the current and the numbers represent the magnitude of the current.

Since current flows from negative to positive, it flows from the conductor(s) with the negative current, through the motor windings, to the conductor(s) with the positive current. The direction of flow through coils and the polarity of the magnetic field and motor poles are established by the motor design.

The current changes in each conductor as it rotates through the complete cycle. As the current changes, the amount of current flowing in each coil changes and the magnetic field around the coil changes. The current varies from maximum positive, through zero, to the maximum negative value. As the current changes direction, the polarity of the magnetic field changes. The three phases add to produce a single rotating magnetic field.

Single-Voltage Motors

The windings must be connected to the proper voltage to avoid damage to the motor. The voltage level is determined by the manufacturer and is stamped on the motor nameplate. A *single-voltage motor* is a motor that operates at only one voltage level. Single-voltage, 3-phase motors are less expensive to manufacture than dual-voltage, 3-phase motors, but are limited to locations having the same voltage as the motor. Typical single-voltage, 3-phase motor ratings are 230 V, 460 V, and 575 V. Less common single-voltage, 3-phase motor ratings are 200 V, 208 V, and 280 V.

Several types of motor connections must be evaluated. The stator can be wired in a wye or delta configuration. In addition, many motors are dual voltage and can be wired to operate at a low voltage or a high voltage. For single-voltage motors, the stator coils may consist of two coils per pole, but the coils are internally connected by the manufacturer so that the motor can only operate at one voltage. The motor nameplate typically has a wiring diagram depicting the proper wiring connections for the desired operation. **See Figure 4-12.**

> **Tech Fact**
>
> *Unlike some types of motors, 3-phase motors are simple in construction, are rugged, and require very little maintenance. It is not uncommon to find 3-phase motors that have run for 10 years or more without a failure.*

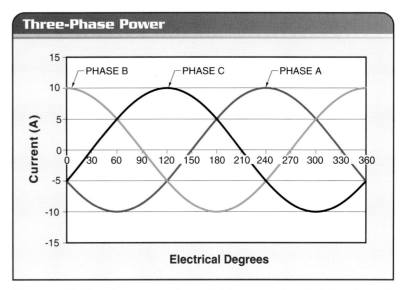

Figure 4-11. The sine curves show −5 A for phase A, +10 A for phase B, and −5 A for phase C. The + and − signs indicate the direction of the current and the numbers represent the magnitude of the current.

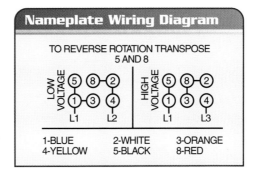

Figure 4-12. The motor nameplate typically has a wiring diagram depicting the proper wiring connections for the desired operation.

> **Definition**
>
> A *single-voltage motor* is a motor that operates at only one voltage level.

Single-Voltage, Wye-Connected Motors. In a wye-connected, 3-phase motor, one end of each of the three phase windings is internally connected to the other phase windings. The remaining end of each phase winding is then brought out externally to form T1, T2, and T3. When connecting to 3-phase power lines, the power lines and motor terminals are connected L1 to T1, L2 to T2, and L3 to T3. **See Figure 4-13.** A 2-pole motor was chosen for this illustration for simplicity. For any other number of poles, the principles are the same, but the illustration would be more complex because there would be more coils.

Single-Voltage, Wye-Connected Motors

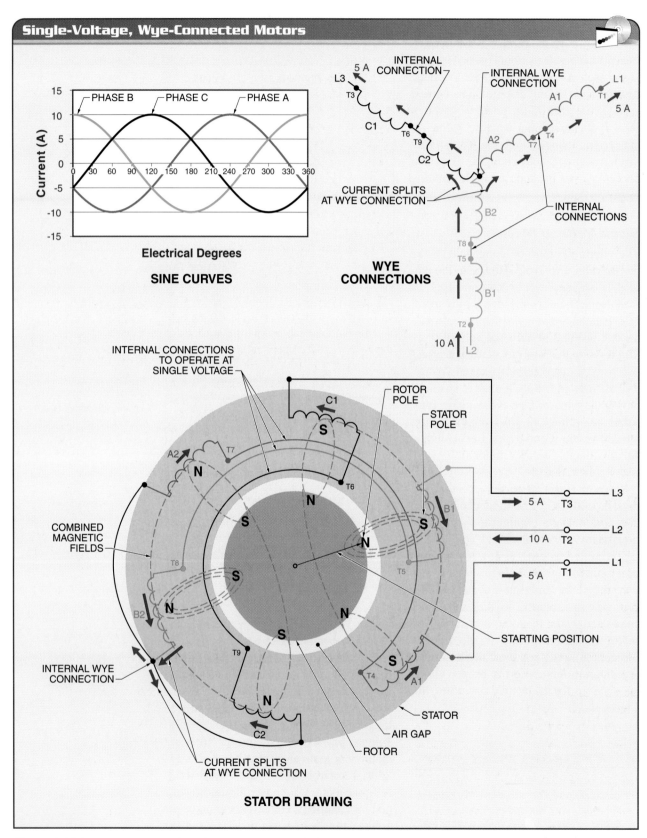

Figure 4-13. In a wye-connected, 3-phase motor, one end of each of the three phase windings is internally connected to the other phase windings. The remaining end of each phase is then brought out externally to form T1, T2, and T3.

The sine-curve graph shows the currents in each of the phases through one cycle. The stator drawing showing the motor coils is the best for showing how the poles are created and showing the location of the strongest fields. The drawing of the wye connections is the best for showing the current flow through the coils. These illustrations work together to describe the operation of a wye-connected motor.

At the beginning at 0 electrical degrees, the sine curves show –5 A for phase A, +10 A for phase B, and –5 A for phase C. Because the stator is wye connected, the current through each coil is equal to the phase current. Looking at the wye connections, the +10 A current from phase B flows through L2 to coils B1 and B2. The current splits equally at the wye connection and 5 A flows through the other coils. The current returns out the T1 and T3 terminals.

The same current can be seen in the stator drawing. The largest current flows through coils B1 and B2, so the strongest magnetic field is also present at those coils, creating the strongest poles. Smaller currents through the other coils create weaker poles. The magnetic fields around like poles combine into one larger field.

The north or south polarity of the poles is determined by the direction of the current flow and the motor design. The strongest north pole in the stator is at coil B2 and the strongest south pole in the stator is a coil B1. The stator magnetic fields induce current in the rotor that creates opposite rotor poles corresponding to the stator poles. The reference starting position is chosen to be centered on coil B1, centered on the combined magnetic field.

The current starts to change as the sine wave rotates through the cycle and the amount of current flowing through each coil changes. As the current changes, the stator poles move to follow the strongest current. **See Figure 4-14.** The magnitude of the current in phase A increases from –5 A toward –10 A, the current in phase B decreases from +10 A toward 0 A, and the current in phase C decreases from –5 A toward 0 A.

At 30 electrical degrees, the current in phase A is –8.66 A, the current in phase B is +8.66 A, and the current in phase C is 0 A. The magnitude of the current flowing through coils A1, A2, B1, and B2 is equal. The poles created in the stator are 30 electrical degrees from the starting position of 0 electrical degrees. Since the motor has a 2-pole stator, the rotor is also rotated by 30° (mechanical degrees) to keep the poles aligned.

The current continues to alternate and the stator poles continue to rotate. At 60 electrical degrees, L2 and L3 are at the same potential. Therefore, no current flows between these conductors. Equal currents enter at T2 and T3 and combine at the internal wye connection. Since the largest current flows through coils A1 and A2, with smaller currents through the other coils, the strongest poles are created at A1 and A2. As a result, the rotor poles also rotate 60 mechanical degrees from the starting point. At 90 electrical degrees, the current flowing through coils A1, A2, C1, and C2 is equal. The poles created in the stator are 90 electrical degrees from the starting position at 0 electrical degrees.

The current continues to alternate through the remainder of the cycle. The stator poles continue to rotate around the stator until the poles have traveled all the way around the stator and the cycle begins again.

Single-Voltage, Delta-Connected Motors. In a delta-connected, 3-phase motor, each phase is wired end-to-end to form a completely closed circuit. At each point where the phases are connected, leads are brought out externally to form T1, T2, and T3. **See Figure 4-15.** T1, T2, and T3 are connected to the three power lines, with L1 connected to T1, L2 connected to T2, and L3 connected to T3. The 3-phase line supplying power to the motor must have the same voltage and frequency rating as the motor.

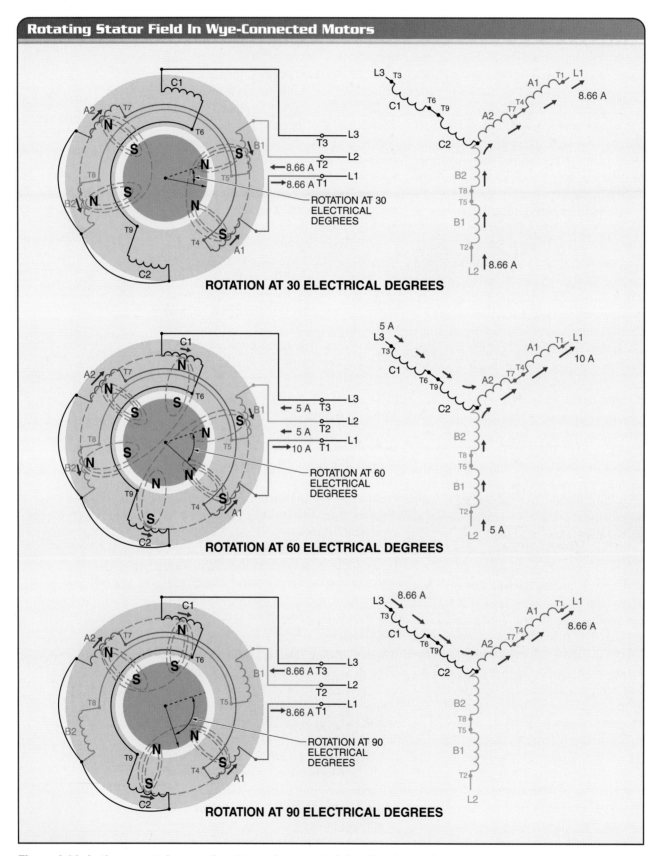

Figure 4-14. As the current changes, the stator poles move to follow the strongest current.

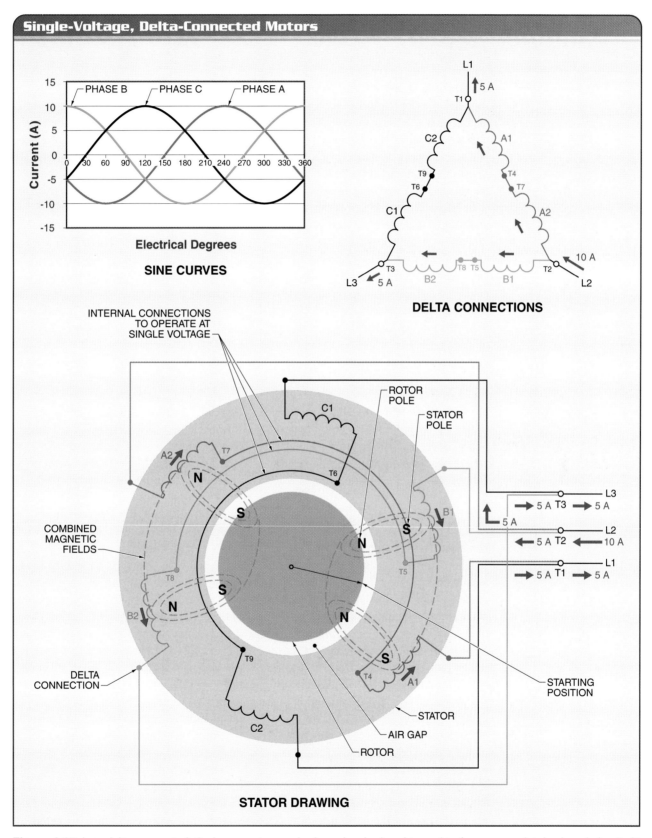

Figure 4-15. In a delta-connected, 3-phase motor, each phase is wired end-to-end to form a completely closed circuit. At each point where the phases are connected, leads are brought out externally to form T1, T2, and T3.

The sine-curve graph shows the currents in each of the phases through one cycle. The drawing of the stator coils is the best for showing how the poles are created and showing the location of the strongest fields. The drawing of the delta connections is the best for showing the current flow through the coils. These drawings work together to illustrate the operation of a wye-connected motor.

At the beginning at 0 electrical degrees, the sine curves show –5 A for phase A, +10 A for phase B, and –5 A for phase C. The + and – signs indicate the direction of current flow and the numbers represent the magnitude of the current flow. Because the stator is delta connected, the current through each coil is not equal to the phase current. The 10-A current from phase B splits equally at T2. A current of 5 A flows through coils A2 to A1 and returns at T1. An equal current of 5 A flows through coils B1 and B2 and returns at T3.

Since equal current flows through coils A1, A2, B1, and B2, the magnetic fields are also equal. The equal magnetic fields create equal poles. The stator magnetic fields induce current in the rotor that creates opposite rotor poles corresponding to the stator poles. The magnetic fields around like poles combine into one larger field. The reference starting position is chosen to be equidistant between coils A1 and B1, at the center of the magnetic field.

The current starts to change as the sine wave rotates through the cycle and the amount of current flowing through each coil changes. As the current changes, the stator poles move to follow the strongest current. **See Figure 4-16.** The magnitude of the current in phase A increases from –5 A toward –10 A, the current in phase B decreases from +10 A toward 0 A, and the current in phase C decreases from –5 A toward 0 A.

At 30 electrical degrees, the current in phase A is –8.66 A, the current in phase B is +8.66 A, and the current in phase C is 0 A. Since there is no current flow through L3, all current flows from L2 to L1. There are two paths for the current to flow. The path through B1, B2, C1, and C2 has twice the impedance of the path through A1 and A2. When the current splits at T2, twice as much current flows through A1 and A2 than through the other path.

The magnitude of the current flowing through coils A1 and A2 is 5.77 A. The magnitude of the current through B1, B2, C1, and C2 is 2.89 A. Since the current is stronger through A1 and A2 than through the other path, the poles created at those coils are stronger than the other poles. These poles in the stator are 30 electrical degrees from the starting position at 0 electrical degrees. Since the motor has a 2-pole stator, the rotor has also rotated by 30° (mechanical degrees) to keep the poles aligned.

The current continues to alternate and the stator poles continue to rotate. At 60 electrical degrees, the current in phase A is –10 A, the current in phase B is +5 A, and the current in phase C is +5 A. As a result, 5 A enters at L2 and 5 A enters at L3. The currents combine at T1 and return at L1. Since equal currents flow through coils A2, A1, C1, and C2, equal poles are formed at these coils. The poles are rotated 60 electrical degrees from the reference starting position. As a result, the rotor poles also rotate 60 electrical degrees from the starting point.

At 90 electrical degrees, the current in phase A is –8.66 A, the current in phase B is 0 A, and the current in phase C is +8.66 A. As a result, 8.66 A enters at L2 and splits. The magnitude of the current flowing through coils C1 and C2 is 5.77 A. The magnitude of the current through B2, B1, A2, and A1 is 2.89 A. Since the current is stronger through C1 and C2 than through the other path, the poles created at those coils are stronger than the other poles. The poles created in the stator are 90 electrical degrees from the starting position at 0 electrical degrees. As a result, the rotor poles also rotate 90 mechanical degrees from the starting point.

The current continues to alternate through the remainder of the cycle. The stator poles continue to rotate around the stator until the poles have traveled all the way around the stator and the cycle begins again.

Chapter 4—Three-Phase Motors

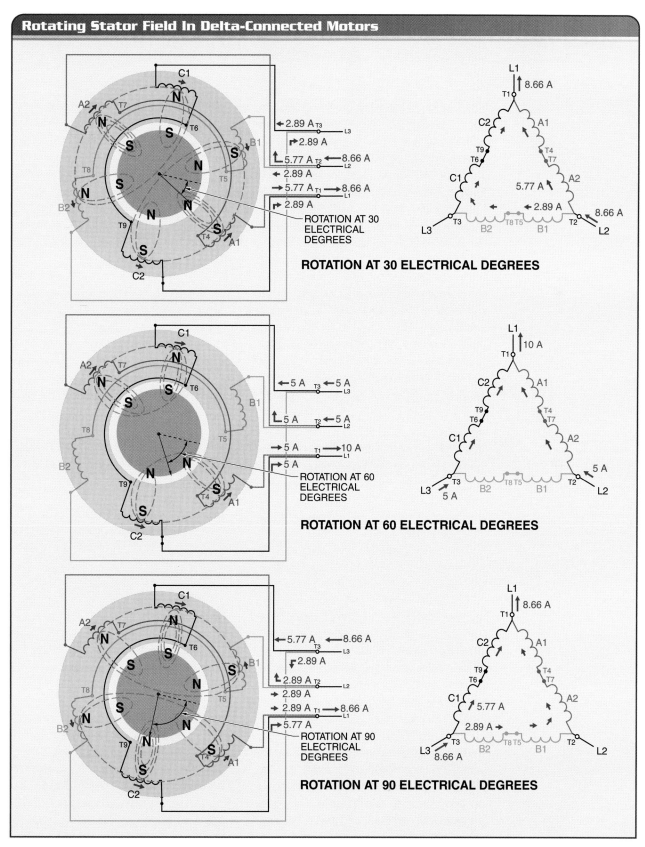

Figure 4-16. As the current changes, the stator poles move to follow the strongest current.

Definition

A *dual-voltage motor* is a motor that operates at more than one voltage level.

Dual-Voltage Motors

A *dual-voltage motor* is a motor that operates at more than one voltage level. Most 3-phase motors are manufactured so that they can be connected for either of two voltages. The purpose in making motors for two voltages is to enable the same motor to be used with two different power line voltages.

A typical dual-voltage, 3-phase motor rating is 230/460 V. Other common dual-voltage, 3-phase motor ratings are 240/480 V and 208–230/460 V. The dual-voltage rating of the motor is listed on the nameplate of the motor. If both voltages are available, the higher voltage is usually preferred because the motor uses the same amount of power, given the same horsepower output, for either high or low voltage. As the voltage is doubled (e.g., 230 V to 460 V), the current drawn on the power lines is cut in half. With the reduced current, the wire size is reduced, and the material cost is decreased.

Dual-Voltage, Wye-Connected Motors. In a dual-voltage, wye-connected, 3-phase motor, each phase coil (A, B, and C) is divided into two equal parts and the coils are connected in a standard wye connection. By dividing the phase coils in two, nine terminal leads are available. These motor leads are marked terminals one through nine (T1 to T9). **See Figure 4-17.**

When a dual-voltage, wye-connected, 3-phase motor is connected for high voltage, L1 to T1, L2 to T2, and L3 to T3 are connected at the motor starter. With wire nuts and tape, T4 is tied to T7, T5 is tied to T8, and T6 is tied to T9. By making these connections, the individual coils in each phase are connected in series and the applied voltage divides equally among the coils.

When a dual-voltage, wye-connected, 3-phase motor is connected for low voltage, L1 to T1 and T7, L2 to T2 and T8, and L3 to T3 and T9 are connected at the motor starter. With a wire nut and tape, T4, T5, and T6 are tied together. By making these connections, the individual coils in each phase are connected in parallel. Since the coils are connected in parallel, the applied voltage is present across each set of coils.

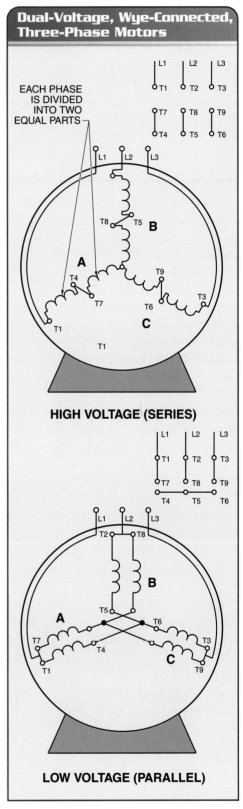

Figure 4-17. Each phase coil (A, B, and C) is divided into two equal parts and the coils are connected in a standard wye connection.

Dual-Voltage, Delta-Connected Motors.
In a dual-voltage, delta-connected, 3-phase motor, each phase coil (A, B, and C) is divided into two equal parts and the coils are connected in a standard delta connection. By dividing the phase coils in two, nine terminal leads are available. These motor leads are marked terminals one through nine (T1–T9). The nine terminal leads can be connected for high or low voltage. **See Figure 4-18.**

When a dual-voltage, delta-connected, 3-phase motor is connected for high voltage, L1 to T1, L2 to T2, and L3 to T3 are connected at the motor starter. With wire nuts and tape, T4 is tied to T7, T5 is tied to T8, and T6 is tied to T9. By making these connections, the individual coils in each phase are connected in series. Since the coils are connected in series, the applied voltage divides equally among the coils.

When a dual-voltage, delta-connected, 3-phase motor is connected for low voltage, L1 to T1, L2 to T2, and L3 to T3 are connected at the motor starter. With a wire nut and tape, T1 is tied to T7 and T6, T2 is tied to T8 and T4, and T3 is tied to T9 and T5. By making these connections, the individual coils in each phase are connected in parallel. Since the coils are connected in parallel, the applied voltage is present across each set of coils.

Twelve-Lead, 3-Phase Motors

Typically, dual-voltage, 3-phase motors have nine leads coming out of the motor box. A 9-lead, wye-connected motor and a 9-lead, delta-connected motor have internal connections made by the manufacturer. However, manufacturers of dual-voltage, 3-phase, wye-connected and delta-connected motors sometimes do not make the internal connections. The internally unconnected motors have 12 leads coming out of the motor box. The three additional leads are labeled T10, T11, and T12. The connections are made externally by the installer. **See Figure 4-19.**

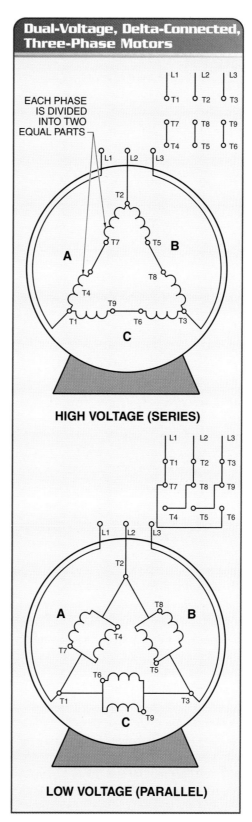

Figure 4-18. Each phase coil (A, B, and C) is divided into two equal parts and the coils are connected in a standard delta connection.

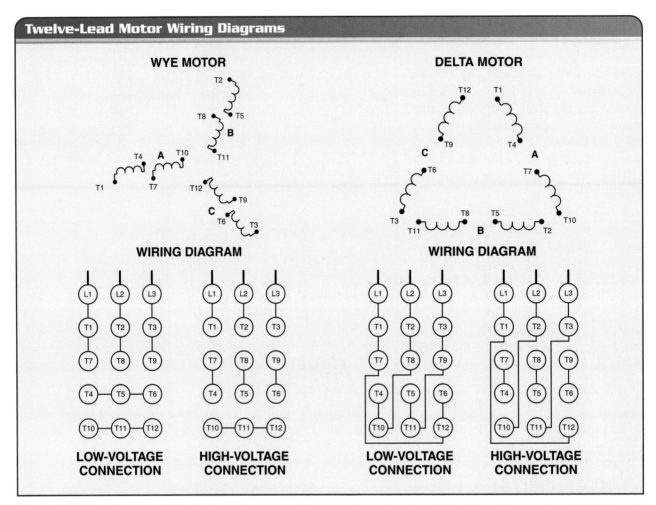

Figure 4-19. Manufacturers of dual-voltage, 3-phase motors sometimes do not make the internal connections. The internally unconnected motors have 12 leads coming out of the motor box labeled T10, T11, and T12. The connections are made externally by the installer.

Reversing Motor Direction

The direction of rotation of 3-phase motors can be reversed by interchanging any two of the 3-phase power lines to the motor. **See Figure 4-20.** Although any two lines can be interchanged, the industrial standard is to interchange T1 and T3. This standard holds true for all 3-phase motors. For example, to reverse the direction of rotation of a delta-connected, 3-phase motor, T1 and T3 are interchanged.

Interchanging T1 and T3 is a standard for safety reasons. When first connecting a motor, the direction of rotation is usually unknown until the motor is started. It is common practice to temporarily connect the motor to determine the direction of rotation before making permanent connections. By always interchanging T1 and T3, T2 can be permanently connected to L2, creating an insulated barrier between T1 and T3.

Motors on 50/60 Hz

The motor nameplate indicates whether a motor is rated to operate at either 50 Hz, 60 Hz, or both. The motor speed will be different at the different line frequencies. A motor operating at 60 Hz operates 20% faster than a motor operating at 50 Hz. Consult the manufacturer if there are any questions about the power and torque characteristics of motors operated at a line frequency other than the design frequency.

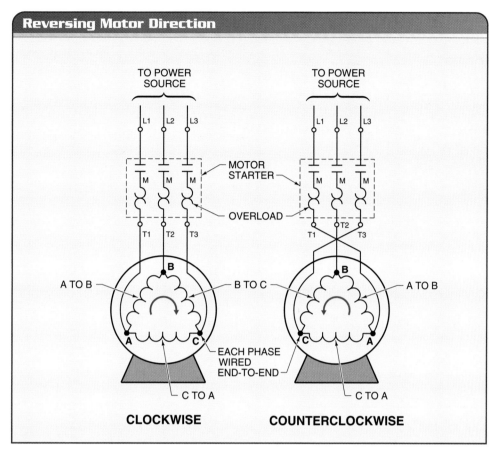

Figure 4-20. The direction of rotation of 3-phase motors can be reversed by interchanging any two of the 3-phase power lines to the motor.

MOTOR LOAD AND TORQUE

Since the function of the motor is to do work, the motor must be able to operate the load to which it is connected. To do this correctly, the motor must be matched to the load that it will drive. When the motor and load are properly matched, the motor should successfully drive the load under all given conditions for a reasonable period and operate economically.

Motor Torque

Torque is a turning or twisting force that causes an object to rotate. The torque required to operate a load from initial startup to final shutdown is considered when determining the type and size of motor required for a given application. Torque is typically measured in foot-pounds (ft-lb), pound-feet (lb-ft), and inch-ounces (in.-oz).

Motor power is rated in watts or horsepower. **See Figure 4-21.** A watt (W) is the base unit of electrical power. Larger motors are rated in kilowatts (kW). A horsepower (HP) is a unit of power equal to 746 watts or 33,000 pound-feet per minute (550 pound-feet per second). The horsepower of a motor represents the rate at which torque is applied. The four most common types of torque related to motors are locked-rotor torque, full-load torque, pull-up torque, and breakdown torque. **See Figure 4-22.**

Locked-rotor torque is the torque a motor produces when the rotor is stationary and full power is applied to the motor. This is the minimum torque that a motor at rest develops for all angular positions of the shaft with rated voltage applied at rated frequency. Locked-rotor torque is also referred to as breakaway or starting torque.

Definition

Torque is a turning or twisting force that causes an object to rotate.

Locked-rotor torque is the torque a motor produces when the rotor is stationary and full power is applied to the motor.

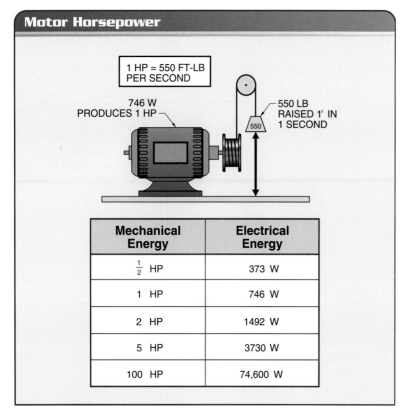

Figure 4-21. Motor power is rated in horsepower or watts.

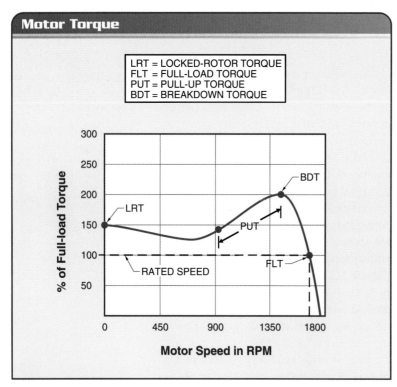

Figure 4-22. The four most common types of torque related to motors are locked-rotor torque, full-load torque, pull-up torque, and breakdown torque.

Full-load torque is the torque required to produce the rated power at full speed of the motor. Once a motor is up to rated speed, full-load torque keeps the load turning.

Pull-up torque is the accelerating torque required to bring a load up to the correct speed. If a motor is properly sized to the load, the time that pull-up torque is applied is brief. If a motor does not have sufficient pull-up torque, the locked rotor torque may start the load turning, but the pull-up torque will not bring it up to rated speed.

Breakdown torque is the maximum torque a motor can provide without an abrupt reduction in motor speed (stalling). As the load on a motor shaft increases, the motor produces more torque. As the load continues to increase, the point at which the motor stalls is reached. This point is the breakdown torque.

During startup, maximum torque is developed when the rotor frequency decreases to the point at which the winding reactance is equal to its resistance. At this point, the current in the rotor lags exactly 45 electrical degrees behind the voltage. The lower frequency in the rotor induces a resistive current high enough to produce a strong field that surrounds the rotor winding and creates a strong pole in the iron.

Matching Motor to Load

Undersized and oversized motors can drive most loads within a given range. However, if a motor is undersized, it may drive the load for a shorter period than is desirable. This is a typical problem with motors that require frequent replacement. To simply oversize a motor is not always the solution. An oversized motor costs more to purchase and operate than a properly sized motor, and requires more space. Nevertheless, if the proper size motor is not available, it is usually better to oversize than to undersize.

To drive a load at a set speed, the motor must produce a certain amount of torque. If the motor's output torque is large enough, the load will be driven. If the motor's output torque is too small, the load will not be driven, or will be driven at a reduced speed.

The torque-speed characteristic of a motor must match the load the motor is to drive. A load may have a definite torque-speed characteristic, such as a pump or fan that has a fixed load. Or the load may have a variable torque-speed characteristic, such as a hoist or conveyor belt used to move loads of varying weights. **See Figure 4-23.**

A *high-inertia load* is a load that has a relatively large amount of momentum. Typical types of high-inertia loads are fans, blowers, and punch presses. These types of loads are hard to accelerate and decelerate, even with a motor that has sufficient torque to drive the load at full speed.

Multiple-Speed Motors

Multiple-speed, 3-phase motors are designed to operate at two, three, or four speeds. The motor's operating speed is dependent upon the number of poles used. The speed of the motor varies inversely with the number of poles.

When the speed of a motor changes, the horsepower and torque required at the new speed may be higher, lower, or the same. The type of load and application determine whether a change in horsepower or torque is required. To meet load requirements, multiple-speed, 3-phase motors are designed with different operating characteristics.

Motor Types

The three motor types available are constant-horsepower, constant-torque, and variable-torque motors. Constant-horsepower motors are designed to give the same maximum horsepower at all speeds. Constant-torque motors are designed to give the same maximum torque at all speeds. Variable-torque motors are designed to produce an increase in torque and horsepower with an increase in speed.

Constant-Horsepower Motors. In a constant-horsepower motor, torque decreases in the same ratio as the speed increases, maintaining a constant horsepower. More current flows at the lower speed, increasing the torque. Less current flows at the higher speed, decreasing the torque.

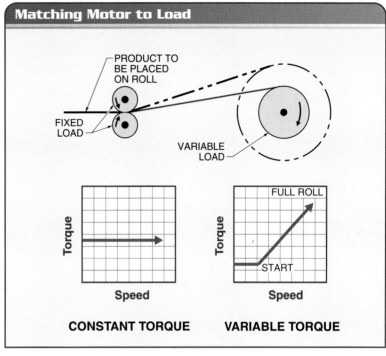

Figure 4-23. The torque-speed characteristic of a motor must match the load the motor is to drive.

Constant horsepower motors are used to drive loads that require the same horsepower output at different speeds. Typical applications include most machine-tool machines, such as boring machines, drilling machines, wheel-driven grinders, lathes, and milling machines. The number of poles in a constant-horsepower motor is effectively changed by changing the direction of current through the motor windings. **See Figure 4-24.**

Constant-Torque Motors. In a constant-torque motor, horsepower changes proportionally to the speed. A proportional change is a change in which factors increase or decrease at the same rate. The horsepower and the line current increase in the same ratio as the motor speed to provide constant torque.

Constant-torque motors are used to drive loads that require a constant torque output at different speeds. Typical applications include rotary and reciprocating compressors, conveyors, displacement fans, and printing presses. The number of poles in the motor is effectively changed by changing the direction of current through the motor windings. **See Figure 4-25.**

Definition

Full-load torque is the torque required to produce the rated power at full speed of the motor.

Pull-up torque is the accelerating torque required to bring a load up to the correct speed.

Breakdown torque is the maximum torque a motor can provide without an abrupt reduction in motor speed (stalling).

A *high-inertia load* is a load that has a relatively large amount of momentum.

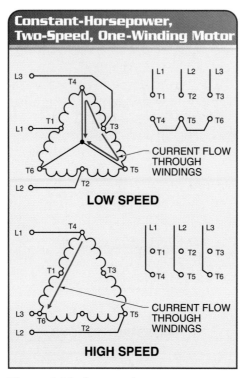

Figure 4-24. Constant-horsepower motors are used to drive loads that require the same horsepower output at different speeds.

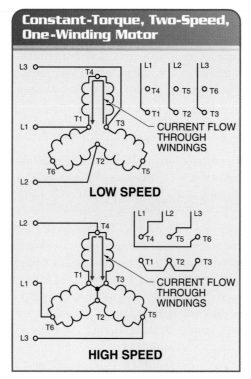

Figure 4-25. Constant-torque motors are used to drive loads that require a constant torque output at different speeds.

Variable-Torque Motors. In a variable-torque motor, torque and horsepower vary directly with the speed. The horsepower varies with the cube of the speed change. Torque and horsepower increase at higher speed and decrease at lower speed. Variable-torque, multiple-speed motors are used to drive fans, pumps, and blowers that require an increase in both torque and horsepower when speed is increased. The number of poles in the motor is effectively changed by changing the direction of current through the motor windings. **See Figure 4-26.**

Tech Fact

The occurrence of voltage sags and swells may indicate a weak power distribution system. In such a system, voltage will change dramatically when a large motor is switched ON or OFF.

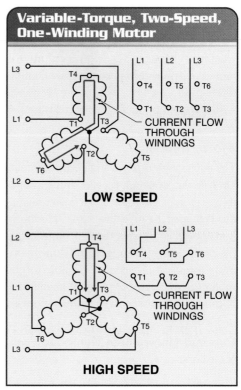

Figure 4-26. Variable-torque, multiple-speed motors are used to drive fans, pumps, and blowers that require an increase in both torque and horsepower when speed is increased.

MOTOR POWER

In a DC circuit, the polarity of the voltage is constant and the current always flows in the same direction. The situation is different with AC circuits. The alternating current causes the magnetic field around an inductor or capacitor to alternately charge and discharge as the current changes direction.

True power is the power, in W or kW, drawn by a motor that produces useful work. True power is used by the resistive part of a circuit that performs the work. True power can be produced only when current and voltage are both positive or both are negative. **See Figure 4-27.** A resistive load consumes true power when the voltage and current are in the same direction (both positive or both negative).

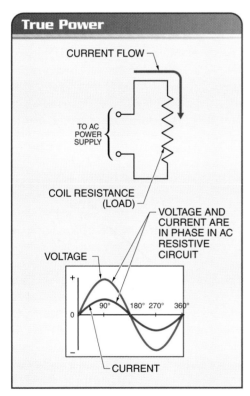

Figure 4-27. True power can be produced only when current and voltage are both are positive or both negative.

Reactive power is the power, in VAR or kVAR, stored and released by the magnetic field around inductors and capacitors. Reactive power is measured in volts-amps-reactive (VAR). In a circuit with reactive components, the voltage and current are out of phase. For purely inductive circuits, the current lags the voltage by 90 electrical degrees. For circuits with mixed inductive and resistive components, the current lags the voltage by a value between 0 electrical degrees and 90 electrical degrees. **See Figure 4-28.** Reactive power flows through the inductor or capacitor when the voltage and current are not in the same direction (one positive and one negative).

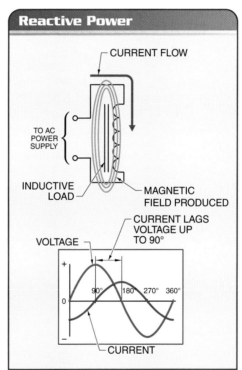

Figure 4-28. For circuits with mixed inductive and resistive components, the current lags the voltage by a value between 0° and 90°.

Apparent power is the power, in VA or kVA, that is the vector sum of true power and reactive power. Apparent power is the product of the total current and voltage in a circuit. A common type of inductive load is an inductive motor. Inductive loads have the current lagging the voltage. Reactive power used to build magnetic fields flows back into the source from the inductors and capacitors. This opposing power affects the power factor of a circuit.

Definition

True power is the power, in W or kW, drawn by a motor that produces useful work.

Reactive power is the power, in VAR or kVAR, stored and released by the magnetic field around inductors and capacitors.

Apparent power is the power, in VA or kVA, that is the vector sum of true power and reactive power.

Definition

Power factor is the ratio of true power, in W or kW, to apparent power, in VA or kVA, in a circuit.

A *power factor correction capacitor* is a capacitor used to improve a facility's power factor by improving voltage levels, increasing system capacity, and reducing line losses.

Motor efficiency is the ratio of useful work performed by a motor to the energy used by the motor to produce the work.

Power Factor

Power factor is the ratio of true power, in W or kW, to apparent power, in VA or kVA, in a circuit. Power factor is sometimes called displacement power factor. The power factor of an installed 3-phase motor is seldom known, but it can be calculated as follows:

$$pf = \frac{hp \times 746}{V \times i \times \sqrt{3} \times \varepsilon}$$

where
pf = power factor
hp = horsepower
V = voltage, in V
i = current, in A
ε = efficiency

For example, a 10-HP, 3-phase motor with an efficiency of 90% is connected to a 480-V supply. The full load current is 12 A. The power factor at full load is calculated as follows:

$$pf = \frac{hp \times 746}{V \times i \times \sqrt{3} \times \varepsilon}$$

$$pf = \frac{10 \times 746}{480 \times 12 \times \sqrt{3} \times 0.9}$$

$$pf = \frac{7460}{8979}$$

$$pf = 0.83$$

Power Factor Correction

Utility companies penalize customers with low power factors. A low power factor indicates that the circuit is drawing more current than is required by the load. Much of the current draw is used to charge the magnetic fields around the inductors. The power factor can be improved by adding capacitance in parallel with the inductance.

A *power factor correction capacitor* is a capacitor used to improve a facility's power factor by improving voltage levels, increasing system capacity, and reducing line losses. The capacitor should have the same amount of reactance as the inductor to cancel out the reactive power of the inductor. Power factor correction capacitors can be placed ahead of an electric motor drive in the AC supply lines but not between the drive and motor. Power factor correction capacitor units with automatic switching must not be used unless specifically recommended by the manufacturer. **See Figure 4-29.**

MOTOR EFFICIENCY

Motor efficiency is the ratio of useful work performed by a motor to the energy used by the motor to produce the work. Motor efficiency is calculated by dividing the output power by the input power. The output power is the rotary mechanical energy of the motor. The input power is the electrical power required to operate the motor. Motor efficiency information is typically provided by the manufacturer. All motors have motor energy losses that reduce motor efficiency.

In general, any motor that is replaced during maintenance should be replaced by another motor of the same size, unless there is evidence that the original motor was undersized. An oversized motor operates less efficiently than a motor that is properly sized.

Motor Energy Losses

There are always motor energy losses that reduce the efficiency of a motor. Any energy losses reduce the efficiency of a motor because the energy is wasted as heat and does not contribute to driving the load. Older motors typically operate at less than 80% efficiency. Newer, high-efficiency motors typically operate at more than 90% efficiency. The five major components of motor energy losses are resistance loss, core loss, bearing loss, windage loss, and sound loss. These losses add up to the total loss of a motor. **See Figure 4-30.**

Tech Fact

A common cause of a low power factor in industrial facilities is underloaded induction motors. The power factor of a motor is much lower at partial load than at full load.

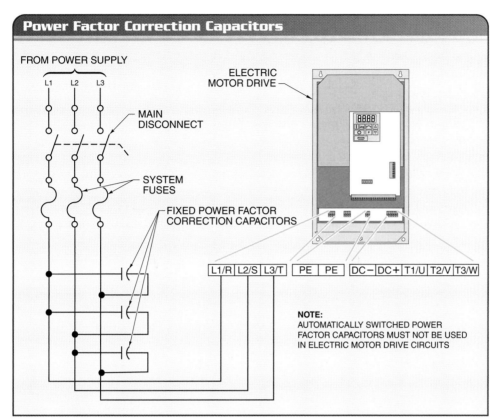

Figure 4-29. Power factor correction capacitors can be placed ahead of an electric motor drive in the AC supply lines but not between the drive and motor.

> **Definition**
>
> *Resistance loss* is the energy loss in a motor due to current flowing through conductors and coils that have resistance.
>
> *Core loss* is the total energy loss in the stator and rotor cores due to circulating currents and to the magnetic field escaping from the core.

Resistance Loss. *Resistance loss* is the energy loss in a motor due to current flowing through conductors and coils that have resistance. Resistance loss is sometimes called copper loss or I^2R loss. Resistance losses vary, depending on the motor load. At low load, the current drawn is small so the resistance losses are small. At full load, the current drawn is relatively high, so the resistance losses are also relatively high.

Resistance losses are found in the stator and the rotor. The loss in the stator due to the resistance of the windings is dissipated in the form of heat. In the rotor, the current draw increases as the load increases. The resistance losses increase the temperature of the motor.

Core Loss. *Core loss* is the total energy loss in the stator and rotor cores due to circulating currents and to the magnetic field escaping from the core. Core losses are fairly constant and depend on the motor design. Core losses are generally not dependant on the load, but do increase with increasing design voltage.

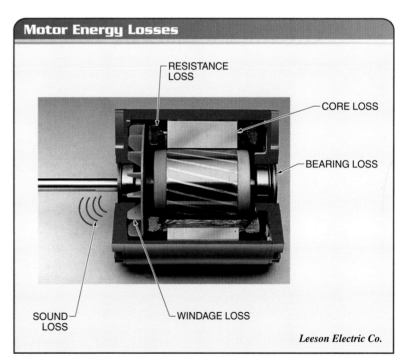

Figure 4-30. The five major components of motor energy losses are resistance losses, core losses, bearing losses, windage losses, and sound losses. These losses add up to the total loss of a motor.

Definition

*An **eddy current** is an undesired current circulating in the stator and rotor core caused by magnetic induction.*

***Hysteresis** is a loss due to the power consumed to realign the magnetic domains in the iron twice every electrical cycle.*

***Saturation** is the loss of magnetic lines of flux out of the core when the core cannot carry any more lines of flux with an increase in current.*

***Flux-linkage loss** is the loss of flux in the air gap because the air gap has increased reluctance compared to the cores.*

***Bearing loss** is any energy lost from friction between the motor shaft, the bearing, and the bearing support.*

***Windage loss** is energy lost by blowing air past a motor to remove heat.*

***Sound loss** is energy lost in a motor when the motor makes noise.*

An *eddy current* is an undesired current circulating in the stator and rotor core caused by magnetic induction. These currents can be minimized by manufacturing the core with thin sheets of stamped metal. The laminations break the potential conductive path in the core into smaller sections and reduce the loss. In addition, the alloys used for the cores usually contain silicon to increase the electrical resistance and reduce the eddy currents.

Hysteresis is a loss due to the power consumed to realign the magnetic domains in the iron twice every electrical cycle. *Saturation* is the loss of magnetic lines of flux out of the core when the core cannot carry any more lines of flux with an increase in current. *Flux-linkage loss* is the loss of flux in the air gap because the air gap has increased reluctance compared to the cores.

Bearing Loss. *Bearing loss* is any energy lost from friction between the motor shaft, the bearing, and the bearing support. The rotating shaft that supports the rotor is suspended in the end bells by the bearings. A bearing is a motor component used to reduce friction and maintain clearance between the stationary parts and the moving shaft. Bearing losses are usually small and fairly constant regardless of the load.

Windage Loss. *Windage loss* is energy lost by blowing air past a motor to remove heat. Almost all motors require cooling, as losses are mostly converted to heat. Blades are often designed into the rotor so that air is blown across the motor while the rotor is turning. In some types of motors, sheet-metal blades are stamped out and installed on the rotor shaft, with some on the inside of the housing and some on the outside with a protective guard over the end bell. Moving air in or across the stator housing opposes rotation, so this process wastes energy and generates more heat. Windage losses are usually small and fairly constant regardless of the load.

Sound Loss. *Sound loss* is energy lost in a motor when the motor makes noise. It takes energy to produce noise. Therefore, any sound produced by a motor is wasted energy. However, sound losses are usually small and fairly constant regardless of the load.

Application—Motor Line and Motor Wiring Diagrams

The two types of diagrams used with motor circuits are line diagrams and wiring diagrams. A line diagram shows only the control circuit. A wiring diagram shows the control circuit and the power circuit.

Motor Line Diagrams

A motor line diagram shows the operational logic of the motor circuit. The motor line diagram shows the electric connections between the components in the control circuit and how the components control the power circuit. For example, the diagram shows that the start and stop pushbuttons are connected by a conductor. When the start pushbutton is pressed, the circuit is completed through the motor starter. The starter's auxiliary contacts are connected in parallel with the start pushbutton. The motor starter remains energized until the stop pushbutton is pressed, or until the overload contacts open. The motor line diagram does not show the location of the components in the circuit.

Motor Wiring Diagrams

A motor wiring diagram shows the actual location of each component used in the control circuit and power circuit. For example, the diagram shows that the start and stop pushbuttons are located in the same pushbutton station and the start pushbutton is located on top. Motor wiring diagrams are useful in troubleshooting because they show the layout and connections of the components. However, motor wiring diagrams can hinder circuit understanding because the conductor connections are often hard to follow.

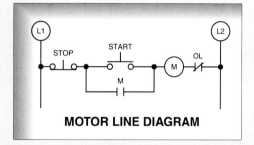

MOTOR LINE DIAGRAM

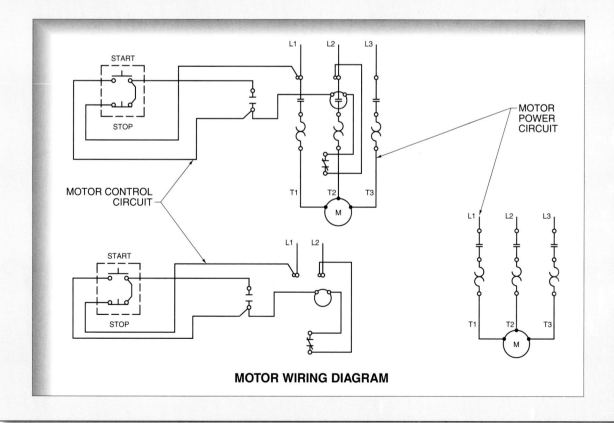

MOTOR WIRING DIAGRAM

Application—Three-Phase Motors

Application—Voltage and Current Unbalance

Multimeters can be used to measure voltage and current unbalance in a 3-phase power system by taking readings at each line. Three separate measurements of voltage and of current must be taken and recorded. Unbalance is determined by calculating the percentage of difference between the largest deviation and the average value.

A 3-phase power quality meter can also be used to measure the voltage and current unbalance. This is the quickest method because the meter is connected to each power line with voltage leads and current clamps and takes readings simultaneously. The meter also makes the calculations required to determine unbalance and displays the result directly.

For example, if the voltage readings are 471 V, 452 V, and 468 V, the average voltage is 463.7 V. The maximum voltage difference from the average is 11.7 V (463.7 − 452 = 11.7). The voltage unbalance is 2.5% (11.7 ÷ 63.7 × 100 = 2.5). In general, voltage unbalance should be 1% or less. A motor with a 2.5% voltage should be not be operated until the cause of the problem is fixed.

If the current readings are 61.5 A, 54.2 A, and 63.4 A, the average current is 59.7 A. The maximum current difference from the average is 5.5 A (59.7 − 54.2 = 5.5). The current unbalance is 9.2% (5.5 ÷ 59.7 × 100 = 9.2). In general, current unbalance should be 10% or less. A motor with a 9.2% unbalance can be operated, but with caution to avoid overheating the motor.

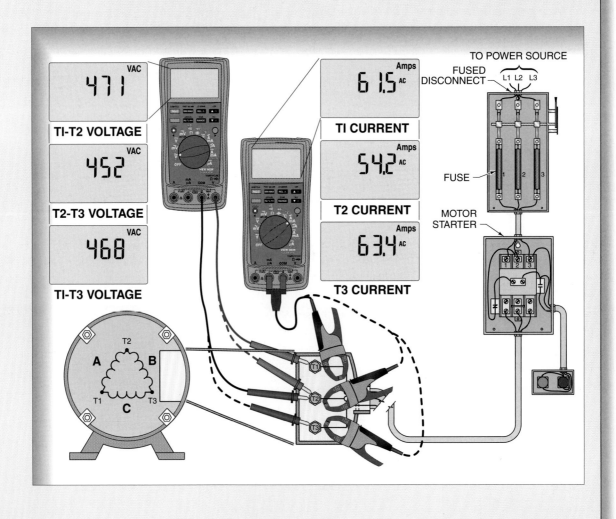

Summary

- A 3-phase motor transfers energy from electric power to the load through the use of a rotating magnetic field.

- A stator consists of a core and windings, or coils. The stator is mounted within a housing.

- The stator core consists of many thin iron sheets laminated together, pressed into a frame, and secured in place.

- The stator windings consist of coils of copper wire placed in slots 120 electrical degrees apart.

- A rotor consists of a core and windings, mounted on a shaft, that is allowed to turn to follow the stator rotating field.

- There are 360 electrical degrees in one cycle of a sine wave. There are 360° (mechanical) in one revolution.

- The synchronous speed of a 60 Hz motor is the calculated by dividing 7200 by the number of poles. For a 4-pole motor, the synchronous speed is 1800 rpm (7200 ÷ 4 = 1800).

- The current changes as it goes through the complete cycle and the amount of current flowing in each coil changes and the magnetic field around the coil changes. This creates the rotating stator field.

- The stator windings are factory wired as wye-connected or as delta-connected and cannot be changed in the field.

- The four most common types of torque related to motors are locked-rotor torque, full-load torque, pull-up torque, and breakdown torque.

- Power factor is the ratio of true power to apparent power.

- The five types of motor energy losses are resistance losses, core losses, bearing losses, windage losses, and sound losses.

Glossary . . .

A **stator** is the fixed, unmoving part of a motor, consisting of a core and windings, or coils, that converts electrical energy to the energy of a magnetic field.

A **rotor** is the rotating, moving part of a motor, consisting of a core and windings, which convert the rotating magnetic field of the stator into the torque that rotates the shaft.

A **salient pole**, or **projecting pole,** is a pole that extends away from the core toward the stator or extends away from the stator toward the rotor.

The **motor shaft** is a cylindrical bar used to carry the revolving rotor and to transfer power from the motor to the load.

The **synchronous speed** is the theoretical speed of a motor based on line frequency and the number of poles of the motor.

The **rotor frequency** is the rate at which the stator magnetic field passes the poles in the rotor.

A **single-voltage motor** is a motor that operates at only one voltage level.

A **dual-voltage motor** is a motor that operates at more than one voltage level.

...Glossary

Torque is a turning or twisting force that causes an object to rotate.

Locked-rotor torque is the torque a motor produces when the rotor is stationary and full power is applied to the motor.

Full-load torque is the torque required to produce the rated power at full speed of the motor.

Pull-up torque is the accelerating torque required to bring a load up to the correct speed.

Breakdown torque is the maximum torque a motor can provide without an abrupt reduction in motor speed (stalling).

A ***high-inertia load*** is a load that has a relatively large amount of momentum.

True power is the power, in W or kW, drawn by a motor that produces useful work.

Reactive power is the power, in VAR or kVAR, stored and released by the magnetic field around inductors and capacitors.

Apparent power is the power, in VA or kVA, that is the vector sum of true power and reactive power.

Power factor is the ratio of true power, in W or kW, to apparent power, in VA or kVA, in a circuit.

A ***power factor correction capacitor*** is a capacitor used to improve a facility's power factor by improving voltage levels, increasing system capacity, and reducing line losses.

Motor efficiency is the ratio of useful work performed by a motor to the energy used by the motor to produce the work.

Resistance loss is the energy loss in a motor due to current flowing through conductors and coils that have resistance.

Core loss is the total energy loss in the stator and rotor cores due to circulating currents and to the magnetic field escaping from the core.

An ***eddy current*** is an undesired current circulating in the stator and rotor core caused by magnetic induction.

Hysteresis is a loss due to the power consumed to realign the magnetic domains in the iron twice every electrical cycle.

Saturation is the loss of magnetic lines of flux out of the core when the core cannot carry any more lines of flux with an increase in current.

Flux-linkage loss is the loss of flux in the air gap because the air gap has increased reluctance compared to the cores.

Bearing loss is any energy lost from friction between the motor shaft, the bearing, and the bearing support.

Windage loss is energy lost by blowing air past a motor to remove heat.

Sound loss is energy lost in a motor when the motor makes noise.

Review

1. Explain why the stator uses thin iron sheets laminated together.

2. Explain why it is necessary to protect the stator windings from damage from movement of the coils.

3. List and describe the parts of a rotor.

4. Explain how electrical degrees are different from mechanical degrees.

5. Demonstrate how to calculate the synchronous speed of a 2-pole motor.

6. Describe how the stator rotating field is created.

7. Describe the difference between the wiring connections of a wye-connected and a delta-connected stator.

8. List and describe four components of motor torque. Sketch a graph showing the components of motor torque.

9. Define power factor. Demonstrate how to calculate the power factor of a 3-phase, 5-hp, 240-V, 94%-efficiency motor that draws a full load current of 12 A.

10. List and describe the five types of motor energy losses.

Refer to the CD-ROM for Quick Quiz® questions related to chapter content.

MOTORS

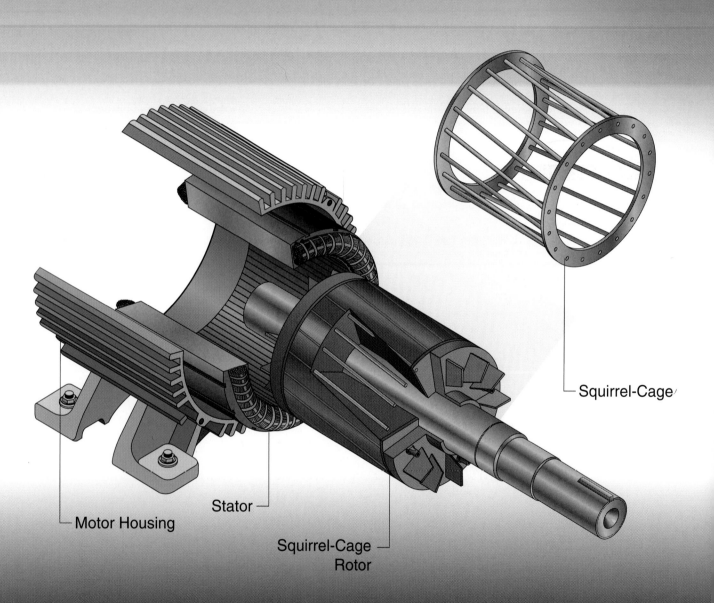

Motor Housing

Stator

Squirrel-Cage Rotor

Squirrel-Cage

Induction Motors

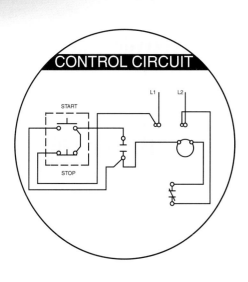

Induction Motor Construction 114
Operating Principles 118
Application—Troubleshooting 3-Phase Motors 120
Application—Troubleshooting Motor Control Circuits 121
Summary 122
Glossary 122
Review 123

OBJECTIVES

- List and describe the parts of a stator.
- List the parts of an induction motor rotor.
- Describe the shorting ring in a rotor.
- List the different types of squirrel-cage rotors (rotor design codes) and explain how their design determines torque.
- Describe how the induced current in the rotor is converted to torque.

Industry today depends on the induction motor to meet its needs. The typical squirrel-cage rotor is a simple, economical design that greatly reduces maintenance on these motors, making the motors very reliable.

114 MOTORS

Definition

*An **induction motor** is an electric motor that uses the principles of mutual induction to develop current and torque in the rotor.*

INDUCTION MOTOR CONSTRUCTION

An *induction motor* is an electric motor that uses the principles of mutual induction to develop current and torque in the rotor. Induction motors consist of a stator and a rotor enclosed within a frame. There are no physical electrical connections between the stator and the rotor. In an induction motor, the rotating magnetic field in the stator induces voltage in the rotor conducting bars that induces an opposing field. **See Figure 5-1.** Induction motors are the most common 3-phase motors used in industry today.

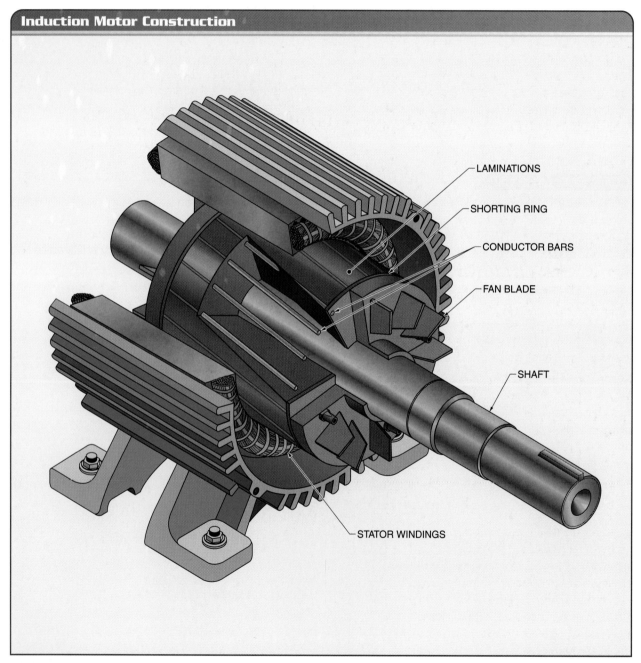

Figure 5-1. Induction motors consist of a stator and a rotor enclosed within a frame, with physical electrical connections between the stator and the rotor.

Stator Construction

The stator is the stationary part of an AC motor. The stator consists of a core and windings, or coils, that convert electrical energy to the energy of a magnetic field. **See Figure 5-2.**

The stator core consists of many thin iron sheets laminated together, pressed into a frame, and secured in place. The iron sheets are electrically separated from each other by an insulating coating. The separation reduces the cross-sectional area of the core and shortens the conduction path for damaging eddy currents.

The stator windings consist of coils of copper wire placed in slots 120 electrical degrees apart. A 3-phase stator is wound with coils that are connected to produce the three separate phases, A, B, and C. The motor nameplate shows the connections required to give a wye configuration or a delta configuration.

Rotor Construction

A rotor is the rotating part of an AC motor. The rotor consists of a core and windings, which convert the rotating magnetic field of the stator into the torque that rotates the shaft. The rotor is mounted on a shaft that is used to transfer the power to the load. The rotor core is made of many individual sheets of iron. These sheets are usually thicker than the stator sheets.

Rotor Windings. The main difference between different types of 3-phase motors is the design of the rotors and windings. In an induction motor, the field windings are in the stator and the rotors are designed to interact with the rotating stator magnetic field. Induction motors all have some method of inducing a current in the rotor windings.

A squirrel-cage rotor is the most common design for induction motor rotor design. A *squirrel-cage rotor* is an induction motor rotor consisting of conductors made from solid bars assembled into a cage frame resembling a squirrel cage. **See Figure 5-3.** The bars may be made from copper and pressed into the slots, or made from aluminum and poured into the slots under pressure in a die-cast machine. Squirrel-cage rotors are usually used with smaller motors.

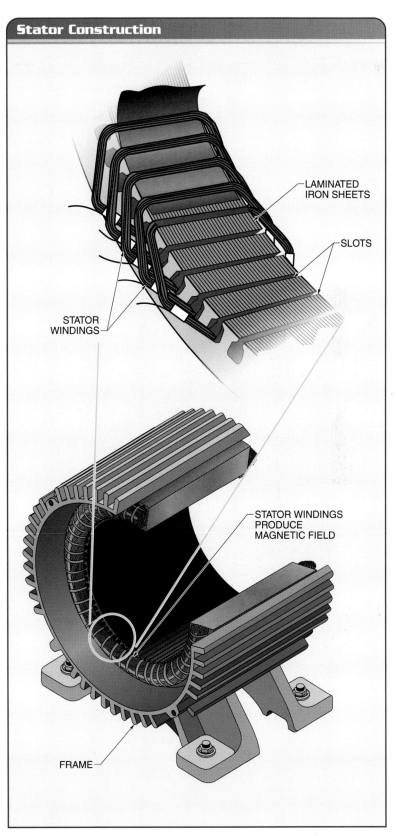

Figure 5-2. The stator consists of a core and windings, or coils, that convert electrical energy to the energy of a magnetic field.

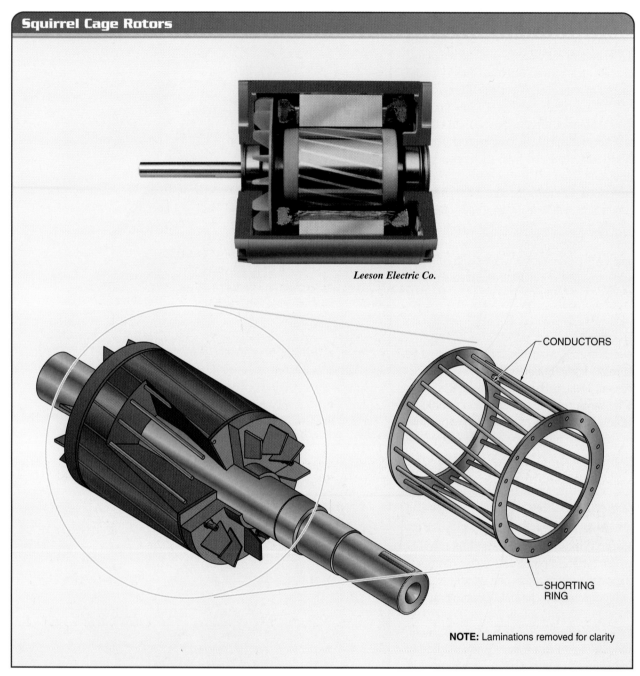

Figure 5-3. A squirrel-cage rotor is an induction motor rotor consisting of conductors made from solid bars assembled into a cage frame resembling a squirrel cage.

Definition

A *squirrel-cage rotor* is an induction motor rotor consisting of conductors made from solid bars assembled into a cage frame resembling a squirrel cage.

The copper or aluminum bars are attached to a connecting ring on each end of the drum. A *shorting ring,* or *conducting ring,* is a metal ring used to electrically connect the bars of a squirrel-cage rotor at the end of the cage frame. The shorting ring provides a short-circuit path for the current to circulate within the rotor.

In many motor styles, the shorting rings contain fan blades that move the air through the motor for cooling. The fan blades are also used for balancing the rotor. At the heaviest point at each end, material is added or removed to balance the rotor. Perfect balance is essential to prevent premature bearing failure.

NEMA Design Code Letters. There are several different standard rotor designs for an induction motor that deliver different starting torques and starting currents. **See Figure 5-4.** The rotor design code is identified on the nameplate. Motor Design A is a design with normal starting torque and normal starting current. The bars (squirrel cage) are placed near the surface of the rotor and have low reactance. The low reactance allows for large current to flow through the bars and a large torque to be developed. The locked-rotor current is about 500% to 1000% of full-load current.

Motor Design B is a fairly common design with normal starting torque and low starting current. This type of design is often used with larger motors that require torque similar to that of Design A, with lower currents. The bars are narrow and are placed deep in the iron, which increases the reactance and lowers the current, while maintaining normal starting torque. Design B motors are designed for a broad variety of applications, such as in HVAC applications providing power for fans and blowers and in industrial applications providing power for pumps.

Motor Design C has high starting torque and low starting current. The rotor has two conductor bars in each slot, with one above the other. The top bar is a high-resistance conductor, while the bottom bar is a high-reactance conductor. This arrangement forces the top bar to carry most of the current during starting. Design C motors are suitable for equipment with high-inertia starts, such as positive-displacement pumps and heavily loaded conveyors.

Motor Design D has high starting torque, high slip, and low starting current. The bars in the rotor have high resistance, are large, and are placed deep in the iron. Design D motors have high slip because of the relatively low starting current. Because of the high resistance and high slip, Design D motors are less efficient than other designs. Design D motors are suitable for equipment with very-high-inertia starts, such as cranes and hoists.

> **Definition**
>
> A ***shorting ring***, or ***conducting ring***, is a metal ring used to electrically connect the bars of a squirrel-cage rotor at the end of the cage frame.

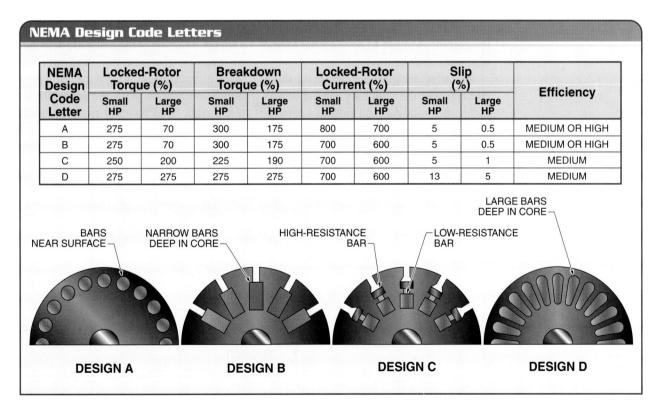

NEMA Design Code Letters

NEMA Design Code Letter	Locked-Rotor Torque (%)		Breakdown Torque (%)		Locked-Rotor Current (%)		Slip (%)		Efficiency
	Small HP	Large HP	Small HP	Large HP	Small HP	Large HP	Small HP	Large HP	
A	275	70	300	175	800	700	5	0.5	MEDIUM OR HIGH
B	275	70	300	175	700	600	5	0.5	MEDIUM OR HIGH
C	250	200	225	190	700	600	5	1	MEDIUM
D	275	275	275	275	700	600	13	5	MEDIUM

Figure 5-4. Different standard rotor designs are used for different applications.

Summary

- The stator consists of a core and windings, or coils, that convert electrical energy to the energy of a magnetic field.

- The stator core consists of many thin iron sheets laminated together, pressed into a frame, and secured in place. The iron sheets are electrically separated from each other by an insulating coating.

- The stator windings consist of coils of copper wire placed in slots 120 electrical degrees apart.

- The rotor consists of a core and windings, which convert the rotating magnetic field of the stator into the torque that rotates the shaft. The rotor is mounted on a shaft that is used to transfer the power to the load.

- A shorting ring, or conducting ring, is a metal ring used to electrically connect the bars of a squirrel-cage rotor at the end of the cage frame. The shorting ring provides a short-circuit path for the current to circulate within the rotor.

- With Motor Design A, the bars are placed near the surface of the rotor and have low reactance. The low reactance allows for large current to flow through the bars and a large torque to be developed.

- With Motor Design B, the bars are narrow and are placed deep in the iron, which increases the reactance and lowers the current, while maintaining normal starting torque.

- With Motor Design C, the rotor has two conductor bars in each slot, with one above the other. The top bar is a high-resistance conductor that carries most of the current during starting.

- With Motor Design D, high-resistance bars are placed deep in the iron, creating low starting current.

- Current flowing through the conducting bars in the rotor creates a magnetic field around the bars. When the current in adjacent bars flows in the same direction, the magnetic fields around the bars combine into a larger field that determines the location of the poles.

- The magnetic field and the current flow interact to create the torque. The direction of the torque can be determined with the right-hand motor rule.

Glossary

An *induction motor* is an electric motor that uses the principles of mutual induction to develop current and torque in the rotor.

A *squirrel-cage rotor* is an induction motor rotor consisting of conductors made from solid bars assembled into a cage frame resembling a squirrel cage.

A **shorting ring**, or **conducting ring**, is a metal ring used to electrically connect the bars of a squirrel-cage rotor at the end of the cage frame.

Review

1. List the parts of the stator.

2. Describe the parts of a stator.

3. List the parts of an induction motor rotor.

4. Explain the function of the shorting ring in a rotor.

5. Summarize how the different types of squirrel-cage rotors create different currents and torque.

6. Explain how torque is developed in an induction motor.

Refer to the CD-ROM for Quick Quiz® questions related to chapter content.

MOTORS

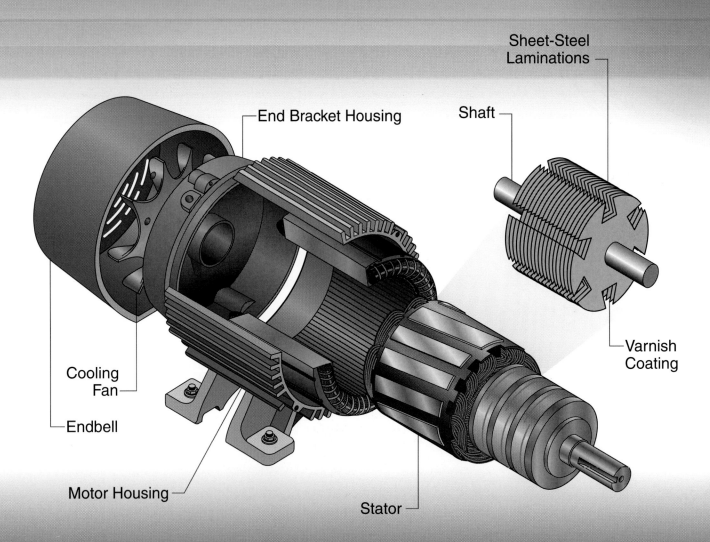

Wound-Rotor Motors

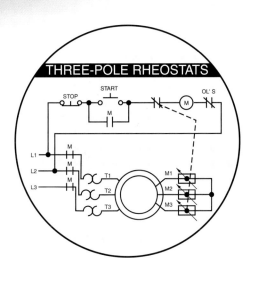

Wound-Rotor Motor Construction	126
Operating Principles	130
Supplemental Topic—Reactance and Phase Angle	133
Application—Crane and Elevator Operation	134
Application—Current-Control Resistance Switching	135
Summary	136
Glossary	137
Review	137

OBJECTIVES

- Describe the differences between the rotors of a wound-rotor motor and a squirrel-cage induction motor.
- Explain why resistance is included in the rotor circuit during startup.
- Demonstrate how resistance in the rotor circuit is used to control motor speed.
- Describe the difference between a starter and a regulator for a wound-rotor motor.

Wound-rotor motors were one of the first types of motors to allow variable-speed operation. In addition, the torque can be varied while the motor is operating at full line voltage. Wound-rotor motors are often used for conveyors, large grinders, crushers, fans, pumps, elevators, and bridge cranes. They are fairly simple to operate and do not require sophisticated control equipment. However, the cost of the motor itself is higher, along with higher maintenance costs, because of the addition of slip rings and an external resistor circuit to the design.

126 MOTORS

Definition

A *wound-rotor motor* is an induction motor with the squirrel-cage conductor bars replaced with coils of wire, and with added slip rings, brushes, and a resistor circuit.

WOUND-ROTOR MOTOR CONSTRUCTION

A *wound-rotor motor* is an induction motor with the squirrel-cage conductor bars replaced with coils of wire, and with added slip rings, brushes, and a resistor circuit. Wound-rotor motors are especially useful because they are able to deliver high starting torque without overloading the electrical supply system.

Stator Construction

The stator of a wound-rotor motor is the same as the stator of a squirrel-cage induction motor. **See Figure 6-1.** The windings are placed in the slots in the stator 120 electrical degrees apart. The windings can be wound in either a wye or a delta configuration and can be either single or dual voltage. They are brought out of the motor and are marked T1 through T9. The terminals are connected to the motor starter.

Figure 6-1. The stator of a wound-rotor motor is the same as the stator of a squirrel-cage induction motor.

Rotor Construction

The rotor of a wound-rotor motor is made from laminations stacked together in the same manner as the rotor of a squirrel-cage induction motor, with an oxide or varnish coating between each lamination. **See Figure 6-2.** However, the rotor of a wound-rotor motor is constructed by placing insulated coils of wire in the slots instead of the solid conductor bars used in the squirrel-cage induction motor. A wound-rotor motor can be distinguished from a squirrel-cage induction motor by the presence of the coils of wire in the winding slots instead of the solid conductor bars, by the presence of the three slip rings on the shaft, and by the presence of an external resistance bank.

Rotor Windings. The rotor windings of a wound-rotor motor consist of coils of insulated copper conductor placed into the slots in the rotor. The windings are placed in slots 120 electrical degrees apart. The rotor is wound so that it has the same number of poles as the stator.

With a squirrel-cage induction motor, the rotor windings are short-circuited by the shorting ring. Wound-rotor stators are wound just as the stator, either wye or delta. The leads are connected to the slip rings. **See Figure 6-3.** This allows the rotor windings to be connected through the brushes to the external wye-connected resistance. The rotor windings are marked M1, M2, and M3.

The NEC® requires overload protection for most circuits. In the case of a wound-rotor motor, the secondary circuit is considered protected against overcurrent by the motor overload protection means.

Tech Fact

Resistance is opposition to the flow of current and is measured in ohms (Ω). External resistors are used to limit the amount of current while starting a wound-rotor motor. Resistors must be sized properly to withstand the voltage and to be able to dissipate the heat generated during starting.

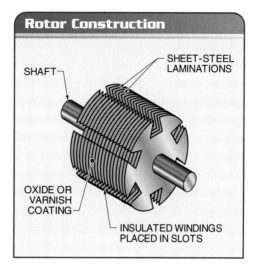

Figure 6-2. The rotor of a wound-rotor motor is constructed by placing insulated coils of wire in the slots, instead of the solid conductor bars used in the squirrel-cage induction motor.

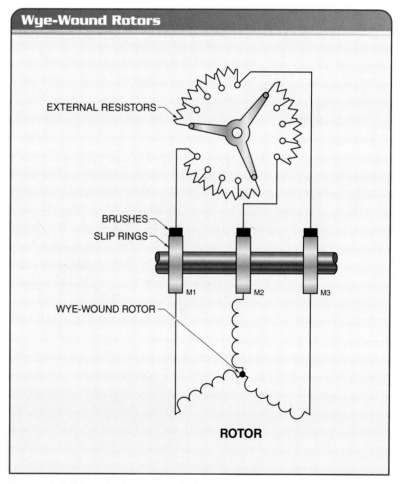

Figure 6-3. Wound-rotor stator windings are wye connected, with the free end of each winding connected to a slip ring.

Definition

*A **slip ring** is a metallic ring mounted on a motor shaft and electrically insulated from the shaft.*

*A **brush** is a sliding contact that rides against a rotating component to provide a connection to a stationary circuit.*

*A **brush rigging** is the entire assembly of the brush, brush holder, insulators, and any wiring included in the assembly.*

*A **pigtail** is an extended, flexible connection or a braided copper conductor.*

Slip Rings and Brushes. A *slip ring* is a metallic ring mounted on a motor shaft and electrically insulated from the shaft. A *brush* is a sliding contact that rides against a rotating component to provide a connection to a stationary circuit. A wound-rotor motor has three slip rings mounted on the shaft. Brushes are used with the slip rings to connect the rotor to an external set of resistors.

A brush is held in a brush holder and is free to move up and down in the holder. **See Figure 6-4.** A *brush rigging* is the entire assembly of the brush, brush holder, insulators, and any wiring included in the assembly. This allows the brush to follow irregularities in the surface of the slip ring. Brush holders are mounted on the motor frame, but they are electrically insulated from it. A spring placed behind a brush forces the brush to make contact with the slip ring. The spring pressure is usually adjustable. A pigtail connects the brush to the power supply. A *pigtail* is an extended, flexible connection or a braided copper conductor.

Brushes must have good conductivity and be soft enough not to damage the slip ring. Brush material is typically composed of a mixture of carbon and graphite for high-voltage machines and of graphite and metallic powder for low-voltage machines. The graphite in the brushes provides lubrication. No other lubrication should be used since it may cause electrical problems and equipment damage.

Brushes can usually be worn down to a fraction of their length before they need to be replaced. Because brushes wear down, constant pressure is applied with an adjustable tension spring so that the brushes can maintain contact with the slip rings.

Resistors. Brushes are used on the slip rings to connect the rotor to an external set of resistors. The resistors are often referred to as secondary resistors, because the rotor is the secondary of the motor. The resistors are switched to add or subtract the resistance from the rotor circuit. **See Figure 6-5.**

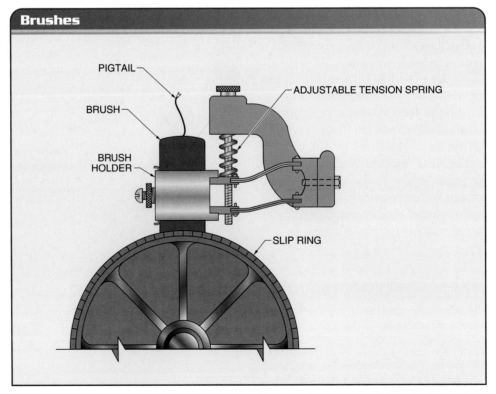

Figure 6-4. A brush is held in a brush holder and is free to move up and down in the holder.

The resistance may be in the form of a resistor wire or cast iron resistor grids, or as electrolytic liquid with metallic electrodes. The resistor usually requires tapping the wire, continuously or in steps, with a rheostat. Cast iron grids can be switched in and out of the circuit with contactors. Electrolytic resistors contain a conductive electrolyte solution. These resistors have a large thermal mass to absorb the heat generated during starting, but are seldom used anymore because other types of resistors are available that can dissipate the heat.

The external resistor circuit carries the entire rotor current. This is very helpful in troubleshooting a wound-rotor motor. The current can be easily measured with an ammeter and differences between the resistor circuits can be detected. Any differences in current in the different phases can indicate a problem with the resistor bank.

A wound-rotor motor uses coils of wire for the rotor conductors and uses three slip rings.

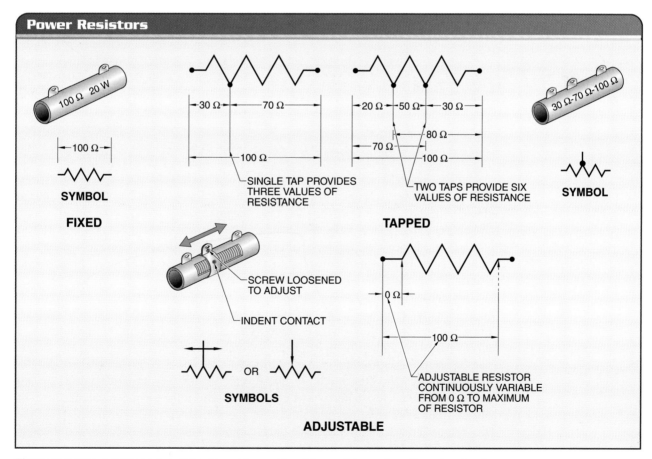

Figure 6-5. The resistors are switched to add or subtract the resistance from the rotor circuit.

The resistors are used to keep the current and voltage in phase as the frequency varies. The secondary resistors are connected through the slip rings and brushes to the terminals M1, M2, and M3. The M1 connection is the outside slip ring, the M2 connection is the middle slip ring, and the M3 connection is close to the motor windings. **See Figure 6-6.**

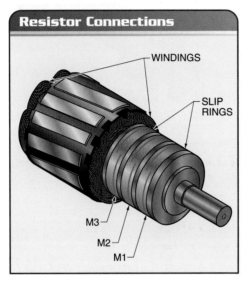

Figure 6-6. The secondary resistors are connected through the slip rings and brushes to the terminals M1, M2, and M3.

OPERATING PRINCIPLES

As with a squirrel-cage induction motor, three-phase power is applied to the stator through the three motor leads, T1, T2, and T3. This establishes the rotating field in the stator. The coils in the rotor have current induced in them in a similar manner to a squirrel-cage induction motor.

A wound-rotor motor is normally started with full resistance in the circuit. As the motor accelerates, resistance is gradually switched out of the circuit. Resistance can be switched out either manually or automatically. When the motor reaches full speed, all the resistance is switched out and the rotor windings are shorted. The rotor windings themselves have only slightly more resistance than the bars in a squirrel-cage rotor. This low resistance results in the same basic characteristics as a 3-phase squirrel-cage induction motor, but with slightly more slip and slightly lower efficiency.

Starting and Torque

The maximum torque is produced when the maximum resistance is connected to the rotor and the induced frequency is at its highest. A high-resistance rotor develops a high starting torque at low starting current. To gain the maximum starting torque, the motor is started with maximum rotor resistance. The resistance is reduced as the motor accelerates in order to shift the point of maximum torque. **See Figure 6-7.**

Curve 1 shows the starting torque when no resistance is applied to the rotor circuit. When no resistance is applied, the motor has the same basic starting torque characteristics as an induction motor. The starting torque is about 125% of the full-load torque (FLT).

Curve 2 shows the starting torque when maximum resistance is applied to the rotor. When maximum resistance is applied, the motor's starting torque is almost equal to the motor's breakdown torque (BDT), or about 200%. This is used for starting a wound-rotor motor.

Curve 3 shows the starting torque when medium resistance is applied to the rotor. When medium resistance is applied, the motor has less starting torque than at maximum resistance, but has a higher torque at approximately 15% to 30% of full speed. The resistance is changed during startup to keep the maximum torque near the actual operating speed throughout the acceleration period.

Once the motor starts, the speed of the rotor increases and the induced frequency decreases, decreasing the induction and induced current in the rotor. The torque also decreases as the induced current decreases. As the torque decreases toward the minimum torque needed, some of the resistance can be removed from the circuit. This can be done manually with a rheostat or automatically with a controller and contactors.

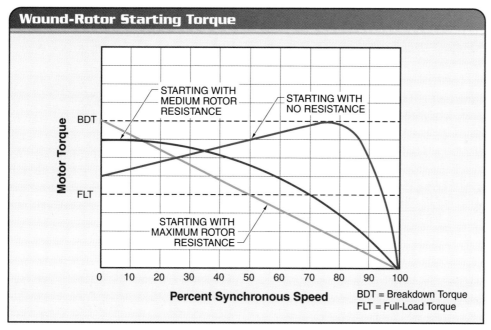

Figure 6-7. Resistance is reduced as the motor accelerates in order to shift the point of maximum torque.

When resistance is switched out of the circuit, the current increases. This increases the torque and allows the motor to continue to accelerate. This can continue in steps, with the number of steps depending on the motor design. For each step, the motor accelerates, the current decreases, and more resistance is removed, until the motor is at full-load speed. At this point, there is no resistance, except for the small amount in the coil conductors, and the motor operates in a manner similar to a standard squirrel-cage induction motor.

Speed Control

Adding resistance improves the starting torque of a wound-rotor motor at low speeds. However, there is an opposite effect at operating speeds. The resistors that were connected to the rotor can be used to reduce the speed of the motor. Increasing the resistance of the rotor while it is running reduces the current in the rotor windings and reduces the speed of the motor. When the speed of the rotor is reduced, the slip increases and more current is induced into the windings of the rotor. This increases the torque and allows the motor to operate at a lower speed and still drive the connected load.

When resistors are used for speed control, they are continuously in the rotor circuit. However, resistors must not be used continuously unless specifically designed for the purpose. Resistors designed for starting are able to dissipate the heat generated during the short time required for starting the motor, but cannot survive the heat generated during continuous operation. Power resistors that are able to dissipate large quantities of heat are required.

When the motor operates at a lower speed, the increased current causes a higher temperature in the motor. Since the motor is operating at a lower speed, normal ventilation is also reduced. Therefore, the speed of the motor is usually reduced no more than 50%.

Starters

A *wound-rotor motor starter* is a device containing low-wattage resistors that is designed to provide rotor circuit resistance during startup and remove that resistance when the motor is up to speed. The resistors used in a wound-rotor motor starter are only able to dissipate small amounts of heat and therefore can only be used during starting. They cannot be used continuously to control the motor speed.

Definition

*A **wound-rotor motor starter** is a device containing low-wattage resistors that is designed to provide rotor circuit resistance during startup and remove that resistance when the motor is up to speed.*

132 MOTORS

Definition

*A **wound-rotor motor regulator** is a device containing high-wattage resistors that is designed to control the speed of a wound-rotor motor and operate in variable-speed mode for as long as needed.*

*A **drum switch** is a rotating control device used to switch resistors in or out of a wound-rotor circuit.*

*A **three-pole rheostat** is a switch with tapped resistors used to switch a set of resistors in or out of a wound-rotor circuit.*

*A **silicon-controlled rectifier (SCR)** is a solid-state device used to switch a set of resistors in or out of a wound-rotor circuit.*

Starters usually have one or more timers used to automatically remove resistance from the rotor circuit. **See Figure 6-8.** When the start button is pressed, the P coil is energized and one P contact latches in the start circuit. Three other P contacts are closed and the stator is placed across the line via L1, L2, and L3. The rotor current flows through all the resistors in the rotor circuit.

The timing contact on the P contactor reaches its preset and closes, energizing contactor S1, and closes three contacts. One contact latches in contactor S1. The other two contacts are in the rotor circuit and shunt a portion of the resistor grid. Because some of the resistors in the rotor circuit are shunted, the resistance in the rotor circuit is decreased and the rotor current is increased. This increases the pole strength and reduces the slip, accelerating the rotor.

Contactor S1 also contains a timing contact. When the timing contact on the S1 contactor reaches its preset and closes, it energizes contactor S2, closing three contacts and opening one contact. One contact latches in contactor S2. One contact opens to remove contactor S2 from the circuit. These contacts ensure that contactor S2 is latched in before contactor S1 is removed from the circuit. The other two contacts shunt the remaining resistors.

Regulators

A *wound-rotor motor regulator* is a device containing high-wattage resistors that is designed to control the speed of a wound-rotor motor and operate in variable-speed mode for as long as needed. During startup, a regulator works the same as a starter. During normal operation, the high-wattage resistors in a regulator allow the resistors to remain continuously in the rotor circuit. Three common types of regulators are drum switches, three-pole rheostats, and silicon-controlled rectifiers.

Drum Switches. A *drum switch* is a rotating control device used to switch resistors in or out of a wound-rotor circuit. **See Figure 6-9.** A drum switch may have contacts large enough to directly switch the rotor current or it may have smaller contacts to pull in contactors that shunt the resistors and create a path for the rotor current. A drum switch often contains a microswitch to pull in the stator contactor.

A drum switch typically does not have low-voltage protection. If the power is lost, the drum switch restarts in the same position in which it was left when the power failed. This lack of protection is generally not a problem because this type of wound-rotor motor usually has an operator at the controls.

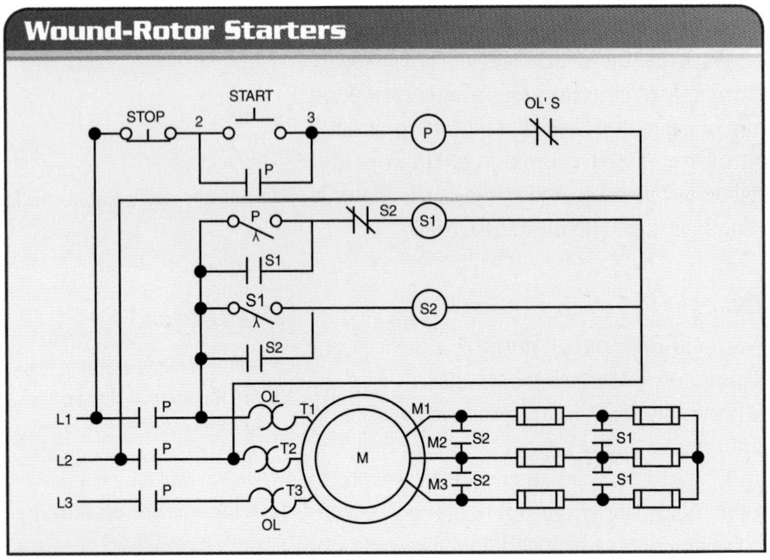

Figure 6-8. Starters usually have one or more timers used to automatically remove resistance from the rotor circuit.

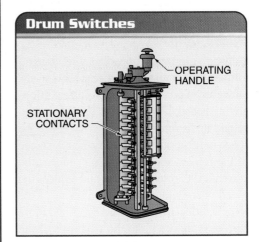

Figure 6-9. A drum switch is a rotating control device used to switch resistors in or out of a wound-rotor circuit.

Three-Pole Rheostats. A *three-pole rheostat* is a switch with tapped resistors used to switch a set of resistors in or out of a wound-rotor circuit. It can be manually controlled or motor driven. A rheostat uses a three-wire control circuit to start the wound-rotor motor. The system should have an interlock to ensure that the rheostat is in the full resistance position when started. **See Figure 6-10.**

Silicon-Controlled Rectifiers. A *silicon-controlled rectifier (SCR)* is a solid-state device used to switch a set of resistors in or out of a wound-rotor circuit. SCRs are rarely used with wound-rotor motors, and then only in applications that require continuous speed variations. This system uses feedback to regulate the speed and can keep the speed constant. It also has the three-wire system to start the wound-rotor motor.

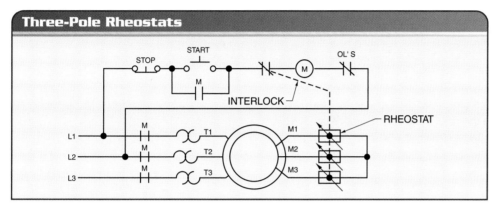

Figure 6-10. A rheostat uses a three-wire control circuit to start the wound-rotor motor.

Reactance and Phase Angle

With a squirrel-cage induction motor, the rotor conductor bars are shorted together and have very low resistance. This makes the reactance large relative to the resistance at startup and creates a reactive circuit. For a reactive circuit, there is a 90° phase angle between the voltage and current in the rotor. Since the voltage in the rotor is already 90° out of phase with the stator, the rotor current is 180° out of phase with the stator current. Since the stator and rotor currents are 180° out of phase, the stator and rotor poles are also 180° out of phase. The stator and rotor poles are far apart and the force between them is relatively weaker than if the poles were closer together.

A wound-rotor motor has resistance added to the rotor circuit during startup. This makes the rotor circuit resistive instead of reactive. For a resistive circuit, the voltage and current are in phase. This means that the rotor current is only 90° out of phase with the stator current.

This places the stator and rotor poles closer together than in a squirrel-cage motor, increasing the force between them. When a wound-rotor motor is operating at full speed and the resistance is removed from the rotor circuit, the rotor circuit becomes reactive instead of resistive.

The rotor winding of a wound-rotor motor has multiple conductors, so it has more inductance than a similar squirrel-cage rotor. Because of the higher inductance, the reactance is higher and the current is lower than in a squirrel-cage motor. This lower current means that it would be very difficult to start a wound-rotor motor without the resistance in the circuit.

With a wound-rotor motor, the resistors in the rotor circuit make the circuit resistive instead of reactive and the power factor improves. This reduces the starting current.

Supplemental Topic

Application—Wound-Rotor Motors

Application—Crane and Elevator Operation

A wound-rotor motor can easily be used to operate a crane or elevator. As the load leaves the ground, maximum resistance is inserted in the rotor circuit to ensure there is enough torque to move the load. As the load begins to accelerate, some of the resistance is removed from the rotor circuit. As the load approaches the required height, the resistance is reinserted, slowing the load and allowing it to stop. The sequence is reversed to lower a load.

In order to implement this scenario, the shunt contactors are open when the stator is placed across the line. This creates a circuit with maximum resistance in series with the rotor windings. This provides very strong torque, but at a slow speed.

The shunt contactors close, removing the resistance from the rotor circuit, and the motor accelerates to the speed allowed by the load. As the load approaches the desired height, the resistance is inserted back into the rotor circuit. This decreases the current in the rotor and reduces the interaction between the rotor and the stator, slowing the rotor. The slower rotor sees an increase in frequency and torque as the result. The slow speed at high slip allows the load to approach the required height at a speed slow enough to stop at the proper position. At this position, the stator is removed from the line and the motor stops.

Speed regulation in the high-slip condition is poor because much of the power consumed in the resistive rotor circuit is wasted as heat and dissipated by the resistor network. The acceleration is very different for a light load than a heavy load.

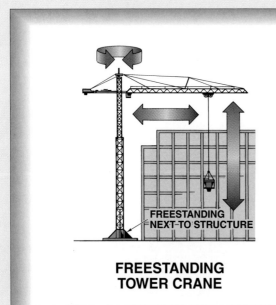

FREESTANDING TOWER CRANE

Portland Cement Association

Application—Current-Control Resistance Switching

There are many applications where a wound-rotor motor must be started up while keeping the torque within a limited range. For example, a large rock crusher has a heavy breaker bar or grinding teeth. Because the load has a lot of inertia and starts slowly, the starting current can overload the electrical supply system. In order to manage the load on the electrical supply system, the torque must be kept between 100% and 150% of full-load torque (FLT) during startup. In this case, a series of contactors are used in the starting circuit to switch the resistors.

A current transformer is used to monitor the rotor current flow through the external circuit. During startup, the brushes pick up the current from the slip rings and send it to the resistor. The resistors are used to limit the rotor current and control the torque. The resistor bank must be carefully designed as part of the control system to keep the torque within the desired range.

As the rotor begins to speed up, the current and torque decrease. The current is monitored through the current transformer. When the current decreases to the point where the torque has decreased from 150% to 100%, the motor controller switches out resistors labeled D by closing contactor S1 to short that section of the resistor bank. This reduces the resistance and allows more to current to flow, and the torque jumps back up to 150% of FLT.

This type of control can be used to reduce the problems caused by an overload situation. If the crusher gets overloaded, the measured current can be used to activate a switch to place some of the resistance back into the rotor circuit and decrease the current. This signal can also be used to turn off a conveyor feeding the crusher to allow the system to clear out any jams.

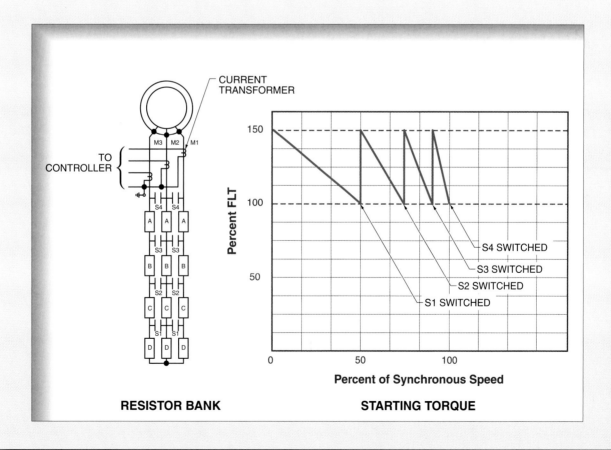

RESISTOR BANK **STARTING TORQUE**

Summary

- The stator of a wound-rotor motor is the same as the stator of a standard squirrel-cage induction motor.

- The rotor windings of a wound-rotor motor consist of coils of insulated copper conductor placed into the slots in the rotor.

- Slip rings and brushes are used to make an electrical connection between the moving rotor and the stationary external resistor circuit.

- Wound-rotor motors use external resistors in the rotor circuit.

- Resistance in the rotor circuit limits the starting current.

- The phase angle in the rotor is the primary reason a wound-rotor motor has a large torque with low stator current.

- As the rotor starts turning during startup, rotor frequency and the current decrease.

- As the current decreases during startup, resistance can be switched out of the circuit to increase the current and torque.

- Increasing the resistance while the motor is running decreases the current. The rotor slows to increase the induced current. This allows the motor speed to be controlled.

- A wound-rotor motor starter contains low-wattage resistors and should only be used during startup. The resistors must be removed from the rotor circuit after the motor is up to speed.

- A wound-rotor motor regulator contains high-wattage resistors and is used for starting and speed control. The resistors can remain in the rotor circuit continuously to allow for speed control.

- Three types of wound-rotor motor regulators are drum switches, three-pole rheostats, and silicon-controlled rectifiers.

Glossary

A **wound-rotor motor** is an induction motor with the squirrel-cage conductor bars replaced with coils of wire, and with added slip rings, brushes, and a resistor circuit.

A **slip ring** is a metallic ring mounted on a motor shaft and electrically insulated from the shaft.

A **brush** is a sliding contact that rides against a rotating component to provide a connection to a stationary circuit.

A **brush rigging** is the entire assembly of the brush, brush holder, insulators, and any wiring included in the assembly.

A **pigtail** is an extended, flexible connection or a braided copper conductor.

A **wound-rotor motor starter** is a device containing low-wattage resistors that is designed to provide rotor circuit resistance during startup and remove that resistance when the motor is up to speed.

A **wound-rotor motor regulator** is a device containing high-wattage resistors that is designed to control the speed of a wound-rotor motor and operate in variable-speed mode for as long as needed.

A **drum switch** is a rotating control device used to switch resistors in or out of a wound-rotor circuit.

A **three-pole rheostat** is a switch with tapped resistors used to switch a set of resistors in or out of a wound-rotor circuit.

A **silicon-controlled rectifier (SCR)** is a solid-state device used to switch a set of resistors in or out of a wound-rotor circuit.

Review

1. Explain how to tell the difference between a wound-rotor motor and a squirrel-cage induction motor.

2. Explain why resistors are included in the rotor circuit during startup of a wound-rotor motor.

3. Describe what happens when resistors are switched out of the rotor circuit during startup of a wound-rotor motor.

4. Describe how changing the resistance in the rotor circuit changes the motor speed while running at nominal speed.

5. Describe the difference between a starter and a regulator in a wound-rotor circuit.

Refer to the CD-ROM for Quick Quiz® questions related to chapter content.

MOTORS

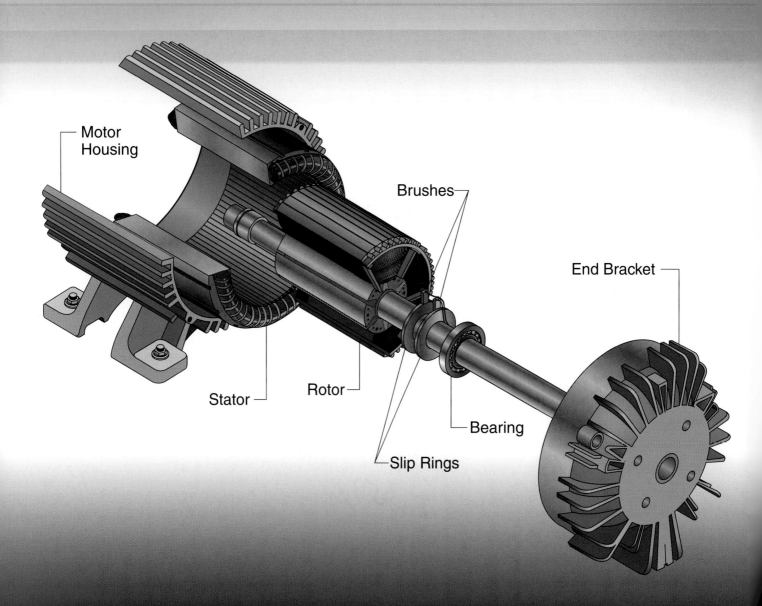

Synchronous Motors

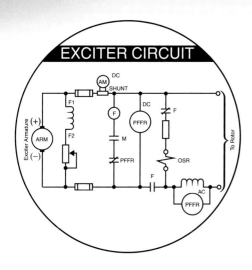

Synchronous Motor Construction	140
Operating Principles	148
Supplemental Topic—Hunting	159
Power Factor	160
Application—Synchronous Motor Power Factor	163
Summary	164
Glossary	164
Review	165

OBJECTIVES

- Explain why a synchronous motor must operate at synchronous speed.
- Describe how the rotor of a synchronous motor differs from the rotor of an induction motor.
- List the different ways that DC excitation current can be applied to a rotor field winding.
- List the types of relays used with synchronous motors.
- Describe synchronous motor starting.
- Explain why and how a discharge resistor is used.
- Explain the differences between pull-in torque, pull-out torque, and torque angle.
- Describe how synchronous motors can be used for power factor correction.

Synchronous motors are used where a very constant speed is required, where power factor correction is required, or for slow-speed machines. Because synchronous motors operate at synchronous speed, they are not subject to the slip found in other polyphase motors. One big advantage of synchronous motors is that they can be used to improve power factor for the location where they are installed. Synchronous motors are generally used to power large equipment such as compressors and pumps.

Definition

A *synchronous motor* is a motor that rotates at exactly the same speed as the rotating magnetic field of the stator.

The *field windings* are magnets or stationary windings used to produce the magnetic field in an alternator or motor.

SYNCHRONOUS MOTOR CONSTRUCTION

A *synchronous motor* is a motor that rotates at exactly the same speed as the rotating magnetic field of the stator. Standard induction motors and wound-rotor motors always run at slower than synchronous speed. Because of their speed characteristics, synchronous motors are used with loads that require constant speed. Synchronous motors are used for applications in which a NEMA Class B motor is designed.

Synchronous motors are similar to induction motors in that both have stator coils that produce a rotating magnetic field. **See Figure 7-1.** Unlike an induction motor, a DC field winding is placed in the rotor of a synchronous motor. The *field windings* are magnets or stationary windings used to produce the magnetic field in an alternator or motor. The field winding is excited by an external DC power source. Because of this external excitation, many existing synchronous motor designs require slip rings and brushes to provide current to the rotor. Newer designs may be brushless.

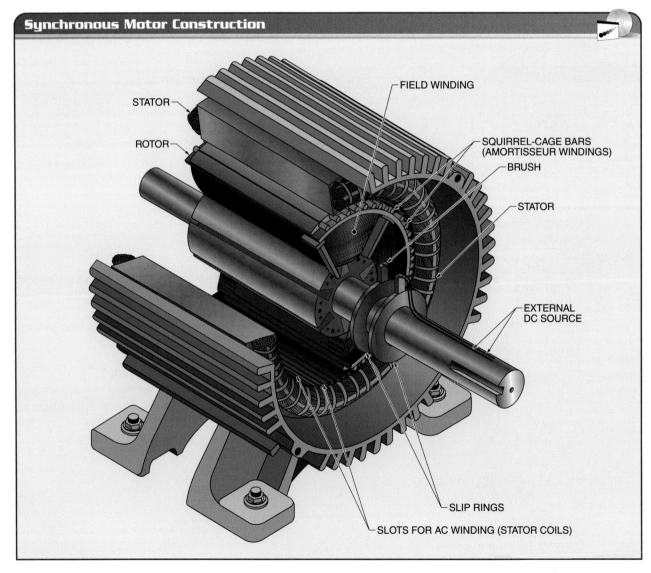

Figure 7-1. Synchronous motors are similar to induction motors in that they both have stator coils that produce a rotating magnetic field. The field winding of a synchronous motor is in the rotor and is excited by an external DC source.

The DC power establishes electromagnets in the rotor with north-south poles that enable the rotor to synchronize with the rotating stator flux. The rotor locks into step with the rotating magnetic field and rotates at exactly synchronous speed. If the synchronous motor is overloaded to the point where the rotor is pulled out of step with the rotating magnetic field, no torque is developed and the motor is taken off-line. The rotating field of the motor runs at the synchronous speed determined by the frequency and the number of poles.

Synchronous motors are very similar to alternators and can be used as such. With a prime mover connected to the shaft and a voltage regulator connected to the output to control the exciter, the motor operates as an alternator.

Stator Construction

The stator of a synchronous motor is constructed like the stator of any polyphase motor, with the field coils placed in slots 120 electrical degrees apart. The stator is constructed of laminated iron sheets to minimize eddy current losses. **See Figure 7-2.** Since synchronous motors range in size from small to very large, the stators also range in size from small to very large. Larger stators may have either a cast iron frame or a welded steel ring with feet welded on to mount the stator to the floor.

An advantage of some larger stators is that the stator coils are individually constructed and connected. The end of each coil is made so that it can be bolted to the next coil. This makes it easier to perform major maintenance, allowing in-field replacements and eliminating the need to pull the stator and completely rewind it.

Tech Fact

Motor windings consist of coils of insulated copper wire, insulated aluminum wire, or heavy, rigid insulated conductors. Since windings are real conductors, all windings have electrical losses while conducting power. These losses create heat that must be removed.

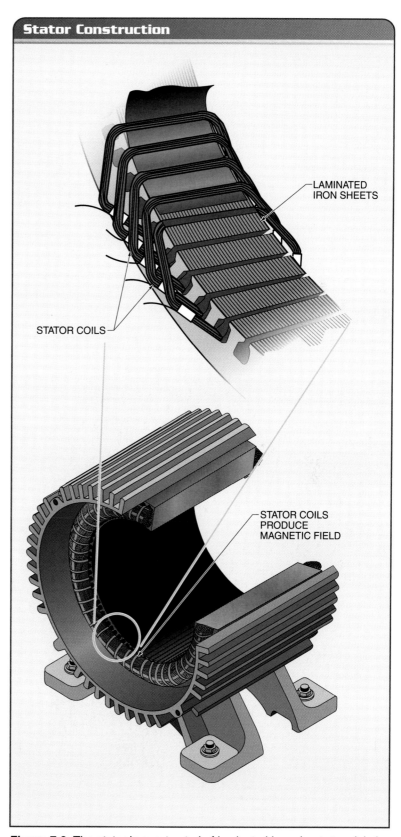

Figure 7-2. The stator is constructed of laminated iron sheets to minimize eddy current losses.

Definition

Amortisseur windings, or *damper windings,* are squirrel-cage conducting bars placed in slots on the pole faces and connected at the ends.

Rotor Construction

A synchronous motor has a two-part rotor. The rotor has an induction-motor rotor section and a wound rotor section. The motor is started the same way as a standard induction motor. Starting a synchronous motor as an induction motor requires amortisseur windings. *Amortisseur windings,* or *damper windings,* are squirrel-cage conducting bars placed in slots on the pole faces and connected at the ends. The number of rotor fields must equal the number of stator field poles.

Each pole is assembled individually. A conducting ring connects all the conducting bars in the rotor. **See Figure 7-3.** The amortisseur winding of a synchronous motor is generally smaller than the squirrel-cage winding of a comparable induction motor because the winding is only used for starting and is not expected to dissipate heat from continuous operation. Because of this, many motor drives require either a minimum amount of cooling time between motor starts or a maximum amount of time in slip operation.

The wound rotor section consists of field windings wrapped around salient poles. The salient poles are attached to a spider ring mounted on the rotor shaft so that the entire assembly rotates. In a common design, all the windings of the rotor poles are connected in series and two leads are brought out to slip rings. The slip rings are mounted on the shaft, but insulated from it. This allows a DC excitation current to be connected to the windings via brushes that ride on the slip rings. The main DC exciter windings on the salient poles are wound around the cores with the wire exposed.

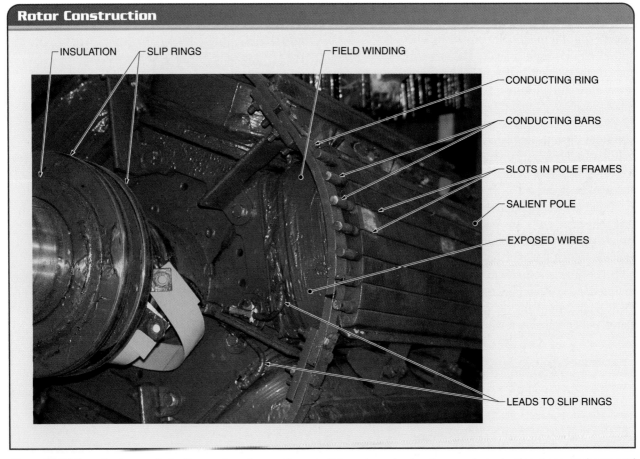

Figure 7-3. The rotor has an induction-motor rotor section consisting of conducting bars and conducting rings, and a wound rotor section consisting of the salient poles and the field windings.

As the rotor approaches synchronous speed, a DC excitation current is applied to the rotor field windings. **See Figure 7-4.** When the DC excitation is applied to the rotor, it creates an electromagnet with north and south poles, which lock into step with the revolving field of the stator. The rotor then rotates in synchronization with the source.

When a synchronous motor is running at synchronous speed, there is no relative motion between the rotor and stator magnetic fields and no eddy currents are induced in the iron. Therefore, when running at synchronous speed, rotor laminations would not be required and a solid iron core would be acceptable. However, from the time the rotor is started in locked rotor until the time the rotor is at synchronous speed, eddy currents are induced in the core. The heat induced into the core as a result of the eddy currents is detrimental to the service life of the motor. For this reason, synchronous rotors are also assembled from iron laminations.

High-Speed Rotor Designs. Synchronous motors that typically run faster than about 450 rpm are considered high-speed synchronous motors. Because of higher speeds and increased centrifugal force compared to a low-speed rotor design, the spider ring is constructed from high-strength steel laminations with punched or machined dovetail grooves into which the poles slide. **See Figure 7-5.** Another less-common design uses a cylindrical rotor with the field windings embedded in slots in the rotor.

For high-speed rotor designs, the rotor has a relatively long axial length compared to the rotor diameter. The bearings are commonly mounted in endbells that are bolted to the motor.

Tech Fact

The rotor of a synchronous motor has a field winding that creates electromagnetic fields. The electromagnetic fields follow the rotating stator field.

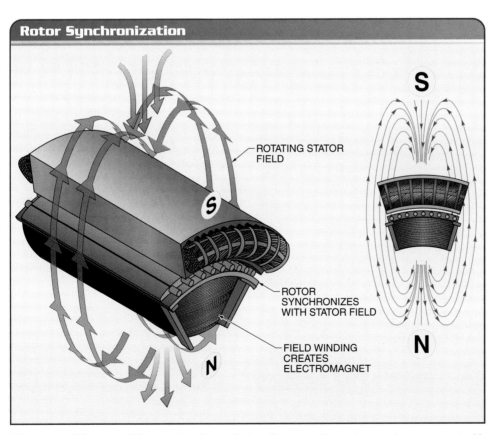

Figure 7-4. When the DC excitation is applied to the rotor, it creates an electromagnet with north and south poles, which lock into step with the revolving field of the stator.

144 MOTORS

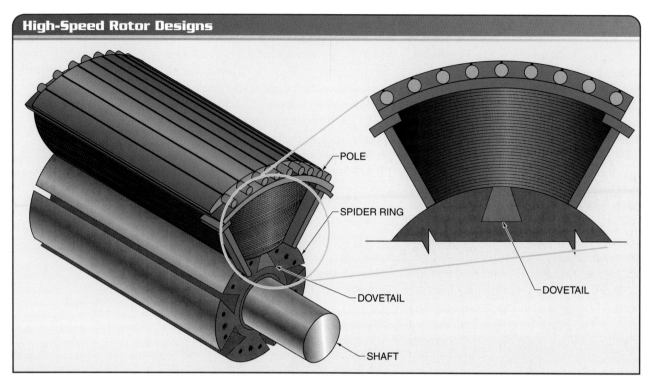

Figure 7-5. For high-speed rotor designs, the spider ring is constructed from high-strength steel laminations with punched or machined dovetail grooves into which the poles slide.

Synchronous motors are usually very large and require careful handling and assembly because of their weight.

Low-Speed Rotor Designs. Synchronous motors that typically run slower than about 450 rpm are considered low-speed synchronous motors. Because of the lower speeds and decreased centrifugal force compared to a high-speed rotor design, the spider ring is a hollow cylinder. The salient poles are bolted to the spider ring. **See Figure 7-6.** The spider ring can also be constructed as a split-ring assembly to make it easy to remove for maintenance. The rotor can even be constructed as a split-rotor assembly. This is especially helpful when there is driven equipment on each end of the drive shaft.

For low-speed rotor designs, the rotor has a relatively shorter axial length compared to the rotor diameter. The bearings are commonly mounted on integral or separate pedestals.

Pony Motor Operation. Some synchronous motor designs for synchronous condensers do not include the amortisseur windings. This type of rotor cannot be started by the application of a rotating magnetic stator field.

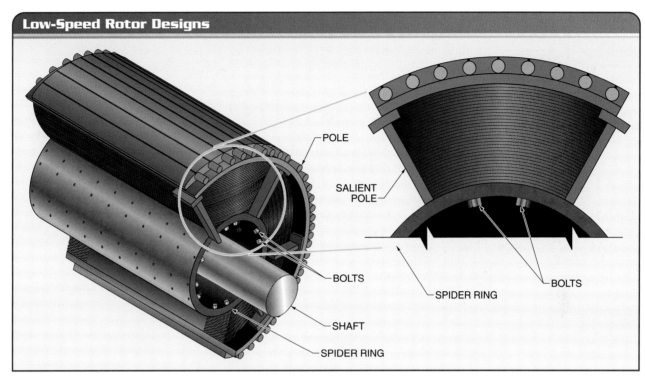

Figure 7-6. For low-speed rotor designs, the spider ring is a hollow cylinder and the salient poles are bolted to the spider ring.

In this case, a pony motor (small DC motor) is used to start the rotor rotation and speed it up to almost synchronous speed. At this point, the DC motor is switched to generator operation, with the synchronous motor acting as the prime mover. The DC generator then acts as a standard DC shaft-mounted exciter.

DC Exciter Generator Construction

When a synchronous motor is started up, it operates as an induction motor until it reaches a speed just below synchronous speed. At approximately 95% of synchronous speed, DC excitation is applied to the wound rotor section. The DC excitation creates electromagnets that polarize the rotor poles. This creates a flux that can synchronize with the rotating stator field so the motor can pull up to synchronous speed. **See Figure 7-7.**

Most often, the DC is applied to the rotor through slip rings mounted on the shaft. The polarity of the slip rings is not critical. The negative-polarity ring typically wears faster than the positive ring due to electrolysis. As part of a preventive maintenance program, the rings should be reversed periodically to equalize the wear.

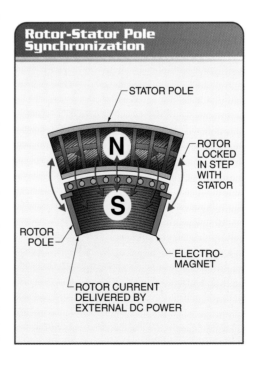

Figure 7-7. The DC excitation creates electromagnets that polarize the rotor poles. This creates a flux that can synchronize with the rotating stator field so the motor can pull up to synchronous speed.

Definition

A motor-generator (M-G) set is a motor and a generator with shafts connected and used to convert one form of power to another form.

DC Shaft-Mounted Exciter Generators. The DC power to synchronous motors is often supplied by an exciter generator mounted on the end of the synchronous motor shaft, with the motor acting as the prime mover for the DC exciter generator. The exciter generator typically is a shunt- or compound-wound DC generator with the armature in parallel with the shunt field and its shunt field rheostat. As the armature is rotated by the synchronous motor, the windings of the armature cut the residual lines of flux in the stationary field. **See Figure 7-8.**

The integrally mounted generator is usually driven by an extension of the synchronous motor shaft. The exciter generator voltage is seldom higher than 250 V and the kW capacity of an exciter generator usually ranges from approximately 1% to 3% of the synchronous motor rating.

Motor-Generator Sets. A synchronous motor may use a motor-generator (M-G) set to produce the required DC power. A *motor-generator (M-G) set* is a motor and a generator with shafts connected and used to convert one form of power to another form. An M-G set for a synchronous motor consists of a small high-speed induction motor that drives a DC generator. An M-G set can be mounted at a remote location from the synchronous motor and can be used with more than one synchronous motor at a time. **See Figure 7-9.**

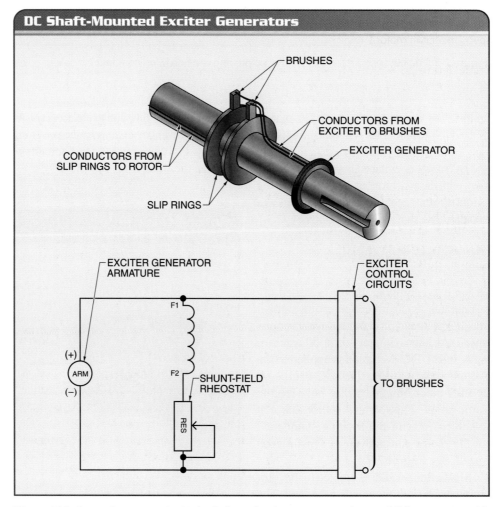

Figure 7-8. An exciter generator typically is a shunt- or compound-wound DC generator with the armature in parallel with the shunt field and its shunt-field rheostat.

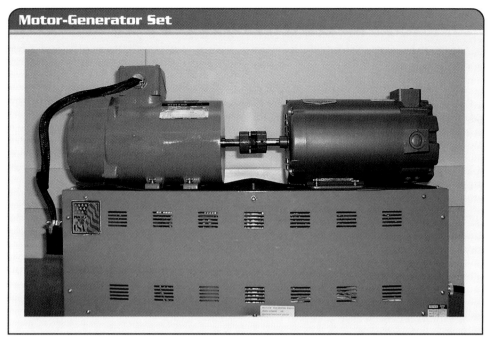

Figure 7-9. A motor-generator (M-G) set for a synchronous motor consists of a small high-speed induction motor that drives a DC generator.

Static Exciters. A static exciter uses an AC source to provide rectified DC power to the field winding. A static exciter is usually mounted near the motor-starting equipment. This type of exciter is becoming more popular because of its reliability and simple design.

Permanent Magnets. There are a few designs of synchronous motors that use permanent magnets for the poles. Since permanent magnets are used, the rotors do not need to be excited. These types of synchronous motors are typically used in variable-speed applications. Permanent-magnet rotors are simpler than excited rotors and have minimal maintenance requirements.

Brushless Exciters. Many new synchronous motors are supplied with brushless exciters. A brushless exciter is an AC generator and rectifier mounted on a motor shaft. A brushless exciter rectifies AC power and provides the resulting DC power to the rotor field windings. Typically, the brushless exciter is mounted on the non-drive end of the motor shaft.

A brushless synchronous motor starts with the assistance of the amortisseur winding, similar to rotors with brushes. The brushless-style main rotor receives its DC power from a rectified single-phase AC supply that rotates on the same shaft as the rotor. **See Figure 7-10.** The control unit monitors the induced frequency and, when the rotor accelerates to about 95% of the rotating field, the AC source to the exciter is energized, providing current through the single-phase bridge rectifier. The bridge DC output produces a current in the exciter field stator coil. The exciter field stator coil is a series of coils surrounding the exciter rotor that is mounted on the motor shaft.

As the exciter rotor spins in the DC field of the exciter field coils, an AC voltage is induced in the rotor. The rotor is a three-phase system, with the leads connected to a three-phase bridge rectifier that rotates on the shaft. The output of the three-phase rectifier is connected to the two leads from the main rotor coils in the synchronous motor. The lead that is connected to the three cathodes on the bridge is positive, while the lead connected to the three anodes is negative.

The rectified output of the three-phase bridge provides the current for the rotor. As a rheostat or an adjustable autotransformer is adjusted, the exciter rotor output to the three-phase bridge is varied. This controls the level of excitation in the main rotor, which in turn controls the power factor.

Brushless exciters are typically used in high-speed applications because of ignition problems from brushes in physical contact with the slip ring. A brushless synchronous motor has reduced maintenance costs because it does not have brushes, collector rings, or exciter commutators. A brushless synchronous motor can be considered for use in hazardous locations because of the lack of sparking from brushes.

OPERATING PRINCIPLES

Because of their steady speed, synchronous motors are often used to power large, sometimes slow-moving machines. Large plant compressors are a popular application of synchronous motors. In addition, fans, pumps, and large industrial grinders are powered by synchronous motors. Steady-speed mills in the steel industry are also often powered with synchronous motors.

Because synchronous motors must operate at synchronous speed, they have several features that are different from those of induction motors. A synchronous motor typically starts up like an induction motor and switches to synchronous operation only after the rotor is accelerated to a speed almost as fast as the rotating stator field. A special relay is required to control this transition from inductive to synchronous operation.

The rotor of a synchronous motor has amortisseur windings as well as synchronous windings. This means that if the motor slips out of synchronous speed, a voltage is induced in the amortisseur windings. This induced voltage is very dangerous because the windings can quickly overheat and damage the motor.

Protective relays are needed to shut down a synchronous motor if it slips out of synchronous speed. The pull-in torque required to get a motor up to synchronous speed and the pull-out torque that can cause the motor to slip out of synchronous speed must be understood.

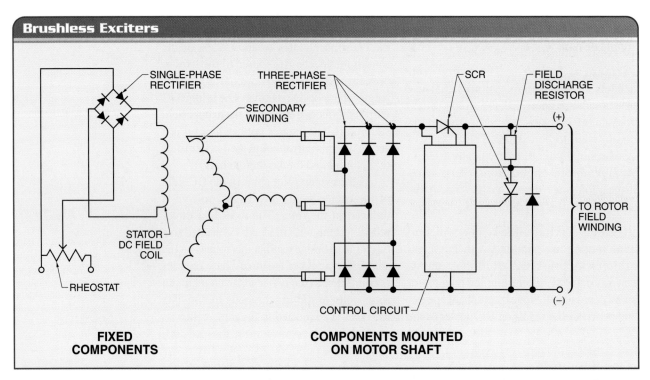

Figure 7-10. A brushless main rotor receives its DC power from a rectified single-phase AC supply that rotates on the same shaft as the rotor.

Since synchronous motors are often used to provide power to large loads, reduced-voltage starting is often required to protect the power circuits in the plant. On the relatively uncommon occasions where synchronous motors are used for fractional-horsepower loads, the motor must be started with methods common to induction electric motors, such as split-phase, capacitor-start, repulsion-start, and shaded-pole starting.

Relays and Accessories

All electric motors normally have protective relays such as overcurrent relays. There are several additional types of relays used with synchronous motors. Three common types of relays used with synchronous motors are polarized field frequency relays, loss-of-excitation (field-failure) relays, and out-of-step relays. In addition, AC and DC ammeters and a discharge resistor are used with synchronous motors.

Tech Fact

Overloads cause overheating and damage to equipment. Overload protection is required by the National Electric Code®.

Polarized Field Frequency Relays. A *polarized field frequency relay (PFFR)* is a relay used to apply current to the DC field windings of a synchronous motor and to remove the discharge resistor from the starting circuit. Polarized field frequency relays are also known as field application relays or polarized frequency relays.

When starting a synchronous motor, the final step is to simultaneously remove the discharge resistor from the circuit and apply the DC voltage to the field windings on the rotor. Since it is very difficult to manually perform this operation, a PFFR is used. Polarized field frequency relays operate by monitoring the frequency and slip of the motor. The relays operate at the correct time to coordinate motor synchronization.

A PFFR consists of an AC coil connected across a reactor in series with the field windings and a DC coil connected to the source of the DC excitation. **See Figure 7-11.** As the synchronous motor starts, the rotor is not moving and the stator magnetic field is rotating at full speed. This induces a high-frequency current in the rotor amortisseur windings.

Definition

*A **polarized field frequency relay (PFFR)** is a relay used to apply current to the DC field windings of a synchronous motor and to remove the discharge resistor from the starting circuit.*

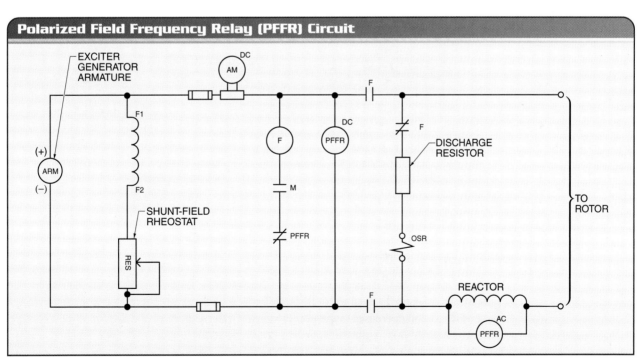

Figure 7-11. A polarized field frequency relay (PFFR) consists of an AC coil connected across a reactor in series with the field windings and a DC coil connected to the source of the DC excitation.

Definition

A loss-of-excitation relay is a relay used to protect a synchronous motor from damage caused by the loss of excitation in the DC winding.

An out-of-step relay (OSR) is an overload relay that is used to protect a synchronous motor from damage from induced currents caused by the rotor falling out of step with the rotating stator field.

The AC power induced in the field windings flows through a reactor in parallel with the PFFR AC coil. The inductive reactance of the PFFR AC coil is lower than the inductive reactance of the reactor, so most of the current flows through the PFFR AC coil. The combined field of the DC and AC coils is strong enough to pull in the armature and open the contact. **See Figure 7-12.**

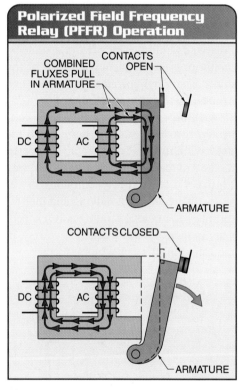

Figure 7-12. The combination of the flux from the AC coil and the DC coil creates a magnetic flux that pulls in the armature and opens the contact.

As the rotor speeds up, the induced frequency decreases in the rotor field windings. The inductive reactance of the reactor and of the PFFR AC coil decreases as the frequency decreases. The flux is proportional to the current. Eventually the current through the PFFR AC coil decreases to the point where the magnetic flux is not sufficient to hold the contacts open. The armature has a hinge that allows it to move away from the coil and close the contact.

When the PFFR closes the contact, relay F is actuated and DC is applied to the rotor. At the same time, DC is still being applied across the DC coil in the PFFR. This polarizes the core material of the relay.

After the DC power is applied from the exciter, the induced AC power in the rotor and the applied DC power are briefly opposed to each other. As the exciter establishes poles, the rotor synchronizes with the rotating stator field. At the same time, the discharge resistor is removed from the circuit. The PFFR has DC flowing through the DC coil. The AC coil does not have any current.

Loss-of-Excitation Relays. A *loss-of-excitation relay* is a relay used to protect a synchronous motor from damage caused by the loss of excitation in the DC winding. The DC power from the exciter is necessary to maintain the excitation of the rotor poles. If the excitation is lost, the rotor falls out of step with the stator field and large, dangerous voltages are induced in the windings in the rotor.

Synchronous motors can be protected from damage caused by the loss of excitation by an undercurrent relay coil in series with the field winding. **See Figure 7-13.** The relay circuit should include a time delay so that momentary fluctuations do not prematurely shut down the motor.

In addition, when the rotor loses synchronization with the stator field, the motor draws excessive VAR power. An impedance relay that activates on excessive VAR power may be used. A power factor relay can also be used to detect the reactive power in a synchronous motor.

Out-of-Step Relays. An *out-of-step relay (OSR)* is an overload relay that is used to protect a synchronous motor from damage from induced currents caused by the rotor falling out of step with the rotating stator field. **See Figure 7-14.** A synchronous motor may lose speed and fall out of step if it is overloaded, if the source voltage sags, or if the rotor field loses excitation. An OSR includes a timer that allows a reasonable amount of time for a synchronous motor to get up to speed before the OSR is switched into the circuit.

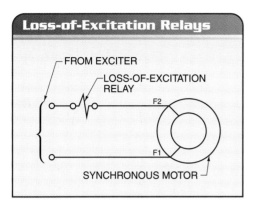

Figure 7-13. Synchronous motors can be protected from damage caused by loss of excitation by an undercurrent relay with the coil in series with the field winding.

A common OSR design is a current-type relay in series with the discharge resistor. This means that the magnetic field that actuates the relay is a result of the rotor circuit current. An OSR includes a time-delay function controlled by an adjustment on the plunger (armature) that controls the field strength needed to pull the plunger up into the core. The plunger trips an NC contact wired in series with the OL contacts.

An OSR has a piston immersed in a viscous fluid that controls the amount of time it takes for the piston to ascend into the core. The piston is connected to the bottom of the plunger and has an adjustable orifice to control the displacement of the fluid from the top of the piston to the bottom. As the orifice is reduced via a movable cover that slides over it to control its size, the time it takes for the piston to ascend in the core is extended.

An OSR is often packaged as part of a field monitor relay. A field monitor relay measures the power factor of the motor and trips the motor offline if synchronization is not achieved within the proper amount of time or if the motor pulls out of step while running.

Discharge Resistors. A *discharge resistor* is a resistor used to discharge any AC potential that builds up in the DC field winding of a rotor in a synchronous motor.

See Figure 7-15. A discharge resistor may also be called a field discharge resistor, external discharge resistor, or external shorting resistor. When a synchronous motor is started or stopped, the rotor is out of synchronization with the stator field, and AC voltage is induced into the DC field windings. The amount of voltage is determined by the difference between the rotational speed of the rotating stator field and the speed of the rotor. Since the field windings are wired in series, high voltages can be induced into the rotor field, creating a dangerous condition.

A discharge resistor is included as part of the starting circuit. The discharge resistor is shunted across the rotor field during starting. When the rotor is up to speed and ready to switch to DC excitation, the discharge resistor is removed from the starting circuit by the F relay. During a shutdown, when the motor slows down, the shorting resistor is placed back in the circuit.

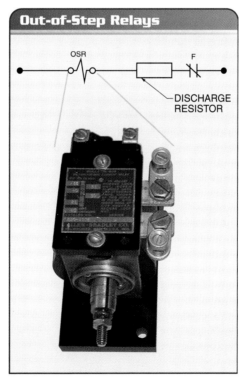

Figure 7-14. A common design for an out-of-step relay circuit is a current-type relay in series with the discharge resistor.

Definition

A *discharge resistor* is a resistor used to discharge any AC potential that builds up in the DC field winding of a rotor in a synchronous motor.

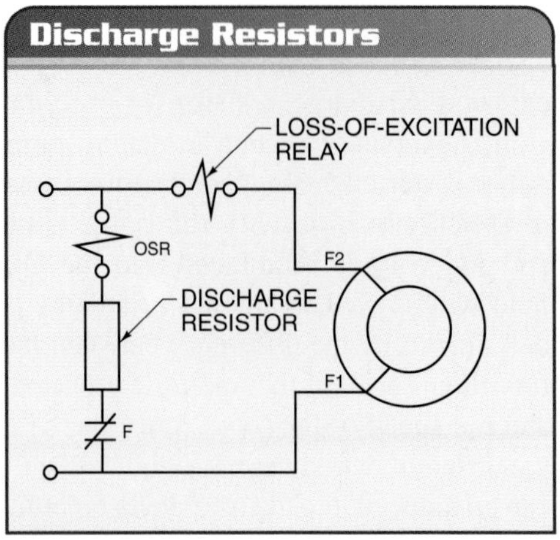

Figure 7-15. A discharge resistor is included as part of the starting circuit. The discharge resistor is shunted across the rotor field winding during starting.

Many older synchronous motors also included a back-up component for times when the discharge resistor failed. Each conducting ring in the rotor was built with two small, sharp-pointed screws with the pointed ends facing each other and leaving a small spark gap. If the discharge resistor fails, the voltage builds up in the windings until the induced voltage causes a flashover between the screw tips and relieves the stress on the DC field insulation.

Synchronous Motor Starting

Induction motors depend on using magnetic fields to induce current in rotor windings. The power is transferred from the rotating field of the stator windings to the squirrel-cage winding of the induction motor. All synchronous motors larger than subfractional sizes use permanent magnets or use an applied DC to create electromagnets that follow the rotating field of the stator windings. The rotor current is transferred directly through the DC exciter circuit.

A rotating magnetic field is produced when 3-phase power is applied to the stator of the motor. This revolving magnetic field produces the torque in the amortisseur winding that causes the rotor to rotate. The rotor starts and accelerates to near-synchronous speed, similarly to a standard induction motor.

The motor speed remains at slightly less than synchronous speed until the DC power is applied to the rotor field. **See Figure 7-16.** When the DC power is applied to the rotor field, the discharge resistor is switched out of the circuit. It is important to switch the DC power at the correct time. The magnetic flux flows through the air gap between the rotor and stator. Since the purpose of applying the DC to the field winding is to create a magnetic field (electromagnet) that synchronizes with the stator field, it is very important to switch the DC when the N and S rotor and stator poles are aligned in order to lock the fields together and minimize the pull-in torque.

Direct On-line Starting. The simplest starting method for a synchronous motor is direct on-line (DOL) starting, where the motor is wired directly across the line conductors. A manual starter can be used as the actual switching device. This method may work satisfactorily for relatively small motors, but is seldom used for large motors. Large motors draw excessive power from the electrical network. This can result in unacceptable voltage drop and high inrush current.

Reduced-Voltage Starting. Low inrush current at starting is often desirable to protect the power system from undesirable high currents. There are many designs for synchronous motors that allow for a low inrush current. However, these designs may not be readily available or may not be cost efficient. Because of the inertia from the size and weight of the rotor, along with the starting currents of the synchronous motor, reduced-voltage starting is often used.

The methods of applying a reduced voltage to a synchronous motor are the same as the methods of applying a reduced voltage to an induction motor. Because of the reduced inrush current and torque during starting, a motor drive normally allows an increased stall time to allow more time for the motor to reach synchronous speed.

> **Tech Fact**
> *The addition of a reactor or choke can create a resonant circuit, especially where power factor correction capacitors are used.*

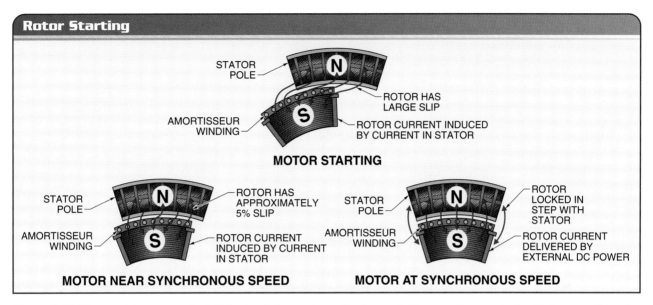

Figure 7-16. The revolving magnetic field in the stator produces the torque in the amortisseur winding that causes the rotor to rotate. The rotor starts and accelerates to near-synchronous speed, similar to a standard induction motor. The motor speed remains slightly less than synchronous speed until the DC power is applied to the rotor field.

Starting Circuit Operation. For synchronous motors started with PFFR relays, DC power needs to be available. If the synchronous motor uses an M-G set to provide DC power, the M-G set needs to be started first. If the synchronous motor uses a DC generator mounted on the motor shaft, the generator starts providing DC power as soon as the rotor starts turning. The starting circuit for a synchronous motor includes a power circuit, a control circuit, and an exciter circuit. **See Figure 7-17.**

When the start button is pressed, relay CR1 is energized in the control circuit. This closes the CR1 holding contact (latching contact) in parallel with the start button and the CR1 contact in series with the coil of the main motor starter. This energizes the motor relay (M) that closes the M holding contact. Both a contact of the M contactor and a contact of the CR1 are in series, with both in parallel with the start button. This requires that both CR1 and M be energized to latch up CR1.

Energizing relay M also closes the three NO contacts in the power circuit. These contacts are in the conductors connecting the line to the stator and allow current to flow to the overloads. The low-impedance coils of the magnetic overload relays sense the current, and the plungers (armatures) are drawn toward the core, closing the OL contacts and allowing current to flow to the stator.

The current transformer that surrounds line 1 provides a current to the AC ammeter. At startup in locked rotor, the meter may read up to about 700% of full-load current, depending on the motor design. The stator is energized and the rotating synchronous field established, rotating around the stator frame.

The rotating stator field is applied to the rotor amortisseur windings, and a voltage and current are induced in the windings. Because the current in the amortisseur windings is 180° out of phase with the current in the stator, the poles are unlike and there is an attraction between the poles of the rotor and the poles of the stator.

The same rotating stator field that induced voltage in the amortisseur windings also induces voltage in the DC field windings. Since the field windings are all in series, the voltages add to produce a very high induced voltage. Because of the high induced voltage, it is imperative that the circuit not be left in an open condition. With no current flow, the voltage would rise to a level at which the insulation would be damaged by the stress from the high voltage.

154 MOTORS

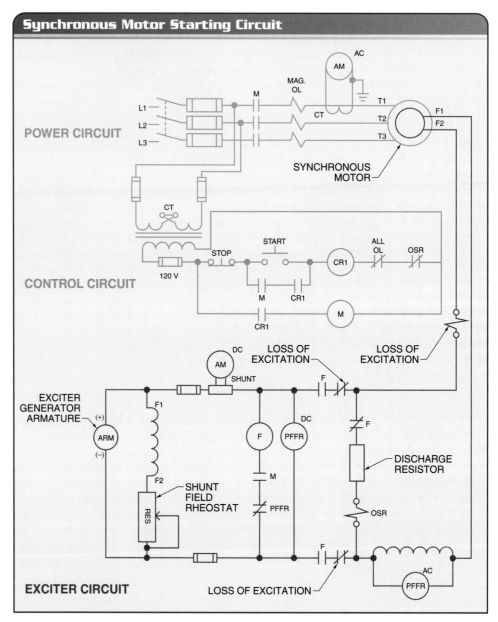

Figure 7-17. The starting circuit for a synchronous motor includes a power circuit, a control circuit, and an exciter circuit.

The exciter circuit contains a discharge resistor that provides a complete path for the current induced in the field windings to flow. This provides a means of discharging the high voltage in the rotor caused by the induced current in the field windings. The low-impedance coil of the OSR is energized by the current and the plunger is drawn toward the core.

If any of the three magnetic overloads or the OSR plungers pulls all the way up into the core and actuates its NC contact in series with CR1, the stator will be disconnected from the line. The OSR has a delay built into it to allow time for the rotor to achieve synchronous speed.

The fifth contact of motor relay M is in series with field contactor relay F in the exciter circuit. Since relay F has not been energized yet, the relay F NO contacts prevent the AC in the field windings from reaching the F coil and keep relay F from being energized. The AC flows through the relay F NC contact, the discharge resistor, the OSR, and the parallel combination of the reactor and the PFFR AC coil. Since the AC frequency is relatively high at startup, the inductive reactance of the reactor is relatively high and most of the current flows through the PFFR AC coil.

Because of the induced voltage in the field winding, the rotor starts to turn. The exciter armature starts to turn with the rotor and produces DC power. The DC coil of the PFFR is energized and flux from the coil fills the iron core. The flux in the PFFR core is a combined flux from the DC coil, created by the DC generator, and from the AC coil, created in the DC windings. The combined flux is enough to pull in the armature and open the PFFR contact. This prevents relay F from energizing and prevents the DC from being applied to the rotor DC field windings.

As the rotor starts to rotate, the induced frequency in the field windings is reduced in proportion to the increase in speed. As the rotor speed increases, the slip is reduced and the current in the rotor is reduced, which reduces the flux in the OSR coil. This reduces the speed at which the OSR plunger moves toward the core. Although the speed is reduced, the plunger will still be drawn up into the core if the rotor does not reach synchronous speed.

The reactive voltage in the reactor is reduced as the frequency decreases. At the optimum point, the frequency is low enough that the PFFR AC coil cannot force the DC flux through the armature. The armature opens, closing the PFFR NC contact in series with the M contact in the exciter circuit. (The M contact closed previously when the M coil was energized.) This energizes relay F, and the contactor closes the two NO contacts, connecting the exciter generator output to the rotor.

Right after the relay F NO contacts close, the relay F NC contact opens, disconnecting the OSR coil and the discharge resistor from the rotor. The contact operation of the make before break of the F contactor contacts is essential to prevent the high-voltage spike that would be induced in the rotor circuit if the circuit were allowed to open for even a short time. This spike would damage the contacts on the F contactor and the rotor coils.

The DC that is applied to the rotor field windings creates an electromagnet that pulls the rotor into synchronous speed with the rotating stator field. The DC ammeter is connected in series with the exciter output. The meter is connected to a meter shunt, providing a voltage for the meter via a voltage drop across the shunt.

The field strength can be adjusted by varying the rheostat that is in series with the shunt field of the exciter. The output of the exciter is varied to control the flux in the rotor poles. The strength of the rotor field determines the power factor correction of the synchronous motor.

A loss-of-excitation relay is placed in the conductors between the rotor field windings and the exciter circuit. This relay is used to monitor the field windings and to shut down the exciter circuit if the field excitation fails.

Many modern synchronous motors are made with brushless exciters. These types of exciters use solid-state electronic circuits to replace the PFFRs used on conventional synchronous motor starting. This results in a slightly different method of applying DC to the rotor, but the same type of power circuit and control circuit are used.

Torque

Torque is a twisting force that is used to rotate a motor shaft. The three types of torque that are important to understand for synchronous motors are reluctance torque, pull-in torque, and pull-out torque. Understanding torque angle is also important.

Reluctance Torque. *Reluctance torque* is torque developed by the salient rotor poles before the poles are excited by the external DC power. Reluctance torque is typically fairly small, but it may be enough to enable a synchronous motor to run at very light loads without DC excitation. In addition, the reluctance torque may be enough to pull the rotor into synchronization when the motor is coupled to a very low inertial load. **See Figure 7-18.**

Definition

Torque is a twisting force that is used to rotate a motor shaft.

Reluctance torque is torque developed by the salient rotor poles before the poles are excited by the external DC power.

Definition

Pull-in torque is the maximum torque required to accelerate a synchronous motor into synchronization at the rated voltage and frequency.

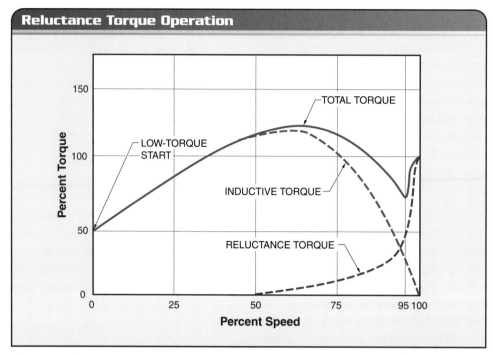

Figure 7-18. Reluctance torque is torque developed by the salient rotor poles before the poles are excited by the external DC power. Reluctance torque is typically small, but it may be enough to enable a synchronous motor to run at very light loads without DC excitation.

Pull-In Torque. *Pull-in torque* is the maximum torque required to accelerate a synchronous motor into synchronization at the rated voltage and frequency. Pull-in torque develops during the transition as the synchronous motor speeds up to the synchronous speed from the maximum speed running as an induction motor. At the point where the rotor is being accelerated to synchronous speed, the slip is very small and only the reluctance torque and the torque provided by the DC exciting the field windings can provide the pull-in torque.

The excitation DC should be applied to the field windings at the point of maximum flux to optimize the change from inductive operation to synchronous operation. The point of maximum flux occurs in the PFFR relay as the AC power crosses through zero from positive to negative. **See Figure 7-19.** The flux creates the strongest magnetic field in the rotor if the field discharge resistor is removed from the circuit at the point of maximum flux. At the same time, the excitation DC is applied and increases the flux. This provides maximum torque for the rotor to pull into synchronization.

Normal pull-in torque typically ranges from about 40% to 100% of full-load torque. For large loads, it takes more torque to accelerate the load to synchronous speed than for a small load. The ability to accelerate the rotor from slip speed to synchronous speed is limited by the strength of the DC field and the timing of the synchronization. Therefore, for large loads the amortisseur winding must bring the load to a higher speed than for smaller loads. This higher speed can be achieved by special designs of the amortisseur winding.

For example, a high-inertia load, such as a blower, requires a very high torque to operate at synchronous speed. It also takes a very high torque to accelerate the load from slip speed to synchronous speed. This application may require an amortisseur winding that accelerates the rotor to at least 97% of synchronous speed to be sure that the motor will pull into synchronization. If the rotor is accelerated to only 95% of full speed, the motor may stall when the DC excitation field is applied. Thus, a synchronous motor should generally be built to a specific application.

Pull-Out Torque. *Pull-out torque* is the torque produced by a motor overload that pulls the rotor out of synchronization. *Synchronous torque* is the torque required to keep the rotor turning at synchronous speed and represents the torque available to drive the load. When the synchronous torque is at maximum, it is equal to the pull-out torque. **See Figure 7-20.**

The torque developed by the amortisseur winding becomes zero at synchronous speed. If the motor pulls out of synchronization, there is slip again and voltage is induced in the amortisseur winding. Pull-out protection is required to prevent the motor from running as an induction motor that draws high current through the rotor windings. Normal pull-out torque typically ranges from about 150% to 200% of full-load torque.

The pull-out torque is not the same thing as the break-down torque of an induction motor. There is no speed change as the load is increased. The motor maintains synchronous speed until the pull-out torque pulls the motor out of step. When the motor pulls out of step, current starts to flow in the amortisseur winding and overheating can occur.

Synchronous torque is provided by the interaction of the stator field and the rotor DC field windings. The amount of torque that can be sustained before the stator and rotor fields pull out of synchronization depends on the strength of the fields. A synchronous motor with a DC excitation source that can be increased, such as an M-G set or an exciter mounted on the shaft, has a pull-out torque that varies with a voltage change. These types of synchronous motors have a variable pull-out torque. If the DC excitation source is a static source of rectified AC, the pull-out torque does not vary unless the line voltage varies. These types of synchronous motors have a constant pull-out torque.

> **Tech Fact**
>
> In an electric motor, torque is normally developed on a current-carrying wire loop moving at a right angle to a magnetic field. Higher torque is developed with more turns of wire, a stronger field, and faster movement.

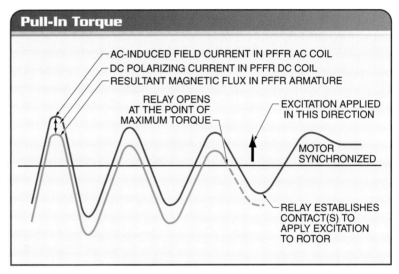

Figure 7-19. The DC excitation current should be applied to the field windings at the point of maximum flux to optimize the change from inductive operation to synchronous operation.

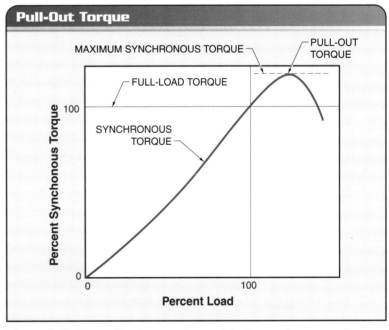

Figure 7-20. The pull-out torque is equal to the maximum synchronous torque.

Definition

Pull-out torque is the torque produced by a motor overload that pulls the rotor out of synchronization.

Synchronous torque is the torque required to keep the rotor turning at synchronous speed and represents the torque available to drive the load.

Torque angle is the angle between the rotor and stator fields as a synchronous motor is running under load.

When a synchronous motor pulls out of synchronous operation while under load, the rotor poles begin to slip relative to the stator poles. This can create very large torque pulsations at the motor shaft and large pulsations in line current. The torque pulsations can create damaging vibrations in the shaft or other mechanical components. The current pulsations can interfere with proper operation of the power system. Power factor is related to the phase angle between voltage and current. Synchronous motors typically operate at unity or leading power factor. Therefore, a power-factor monitor or relay can be used to protect a motor from pole slippage.

Torque Angle. *Torque angle* is the angle between the rotor and stator fields as a synchronous motor is running under load. As the rotor of a synchronous motor synchronizes with the rotating stator field under no load, the rotor and stator become aligned. As load is placed on the motor, the torque angle increases and the poles of the rotor begin to lag behind the stator. **See Figure 7-21.** The counter voltage is affected and the stator current increases, providing a stronger pole on the stator. This is not the same thing as slip. The poles lag, but they maintain a synchronous speed.

Discharge Resistors. The amount of resistance in the discharge resistor influences the amount of torque available to bring the slip speed of the rotor up to a value close to synchronous speed. At startup, the slip is very high and therefore the rotor frequency is very high. A high frequency causes a high reactance in the DC field windings and a high starting torque. As the rotor speeds up, the slip decreases and the starting torque decreases.

As with an induction motor, the rotor circuit impedance determines the starting torque. At low speeds, the DC field windings respond to the high frequency created in the rotor. Therefore, the impedance of the field windings and exciter circuit is very reactive. The discharge resistor has little effect on the impedance and starting torque because the resistance is only a small part of the total impedance. At high speeds, the field windings respond to the lower frequency created in the rotor. Therefore, the impedance of the field windings is less reactive and the discharge resistor has more effect on the impedance. At this point, the field windings start to contribute significantly to the starting torque.

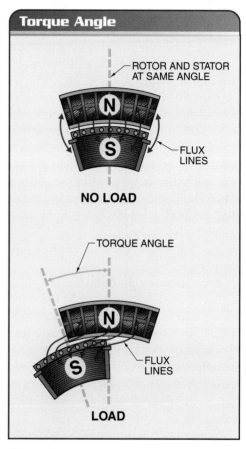

Figure 7-21. Torque angle is the angle between the rotor and stator fields as a synchronous motor is running under load.

A higher value of resistance changes the ratio of reactive impedance relative to resistive impedance and allows a higher torque and higher slip speed. A low value of resistance results in a lower torque and lower slip speed. **See Figure 7-22.**

Tech Fact

A meter with a MIN/MAX function that holds readings over time can be used to help identify intermittent problems such as loose connections.

It may seem desirable to modify the value of the discharge resistor to increase the slip speed and make it easier to bring a load to synchronous speed. However, this can be very dangerous. While it is true that a larger value of resistance allows a higher slip speed and easier synchronization, it also allows a higher voltage to exist in the windings. The winding insulation must be capable of withstanding the higher voltage.

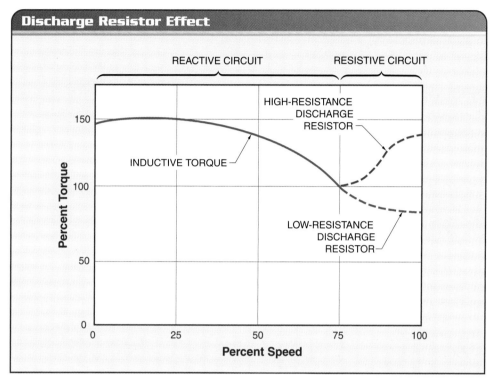

Figure 7-22. The amount of resistance in the discharge resistor influences the amount of torque available to bring the slip speed of the rotor up to a value close to synchronous speed.

Hunting

Hunting is the oscillation of the phase angle between the stator pole and the rotor pole of a synchronous motor. Synchronous motors must maintain synchronization with the stator field. However, when the load increases, the rotor slows down and the phase angle between the stator pole and the locked rotor pole changes in order to provide the proper torque. The motor remains synchronized, but the rotor lags behind the stator field.

As the phase angle increases, the motor draws more current. As more current flows, the torque increases and the phase angle decreases, often overshooting the equilibrium. As a result, the phase angle oscillates back and forth around the equilibrium point. The oscillations must be dampened to minimize the hunting.

The amortisseur windings help to dampen power-angle hunting that occurs when the motor is running at synchronous speed. The amortisseur windings are used to help maintain consistent position and magnitude of the magnetic field connecting the poles. When hunting occurs, the armature reaction on the field flux causes a shift of flux across the pole shoe. Circulating currents are induced in the damper winding by the shifting flux. The kinetic energy of the oscillations is converted into heat energy in the windings, thus eliminating the oscillations.

Supplemental Topic

POWER FACTOR

A leading power factor is a circuit with current leading voltage by some angle between 0° and 90°. A lagging power factor is a circuit with current lagging behind voltage by some angle between 0° and 90°. A leading or lagging power factor causes power losses in the system because voltage and current are not in phase. **See Figure 7-23.**

Synchronous-Motor Power Factor

The power factor of a synchronous motor is determined by the amount of DC power applied to the rotor. Varying the power varies the power factor. A synchronous motor can operate at unity power factor, a lagging power factor (underexcited), or a leading power factor (overexcited).

Unity. Unity power factor is where the AC is in phase with the applied voltage. Except for normal motor losses, all of the power is delivered to the load as mechanical power. This is the most efficient operating condition of a synchronous motor.

> **Tech Fact**
> A common cause of a low (lagging) power factor in industrial facilities is induction motors. The power factor can be increased by installing synchronous motors (leading power factor) or static capacitors across the line.

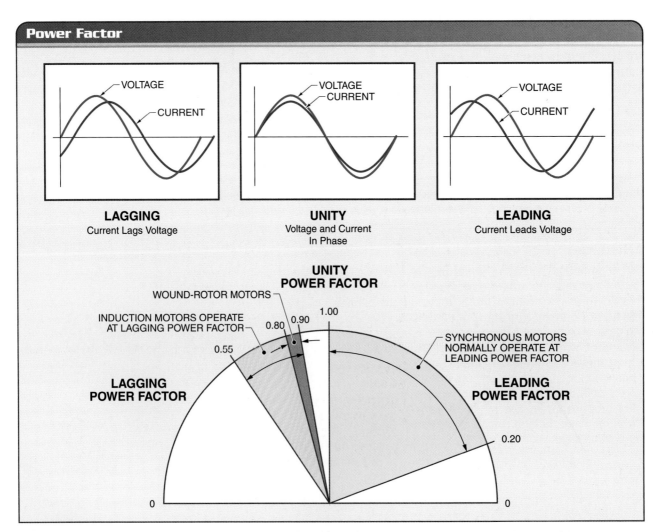

Figure 7-23. A leading or lagging power factor causes power losses in the system because voltage and current are not in phase.

Underexcited. With weak DC power applied to the rotor, the power factor is less than unity. The current lags the applied voltage by some angle. When underexcited, the synchronous motor operates with the same type of lagging power factor as induction and wound-rotor motors. Being underexcited is not the normal operating condition of a synchronous motor and typically only happens when the motor is overloaded and about to slip out of synchronization.

Overexcited. With high DC power applied to the motor, the power factor is also less than unity. In this case, the current leads the applied voltage and produces a leading power factor. The leading power factor of the synchronous motor offsets the lagging power factor of other loads. Being overexcited is the normal operating condition of a synchronous motor. It is not the most efficient operating condition for the motor, but it is the most efficient operating condition for the entire distribution system.

Power Factor Correction

Synchronous motors are not only used to produce mechanical power, such as running a plant compressor, but are also used for power factor correction at the same time. Power factor correction is possible when the synchronous motor is not fully loaded. A *synchronous condenser* is a synchronous motor operated at no load in order to provide power factor correction. A synchronous motor usually operates with a leading power factor, which is the opposite of induction and wound-rotor motors. This leading power factor can be used to counteract the effect of the lagging power factor caused by other motor types.

Synchronous motors are generally available with power factor ratings of unity (1.0) or 0.8 leading. Other ratings are available. A leading power factor can provide substantial electrical power savings by supplying reactive power to counteract the lagging power factor typically caused by inductive loads.

A synchronous motor operates at a constant speed. Variation in DC excitation current does not affect the speed. However, the excitation level changes the power factor. The exciter circuit has a rheostat that changes the amount of DC power available to excite the field windings. As the rheostat on the exciter field is adjusted, the exciter rotor output to the three-phase bridge is varied. This controls the level of excitation in the field windings in the rotor, which in turn controls the power factor.

By varying the rotor current, the pole can be either just strong enough to keep the rotor synchronized with the rotating field, or it can be overexcited, cutting the stator coils ahead of the coils that are excited by the source. Overexcitation provides a leading power factor on the line, which affects the supply system power factor. If the supply system has a lagging power factor, raising the field current overexcites the rotor and the power factor is improved.

The rotor voltage supplied by the exciter, E_R, provides a field that is sufficient to maintain synchronization between the rotor and the rotating stator field. **See Figure 7-24.** The stator current, I, lags the source stator voltage, E_A. As a result of the inductive reactance of the stator, the stator current lags 90° behind the rotor voltage.

The torque angle remains in the same position as long as there is no change in load. If more load is added, the torque angle will increase and the stator current will move in a clockwise direction and lag the stator voltage at a greater angle. This increase in the torque angle increases the current in the stator and allows for the heavier load.

The power circuit is inductive because the stator current is lagging the stator voltage. Therefore, the power factor is lagging. When the rotor current is increased, the strength of the field surrounding the poles in the rotor is also increased. This expanded field cuts the coils ahead of the source and induces a power in the stator that leads the source. This leading power produces a leading power just as a capacitor does. This corrects the power factor on the system. Modern synchronous motors often have automatic controls that monitor the power factor and adjust the rotor field strength as needed to maintain the desired power factor.

Definition

*A **synchronous condenser** is a synchronous motor operated at no load in order to provide power factor correction.*

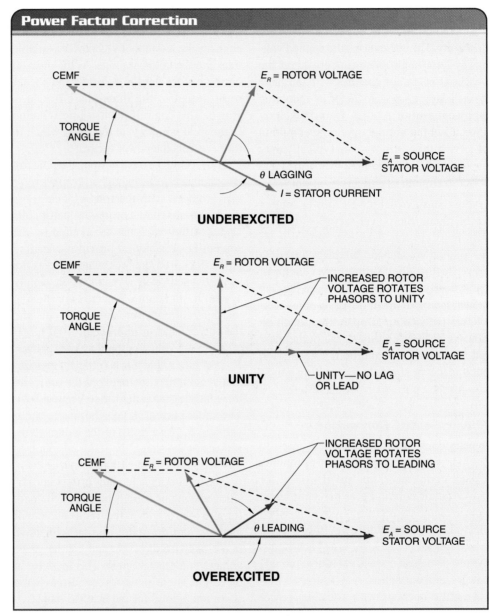

Figure 7-24. The phasor diagrams show how the phase angle between the stator current and voltage can change with the amount of applied DC power to the rotor.

Application—Synchronous Motor Power Factor

When a synchronous motor is at unity power factor, the current in the stator power circuit is at a minimum level. If the current in the rotor winding is increased (overexcited), the current in the stator increases and the motor develops a leading power factor. If the current in the rotor winding is decreased (underexcited), the current in the stator increases and the motor develops a lagging power factor.

To measure stator current, an AC ammeter connected to a current transformer (CT) is added to the stator power circuit. A CT with a typical value of 50:5 up to about 400:5, depending on motor size, is used to measure the current from the main power lines. To measure the rotor current, a DC ammeter connected to a DC shunt is added to the exciter circuit. A rheostat or other resistance is used to vary the rotor current.

A synchronous motor operating a blower is used to move air as well as correct the system power factor. The following table gives the measured currents. The currents are plotted on the graph. A rotor current greater than 18 A results in an overexcited rotor, which provides leading power factor to correct for other lagging loads. This type of graph can easily be used to help operators understand how to control synchronous motors to deliver power factor correction.

Baldor

Ammeter Readings

Rotor Current	Stator Current
7	195
12	163
14	157
16	155
18	150
20	155
22	170
24	180

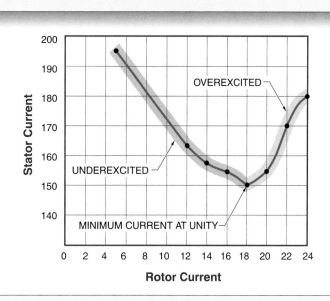

Summary

- A synchronous motor rotates at exactly the same speed as the rotating stator field. External DC excitation creates an electromagnetic field that synchronizes with the rotating stator field.

- A synchronous motor has a two-part rotor. The rotor has an induction-motor rotor section consisting of amortisseur windings and a wound rotor section consisting of field windings.

- Common methods of applying a DC exciting current to the rotor windings include DC shaft-mounted exciters, motor-generator (M-G) sets, static exciters, and brushless exciters. In addition, some rotors are manufactured with permanent magnets to eliminate the need for excitation.

- Common relays and accessories include polarized field frequency relays (PFFRs), loss-of-excitation relays, out-of-step relays (OSRs), and discharge resistors.

- The starting circuit of a synchronous motor is very similar to the starting circuit of an induction motor, with the addition of an exciter circuit. The exciter circuit includes relays or a control module that switches the motor from inductive operation to synchronous operation.

- The discharge resistor provides a conductive path for the current induced in the field windings during startup. In addition, the value of the discharge resistor affects the amount of torque available to bring the slip speed of the rotor up to a value close to synchronous speed.

- Torque is a twisting force that is used to rotate a motor shaft. Reluctance torque is torque developed by the salient rotor poles that occurs before the poles are excited by the external DC power. Pull-in torque is the maximum torque required to accelerate a synchronous motor into synchronization at the rated voltage and frequency. Pull-out torque is the torque produced by a motor overload that pulls the rotor out of synchronization. Torque angle is the angle between the rotor and stator fields as a synchronous motor is running under load.

- The power factor of a synchronous motor is determined by the amount of DC power applied to the rotor. As the rheostat on the exciter field is adjusted, the level of excitation in the field windings changes, changing the power factor.

Glossary...

A **synchronous motor** is a motor that rotates at exactly the same speed as the rotating magnetic field of the stator.

Field windings are magnets or stationary windings used to produce the magnetic field in an alternator or motor.

Amortisseur windings, or **damper windings,** are squirrel-cage conducting bars placed in slots on the pole faces and connected at the ends.

A **motor-generator (M-G) set** is a motor and a generator with shafts connected and used to convert one form of power to another form.

A **polarized field frequency relay (PFFR)** is a relay used to apply current to the DC field windings of a synchronous motor and to remove the discharge resistor from the starting circuit.

A **loss-of-excitation relay** is a relay used to protect a synchronous motor from damage caused by loss of excitation in the DC winding.

...Glossary

An **out-of-step relay (OSR)** is an overload relay that is used to protect a synchronous motor from damage from induced currents caused by the rotor falling out of step with the rotating stator field.

A **discharge resistor** is a resistor used to discharge any AC potential that builds up in the DC field winding of a rotor in a synchronous motor.

Torque is a twisting force that is used to rotate a motor shaft.

Reluctance torque is torque developed by the salient rotor poles before the poles are excited by external DC power.

Pull-in torque is the maximum torque required to accelerate a synchronous motor into synchronization at the rated voltage and frequency.

Pull-out torque is the torque produced by a motor overload that pulls the rotor out of synchronization.

Synchronous torque is the torque required to keep the rotor turning at synchronous speed and represents the torque available to drive the load.

Torque angle is the angle between the rotor and stator fields as a synchronous motor is running under load.

A **synchronous condenser** is a synchronous motor operated at no load in order to provide power factor correction.

Review

1. Explain why a synchronous motor rotates at exactly the same speed as the rotating stator field.

2. Describe the two parts of a rotor in a synchronous motor.

3. List and describe the typical methods of applying a DC excitation current to the field windings of a synchronous motor.

4. List and describe the common relays and accessories used with a synchronous motor.

5. Describe the operation of the exciter circuit of a synchronous motor.

6. Explain the purpose and operation of a discharge resistor.

7. Describe the types of torque that apply to a synchronous motor.

8. Describe how a synchronous motor is used to correct power factor.

Refer to the CD-ROM for Quick Quiz® questions related to chapter content.

MOTORS

OUT-OF-PHASE CURRENT PRODUCES A ROTATING MAGNETIC FIELD

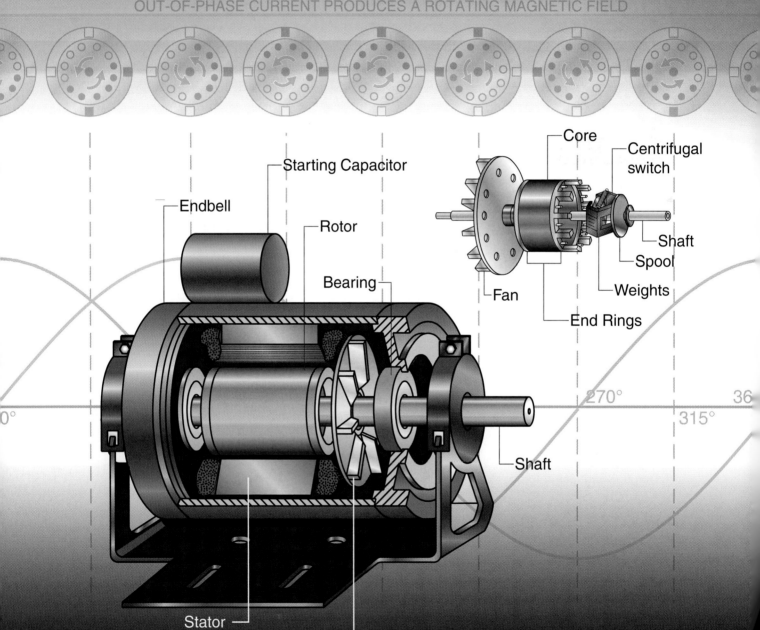

Single-Phase Motors

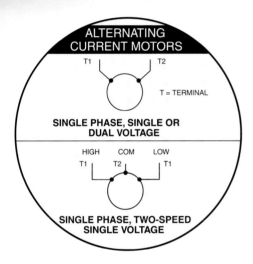

Single-Phase Motors	168
Shaded-Pole Motors	172
Split-Phase Motors	175
Capacitor Motors	176
Supplemental Topic—Repulsion Motors	182
Application—Centrifugal Switches	183
Application—Resistors	183
Application—Single-Phase AC Motor Terminal Designations	184
Application—Motor Winding Lead Terminations	185
Summary	186
Glossary	187
Review	187

OBJECTIVES

- Explain why there is a difference in current and phase between the start winding and the run winding.
- Describe the operation of a shaded-pole motor.
- Describe the operation of a split-phase motor.
- Describe the operation of a capacitor motor.
- Explain the operating characteristics of the different types of single-phase motors and why a certain type of motor would be chosen for a certain application.

More single-phase, AC motors are used for residential applications than any other type of motor. Single-phase, AC-motor-driven appliances and devices in a typical dwelling include furnaces, air conditioners, refrigerators, washing machines, dryers, ovens, microwave ovens, clocks, and cooling fans for computers and stereos. Single-phase motors do not naturally have a rotating magnetic field and are named for the method used to develop the rotating magnetic field. Single-phase motors include shaded-pole, split-phase, and capacitor motors.

Definition

A *stator* is the fixed, unmoving part of a motor, consisting of a core and windings, that converts electrical energy to the energy of a magnetic field.

A *rotor* is the rotating, moving part of a motor, consisting of a core and windings, that converts the rotating magnetic field of the stator into the torque that rotates the shaft.

SINGLE-PHASE MOTORS

Single-phase power is taken from a utility transformer. Single-phase, 120 V/240 V power has three conductors, with a potential of 240 V between conductors A and B. **See Figure 8-1.** There is a potential of 120 V between conductors A-C and B-C. During each cycle, voltage A-C and voltage B-C are exactly the same and are equal to half of the total voltage A-B.

No rotation is produced because the fields are alternating. As the voltage rises from zero to peak and falls back to zero, the unlike poles that are formed are attracted to each other and resist rotation. A single-phase motor must have a method to create a rotating magnetic field. Single-phase motor types are named for the method used to develop the rotating magnetic field.

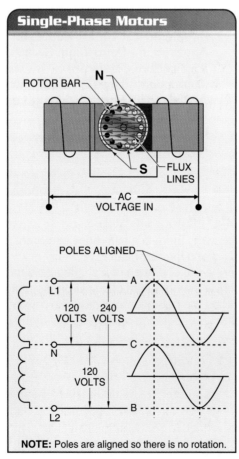

Figure 8-1. Single-phase motors resist rotation without some method of creating a rotating magnetic field.

Single-Phase Motor Construction

All motors deliver work by converting electrical energy to mechanical torque. Single-phase motors, like all motors, have a stator and a rotor. **See Figure 8-2.** A *stator* is the fixed, unmoving part of a motor, consisting of a core and windings, that converts electrical energy to the energy of a magnetic field. A *rotor* is the rotating, moving part of a motor, consisting of a core and windings, that converts the rotating magnetic field of the stator into the torque that rotates the shaft. Single-phase motors also typically have a frame that supports the endbells and bearings. They may also have a fan connected to the motor shaft.

Single-Phase Motor Stator Construction. A single-phase stator is nearly identical to a three-phase stator, with the exception that some method of creating a rotating field must be added to the motor. In many cases, a single-phase motor is constructed with both a start and a run winding in the stator. Both start and run windings are in parallel, connected to the same source. In other cases, one or more capacitors are added to the windings, or a shading ring is used to cause a rotating magnetic field.

The stator core is constructed of many notched thin sheets (laminations) of steel to limit eddy currents. When the sheets are stacked and pressed into the stator frame, the notches become slots. The slots hold the windings in the iron, which become the poles that interact with the rotor. The start winding is placed 90 mechanical degrees from the run windings, allowing the motor to start in either direction depending on the direction of current in one with respect to the other. **See Figure 8-3.**

Single-Phase Motor Rotor Construction. The rotor of a single-phase motor is the same as in an induction squirrel-cage motor. The rotor core consists of stacked laminations with the slots at the outside of the core. The shaft is pressed through a hole in the center of the assembly. In many cases, the conductor bars are made from molten aluminum forced through the slots under pressure.

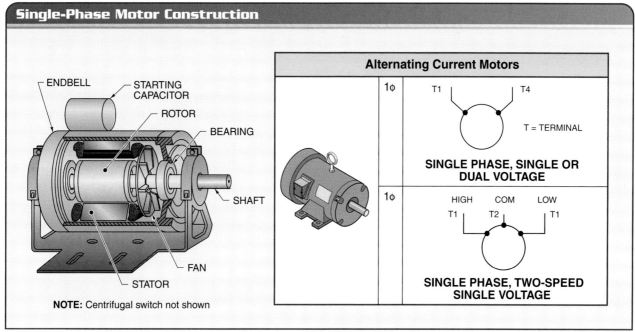

Figure 8-2. Single-phase motors have a stator and rotor, just like three-phase motors.

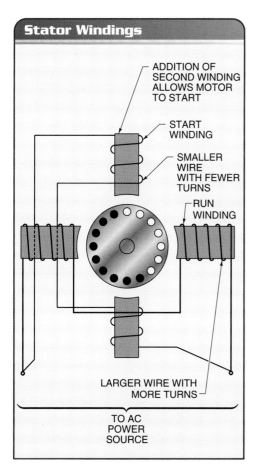

Figure 8-3. The start winding is placed at a 90° angle to the run winding.

Single-Phase Motor Operating Principles

A relationship exists between the stator and rotor of a single-phase motor. **See Figure 8-4.** When electrical power is applied to the stator, a magnetic field is produced by the stator windings. The magnetic field induces a voltage into the rotor bars. This induced voltage causes current to flow in each rotor bar. The current flow produces a magnetic field in the rotor. The magnetic field produces poles on the rotor, which are opposite in polarity to the stator poles.

Regardless of the position the rotor is in when power is applied, the rotor may still not rotate. The stator field only alternates back and forth, and does so at such a rapid rate (60 times per second for standard 60 Hz AC) that the rotor's magnetic field may simply be locked in step with the stator's alternating magnetic field and not rotate.

Locked in step is the lack of rotation when the stator's field and the rotor's field are parallel to one another. Rotation is not possible because the magnetic repulsion between the two fields is equal for both directions of rotation. The force rotating the motor clockwise is equal to the force rotating

Definition

Locked in step is the lack of rotation when the stator's field and the rotor's field are parallel to one another.

the motor counterclockwise. If the rotor is given a spin in one direction or the other, it will continue to rotate in the direction of the spin. The rotor will then quickly accelerate until it reaches a speed slightly less than the rated synchronous speed of the motor.

In order to start automatically, some single-phase motors use a start winding. The start winding, when present, has smaller wire and fewer turns than the run winding. It is more resistive and less inductive than the run winding. The more resistive the circuit, the closer the current is to being in phase with the voltage. This allows current to flow in the start winding ahead of the run winding. This creates a phase shift where the current in the start winding leads the current in the run winding. This produces a rotating magnetic field.

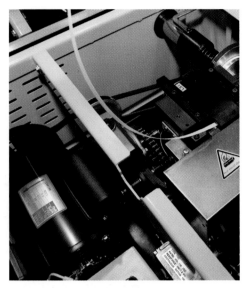

Single-phase motors are typically used in commercial applications, such as in banding machines.

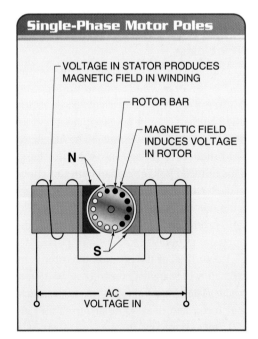

Figure 8-4. Voltage in the stator produces a magnetic field in the winding, which induces a voltage in the rotor.

Tech Fact

A run winding has many turns of heavy copper wire to develop a high reactance. The start winding is smaller and has higher resistance. This creates a phase difference between the two windings.

Dual-Voltage Single-Phase Motors. Many single-phase motors can be connected for either of two voltages. The purpose in making motors for two voltages is to enable the same motor to be used with two different power line voltages. Usually the dual-voltage rating of single-phase motors is 115/230 V. The nameplate of the motor should always be checked for proper voltage ratings.

In a typical dual-voltage motor, the run winding is split into two sections. **See Figure 8-5.** The two sections may be connected in series or in parallel. The series connection is used for high voltage. The parallel connection is used for low voltage. The start winding is connected across one of the run winding sections. Electrically, all windings receive the same voltage when wired for low- or high-voltage operation.

When there is a choice of which voltage to use for a dual-voltage motor, the higher voltage is preferred. The motor develops the same amount of power for either voltage. However, as the voltage is doubled (115 V to 230 V), the current is reduced by half. With half the current, the ampacity of conductors is reduced and the installation is more cost-efficient.

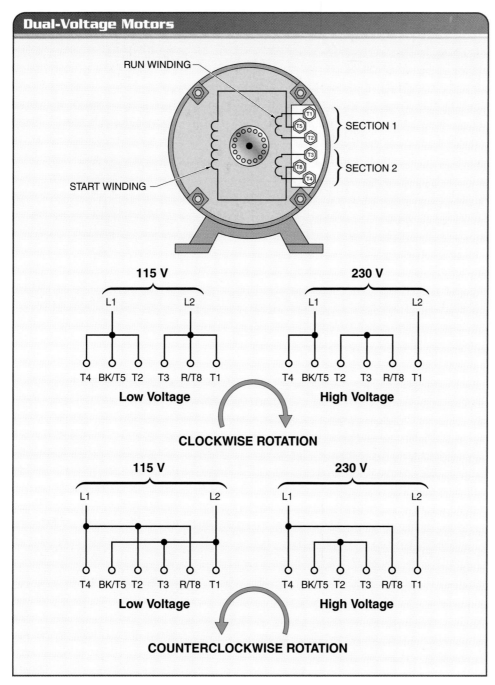

Figure 8-5. Dual-voltage motors have the run winding split into two sections.

Reversing Single-Phase Motors. To reverse the direction of most single-phase motors, all that is necessary is to switch the two start leads in the circuit. These leads will be colored black and red or marked T5 and T8. Interchanging these leads reverses the current and the poles in the rotor. This causes the rotor to turn in the opposite direction. **See Figure 8-6.**

Tech Fact

A universal motor is a motor that can be operated on either single-phase AC or DC power. Universal motors are the most common motors found in residences. The main advantages of universal motors are high torque, high speed, and small size when compared to other AC motors.

Definition

A *shaded-pole motor* is an AC motor that uses a shaded stator pole for starting.

A *shaded pole* is a short-circuited winding, consisting of a single turn of copper wire, that acts on only a portion of the stator windings.

A *shading coil* is a single turn of copper wire wrapped around part of the salient pole of a shaded-pole motor.

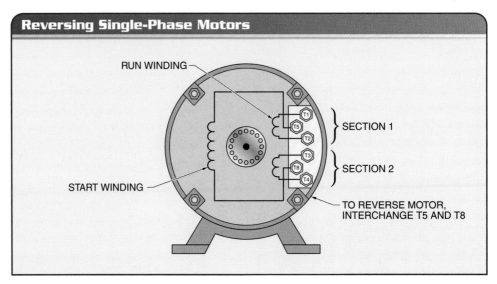

Figure 8-6. To reverse the motor direction, all that is needed is to interchange the T5 and T8 leads.

SHADED-POLE MOTORS

A *shaded-pole motor* is an AC motor that uses a shaded stator pole for starting. A *shaded pole* is a short-circuited winding, consisting of a single turn of copper wire, that acts on only a portion of the stator windings. Although the shaded pole produces a rotation effect on the rotor, it produces low starting torque. For this reason, shaded-pole motors are used primarily in applications requiring very small loads to be driven.

Shaded-pole motors are commonly 1/20 HP or less and have low starting torque. The most common application of shaded-pole motors is for use as the cooling fan in small appliances, such as computers and stereos. The only load the motor must turn at startup in such applications is the fan blade.

Shaded-Pole Motor Construction

The stator of a shaded-pole motor is constructed of thin sheets of iron, just as with other types of motors. For a shaded-pole motor, the stator has two or more salient poles that are divided into parts. **See Figure 8-7.** A *shading coil* is a single turn of copper wire wrapped around part of the salient pole of a shaded-pole motor. The shading coil creates the shaded pole.

The field coil is placed around part of the stator to create a magnetic circuit. With a shaded-pole motor, there is no start winding as there is with other types of single-phase motors. The shaded pole creates the rotating magnetic field.

The rotor of a shaded-pole motor is the same as the rotor of a three-phase induction motor. It is constructed of thin sheets of iron with conductor bars filling the slots. The conductor bars are shorted at the ends, creating the rotor circuit. The simplicity of the design makes the shaded-pole motor a reliable and dependable means to power small-horsepower loads.

Shaded-Pole Motor Operating Principles

The function of the shaded pole is to delay the magnetic flux in the area of the pole that is shaded. Shading causes the magnetic flux at the pole area to be about 90 electrical degrees behind the magnetic flux of the main stator pole. The two main stator poles are 180° apart, so the shaded poles are between the main poles. **See Figure 8-8.**

The shaded-pole motor is unique in that the power is applied to the main winding to produce flux in the iron core, and the shading coil receives its power as a result of transformer action by the motion of the flux in the iron. The main pole is at its strongest when the current in the main pole is at its maximum.

The shaded pole is its strongest when the flux is at its maximum. This is at 90° behind the maximum current in the main pole. It occurs at the moment when the current in the main windings is switching from positive to negative or from negative to positive.

Tech Fact

Residential and light commercial wiring uses two conductors and a neutral created from a center grounded tap between two phases. This type of circuit can supply 120 V power for small single-phase motors and 240 V power for larger single-phase loads.

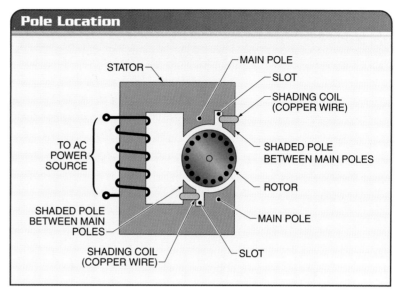

Figure 8-8. The two main stator poles are 180° apart, with the shaded poles between the main poles.

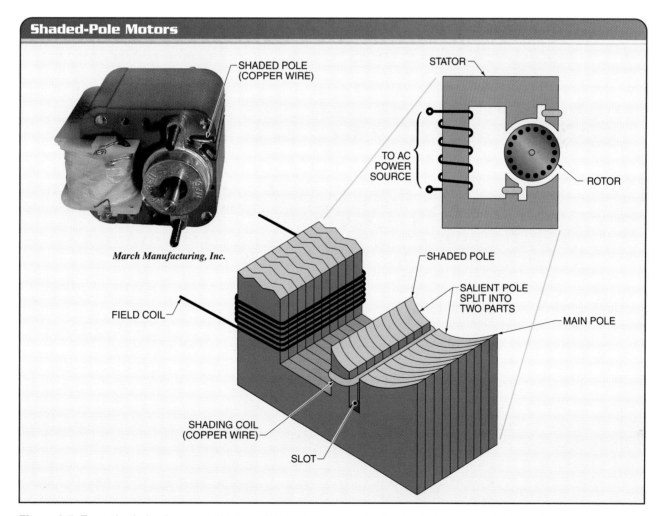

Figure 8-7. For a shaded-pole motor, the stator has two or more salient poles that are divided into parts.

As a result of the shaded pole being 90° behind the main pole, a rotating magnetic field is created. As the magnetic field alternates, the shaded poles follow the main poles and create north and south poles that are offset from the main poles. As a result of the offset field, the rotor moves from the main pole toward the shaded pole. **See Figure 8-9.**

Reversing Shaded-Pole Motors. Reversing a shaded-pole motor is very different from reversing the usual single-phase motor. In many cases, the rotor must be removed from the housing and turned around so that the shaft exits the opposite side of the housing. This procedure requires switching the endbells at the same time. In other cases, some shaded-pole motors are wound with two main windings that reverse the direction of the field. One of the main windings must be deenergized and the other main winding energized to reverse the motor.

Speed Control of Shaded-Pole Motor. Speed control of shaded-pole motors is accomplished by adding windings in series with the main winding. As the windings are added in series, the source voltage is divided across each winding, and the current is reduced. This lowers the flux in the iron core, and the slip increases. Slip in the shaded pole is not a problem, as the current in the stator is not controlled by a countervoltage determined by the rotor speed, as in other types of single-phase motors. The greater the number of windings inserted in series with the main winding, the greater the slip and the slower the speed.

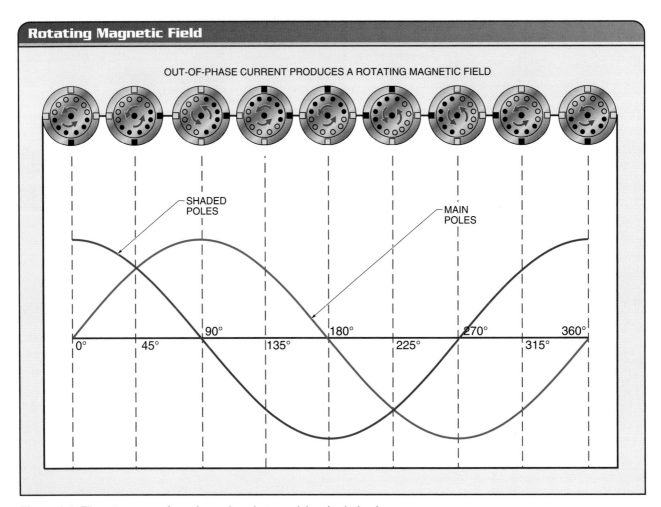

Figure 8-9. The rotor moves from the main pole toward the shaded pole.

SPLIT-PHASE MOTORS

A *split-phase motor* is a single-phase AC motor that includes a run winding and a resistive start winding that creates a phase-shift for starting. The two windings are placed in the stator slots and spaced 90° apart. The split-phase motor is one of the oldest and most common motor types. It is used in applications such as fans, business machines, machine tools, and centrifugal pumps, where starting is easy. Split-phase motors are generally available in sizes ranging from 1/30 HP to 1/2 HP.

Split-Phase Motor Construction

A split-phase motor has a stator constructed with a run winding and a start winding. **See Figure 8-10.** A split-phase motor has a rotor similar to a squirrel-cage rotor. In addition, a centrifugal switch is located inside the motor. The centrifugal switch is used to disconnect the start winding as the motor approaches full-load speed.

Split-Phase Motor Windings. The run winding is made of larger wire and has a greater number of turns than the start winding. When the motor is first connected to power, the reactance of the run winding is higher and the resistance is lower than the start winding. *Inductive reactance* is the opposition to the flow of AC in a circuit due to inductance.

The start winding is made of relatively small wire and has fewer turns than the run winding. When the motor is first connected to power, the reactance of the start winding is lower and the resistance is higher than in the run winding.

Centrifugal Switches. To minimize energy loss and prevent heat buildup in the start winding once the motor is started, a centrifugal switch is used to remove the start winding when the motor reaches a set speed. A *centrifugal switch* is a switch that opens to disconnect the start winding when the rotor reaches a certain preset speed and reconnects the start winding when the speed falls below a preset value. In most motors, the centrifugal switch is located inside the enclosure on the shaft. **See Figure 8-11.** For some motors, the centrifugal switch is located outside the enclosure for easier repair. For example, switches for motors used on washers and dryers are located outside the enclosure.

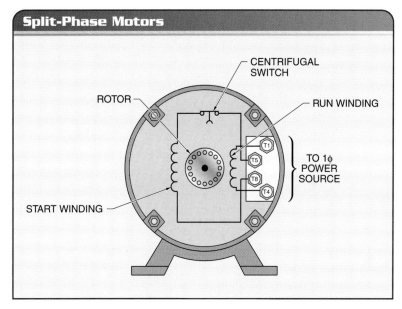

Figure 8-10. A split-phase motor has a stator constructed with a start winding and a run winding.

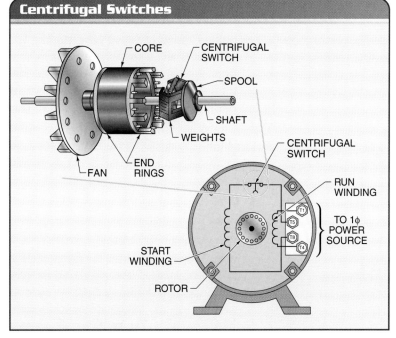

Figure 8-11. A centrifugal switch opens the start winding when the motor reaches a preset speed.

Definition

A split-phase motor is a single-phase AC motor that includes a run winding and a resistive start winding that creates a phase-shift for starting.

Inductive reactance is the opposition to the flow of AC in a circuit due to inductance.

A centrifugal switch is a switch that opens to disconnect the start winding when the rotor reaches a certain preset speed and reconnects the start winding when the speed falls below a preset value.

A capacitor motor is a single-phase motor with a capacitor connected in series with the start windings to produce phase displacement in the start winding.

A capacitor is a device that stores an electric charge.

Split-Phase Motor Operating Principles

When power is first applied, both the run winding and the start winding are energized. Because of their different inductive reactances, the run winding current lags the start winding current. This produces a phase difference between the start and run windings. A 90° phase difference is required to produce maximum starting torque, but the phase difference is commonly much less. A rotating magnetic field is produced because the two windings are out of phase.

When the motor reaches approximately 75% of full speed, the centrifugal switch opens, disconnecting the start winding from the circuit. This allows the motor to operate using the run winding only. When the motor is turned OFF (power removed), the centrifugal switch recloses at approximately 40% of full-load speed.

Reversing Split-Phase Motors. To reverse the direction of rotation of a split-phase motor, the connections of either the start or the run windings are reversed. If both the start and the run windings are reversed, the motor will not change direction of rotation. Reversing the start winding is the standard method used in industry for reversing the direction of rotation of a single-phase motor.

Speed Control of Split-Phase Motors. The speed at which a split-phase motor runs is determined by the number of poles in the motor and the frequency of the supply voltage. Since the frequency of split-phase motors is almost always fixed, the speed is usually changed by changing the number of poles. To change the frequency of the supply voltage, an expensive motor drive is required.

In a typical two-speed, split-phase motor, two run windings and one start winding are used to develop two separate speeds. As the number of poles is increased, the speed of the motor is decreased. As the number of poles is decreased, the speed of the motor is increased. **See Figure 8-12.**

A two-speed, split-phase motor is commonly wound to run on either six or eight poles. When the motor is connected to six poles, the synchronous speed is 1200 rpm, and the actual speed is about 1152 rpm. When the motor is connected to eight poles, the synchronous speed is 900 rpm and the actual speed is about 864 rpm. A double-contact centrifugal switch is used in the circuit. The motor starts on the high-speed run winding, regardless of which speed is selected by the starting switch. However, when the starting switch is set for low speed, the centrifugal switch disconnects the high-speed run winding and connects the low-speed run winding after the motor has reached a set speed.

The number of poles may be changed, providing two or three different possible speeds. Two-speed motors are the most common. However, three-speed motors are often used for washing machines and fans.

CAPACITOR MOTORS

A *capacitor motor* is a single-phase motor with a capacitor connected in series with the start windings to produce phase displacement in the start winding. A *capacitor* is a device that stores an electric charge. A capacitor motor introduces capacitance into an AC circuit to create a phase shift between the start and the run windings.

Capacitor Motor Construction

A capacitor motor is similar in design to a split-phase motor. Both of these motors have a start and run winding, but the capacitor motor has a capacitor connected in series with the start winding. The capacitor is added to provide a higher starting torque at lower starting current than is delivered by the split-phase motor. Typical applications of capacitor motors include refrigerators, air conditioners, air compressors, and some power tools. These loads are harder to start and require more starting torque than a split-phase motor produces.

Capacitor motors have a winding that is more reactive and less resistive than a split-phase motor. It is wound with larger wire and more turns, which will produce a much stronger pole in the rotor. The reactive start winding then is connected to the capacitor to provide the phase shift.

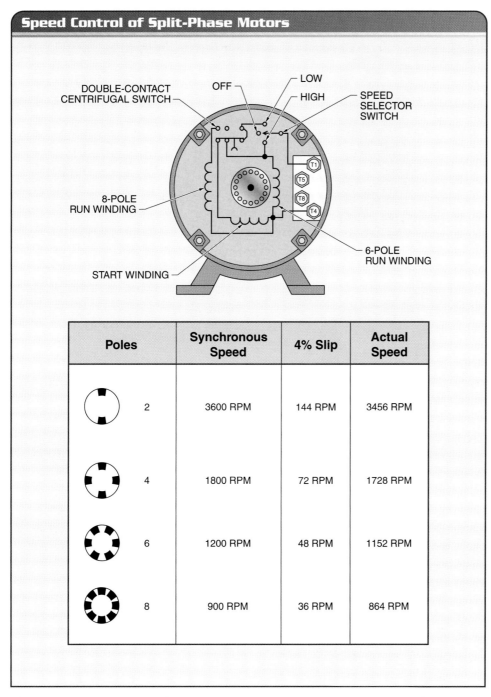

Figure 8-12. As the number of poles is increased, the motor speed is decreased.

The capacitors used to start motors are electrolytic capacitors. **See Figure 8-13.** The large value of capacitance in a small package means that starting capacitors cannot be left in circuit after startup and a centrifugal switch is used to disconnect the capacitor. In some motor designs, capacitors are designed to remain in the circuit during operation. These capacitors are oil-filled and have a low value of capacitance when compared to the start capacitor. If a capacitor is replaced during maintenance, the replacement must be the same capacitance as the original.

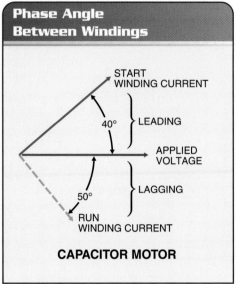

Figure 8-14. In a capacitor motor, the start current and the run current are about 90° out of phase with each other.

Figure 8-13. Electrolytic capacitors are used as starting capacitors. Oil-filled capacitors are used as running capacitors.

Tech Fact

The simplest single-phase motors for industrial use are split-phase motors. Split-phase motors have a simple design with a start winding and a run winding, but generally are not used for applications with high starting torque. Capacitor motors are more expensive, but typically have more starting torque and can be used in many industrial applications.

Capacitor Motor Operating Principles

In a capacitor motor, the capacitor causes the start winding current to lead the applied voltage by about 40°. The run winding current lags the applied voltage by about 50° because of the high inductance of the coil. **See Figure 8-14.** Since the start and run windings are about 90° out of phase, the motor's operating characteristics are improved. A higher starting torque is produced, and the motor has a better power factor with a lower current draw. Three types of capacitor motors are the capacitor-start motor, capacitor-run motor, and the capacitor start-and-run motor.

Capacitor-Start Motors

The capacitor-start motor is the most common type of capacitor motor. With the capacitor in the circuit, the capacitor-start motor develops considerably more locked-rotor torque per ampere than the split-phase motor. A capacitor-start motor operates much the same as a split-phase motor, in that it uses a centrifugal switch that opens at approximately 60% to 80% of full-load speed. **See Figure 8-15.** In a capacitor-start motor, the start winding and the capacitor are removed when the centrifugal switch opens. The larger wire used in the start winding gives a capacitor-start motor high starting torque.

Capacitor-Start Motors

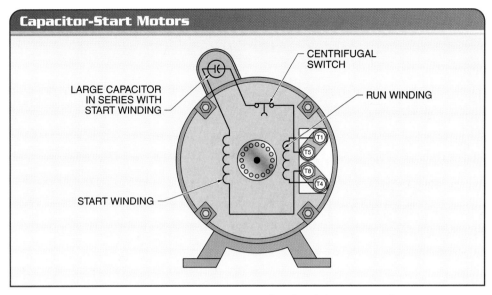

Figure 8-15. In a capacitor-start motor, the centrifugal switch opens and removes the capacitor and start winding from the circuit.

Although a capacitor-start motor develops more starting torque than a split-phase motor, both motor types are the same when running because the capacitor and start windings are removed from the circuit when the motor comes up to speed. For this reason, the capacitor-start motor is also known as a capacitor-start induction run motor. Capacitor-start motors are typically used to drive power tools, pumps, and small machines.

Reversing Capacitor-Start Motors. A capacitor-start motor is reversed in the same way as a split-phase motor, with the current in the start winding and the run winding being reversed with respect to each other. If the motor is to be reversed often as part of its operation, then a drum switch or foot switch can be used to reverse the current through the start with respect to the run windings.

Speed Control of Capacitor-Start Motors. A two-speed capacitor-start motor functions the same as a split-phase multi-speed motor, except that the capacitor that is in series with the start winding provides more torque. If this motor has two run windings and one start winding, it will start on the high speed even though the selector switch is set to the low speed. When the motor gets up to speed and the centrifugal switch opens, the low-speed winding will be energized and the motor will run at the slower speed.

Capacitor-Run Motors

A capacitor-run motor has the start winding and capacitor connected in series at all times. **See Figure 8-16.** A capacitor-run motor is also known as a permanent split-capacitor motor. A smaller capacitor is used in a capacitor-run motor than in a capacitor-start motor because the capacitor remains in the circuit at full-load speed.

Since there is no centrifugal switch, the start winding is not removed as the motor speed increases. The capacitor-run motor has a lower full-load speed than the capacitor-start motor because the capacitor remains in the circuit at all times. The advantage of leaving the capacitor in the circuit is that the motor has more running torque than a capacitor-start motor or split-phase motor. This allows a capacitor-run motor to be used for loads that require a higher running torque, such as to drive shaft-mounted fans and blowers ranging in size from $1/16$ HP to $1/3$ HP.

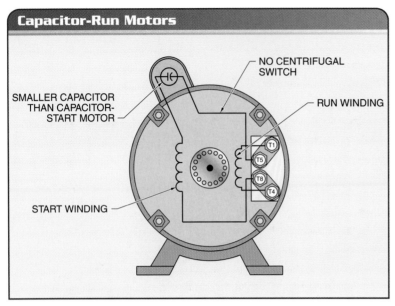

Figure 8-16. In a capacitor-run motor, the capacitor and start winding remain in the circuit at all times.

Tech Fact

Capacitor-run motors have the start winding and run winding connected at all times. They do not need a centrifugal switch to remove the start winding, thus increasing the running torque.

Speed Control of Capacitor-Run Motors.
A multispeed capacitor-run motor does not change speed by changing the number of poles, but rather by adding impedance to the run winding and increasing the slip. Capacitor-run motors are available as two- or three-speed motors with the addition of auxiliary windings that are placed in series with either the start or the run windings. The two-speed motor has a selector switch that places the auxiliary winding in series with the start winding when it is in the high-speed position. **See Figure 8-17.** This places the source voltage across the run and the combination of the start and the auxiliary winding, which allows the rotor to turn at the synchronous speed minus the slip.

When the selector is placed in the low position, the auxiliary winding is placed in series with the run winding. This places the source voltage across the start and the combination of the auxiliary and the run winding. This reduces the current through the run winding. The reduced current lowers the flux in the stator and reduces the pole strength, increasing the slip and decreasing the speed of the rotor.

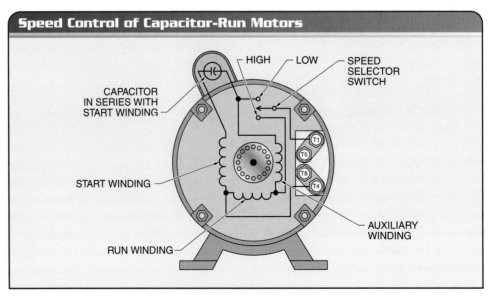

Figure 8-17. In a two-speed capacitor-run motor, an auxiliary winding can be placed in series with the start winding for high-speed operation.

The addition of a second auxiliary winding creates a three-speed version of the motor. **See Figure 8-18.** This further reduces the stator flux and increases the slip. With the selector placed in the high-speed position, the source voltage is placed across the run and the combination of the two auxiliary windings and the start winding. This allows the rotor to run at the synchronous speed minus the slip. As the auxiliary windings are placed in series with the run windings, the rotor slows, with the added impedance reducing the stator current. The slow speed places both of the auxiliary windings in series with the run winding, and the motor runs at its slowest speed.

Reversing Capacitor-Run Motors. To reverse a capacitor-run motor, the capacitor is switched from one winding to the other. The start and the run windings are identical to each other in this type of motor. Whichever winding has the capacitor connected to it is the start winding, as current flows in this winding first and controls the direction.

Capacitor Start-and-Run Motors

A capacitor start-and-run motor uses two capacitors. It is also known as a dual-capacitor motor. A capacitor start-and-run motor has two capacitors in parallel with each other and both in series with the start winding. **See Figure 8-19.** A larger capacitor is used for starting and a smaller capacitor is used for running. A capacitor start-and-run motor has a centrifugal switch that disconnects the large capacitor that is in series with the start winding. The start winding is left in the circuit to increase the running torque.

A capacitor start-and-run motor has the same starting torque as a capacitor-start motor. However, the capacitor start-and-run motor has more running torque than the capacitor-start or capacitor-run motor. This is because the start winding is left in the circuit. Capacitor start-and-run motors are typically used to run refrigerators and air compressors.

Even though the starting torque of a capacitor-run motor is less than the starting torque in a capacitor start-and-run motor, its lower cost makes this type an attractive choice as long as operational requirements are satisfied.

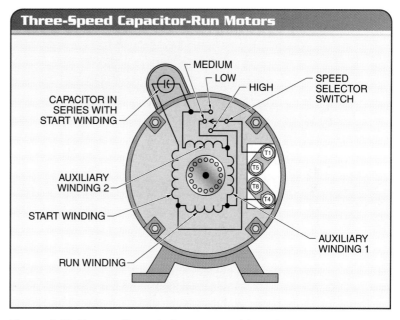

Figure 8-18. Two auxiliary windings are used for a three-speed capacitor-run motor.

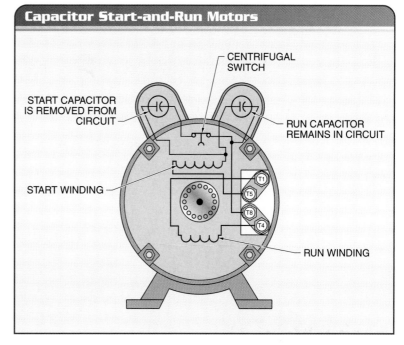

Figure 8-19. A capacitor start-and-run motor has one capacitor in series with the start winding and one capacitor in series with the run winding.

Supplemental Topic

Repulsion Motors

A repulsion motor is a single-phase motor without a rotating magnetic field, where the rotor winding is connected to a commutator with two short-circuited brushes. In a squirrel-cage induction motor, an induced rotor pole rotates in order to "catch up" with the unlike rotating stator pole. Repulsion motors use a wound rotor instead of a squirrel-cage rotor. The stator has salient poles that produce the required magnetic poles to induce rotor current. The induced rotor current produces like poles relative to the stator poles. Since the poles in the rotor and stator are alike, they are repelled, or repulsed, from each other.

The rotor includes brushes and a commutator. The brushes are not used to conduct line power to the rotor. The brushes are shorted together to provide a conducting path between selected commutator segments. This is similar to the operation of the shorting ring of a squirrel-cage rotor. In some designs, the brushes are permanently in contact with the commutator. In other designs, the rotor accelerates and, when it reaches about 75% of its synchronous speed, a centrifugal device lifts the brushes off the commutator.

A hard neutral position is an orientation in a repulsion motor where the brushes are aligned directly with stator poles. Current is induced in the armature windings by the changing stator field and produces like poles between the rotor and stator. A rotating magnetic field is not present. Since the poles are exactly aligned, the rotor cannot rotate from the hard neutral position.

A soft neutral position is an orientation in a repulsion motor where the brushes are aligned 90° from the stator poles. There is no induced current because the stator and rotor poles are at right angles to each other. Since there is no induced current, the rotor does not rotate.

If the brushes shift about 15° to 17° from the hard neutral position, unlike poles of the stator and rotor repel each other, and the rotor rotates due to the repulsion between the magnetic fields. Markings on the endbell indicate the correct location for the brushes. The rotor rotates in the direction of the brush shift. If the brush is shifted counterclockwise from hard neutral position, the rotor rotates counterclockwise.

The motor can also start from the soft neutral position by shifting the brushes. However, starting torque is greatly reduced since the repelling magnetic fields are further apart. Repulsion motors should be started at an angle relative to the hard neutral position.

Due to the use of a commutator, brushes, and centrifugal mechanism to short-circuit the commutator and lift the brushes, repulsion motors are more expensive and require greater maintenance than other types of motors. Because of the higher cost for repulsion motors, capacitor-start motors are typically chosen over repulsion motors when high starting torque is required.

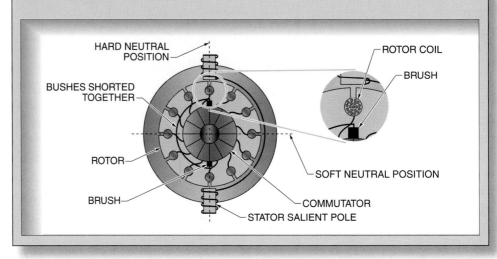

Application—Centrifugal Switches

A centrifugal switch removes the capacitor from the circuit after the motor accelerates in a capacitor-start motor. In normal applications, the charged capacitor slowly discharges as the motor runs. In rapid cycling applications, the capacitor is added back to the circuit before it is discharged. The charged capacitor causes arcing and welding on the contacts of the centrifugal switch. In rapid cycling applications, a bleeder resistor is connected across the terminals of the starting capacitor to bleed off the electric charge that is stored in the capacitor.

Application—Resistors

The value of the resistor must be low enough to bleed the stored charge, but high enough not to interfere with starting the motor. Most bleeder resistors are between 12,000 Ω and 20,000 Ω, and are rated at 2 W. If a motor manufacturer does not specify a value, a 15,000-Ω resistor is used. If the motor application requires fast cycling, a 12,000-Ω resistor is used.

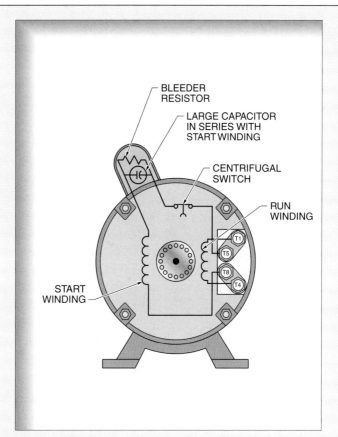

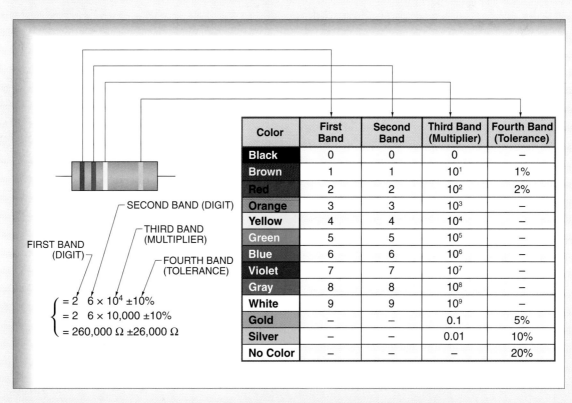

$= 2 \quad 6 \times 10^4 \pm 10\%$
$= 2 \quad 6 \times 10,000 \pm 10\%$
$= 260,000 \, \Omega \pm 26,000 \, \Omega$

Color	First Band	Second Band	Third Band (Multiplier)	Fourth Band (Tolerance)
Black	0	0	0	—
Brown	1	1	10^1	1%
Red	2	2	10^2	2%
Orange	3	3	10^3	—
Yellow	4	4	10^4	—
Green	5	5	10^5	—
Blue	6	6	10^6	—
Violet	7	7	10^7	—
Gray	8	8	10^8	—
White	9	9	10^9	—
Gold	—	—	0.1	5%
Silver	—	—	0.01	10%
No Color	—	—	—	20%

Application—Single-Phase AC Motor Terminal Designations

Manufacturers use different terminal and wire designations for single-phase motors. Most single-phase motors and their auxiliary devices, such as capacitors, starting switches, and thermal protection, placed in operation over the past few years use standard letters, numbers, and colors. Capacitors are labeled with the letter J and a number, the supply line with the letter L and a number, the stator terminations with the letter T and a number, and thermal protection devices with the letter P and a number.

- Capacitor— J1, J2, J3, J4, etc.
- Supply line— L1, L2, L3, L4, etc.
- Stator— T1, T2, T3, T4, etc.
- Thermal protection device— P1, P2, P3, P4, etc.

The wires used in single-phase motors are assigned specific colors:

T1—	Blue	T5—	Black
T2—	White	T8—	Red
T3—	Orange	P1—	No color assigned
T4—	Yellow	P2—	Brown

Single-Voltage Designations

On single-voltage, single-phase motors, T1 and T4 are assigned to the run winding, and T5 and T8 are assigned to the start winding (if present).

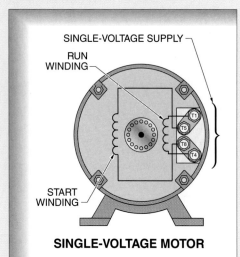

SINGLE-VOLTAGE MOTOR

Dual-Voltage Designations

On dual-voltage motors in which the run winding is divided into two parts, one half of the winding is designated T1 and T2, and the other half is designated T3 and T4. One half of the start winding is designated T5 and T6, and the other half is designated T7 and T8 (if used). If only two wires are used for the start winding, they are designated T5 and T8.

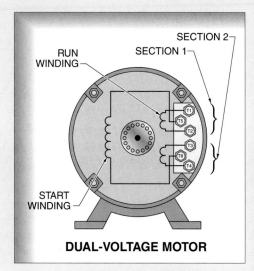

DUAL-VOLTAGE MOTOR

Mixed Numbers and Colors

Some manufacturers mix numbers and colors. Typically, the run winding is designated T1, T2, etc., and the start winding is designated with a color. Red and black are commonly used to designate start windings.

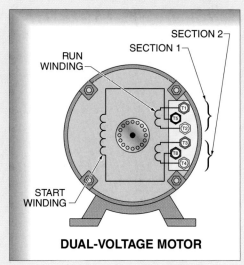

DUAL-VOLTAGE MOTOR

Application—Motor Winding Lead Terminations

Motor winding leads are terminated outside a motor or in a terminal box on a motor. Motors with winding leads terminated outside the motor are used in machines that include the motor as part of other equipment. Applications include central air conditioners, furnaces, and portable equipment. The advantage is that the motor is connected to other devices through a terminal strip. The terminal strip includes other wiring, such as the control circuit.

Motors with winding leads terminated inside a terminal box mounted on the motor are the most common. The power lines are brought to the motor through conduit or cable connected to the terminal box. Wiring methods inside a terminal box include direct wire, terminal post, and terminal post with links.

Direct Wire

The motor winding leads are connected to the power lines using wire connectors (nuts) in the direct wire method. The wires are then pushed inside the terminal box.

Terminal Post

The terminal post method is the most common terminal box wiring method. The ends of the motor winding leads and power lines have terminal bolt or spade lugs. The ends of motor winding leads that require periodic changing have spade lugs. The terminal bolt lugs are connected to terminal posts. The terminal bolt lugs are placed under the bolt on the terminal post and tightened. The spade lugs are connected to spade posts. The spade lugs are designed to be easily connected to the spade post, which permits easy motor direction changes.

Terminal Post with Links

The motor winding leads from the motor are connected to terminal posts inside the terminal box. The terminal posts are evenly spaced to allow connection of predrilled links. The links are used as jumpers to change motor direction. The links are added by removing the terminal bolt, placing the link over the terminal post, and reconnecting the terminal bolt.

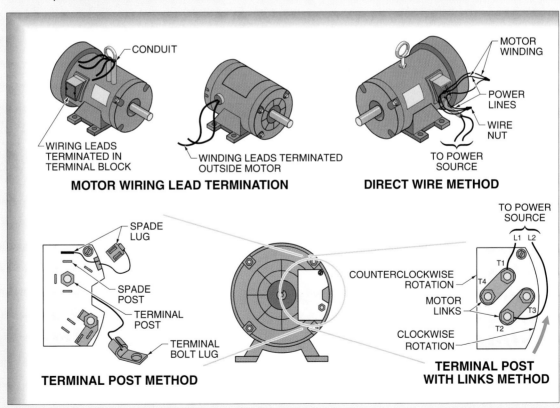

Summary

- A start winding has higher resistance and lower reactance than a run winding. The start winding is resistive at startup. The run winding is reactive at startup.

- A resistive circuit has no phase shift between voltage and current. A reactive circuit has a phase shift between voltage and current. This creates a phase shift between the current in the two windings.

- Shaded-pole motors use a shading coil to act as a transformer and develop a rotating stator magnetic field.

- Shaded-pole motors develop very low starting torque. They are commonly used in very easy-to-start loads such as small fans and blowers.

- Split-phase motors use a start winding and a run winding to develop a rotating stator magnetic field.

- Split-phase motors develop low starting torque. They are commonly used in easy-to-start loads such as fans, business machines, machine tools, and centrifugal pumps.

- Split-phase motors use a centrifugal switch to remove the start winding from the circuit after the motor gets up to speed.

- Capacitor motors use a start winding, run winding, and a capacitor to develop a rotating stator magnetic field.

- Capacitor motors develop medium to high starting torque. They are commonly used in hard-to-start loads such as refrigerators, air conditioners, air compressors, and some power tools.

- Capacitor-start motors use a centrifugal switch to remove the capacitor from the circuit after the motor gets up to speed. They develop a high starting torque because of the large capacitance in the start winding. They develop a medium operating torque because the starting capacitor is removed after startup.

- Capacitor-run motors do not remove the capacitor from the circuit during operation. They develop a medium starting torque because a smaller capacitor is used than in other types of capacitor motors. They develop a high operating torque because the starting capacitor is left in the circuit at all times.

Glossary

A **stator** is the fixed, unmoving part of a motor, consisting of a core and windings, that converts electrical energy to the energy of a magnetic field.

A **rotor** is the rotating, moving part of a motor, consisting of a core and windings, that converts the rotating magnetic field of the stator into the torque that rotates the shaft.

Locked in step is the lack of rotation when the stator's field and the rotor's field are parallel to one another.

A **shaded-pole motor** is an AC motor that uses a shaded stator pole for starting.

A **shaded pole** is a short-circuited winding, consisting of a single turn of copper wire, that acts on only a portion of the stator windings.

A **shading coil** is a single turn of copper wire wrapped around part of the salient pole of a shaded-pole motor.

A **split-phase motor** is a single-phase AC motor that includes a run winding and a resistive start winding that creates a phase-shift for starting.

Inductive reactance is the opposition to the flow of AC in a circuit due to inductance.

A **centrifugal switch** is a switch that opens to disconnect the start winding when the rotor reaches a certain preset speed and reconnects the start winding when the speed falls below a preset value.

A **capacitor motor** is a single-phase motor with a capacitor connected in series with the start windings to produce phase displacement in the start winding.

A **capacitor** is a device that stores an electric charge.

Review

1. Compare a start winding to a run winding.

2. Describe how a phase shift is created when a start winding is used along with the standard run winding.

3. Describe how a shading coil creates a phase shift in a shaded-pole motor.

4. Explain why a centrifugal switch is used in single-phase motors.

5. Explain why a capacitor is used in addition to a start winding in capacitor motors.

6. Describe the differences in starting torque in shaded-pole, split-phase, and all three types of capacitor motors.

7. Give two examples of loads that can be started with a split-phase motor.

8. Describe the difference between a capacitor-start and a capacitor-run motor.

Refer to the CD-ROM for Quick Quiz® questions related to chapter content.

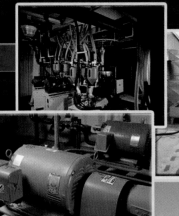

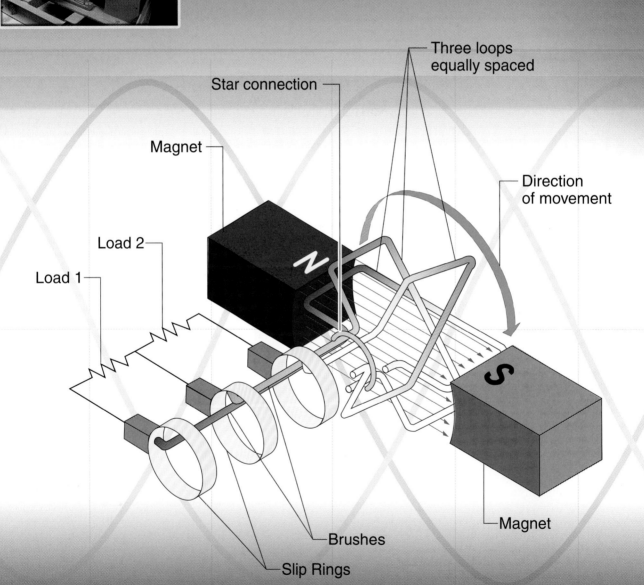

MOTORS

AC Alternators

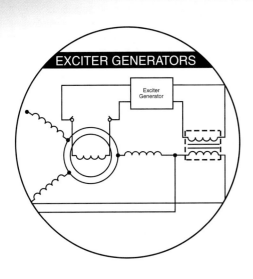

Alternator Construction	190
Operating Principles	192
Alternator Ratings	198
Paralleling Alternators	198
Application—Selecting Portable Generators	202
Summary	204
Glossary	205
Review	205

OBJECTIVES

- Describe the construction of an alternator.
- Describe the differences between a revolving-field alternator and a revolving-rotor alternator.
- Describe how the output voltage and frequency are controlled in an alternator.
- List and explain the steps in producing AC power from an alternator.
- List the parameters to match and explain how to parallel two or more alternators.

The rotating field in the three-phase motor originates at an alternator in a generating station that is connected to a grid system. This system covers hundreds of square miles and has many generating plants connected to provide the necessary power for a large area. All of the alternators are synchronized so that all of the power has the same phasing and peaks at the same time.

Definition

*An **alternator** is a synchronous machine that produces alternating current (AC).*

*A **stator** is the fixed unmoving part of a generator, consisting of a core and windings, that converts the energy of a magnetic field into electrical energy.*

ALTERNATOR CONSTRUCTION

An *alternator* is a synchronous machine that produces alternating current (AC). Strictly speaking, rotors and stators are found in AC equipment, and field frames and armatures are found in DC equipment. Alternators produce AC power and generators produce DC power. However, in common usage, portable AC alternators are often called generators, and utilities operate power generators. The rotor in an AC alternator may also be called an armature, even though the term rotor is more correct.

The magnetic field in an alternator can be produced by permanent magnets or electromagnets. Most alternators use electromagnets, which must be supplied with current. When the rotor is rotated, a voltage is generated in each half of a coil. The rotor may consist of many coils.

A common alternator design consists of a stator with field windings, a rotor, slip rings, and brushes. **See Figure 9-1.** Similar to an AC motor, a stator is the fixed part of an alternator that supplies a magnetic field through the field windings. The rotor is the rotating coils of wire in the center of an alternator that rotate through the magnetic field. Alternators range in size from very small units delivering only a few watts to very large utility generators delivering megawatts of power.

Stator Construction

A *stator* is the fixed unmoving part of a generator, consisting of a core and windings, that converts the energy of a magnetic field to electrical energy. **See Figure 9-2.** The stator is enclosed within a housing made of cast iron, rolled steel, or cast aluminum. The core consists of many thin iron sheets laminated together, pressed into a frame, and secured in place. An insulating coating electrically separates the iron sheets from each other. The separation reduces the cross-sectional area of the core and reduces the conduction path for damaging eddy currents.

The windings consist of coils of copper wire placed in slots 120 electrical degrees apart. The windings are well insulated and can tolerate high voltage and high current. For smaller stators, the stator coils are assembled into one component. During maintenance, the entire motor must be removed and replaced. For larger stators, the coils are individually constructed and connected. These coils are form wound from rectangular wire. The end of each coil is made so that it can be bolted to the next one. This facilitates in-field replacements by eliminating the need to remove the stator and completely rewind it.

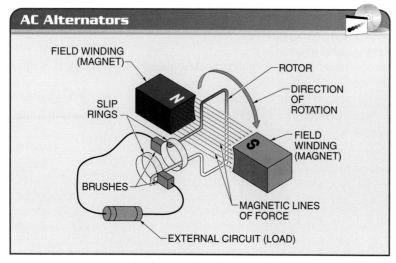

Figure 9-1. A common alternator design consists of a stator with field windings, a rotor, slip rings, and brushes.

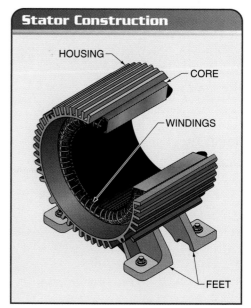

Figure 9-2. The stator consists of a core and windings. The stator is enclosed in a housing that includes feet for mounting.

Larger stators may be mounted within a cast-iron frame or a welded steel ring. Feet are typically welded to the frame to provide anchoring points that allow the alternator to be mounted to the floor.

Rotor Construction

A *rotor* is the rotating moving part of an alternator or generator, consisting of a core and windings, that convert torque to magnetic energy. Like the stator, the rotor core consists of many thin iron sheets laminated together. *Torque* is a turning or twisting force that causes an object to rotate. In a common alternator design, torque is applied to the rotor to move the rotor windings and their magnetic fields past the stator windings. This provides the relative motion that is needed to induce AC current in the stator windings. **See Figure 9-3.**

A rotor uses brushes and slip rings to allow current to flow through the rotor circuit. A brush is held in a brush holder and is free to move up and down in the holder. This allows the brush to follow irregularities in the surface of the slip ring. Brush holders are mounted on the motor frame, but they are electrically insulated from it. A spring placed behind a brush forces the brush to make contact with the slip ring. The spring pressure is usually adjustable. A pigtail connects the brush to the power supply. Some designs do not use brushes and slip rings. These designs use a brushless exciter.

Centrifugal Force. Whenever rotating objects such as motors and generators are designed, centrifugal force must be taken into account. Centrifugal force is larger for high-speed rotation and large diameters and is smaller for low-speed rotation and small diameters.

For large generators rotating at less than about 450 rpm, the centrifugal force is relatively small. Large, slow-moving, multipole generators are typically constructed with salient poles. The salient poles are mounted onto a spider ring that is attached to the rotor shaft so that the entire assembly rotates. The poles are bolted to the ring, which is usually made of cast iron.

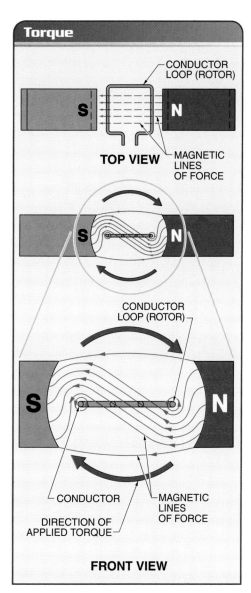

Figure 9-3. Torque is applied to the rotor to provide the relative motion between the conductor and the magnetic lines of force.

Definition

*A **rotor** is the rotating moving part of an alternator or generator, consisting of a core and windings, that convert torque to magnetic energy.*

***Torque** is a turning or twisting force that causes an object to rotate.*

At speeds higher than about 450 rpm, the increased centrifugal forces become more significant and the spider ring has a dovetail groove into which the salient poles slide. **See Figure 9-4.** The dovetail provides a stronger attachment than bolts. For high-speed motors rotating faster than about 1800 rpm, rotors are designed without salient poles. In addition, a reduced diameter is used for the core. The reduced diameter is necessary because of the increased centrifugal force developed at higher speeds.

Rotor Construction

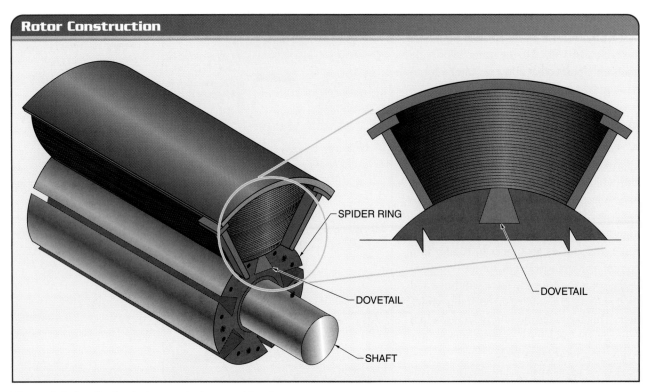

Figure 9-4. A salient pole is mounted on a spider ring.

Definition

*A **revolving-field alternator** is an alternator where a magnetic field is created in the rotor, which turns within the fixed stator windings, and AC power is supplied through the stator windings.*

*An **exciter generator** is an assembly consisting of a small three-phase alternator used to supply current to an alternator rotor.*

*A **brushless exciter** is a rectifier assembly mounted on the main rotor shaft along with the exciter generator.*

OPERATING PRINCIPLES

All alternators operate on the same principle of induction as is used in an induction motor or a transformer. The three requirements for induction are a conductor, a magnetic field, and relative motion between the conductor and the magnetic field.

Revolving-Field Alternators

A *revolving-field alternator* is an alternator where a magnetic field is created in the rotor, which turns within the fixed stator windings, and AC power is supplied through the stator windings. Revolving-field alternators are commonly used by utilities that generate high-voltage power. Fixed stator wiring overcomes the disadvantage of using brushes and slip rings, which limit power output. Excessive arcing occurs at the brushes and slip rings when high levels of current are present.

The magnetic field is developed by supplying DC to the rotor through the brushes and slip rings or with a brushless exciter. As current is supplied to the rotor, north and south fields are set up that expand outward into the stator windings and cut the stator conductors. When poles of different polarity pass the stationary stator conductors, an AC voltage is induced into the stator windings.

An exciter generator is mounted on the shaft. **See Figure 9-5.** An *exciter generator* is an assembly consisting of a small three-phase alternator used to supply current to an alternator rotor. The AC power from the exciter is rectified and supplied to the main rotor fields. A brushless exciter is a rectifier assembly mounted on the main rotor shaft along with the exciter generator. The exciter rotor spins in a DC field supplied by a stationary set of magnets that are mounted in the end bell and surround the exciter rotor. These coils are fed from the voltage regulator, so the output voltage of the alternator can be controlled by varying the current to the exciter.

Tech Fact

Alternating current is used because it allows efficient transmission of electrical power between power generating stations and end users located far apart.

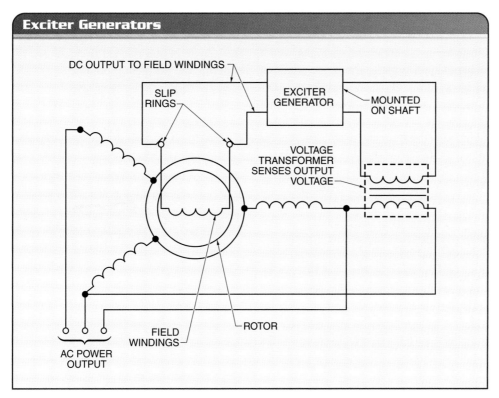

Figure 9-5. An exciter generator supplies current to the rotor to control the strength of the magnetic field.

Revolving-Rotor Alternators

A *revolving-rotor alternator* is an alternator where a fixed magnetic field is created in the stator, with the rotor turning within the stator, and AC power is supplied through the rotor slip rings and brushes. In the past, revolving-rotor alternators were commonly designed for low-voltage alternators. This design is not as common today. Revolving-rotor alternators work well for lower voltages and currents but cannot handle large voltages and currents because of the difficulty of insulating the slip rings, brush holders, and brushes from flashover at higher voltages. AC power is supplied to the load through slip rings and brushes mounted on the shaft. The brushes are held in place and supported by brush holders.

AC Alternator Output Voltage

The voltage output of an AC alternator must be kept constant. Most loads are designed to operate at the specified voltage of the power system. A voltage that is too low can cause loads to draw excessive current. A voltage that is too high can cause damage to equipment. Output voltages of commercial AC generators are typically in the range of 2 kV through 4 kV, but can have output voltages as high as 22 kV.

The three factors controlling the output voltage of an alternator are the length of wire, the speed of the rotor, and the strength of the magnetic field. The wires are wound around the pole and can be very long. These wires are cut by magnetic lines of force and each turn adds to the voltage. In other words, the voltage is proportional to the number of turns.

When an alternator is designed, its speed and coil geometry are fixed. The only factor that can be controlled in the field is the flux density in the rotor. At any given speed, an increase in flux density produces an increase in voltage output.

The *field windings* are magnets or stationary windings used to produce the magnetic field in an alternator or motor.

Definition

A *revolving-rotor alternator* is an alternator where a fixed magnetic field is created in the stator, with the rotor turning within the stator, and AC power is supplied through the rotor slip rings and brushes.

The *field windings* are magnets or stationary windings used to produce the magnetic field in an alternator or motor.

Tech Fact

Most of the alternators in service today are self-excited, with the output from the stator connected to the voltage regulator. The output is controlled this way to protect the alternator from overloads. For a separately excited alternator, the load does not affect the exciter output and the alternator could be damaged by a large overload.

The magnetic field can be produced by permanent magnets or by electromagnets. Most alternators use electromagnets, which must be supplied with current. In a common design of a revolving-field alternator, DC is supplied to the rotor windings through the brushes and slip rings. As current is supplied to the rotor, north and south poles are created. The field at the poles expands outward into the stator windings and cuts the stator conductors. This induces AC voltage and current in the stator conductors.

The current from the exciter generator can be varied to change the strength of the north and south poles in the rotor. This changes the flux density and the output voltage from the stator. A voltage regulator is used to control the main alternator voltage. The voltage regulator senses the output from the main alternator and varies the field supply of the small alternator that supplies current to the electromagnets.

AC Alternator Output Frequency and Rotor Speed

The frequency output of an AC alternator must be kept constant. Many loads are designed to operate at the specified frequency of the power system. A frequency that is too low can cause electric motors to run slow and draw excessive current. A frequency that is too high can cause electric motors to run too fast and overheat. Some devices, such as transformers, draw high current when the frequency is reduced, thereby lowering their efficiencies and reducing their ability to deliver rated capacities. In North America, electrical systems are designed for 60 Hz. In Europe and many other parts of the world, electrical systems are designed for 50 Hz.

The two factors controlling the output frequency of an alternator are the number of poles and the speed of the rotor. The number of poles is determined when the alternator is designed and usually cannot be changed in the field. The speed of the rotor must be precisely controlled to control the output frequency.

With a two-pole AC alternator, each mechanical degree of rotation equals one electrical degree. **See Figure 9-6.** One complete revolution generates one AC cycle. AC generator rotor speed is calculated by applying the following formula:

$$S = \frac{120 \times f}{P}$$

where
S = speed of rotor rotation (in rpm)
f = frequency (in Hz)
P = number of poles

A conversion factor of 120 is used to convert to frequency in hertz. For example, the speed of a two-pole AC generator operating at 60 Hz is calculated as follows:

$$S = \frac{120 \times f}{P}$$

$$S = \frac{120 \times 60}{2}$$

$$S = \frac{7200}{2}$$

$$S = \mathbf{3600\,rpm}$$

Alternator Frequency and Rotor Speed

Poles per Phase	Rotor Speed to Deliver 60 Hz*
 2	$S = \frac{120 \times f}{P}$ $S = \frac{120 \times 60}{2}$ $S = 3600$ rpm
 4	$S = \frac{120 \times f}{P}$ $S = \frac{120 \times 60}{4}$ $S = 1800$ rpm
 6	$S = \frac{120 \times f}{P}$ $S = \frac{120 \times 60}{6}$ $S = 1200$ rpm

* in rpm

Figure 9-6. The alternator frequency is determined by the rotor speed and number of poles.

Single-Phase AC Alternators

Single-phase alternators are generally small units that are used for backup or standby residential power or on a job site. A single-phase AC alternator consists of a stationary magnet and a single winding that is rotated in the field. When the rotor is turned in a magnetic field, a voltage is induced into a loop of wire. Each rotation in a single-phase AC alternator produces one phase of electricity. **See Figure 9-7.**

In a revolving-rotor alternator design, the rotor turns in the magnetic field and AC power is brought out through the slip rings and brushes. Each half of the rotor coil cuts the magnetic lines of force in a different direction. For example, as the rotor rotates, one half of the coil cuts the magnetic lines of force from the bottom up to the left, while the other half of the coil cuts the magnetic lines of force from the top down to the right. The voltage induced in one side of the coil is opposite to the voltage induced in the other side of the coil. The voltage in the lower half of the coil enables current flow in one direction, and the voltage in the upper half enables current flow in the opposite direction.

However, since the two halves of the coil are connected in a closed loop, the voltages add to each other. The result is that the total voltage of a full rotation of the rotor is twice the voltage of each coil half. This total voltage is obtained at the brushes connected to the slip rings, and may be applied to an external circuit. The slip rings do not reverse the polarity of the output voltage produced by the generator. The result is an alternating sine wave output.

In position A, before the rotor begins to rotate, there is no voltage and no current in the external load circuit because the rotor is not cutting across any magnetic lines of force (0° of rotation). At the instant movement begins, the rotor moves parallel with the magnetic field. Because the rotor does not cut the magnetic lines of force while moving parallel to the field, no voltage is induced.

As the rotor rotates from position A to position B, each half of the rotor cuts across the magnetic lines of force, producing current in the external circuit. The current increases from zero to its maximum value in one direction. The maximum voltage is generated at 90° because the movement is crossing the greatest number of magnetic lines of force. This changing value of current is represented by the first quarter (0° to 90° of rotation) of the sine wave.

As the rotor rotates from position B to position C, current continues in the same direction. The current decreases from its maximum value to zero. Zero voltage is generated at 180° because the movement is parallel to the magnetic lines of force. This changing value of current is represented by the second quarter (90° to 180° of rotation) of the sine wave.

As the rotor continues to rotate from position C to position D, each half of the coil cuts across the magnetic lines of force in the opposite direction. This changes the direction of the current. During this time, the current changes from zero to its maximum negative value. The maximum voltage is generated at 270° because the movement is crossing the greatest number of magnetic lines of force. This changing value of current is shown by the third quarter (180° to 270° of rotation) of the sine wave.

As the rotor continues to rotate from position D to position E (position A), the current changes from its maximum negative value to zero. Zero voltage is generated at 360° because the movement is parallel to the magnetic lines of force. This changing value of current is represented by the fourth quarter (270° to 360° of rotation) of the sine wave. This completes one 360° cycle of the sine wave.

Tech Fact

A sine wave is used to represent electric current. The magnitude of the sine wave represents the magnitude of the current. The sign (positive or negative) of the sine wave represents the direction of the current.

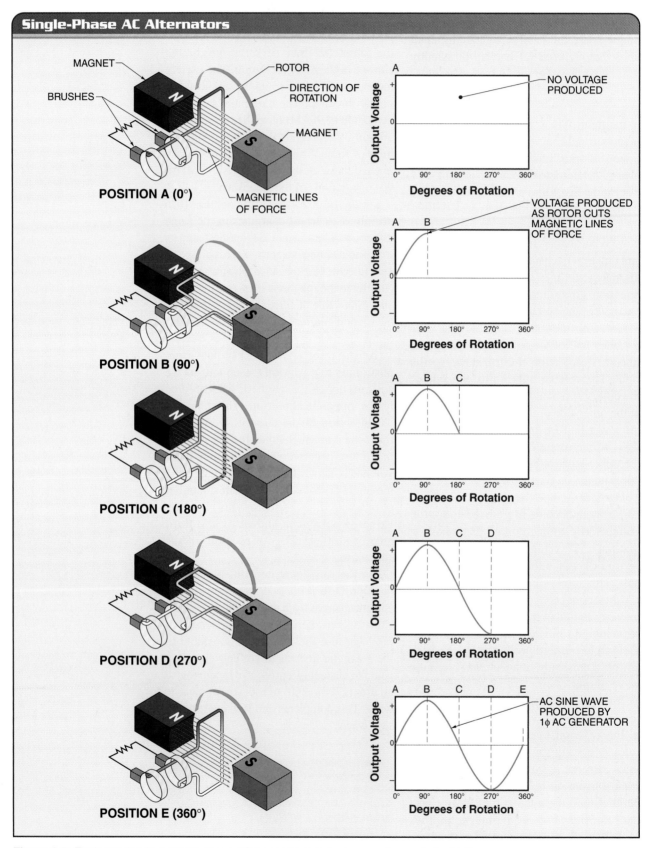

Figure 9-7. Each rotation in a single-phase AC generator produces one phase of electricity.

Three-Phase AC Alternators

Three-phase alternators are normally used by utilities, large industrial facilities, or medical facilities that normally draw 3-phase power. Three-phase AC power is generated by rotating three coils through the magnetic field. Each rotation of the rotor produces three separate AC cycles that are 120° out of phase.

To obtain three phases, 3-phase alternators have three sets of rotor windings that are located mechanically 120° apart. Generation of 3-phase power with a revolving-rotor alternator design is very similar to generation of single-phase power. Each rotation in a three-phase AC alternator produces all three phases. **See Figure 9-8.** As the rotor turns, each phase is brought out to its own set of slip rings. The phases are generated so that phase 1 leads phase 2 by 120° and phase 3 by 240°.

With high-power AC generators that have high current output, revolving-field alternator designs are used. A field is created in the rotor, either through the use of brushes and slip rings or through the use of brushless exciters. As the rotor turns, the field rotates and cuts the stator windings. AC power is transferred through fixed wires connected to the windings.

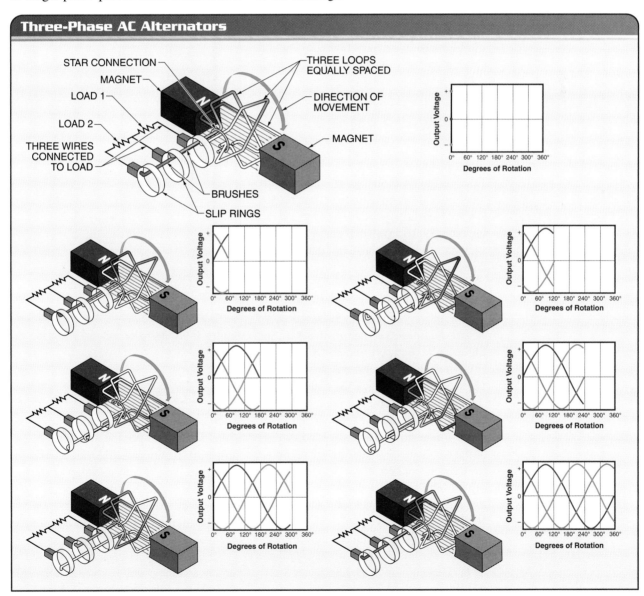

Figure 9-8. Each rotation in a three-phase AC generator produces all three phases.

Definition

A *prime mover* is the power source used to create the relative motion between the coil and the magnetic field.

The *overload rating* of an alternator is the load above the normal load that can be carried for a specified period.

Prime Movers

A *prime mover* is the power source used to create the relative motion between the coil and the magnetic field. Large alternators are often powered by steam turbines, which use natural gas, nuclear energy, coal, or oil as fuel to generate steam. Waste landfills are often a source of energy, providing combustible methane gas to operate turbines that would otherwise be run using fossil fuels. Hydropower at large dams can also be used to drive turbines.

Standby or backup power units are remote units that provide emergency power to critical facilities when there is a problem with the utility power supply. They are usually powered by engines fueled by natural gas, diesel fuel, or, in the case of small units, gasoline.

ALTERNATOR RATINGS

Alternators are rated according to the load they can maintain on a continuous basis. The *overload rating* of an alternator is the load above the normal load that can be carried for a specified period. The overload rating is based on the internal temperature the alternator can withstand. Current flow is the largest cause of heat rise and an alternator's rating is closely tied to its temperature-rise limitations.

Alternators are rated in kilowatts (kW) at a standard power factor of 80%. **See Figure 9-9.** Therefore, the kilovolt ampere (kVA) rating of the alternator is equal to 125% of the kW rating. This needs to be accounted for when considering intermittent overloads and available fault currents. AC alternators come in many sizes, depending on the power requirements to be met. Small residential backup power systems may deliver only about 1000 watts, while large utility alternators may deliver up to about 1000 megawatts.

Alternators have a maximum allowed current. The maximum current is limited by the heat buildup in the alternator. Small AC alternators can be cooled by air flowing through openings in the stator. Large AC alternators can be cooled by air, water, or hydrogen. Hydrogen is expensive, but it is sometimes used because of its superior heat transfer properties. The alternator is sealed to prevent the loss of the hydrogen. Because hydrogen is much less dense than air, windage losses on the rotor are also much lower.

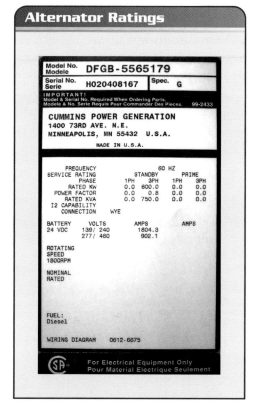

Figure 9-9. Alternator ratings are given on the nameplate.

PARALLELING ALTERNATORS

Alternators are placed in parallel to carry added load when needed, to enhance reliability by having multiple alternators in operation, and to provide power during maintenance of other alternators. Starting up an electricity-generating power plant is much more complex than starting up a machine in an industrial plant.

All the alternators must be connected in parallel and the power from each alternator must be at exactly the right phase, frequency, and voltage before it can be connected to the busbars. If the voltage phases are not matched, large opposing voltages are developed. This causes high currents to flow that can damage the alternators. Phase voltages must be connected A to A, B to B, and C to C. If there is a frequency mismatch, an alternator turning more slowly is a load to the other alternators. If the voltages are not equal, one of the alternators could become a reactive load to the other generators. This causes high currents to circulate between the alternators and can cause extensive damage.

Manual Synchronization

Parallel alternators are often started up manually. In order to manually synchronize an alternator, there must be a way to ensure that the alternator phase, frequency, and voltage match the busbars before the alternator is switched into the circuit. In actual practice, the alternator should have a slightly higher frequency than the power on the busbars. It is easier for the alternator rotor to slow down than to speed up. Common methods of manual synchronization include the use of a synchroscope or the use of synchronizing lights.

Synchroscopes. A synchroscope can be used to help match the frequency and phase of an alternator to the busbar, or to match two alternators. **See Figure 9-10.** A *synchroscope* is a device that indicates whether two AC sources to be connected in parallel are in the correct phase relationship. A synchroscope is connected to both potentials. When the needle on the synchroscope is stationary at the twelve o'clock position, the frequencies and phases are matched. When the needle turns clockwise, the incoming frequency is higher than the operating frequency. When the needle turns counterclockwise, the incoming frequency is lower than the operating frequency. When the synchroscope shows that the units are operating at the correct frequency, a switch can be closed that brings the oncoming unit on-line at the right time.

Lights-Out Methods. Other methods of synchronizing generators are the various lights-out methods. These methods connect a lamp across each of the phases of two alternators or between an alternator and the power busbars. **See Figure 9-11.** The voltage across the lamps is the potential difference between the phases of the alternators. When the phase rotation between the alternators is not matched, the lamps flicker on and off, but not in unison. This means that phase connections on the incoming source are incorrect and must be reversed.

When the peak voltages of the two alternators match, the lamps go out because there is no longer any difference in potential between the two alternators. Two common lights-out methods are the three-lights-out method and the one-light-out method.

In the three-lights-out method, a lamp is connected as a load across each phase of the two sources, A to A, B to B, and C to C. When all three lamps are dark, the potentials of both sources are the same and the second source can be brought on-line. The phase connection is incorrect if the lights do not all turn on and off at the same time. The one-light-out method is very similar except that the phases are connected A to B, B to A, and C to C. When the C to C lamp is dark and the other two are brightly lit with no flickering, the alternators are synchronized and the second source can be brought on-line.

> **Definition**
>
> A *synchroscope* is a device that indicates whether two AC sources to be connected in parallel are in the correct phase relationship.

Figure 9-10. A synchroscope can be used to help match the frequency and phase of an alternator to the busbar.

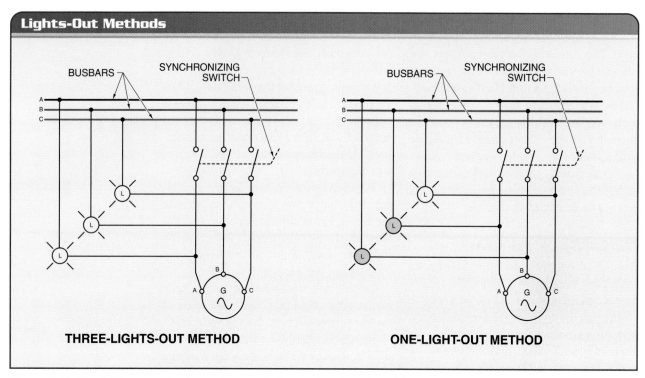

Figure 9-11. The lights-out methods use lamps to indicate when there is a difference in potential between an alternator and the busbar.

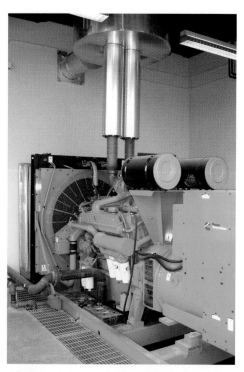

AC alternators are often used as backup power sources. They can be equipped with automatic synchronization to allow the local electric utility to start the alternator during times of high electricity demand.

Automatic Synchronization

Manual synchronization requires the presence of skilled operators to perform the startup. This is possible at a utility generating station, but it is very difficult for a facility with an emergency back-up power supply. For emergency back-up power supplies, automatic synchronization is the simplest and least expensive option for synchronizing alternators.

With automatic synchronization, a controller monitors the frequency, phase angle, and voltage of the incoming source and the busbar. The controller measures the parameters of the incoming source and the busbar and adjusts the incoming alternator to match the busbar. **See Figure 9-12.** An automatic synchronizer energizes a relay and closes a contact when the incoming source and the busbar match. In actual operation, the synchronizer takes into account the actual relay closing time. Therefore, the synchronizer anticipates when the incoming source will match and closes the relay at the appropriate time.

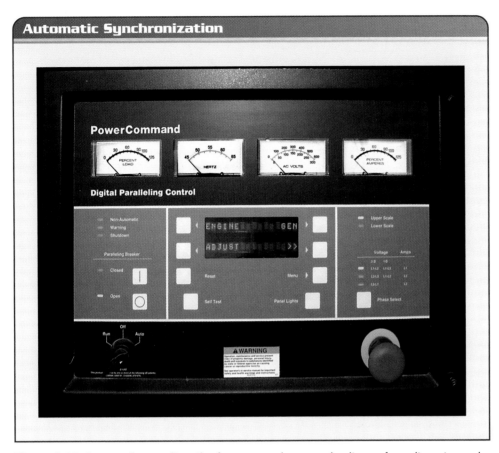

Figure 9-12. A controller monitors the frequency, phase, and voltage of an alternator and a busbar. The controller adjusts the alternator until the parameters match and then switches the alternator into the circuit.

Application—AC Alternators

Application—Selecting Portable Generators

Portable generators are used to supply electrical power in emergencies, on construction sites, for recreation, and as a backup for utility power feeds. A properly sized portable generator should deliver enough power (in watts) at the correct voltage (12 VDC, 120 VAC, 120/240 VAC) to operate all loads connected to the generator. An undersized generator cannot operate the loads as required, which can damage the load and generator. An oversized generator supplies enough power to the loads, but is not energy efficient.

Portable generators are rated by their maximum power output, surge power output, and voltage output(s). For example, a portable generator may be rated for 7500 W (7.5 kW) maximum power output and 10,750 W (10.75 kW) surge power output at 120 VAC. The surge power output rating is used to select an alternator that has enough power to handle loads that include motors with a higher starting power than running power. For example, a ½-HP air compressor requires approximately 1500 W when running and approximately 5500 W when starting.

Backup generators can supply DC, single-phase AC, or 3-phase AC. In the United States and Canada, AC power is 60 Hz. In Europe and many other parts of the world, AC power is 50 Hz. Some generators can supply 60 Hz and 50 Hz power.

Determining Generator Size

To determine the correct size of generator for an application, apply the following procedure:

1. List the electrical devices that must be operated by the generator. This list should include devices likely to be operated (lamps, power tools, etc.) as well as devices that may be added (refrigeration units during prolonged power outages, sump pumps during heavy rains, etc.). The list of devices that must be operated by an alternator determines the absolute minimum generator size that can be selected for an application. By also listing devices that may be added, the generator size will be the most appropriate for the application.

2. Determine the voltage requirements of the loads connected to the generator. The voltage output of an alternator must be within +5%/−10% of the load(s) connected to the generator. Lamps, portable power tools, and most small appliances are rated at 115/120 VAC. Most large power-consuming devices (electric stoves/heaters, air compressors, etc.) are rated at 230/240 VAC. If a device is rated for only one voltage, it cannot be connected to any other voltage. For example, a motor rated for 230 VAC must be connected to a 230-VAC (+5%/−10%) power supply. However, a motor rated for 115/230 VAC may be connected to either a 115-VAC or 230-VAC power supply.

Baldor Electric Co.

Maximum Power*	Surge Power*	Voltage Output
3500	7500	120 VAC GFCI Duplex Outlet
4500	9000	120 VAC GFCI Duplex Outlet 120/240 VAC Twistlock Outlet
7500	10,750	12 VDC (2) 120 VAC GFCI Duplex Outlets 120/240 VAC Twistlock Outlet
10,000	17,000	12 VDC (3) 120 VAC GFCI Duplex Outlets 120/240 VAC Twistlock Outlet
12,500	21,000	12 VDC (4) 120 VAC GFCI Duplex Outlets 120/240 VAC Twistlock Outlet

* in W

A 115/230-VAC motor (or other load) draws the same amount of power when connected to 115 VAC or 230 VAC. However, the current draw at the higher voltage (230 VAC) is half that of the current draw at the lower voltage (115 VAC) because power (P) is equal to voltage (E) times current (I). Thus, wiring a dual-voltage load at a higher voltage reduces the required wire size and increases the permissible wire length, but does not reduce (or increase) the power output.

3. Determine the operating power requirements of each load and the total power requirements of all the loads that are connected to the generator. Some devices, such as lamps and heating elements, have a power listing in watts (100 W, 500 W, etc.). Other loads such as motors, some tools, and appliances do not have a power rating listed but have a current rating listed. The power rating of a load that has a listed voltage and current rating can be determined by multiplying the rated voltage by the rated current.

Identify all the loads to be connected to the generator that have a starting power draw higher than the operating power. This usually includes all loads that contain a motor. In general, any load that includes a motor that is connected directly to a load can be calculated as having a high starting-power requirement. This includes motors that are directly coupled through belts, chains, and gears to loads such as fans, pumps, and tools. However, small motors such as those used in portable cooling fans and cooling motors in computers, etc., need not be considered as having a high starting-power draw.

Problems occur if the high starting-power requirements of loads are not factored into the generator size requirements. The problem occurs when a large amount of starting power is drawn from the generator, as with air compressor starting, etc. In this case, the total voltage output of the generator drops if the generator is fully loaded. The larger the overload, the larger the voltage drop. Even a temporary low voltage on computers, appliances, tools, etc., can cause damage to the loads and generator.

4. The wattage rating listed on a load nameplate can be used to determine the total wattage requirements of an alternator. When load nameplate ratings are not available, average wattage requirement guides can be used to determine total wattage requirements.

5. Once the total power requirements are determined, an alternator can be selected. The minimum generator size should be based on the maximum possible power draw for the given application and the amount of extra (spare) power desired. Typically, 25% to 50% additional power is recommended as a minimum amount of extra power. This ensures that an alternator does not operate at 100% capacity, which shortens generator life and/or lowers the generator voltage output, and there is extra power for the times that may require additional power.

If the total power required is high for any one application, such as a large construction site, etc., it is better to use more than one generator and divide the loads between the generators. This also ensures that there is available power if one of the generators malfunctions.

Electrical Load	Operating Power*	Starting Power*
Construction site		
Air Compressor (½ HP)	1500	5500
Air Compressor (1 HP)	3000	11,000
Electric Welder (200 A)	9000	9000
High-Pressure Washer (1 HP)	1200	3600
Circular Saw (7¼″)	1400	2300
Table Saw (10″)	1800	4500
Hand Drill (½″)	600	800
Grinder (4½″)	750	950
Grinder (6″)	1000	1300
Grinder (9″)	2300	3000
Hand Jigsaw	650	850
Reciprocating Saw (7″ blade)	1150	1600
Sander (⅓ sandpaper sheet size)	350	550
Sander (½ sandpaper sheet size)	450	650
Battery Charger (15 A, no boost)	375	375

* in W

Summary

- An alternator converts mechanical torque into AC power.

- The two types of alternators are the revolving-field alternator and the revolving-rotor alternator.

- As the rotor turns, moving coils of conductor pass through a magnetic field or a moving magnetic field cuts through the conductors, depending on the design.

- A revolving-field alternator has the field in the rotor and brings AC power out to the load through the fixed stator windings. These are usually used in high-power applications.

- A revolving-rotor alternator has the field in the stator and brings AC power out to the load through the slip rings and brushes. These are usually used in low-power applications.

- The output voltage of an alternator is controlled by a voltage regulator. The voltage regulator varies the current through the field windings, which changes the output voltage.

- The output frequency of an alternator is determined by the rotor speed.

- In a single-phase alternator, a single coil is used to generate the single phase. In a three-phase alternator, three coils are used to generate the three phases.

- Alternators are rated in kilowatts (kW) at a standard power factor of 80%. The kilovolt ampere (kVA) rating is 125% of the kW rating.

- The maximum current is limited by the heat buildup in the alternator.

- When paralleling alternators, each alternator must be at the right phase, frequency, and voltage before being switched into the circuit.

- When manually synchronizing alternators, a synchroscope can be used to compare the phases of an alternator and the busbar.

- The lights-out methods of synchronizing alternators use lamps to compare the phase of the alternator to the busbar. When the two sources are in phase, a darkened lamp shows that there is no difference in potential.

- Automatic synchronizers use a controller to monitor the frequency, phase angle, and voltage. Adjustments can be made to the alternator before it is brought on-line.

Glossary

An **alternator** is a synchronous machine that produces alternating current (AC).

A **stator** is the fixed unmoving part of a generator, consisting of a core and windings, that converts the energy of a magnetic field to electrical energy.

A **rotor** is the rotating moving part of an alternator or generator, consisting of a core and windings, that convert torque to magnetic energy.

Torque is a turning or twisting force that causes an object to rotate.

A **revolving-field alternator** is an alternator where a magnetic field is created in the rotor, which turns within the fixed stator windings, and AC power is supplied through the stator windings.

An **exciter generator** is an assembly consisting of a small three-phase alternator used to supply current to an alternator rotor.

A **brushless exciter** is a rectifier assembly mounted on the main rotor shaft along with the exciter generator.

A **revolving-rotor alternator** is an alternator where a fixed magnetic field is created in the stator, with the rotor turning within the stator, and AC power is supplied through the rotor slip rings and brushes.

The **field windings** are magnets or stationary windings used to produce the magnetic field in an alternator or motor.

A **prime mover** is the power source used to create the relative motion between the coil and the magnetic field.

The **overload rating** of an alternator is the load above the normal load that can be carried for a specified period.

A **synchroscope** is a device that indicates whether two AC sources to be connected in parallel are in the correct phase relationship.

Review

1. List the parts of an alternator stator.

2. List the parts of an alternator rotor.

3. Explain how centrifugal force influences the rotor design.

4. Describe the difference between a revolving-field alternator and a revolving-rotor alternator.

5. Explain how a voltage regulator works to control the voltage in an alternator.

6. Explain what happens as a rotor turns in an alternator. Describe the four parts of the rotor movement during one rotation. Explain how the rotation results in AC power.

7. Explain how a synchroscope is used in paralleling alternators.

8. Describe how the lights-out methods are used in paralleling alternators.

9. Describe how automatic synchronization is used in paralleling alternators.

Refer to the CD-ROM for Quick Quiz® questions related to chapter content.

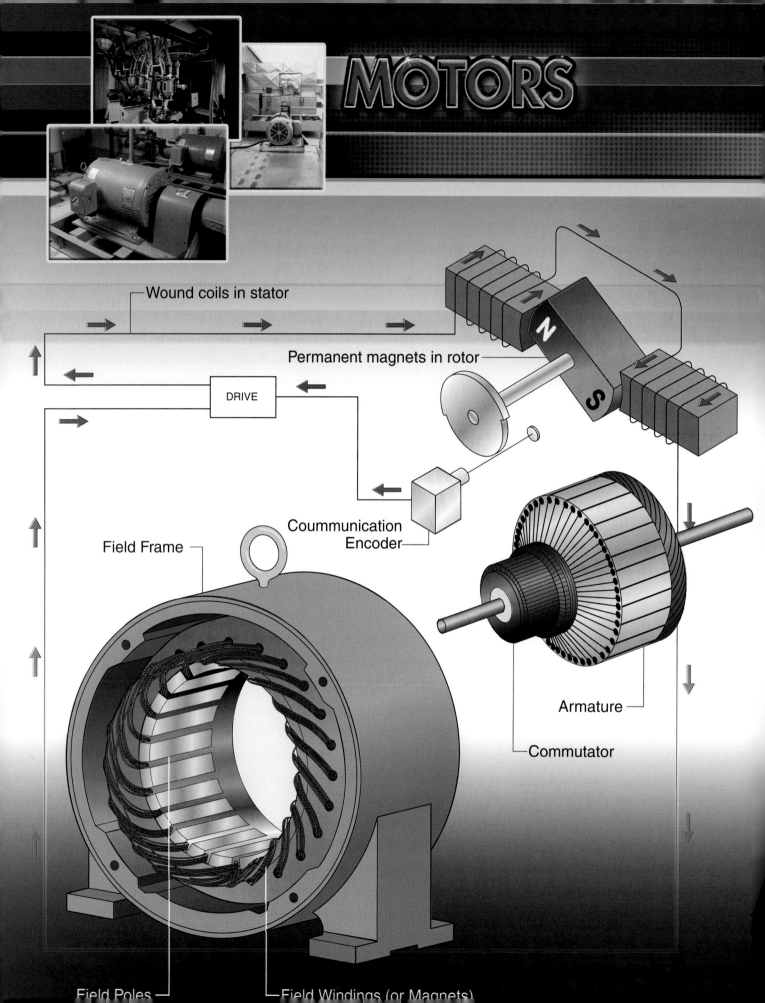

10

DC Motors and Generators

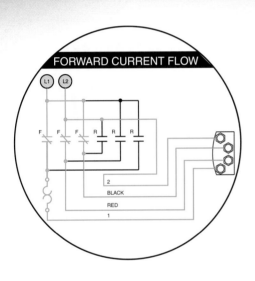

DC Motors and Generators 208
DC Motor Types 217
DC Generators 224
DC Generator Types 227
Supplemental Topic—DC Brushless Motors 229
Supplemental Topic—DC Coreless Motors 230
Application—DC Motor Reduced-Voltage Starting 231
Application—Reversing DC Series Motors 232
Summary 233
Glossary 234
Review 235

OBJECTIVES

- Explain how a DC motor and a DC generator are similar and how they are different.
- Explain how a commutator and brushes work to develop torque.
- Describe the construction of a DC motor armature.
- Describe armature reaction and explain how it moves the neutral plane.
- Describe how the differences in winding connections result in the different types of DC motors.
- Describe the torque curves of the four types of DC motors. Use these curves to explain why certain types of motors are used in particular types of applications.

DC motors and DC generators are very similar to one another. A DC motor converts a DC power source into torque that is used to drive a load. A DC generator reverses the process and converts torque provided by a prime mover into DC power. The armature and field windings can be wired in different ways, resulting in the different types of DC motors and generators. Each type of DC motor has different torque and speed characteristics.

DC MOTORS AND GENERATORS

A *direct current motor,* or *DC motor,* is a machine that uses DC connected to the field windings and armature to produce shaft rotation. A *direct current generator,* or *DC generator,* is a power source that supplies DC when the armature is rotated. DC motors and generators consist of field windings in the field frame, an armature, and a commutator and brushes on the shaft. **See Figure 10-1.**

Connecting voltage directly to the field and armature of a DC motor allows the motor to produce higher torque in a smaller frame than with an AC motor. DC motors provide excellent speed control for acceleration and deceleration, with effective and simple torque control. DC motors perform better than AC motors in most traction equipment applications. DC motors do require more maintenance than AC motors because they have brushes that wear. DC motors are used as drive motors in mobile equipment such as golf carts, quarry and mining equipment, and locomotives. In addition, DC motors were used for many years in applications where speed control was needed. In many of these applications, they have been replaced with AC motors and adjustable-speed drives.

The armature and the field can be wired in different ways, resulting in the different types of DC motors. These different types of motors are used to provide different amounts of torque or different types of speed control.

> **Definition**
>
> A *direct current motor,* or *DC motor,* is a machine that uses DC connected to the field windings and armature to produce shaft rotation.
>
> A *direct current generator,* or *DC generator,* is a power source that supplies DC when the armature is rotated.

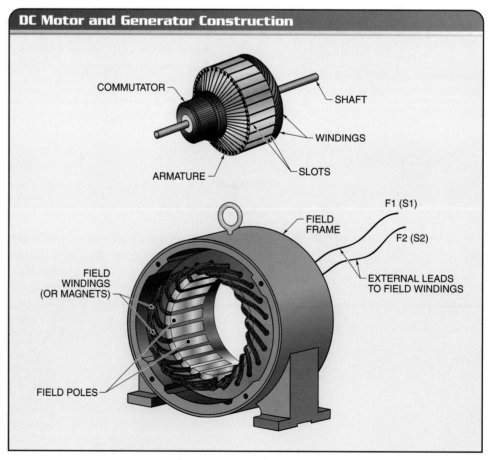

Figure 10-1. DC motors consist of field windings in the field frame, an armature, and a commutator and brushes on the shaft.

Field Frame Construction

A *field frame* is the stationary part in a DC motor or generator. The stationary part is called a stator in an AC motor, but is called a field frame in a DC motor. The *field poles* are metal pieces mounted to the field frame that are used as field windings. The field poles are constructed of thin sheets of steel laminated together, similar to the construction of AC motors. **See Figure 10-2.** The *field windings* are magnets or stationary windings used to produce the magnetic field in an alternator or motor. In most cases, the field windings are made by coiling wire around the field poles. In some cases, interpoles are added between the main poles.

Interpoles. *Interpoles* are auxiliary poles placed between the main field poles of the motor. **See Figure 10-3.** The interpoles are connected in series with the armature windings, with one terminal of the interpole connected to the brushes and one brought out to connect to the DC power supply. The interpoles are made with larger size wire than the main field poles, in order to carry armature current. They are smaller in overall size than the main field poles because they require fewer windings. Interpoles are also known as commutating field poles.

Interpoles are used to reduce sparking at the brushes of larger DC motors. They are used with shunt and compound DC motors of one-half HP or more. The interpoles reduce sparking at the brushes by helping to overcome the effect of armature reaction.

In motors, an interpole must be of the same polarity as the main pole preceding it in the direction of rotation. In generators, an interpole must be of the opposite polarity as the main pole preceding it in the direction of rotation. In a motor, the interpole polarity provides flux of the same direction as the main pole, but in a direction opposite to that of the flux produced by the self-induced reactance. This cancels the flux in the area where the commutated coil is moving at the neutral plane.

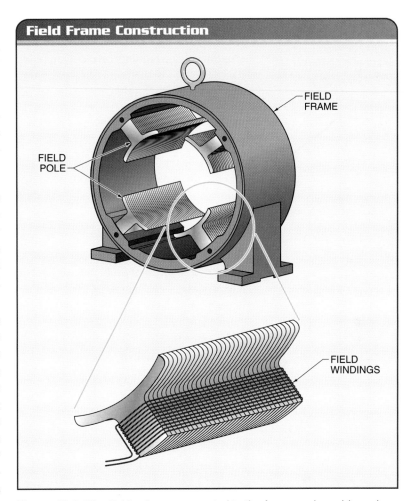

Figure 10-2. The field poles are mounted to the frame and provide a place for the field windings.

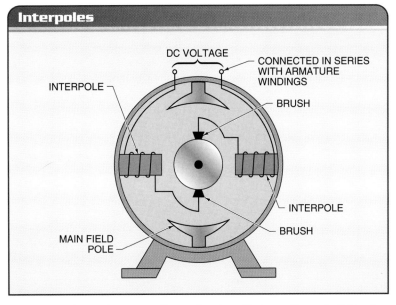

Figure 10-3. Interpoles are auxiliary poles placed between the main field poles.

Definition

A *field frame* is the stationary part in a DC motor or generator.

The *field poles* are metal pieces mounted to the field frame that are used as field windings.

The *field windings* are magnets or stationary windings used to produce the magnetic field in an alternator or motor.

Interpoles are auxiliary poles placed between the main field poles of the motor.

Compensating windings, or *pole-face windings*, are field windings placed in slots on the main poles.

An *armature* is the rotating part of a DC motor, consisting of the laminated core with slots for the coils, the main shaft, and the commutator and brushes.

Compensating Windings. *Compensating windings*, or *pole-face windings*, are field windings placed in slots on the main poles. The purpose of the compensating windings is to help interpoles maintain neutral position on the commutator. Compensating windings look like rotor bars in an induction motor. The compensating windings are in series with the armature, the series field, and the interpoles. They are found in large motors that have a wide range of speeds.

Armature Construction

An *armature* is the rotating part of a DC motor, consisting of the laminated core with slots for the coils, the main shaft, and the commutator and brushes. **See Figure 10-4.** A magnetic field is produced in the armature by current flowing through the armature windings. The armature magnetic field interacts with the DC produced by the field windings. The interaction of the magnetic fields causes the armature to rotate.

Commutators. A *commutator* is a ring made of insulated segments that keep the armature windings in the correct polarity to interact with the main fields. Each armature winding has two ends, with each end connected to one section of the commutator. This means that an armature with 8 coils has 16 commutator segments.

The commutator is mounted on the same shaft as the armature and rotates with the shaft. Each section of the commutator is isolated from the adjacent sections by mica insulation. A mica clamping flange holds the segments together to form a cylinder on the shaft. In a common design, each vertical section of the commutator has a riser where the leads from the armature are attached.

Brushes. A *brush* is the sliding contact that rides against the commutator segments and is used to connect the armature to the external circuit. **See Figure 10-5.** Brushes are made of carbon or graphite material and are held in place by brush holders. A *pigtail* is an extended, flexible connection or a braided copper conductor. The *brush rigging* is the entire assembly of the brush, brush holder, insulators, and any wiring included in the assembly. The brush rigging is mounted on the rear endbell. On some motors there may be as many brush holders as there are main field poles.

Pigtails are used to connect a brush to either the positive or the negative terminal of the DC source. Since each DC source has one positive and one negative terminal, there must be at least two brushes and two brush holders. Each brush keeps the same polarity at all times. For each pair of brushes, one brush is always positive and one brush is always negative.

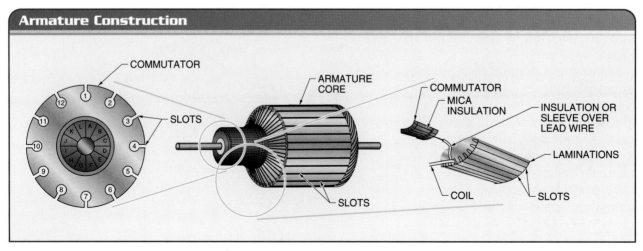

Figure 10-4. The armature consists of the laminated core, the armature coils, the main shaft, and the commutator and brushes.

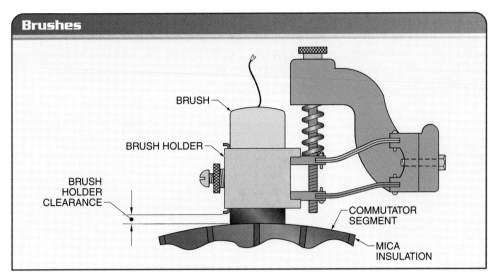

Figure 10-5. A brush is a sliding contact that rides against the commutator.

Definition

*A **commutator** is a ring made of insulated segments that keep the armature windings in the correct polarity to interact with the main fields.*

*A **pigtail** is an extended, flexible connection or a braided copper conductor.*

*The **brush rigging** is the entire assembly of the brush, brush holder, insulators, and any wiring included in the assembly.*

Brushes are free to move up and down in the brush holder. This freedom allows the brush to follow irregularities in the surface of the commutator. A spring placed behind the brush forces the brush to make contact with the commutator. The spring pressure is normally adjustable, as is the entire brush holder assembly. The brushes make contact with successive copper bars of the commutator as the shaft, armature, and commutator rotate.

Brushes should be large enough to carry the armature current without heating up. Three styles of brushes are radial, trailing-edge, and leading-edge brushes. Radial brushes point in to the commutator at 90° and are used for DC motors that may need to be reversed often. Leading-edge brushes are at an angle to the commutator and have less restricted up-and-down movement. Trailing-edge brushes hold the long side of the brush tight against the brush holder. Any maintenance of brushes should use an identical replacement brush. It is best to check with the manufacturer about the type of brush to be used when it is necessary to change a brush.

Tech Fact

Commutators build up a film on the copper surface. The film acts as a low-friction layer between the commutator and the brushes and helps minimize wear.

Operating Principles

Any current-carrying wire is surrounded by a magnetic field. The left-hand conductor rule gives the direction of the magnetic field. When this wire is placed into a stationary magnetic field, the two fields interact and the wire moves at right angles to the stationary magnetic field. The right-hand motor rule gives the direction of motion. In a DC motor, the current-carrying wire is the armature conductor. The stationary magnetic field is the field frame. The left-hand generator rule gives the direction of current flow when the armature is turned within a magnetic field. **See Figure 10-6.**

When voltage is applied to the motor, current flows through the field winding. This sets up a magnetic field in the field winding. At the same time, current flows through the brushes, commutator, and armature windings. The current flows through the armature windings to another commutator segment on the opposite side of the first segment, out a brush opposite to the first, and back to the source. This sets up a magnetic field in the armature.

There are at least two poles in the field frame and an equal number in the armature. The armature begins to rotate as the unlike poles in the field frame and armature attract each other and the like poles repel each other.

Definition

A *countervoltage*, or *counter EMF (CEMF)*, is a voltage induced in the windings that is opposite in polarity to that of the power supply.

Armature reaction is the distortion of the magnetic fields that happens when a current-carrying wire is placed within a fixed magnetic field.

As the armature begins to rotate, a countervoltage is induced in the armature windings. A *countervoltage,* or *counter EMF (CEMF),* is a voltage induced in the windings that is opposite in polarity to that of the power supply.

Armature Reaction. When power is applied to a DC motor, both the armature windings and the field windings generate magnetic fields that become distorted when they interact. *Armature reaction* is the distortion of the magnetic fields that happens when a current-carrying wire is placed within a fixed magnetic field. See **Figure 10-7.** The distorted magnetic fields apply torque to the armature and cause movement. Interpoles are used to cancel the flux in the commutated coil and to control armature reaction.

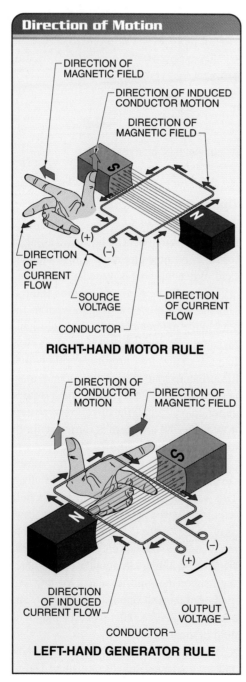

Figure 10-6. The right-hand motor rule and the left-hand generator rule give the direction of armature motion.

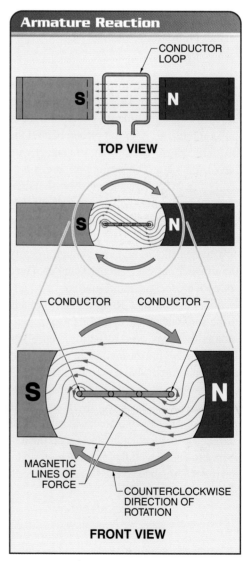

Figure 10-7. The magnetic fields become distorted when one magnetic field is placed within another.

Neutral Plane and Brush Neutral. The *neutral plane* is a line through the armature cross section that is perpendicular to the maximum amount of magnetic flux. At the neutral plane, the armature conductors are parallel to the magnetic flux and no voltage is induced in the armature. The neutral plane can be moved by the reaction of the main fields with the armature fields as a result of armature reaction. When there is no current in the armature, the magnetic field between the field poles is undistorted and the neutral plane is at right angles to the field flux. When there is current in the armature, the magnetic field between the field poles is distorted and the neutral plane is at an angle to the original position. **See Figure 10-8.**

Brush neutral is the position of the brushes where commutation can occur with minimal induced voltage in the armature coils. Brush neutral occurs where the conductors in the armature are operating parallel to the flux of the main fields and see the fewest lines of flux. When a motor or generator has been disassembled for repair, part of the reassembly must be to reset the brush neutral. This adjustment is possible when the brush rigging is movable. On small motors, the brush rigging sometimes is set and cannot be changed. On larger motors, the brush rigging is movable and can be the cause of motor faults.

Torque. The armature windings, commutator, and brushes are arranged so that the flow of current is in one direction in the loop on one side of the armature, and the flow of current is in the opposite direction in the loop on the other side of the armature. Torque is exerted on the armature when it is positioned so that the plane of the armature loop is parallel to the field, and the armature loop sides are at right angles to the magnetic field. **See Figure 10-9.**

Definition

*The **neutral plane** is a line through the armature cross section that is perpendicular to the maximum amount of magnetic flux.*

***Brush neutral** is the position of the brushes where commutation can occur with minimal induced voltage in the armature coils.*

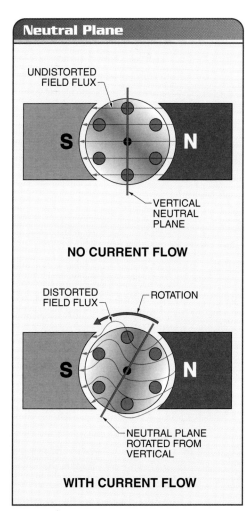

Figure 10-8. Armature reaction moves the neutral plane.

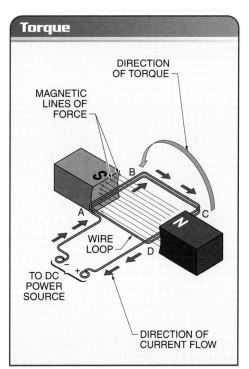

Figure 10-9. Torque is exerted on the armature loop when the plane of the loop is parallel to the magnetic flux.

Definition

Commutation *is the process where the armature current is periodically reversed in order to keep the motor torque in the same direction during the entire armature rotation.*

No movement takes place if the armature loop is stopped in the vertical (neutral) position. In this position, no further torque is produced because the forces acting on the armature are upward on the top side of the loop and downward on the lower side of the loop. Because of inertia, the armature does not stop. The armature continues to rotate for a short distance. The commutator reverses the polarity of the brushes and armature wiring. This keeps the armature rotating. As a practical matter, most motors have many poles. This helps even out the torque and speed and prevents the armature from stopping in the neutral position.

Armature windings are made of many turns of wire with the ends connected to the commutator. The multiple conductors in the slots increase the number of magnetic lines of flux that interact with the stationary flux field, producing more torque. The torque produced in the motor is the total force developed by all the conductors in the armature acting through the radius of the motor shaft. The effective torque is produced at a right angle at a radial distance from the center rotation of the shaft. This torque is measured in pound-inches or pound-feet.

Commutation. *Commutation* is the process where the armature current is periodically reversed in order to keep the motor torque in the same direction during the entire armature rotation. In other words, commutation reverses the current in the armature coils as they leave the influence of one field pole and enter the influence of another, opposite field pole, establishing unidirectional torque.

Commutation can be electronic, as when the current flowing through the armature coils is switched by a brushless exciter or other electronic devices. Electronic commutation can be controlled by the motor drive. Commutation can also be mechanical, as when the current through the armature coils is sequentially switched by the brushes as the commutator and armature rotate. The commutator segments rotate as the armature rotates.

With mechanical commutation, a commutator segment becomes positive when it is touching a positive brush and becomes negative when it is touching a negative brush. As the armature rotates and the brushes move from one commutator segment to another, the commutator segments change polarity from positive to negative and back to positive. The brushes momentarily contact more than one commutator segment.

The fields in a DC motor never change in polarity. The armature windings must have the same polarity as the poles in the field frame in order for the rotor to turn. The purpose of a commutator is to change the direction of current flow as the coils leave one flux field and enter another of opposite polarity. Therefore, the current direction in the armature windings reverses direction and the poles change from positive to negative and back to positive.

For example, brush 2 breaks contact with side B of the commutator and makes contact with side A. **See Figure 10-10.** The flow of current through the commutator reverses because the flow of current is at the same polarity on the brushes at all times. This allows the commutator to rotate another 180° in the same direction. After the additional 180° rotation, brush 1 breaks contact with side B of the commutator and makes contact with side A. Likewise, brush 2 breaks contact with side A of the commutator and makes contact with side B. This reverses the direction of current in the commutator again and allows for another 180° of rotation. The armature continues to rotate as long as the commutator winding is supplied with current and there is a magnetic field.

Most commonly, the brushes contact more than one segment on the commutator. The ends of each coil are connected to adjoining segments of the commutator. As these segments pass under the brush, they are shorted together. The current should be commutated at the point where the fewest lines of flux are being cut. This is the point at which the armature conductor is moving parallel to the flux lines and the smallest amount of generated countervoltage is induced in the coils.

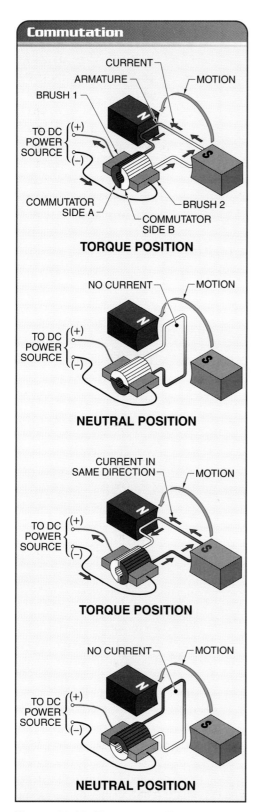

Figure 10-10. The current direction in the armature windings reverses direction and the poles change polarity when the commutator rotates under the brushes.

Even though the armature coil being commutated carries no current because it is not cutting any lines of flux, there is still a collapsing field in that coil that can induce a voltage in the coil. This voltage can cause damaging sparking at the brushes. A countervoltage in the interpoles between the main poles opposes the voltage in the armature coil and minimizes sparking.

Reversing DC Motors

A DC motor can be reversed by changing the polarity of either the armature winding or the field winding. A DC series motor cannot be reversed by changing the polarity of the source power. One of the windings must be reversed relative to the other winding. The armature winding is typically reversed because the terminals are readily available at the brush rigging.

A reversing motor starter can easily be used to reverse the motor. **See Figure 10-11.** When the forward pushbutton is pressed, the F contactor is energized, opening the normally closed F contacts and closing the normally open F contacts. This creates an interlock in the F contactor and routes the current through the armature from A1 to A2.

In order to reverse the motor, the stop pushbutton must be pressed to release the interlock. When the reverse pushbutton is pressed, the R contactor is energized, opening the normally closed R contacts and closing the normally open R contacts. This creates an interlock in the R contactor and routes the current through the armature from A2 to A1. In both forward and reverse directions, current flows through the field winding from S1 to S2. In the forward direction, current flows through the armature from A1 to A2. In the reverse direction, current flows from A2 to A1.

Most compound motors are not considered for reversing applications because as the fields are reversed, the compounding is also reversed. This situation can be overcome by using an elaborate system of contactors, or through the installation of a bridge rectifier feeding the series field.

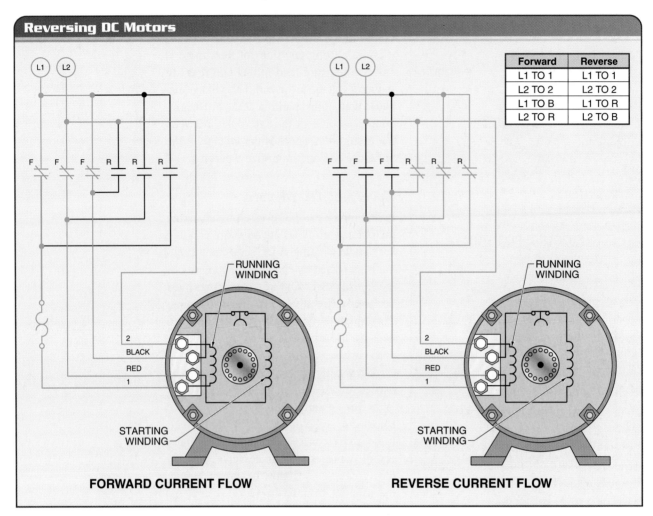

Figure 10-11. A reversing motor starter can be used to reverse a motor.

Comparing Motors and Generators. DC motors can be made into DC generators and DC generators into DC motors. If the unit worked well as a motor, it can be run as a generator with the same connections, but it must run in the opposite direction. This can be done by interchanging A1 and A2.

A DC motor can be made into a DC generator if an external prime mover is coupled to the motor shaft. The motor is driven by an external prime mover and produces an output voltage and current at its terminals. The prime mover must drive the machine that was once used as a motor in the opposite direction from the direction it ran as a motor. The polarity of the generator can be controlled by reversing the armature leads. On many motors and generators, interpoles are part of the armature circuit. Changing the armature circuit changes the polarity of the interpoles as well.

In a split-phase motors, all the leads are brought out. When working on a split-phase motor, care must be taken to ensure correct interpole polarity. Reversing current in only one side of the interpoles causes damage to the armature.

Armature current in a generator flows in the same direction as the generated voltage, but armature current in a motor is forced to flow in the opposite direction to that of the countervoltage. The armature in the generator is being turned by the prime mover, so all the voltage and current in the armature are the result of the conductors cutting the flux in the main poles. The armature in the motor is rotated by connecting the armature to a source and the conductor is repelled out of the main field by the interaction between the main field and armature field. As the armature is repelled out of the main field, the three requirements for induction are met, just as they were in a generator.

When the armature in a DC motor is rotating in a magnetic field, the countervoltage opposes the motion and acts like a generator. When the armature in a DC generator is rotating in a magnetic field, the countervoltage opposes the motion and acts like a motor. The countervoltages in each application oppose the action of the machine.

DC MOTOR TYPES

The four basic types of DC motors are DC series motors, DC shunt motors, DC compound motors, and DC permanent-magnet motors. These DC motors have similar external appearances, but are different in their internal construction and output performance. DC series, shunt, and compound motors have wound poles. DC permanent-magnet motors use permanent magnets as the field poles.

DC Series Motors

A *series motor* is a DC motor that has the field winding connected in series with the armature. The field winding has relatively few turns of heavy-gauge wire and must carry the same load current that passes through the armature. The wires extending from the series coil are marked S1 and S2. The wires extending from the armature are marked A1 and A2. **See Figure 10-12.**

DC Series Motor Operating Principles. Since the field winding and the armature winding are connected in series, power is applied to one end of the series field winding and to the other end of the armature winding at the brush. Current flows from the supply terminals through the series windings. The only resistance in the circuit is the small amount of resistance in the winding wires. This causes the motor to draw a very high starting current. Since the current is very large, a very strong pole is created. The strength of the magnetic poles provides a very high starting torque.

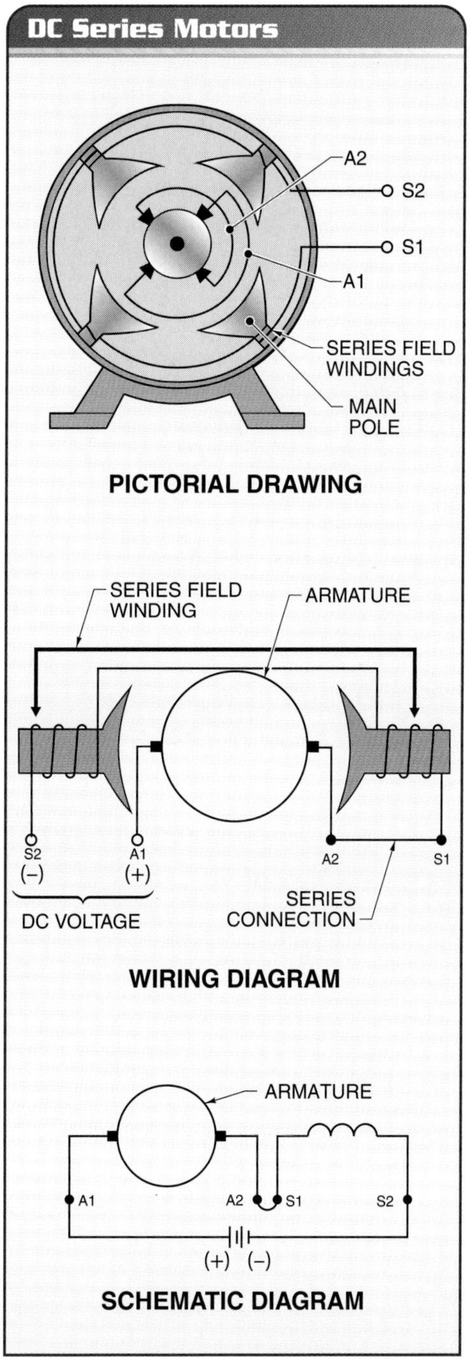

Figure 10-12. A DC series motor has the field winding connected in series with the armature windings.

Definition

*A **series motor** is a DC motor that has the field winding connected in series with the armature.*

As the armature begins to rotate, a countervoltage is induced in the armature windings. This countervoltage is opposite in polarity to that of the power supply. The armature windings see only the net applied voltage, so the effect of the CEMF is that the net applied voltage is reduced. This means that the armature current and the torque are also reduced.

Torque in DC Series Motors. DC series motors can develop about 500% to 800% of full-load torque upon starting. **See Figure 10-13.** Because of the way the current and torque are reduced as the motor speed increases, this type of motor is most suitable for heavy loads that are difficult to start. This is the highest torque produced by any of the DC motors, but the motor does not run at a fixed speed. Typical applications include traction bridges, hoists, winches, gates, and automobile starters. DC series motors are unsuitable to loads where the torque needs to increase with increasing speed, such as fans and blowers.

Speed Regulation of DC Series Motors. The speed regulation of a DC series motor is poor. As the mechanical load on the motor is increased, the motor slows and a simultaneous increase in current occurs in the field and the armature. If the load is removed, the armature speeds up and more countervoltage is induced in the armature. The armature then accelerates to the point at which the countervoltage produces a strong enough unlike pole to limit the speed. Unfortunately, this speed can be higher than the speed at which the armature can spin without failure. In this case, the centrifugal force can cause catastrophic damage. The armature and commutator can come apart with enough force to destroy the field frame. Centrifugal switches are often used to open series motors if the speed gets too high.

Tech Fact
DC series motors can speed up enough to cause damage when the load is removed.

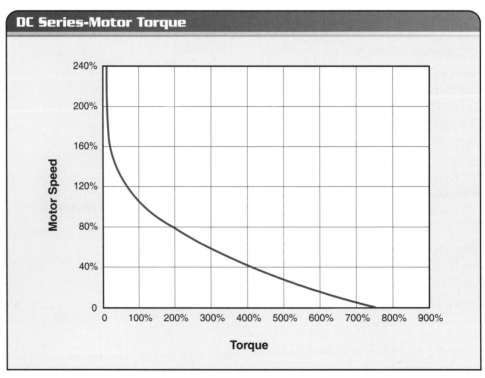

Figure 10-13. A DC series motor has high starting torque and low running torque.

DC Shunt Motors

A *shunt motor* is a DC motor that has the field wiring connected in parallel with the armature. A parallel circuit is often called a shunt. In a shunt motor, the field wiring is a shunt. DC shunt motors are used where constant or adjustable speed is required and starting conditions are moderate. Typical applications include fans, blowers, centrifugal pumps, conveyors, elevators, woodworking machinery, and metalworking machinery.

The field terminal wires extending from the shunt field of a DC shunt motor are marked F1 and F2. The armature windings are marked A1 and A2. **See Figure 10-14.**

DC Shunt Motor Operating Principles. The shunt winding has numerous turns of wire, and the current in the field can be independent of the armature, providing the DC shunt motor with excellent speed control. The large number of turns around the coil means that the coil can produce a strong magnetic field.

The shunt field may be connected to the same power supply as the armature or may be connected to another power supply. A *self-excited shunt field* is a shunt field connected to the same power supply as the armature. A *separately excited shunt field* is a shunt field connected to a different power supply than the armature.

Definition

*A **shunt motor** is a DC motor that has the field wiring connected in parallel with the armature.*

*A **self-excited shunt field** is a shunt field connected to the same power supply as the armature.*

*A **separately excited shunt field** is a shunt field connected to a different power supply than the armature.*

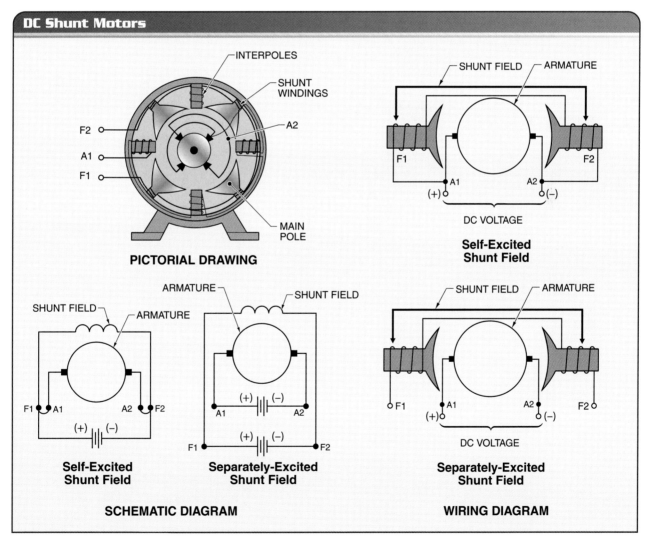

Figure 10-14. A DC shunt motor has the field windings connected in parallel with the armature windings.

Definition

A *stabilizing field winding* is a small series field winding placed over the top of a shunt field winding that improves stability of the fields while running with reduced current in the field circuit.

When voltage is applied to a DC shunt motor, the resistance of the shunt winding limits the overall current flow. The armature draws current to produce a magnetic field that causes rotation. The countervoltage causes the armature current to decrease. At full-load speed, a larger load draws more current than a smaller load and the speed remains fairly constant.

Torque in DC Shunt Motors. DC shunt motors can develop about 250% to 300% of full-load torque upon starting. **See Figure 10-15.** The shunt winding is made of small-gauge wire with many turns around the coil. The small-gauge wire means that the shunt winding has a lower ampacity than a DC series motor. This means that a DC shunt motor has lower starting torque than a DC series motor. However, a DC shunt motor has a very flat speed-torque curve, maintaining the same torque as the load accelerates. This means that a DC shunt motor delivers steady torque at normal operating speeds.

Speed Regulation of DC Shunt Motors. With a DC series motor, the armature accelerates to the point at which the countervoltage produces a strong enough unlike pole to limit the speed. With a DC shunt motor, the armature slows down when a load is placed on the shaft. With the drop in speed, the armature conductors move more slowly through the magnetic flux, and the countervoltage is reduced. This allows the armature current to rise and the torque increases to supply the load. As a result, a shunt motor has very good speed regulation. Typical applications include fans and blowers, centrifugal pumps, conveyors, and machine tools such as lathes.

With a self-excited shunt field, the speed can be varied by varying the field current, the armature current, or both. A rheostat can be placed in series with the field to vary the field current and field strength. Additionally, the supply voltage can be controlled with another rheostat. This allows the motor speed to be controlled over a wide range.

With a separately excited shunt field, there is direct control of the shunt current independent of the armature current. This provides good speed control. A DC shunt motor can be run above its full-voltage base speed by reducing the current in the shunt fields. As the field is reduced, the reduced flux results in a weakened unlike pole, allowing the motor to accelerate. The speed increases to the point at which the armature cuts the same amount of flux as before the current was reduced in the shunt fields.

A *stabilizing field winding* is a small series field winding placed over the top of a shunt field winding that improves stability of the fields while running with reduced current in the field circuit. The motor is labeled a stab/shunt on the nameplate. This winding sees the same current as the armature and has the same polarity as the shunt field, aiding the shunt field if that field is weakened too much by seeking an increase in speed. With a weakened field, the countervoltage decreases and the current increases. The stabilizing field winding adds flux to the fields to stabilize the shunt field.

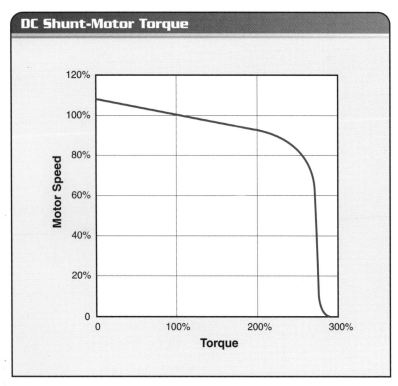

Figure 10-15. A DC shunt motor has moderate starting torque, with steady torque as the motor speeds up.

DC Compound Motors

A *compound motor* is a DC motor with the field connected in both series and shunt with the armature. The field winding is a combination of the series field (S1 and S2) and the shunt field (F1 and F2). **See Figure 10-16.** The series field is connected in series with the armature and both have the same current. The shunt field is connected in parallel with the series field and armature combination and may be powered by a separate source. This arrangement gives a compound motor the high torque of a DC series motor and the constant speed of a DC shunt motor.

DC compound motors have some of the characteristics of a series motor and some of the characteristics of a shunt motor. DC compound motors are used when high starting torque and constant speed are required. Typical applications include punch presses, shears, bending machines, and hoists.

Tech Fact
According to the NEC®, constant-voltage generators shall be protected from overloads by inherent design, circuit breakers, or fuses, or other means suitable for the conditions.

Definition
*A **compound motor** is a DC motor with the field connected in both series and shunt with the armature.*

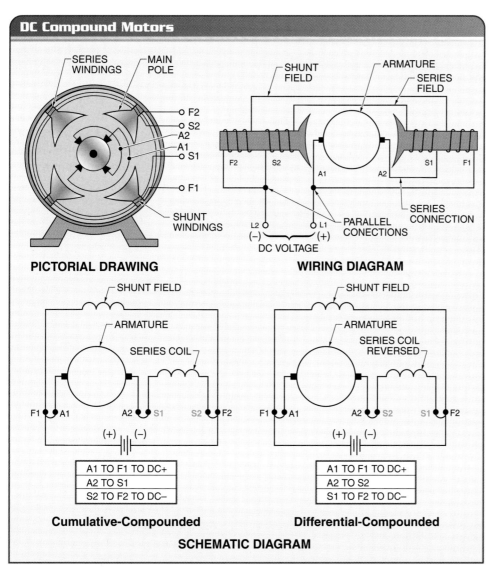

Figure 10-16. A DC compound motor has the field windings connected in series and in parallel with the armature windings.

Definition

A DC cumulative-compounded motor, or *DC overcompounded motor,* is a motor where current flows in the same direction in the series and shunt coils and the flux surrounding the coils adds.

A DC differential-compounded motor, or *DC undercompounded motor,* is a motor where the current flows in the opposite direction in the series and shunt coils and the resulting net flux is the difference between the two fluxes.

DC Compound Motor Operating Principles. The shunt field prevents a DC compound motor from running too fast at light loads and the series field carries the same current as the armature, giving the motor enough torque for heavy loads. DC compound motors are used mostly for special applications where the speed needs to be constant under load, yet high starting torque is required.

The windings of the shunt and series fields are wound over each other on the same pole piece. A *DC cumulative-compounded motor,* or *DC overcompounded motor,* is a motor where current flows in the same direction in the series and shunt coils and the flux surrounding the coils adds. A *DC differential-compounded motor,* or *DC undercompounded motor,* is a motor where the current flows in the opposite direction in the series and shunt coils and the resulting net flux is the difference between the two fluxes.

Torque in DC Compound Motors. DC series motors can develop about 400% to 500% of full-load torque upon starting. **See Figure 10-17.** Since the flux in a cumulative-compounded motor adds together, the motor has a strong starting torque. The total field flux in a differential-compounded motor is the net difference between the two windings, so the total flux is relatively low and gives a constant speed. Under load, the series flux opposes the shunt flux and reduces the total field flux so the motor can speed up. However, if a large enough load is placed on a differentially compounded motor, the series flux cancels all the shunt flux. This places the polarity of the series flux opposite that of the shunt flux and can unexpectedly reverse the motor.

Speed Regulation of DC Compound Motors. When a load is placed on a DC cumulative-compounded motor, the armature slows and cuts fewer lines of flux. The countervoltage is reduced and the armature current increases, increasing the current through the series field winding. This strengthens the flux produced in the main fields, and the countervoltage is increased. The increase in countervoltage opposes rotation and the motor slows.

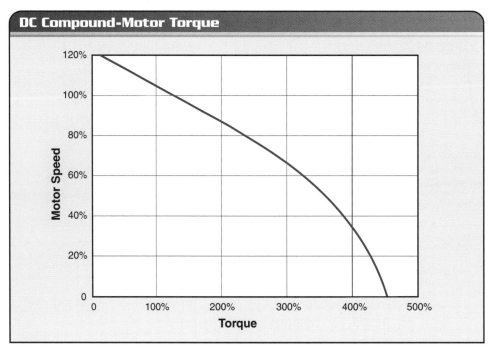

Figure 10-17. A DC compound motor has moderate starting torque that drops off as the motor speeds up.

When a load is placed on a DC differential-compounded motor, the armature also slows and cuts fewer lines of flux. The overall flux is less than in a cumulative-compounded motor because the overall flux is the difference between the two individual fluxes. The countervoltage is smaller than in a cumulative-compounded motor and the force opposing rotation is proportionally smaller. This type of motor also slows under load, but less than with a cumulative-compounded motor. However, a heavy load can stall or even reverse the motor. DC differential-compounded motors are seldom used in industry.

DC Permanent-Magnet Motors

A *DC permanent-magnet motor* is a motor that uses magnets, not a coil of wire, for the field windings. DC permanent-magnet motors have molded magnets mounted into a steel shell. The permanent magnets are the field coils. The armature of a DC permanent-magnet motor is conventionally wound to resemble other DC motors. DC power is supplied only to the armature. **See Figure 10-18.**

DC permanent-magnet motors are typically made in small sizes, such as fractional-horsepower motors. Larger sizes are usually not made because of the high cost of the larger permanent magnets needed in larger motors. They are used where small motors with high starting torque are needed, such as in automobiles to control power seats, power windows, and windshield wipers.

DC permanent-magnet motors produce relatively high torque at low speeds and provide some self-braking when removed from power. Not all DC permanent-magnet motors are designed to run continuously because they overheat rapidly and destroy the magnets.

DC Permanent-Magnet Motor Operating Principles. Modern permanent-magnet motors use strong rare-earth magnets as a replacement for the ferrite magnets used in early motor designs. Rare-earth magnets give a permanent-magnet motor a power-to-weight ratio that ranges from 50% to 200% more than that of a conventional permanent-magnet motor with the same dimensions. Because of the permanent magnets, operating losses are lower and efficiency is better than in other types of DC motors.

> **Definition**
>
> A *DC permanent-magnet motor* is a motor that uses magnets, not a coil of wire, for the field windings.

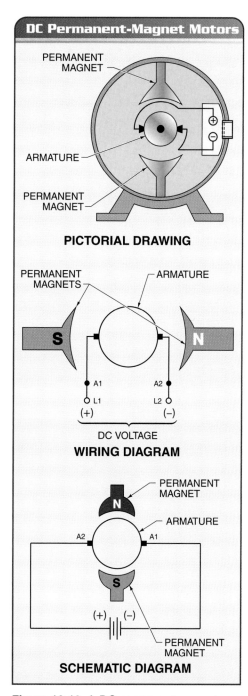

Figure 10-18. A DC permanent-magnet motor uses permanent magnets instead of field windings.

Definition

*A **generator** is a machine that converts mechanical energy into electrical energy by means of electromagnetic induction.*

Torque in DC Permanent-Magnet Motors. DC permanent-magnet motors can develop about 600% to 700% of full-load torque upon starting. **See Figure 10-19.** DC permanent-magnet motors have good starting torque that falls off rapidly as the motor speed increases. DC permanent-magnet motors with rare-earth magnets have a permanent-magnet motor a torque-to-weight ratio that ranges from 40% to 90% more than that of a conventional permanent-magnet motor with the same dimensions.

Speed Regulation of DC Permanent-Magnet Motors. Speed regulation in DC permanent-magnet motors is fairly good, but not as good as with DC shunt motors. There are several motor designs with better speed control than the standard permanent-magnet motors, but with poorer torque.

DC GENERATORS

A *generator* is a machine that converts mechanical energy into electrical energy by means of electromagnetic induction. DC generators operate on the principle that when a coil of wire is rotated in a magnetic field, a voltage is induced in the coil. The amount of voltage induced in the coil is determined by the rate at which the coil is rotated in the magnetic field, the strength of the magnetic field, and the number of turns in the coil. When a coil is rotated in a magnetic field at a constant rate, the voltage induced in the coil depends on the number of magnetic lines of force in the magnetic field at each given instant of time. DC generators consist of field windings, an armature, a commutator, and brushes. **See Figure 10-20.**

DC generators have been used for variable speed control for many years. A generator's output can be varied by changing either the speed of the armature or the strength of the main fields. In a typical generator set, an AC motor is used to power a self-excited DC generator. The field strength can be varied by a rheostat in series with the coil. This provides control of the output of the generator, and a varying voltage to a DC motor.

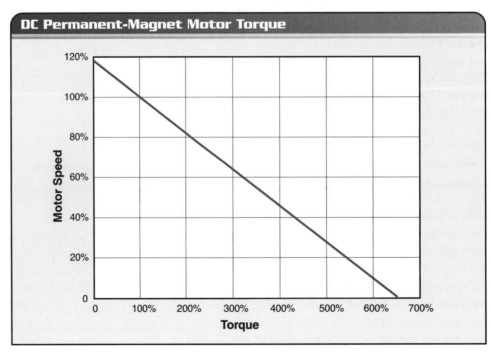

Figure 10-19. A DC permanent-magnet motor has high starting torque that drops off as the motor speeds up.

DC Generators

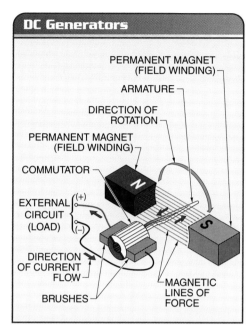

Figure 10-20. A DC generator consists of field windings in the field frame, an armature, and a commutator and brushes on the shaft.

DC Generator Construction

The construction of a DC generator is the same as the construction of a similar DC motor, with the addition of some kind of prime mover. A DC generator has a stationary field frame that holds the poles on which the field windings are wound. The frame can also support the interpoles. The armature consists of a main shaft, a laminated core with slots for the armature windings, and a commutator. The two endbells hold the bearing housings for the armature and shaft; one endbell has the brush holders mounted in it. The brush holders contain the brushes and keep them aligned on the commutator.

DC Generator Operating Principles

As with a DC motor, the field windings of a DC generator are magnets used to produce the magnetic field. If the current for the field windings is supplied by an outside source (a battery or another generator), the generator is separately excited. If the generator itself supplies current for the field windings, the generator is self-excited. DC generators are usually self-excited.

As with a DC motor, the armature is the movable coil of wire in a generator that rotates through the magnetic field. **See Figure 10-21.** A DC generator always has a rotating armature and a stationary field (field windings). A commutator is a ring made of segments that are insulated from one another. Each end of a coil of wire is connected to a segment. The commutator segments reverse the connections to the brushes every half cycle. A brush is the sliding contact that rides against the commutator segments and is used to connect the armature to the external circuit. The resulting output voltage of a DC generator is a pulsating DC voltage. The pulsations of the output voltage are known as ripples. A practical DC generator always has multiple armature windings to smooth out the ripples.

DC Generator Voltage Regulation. *Voltage regulation* is the ability of a source to vary the output voltage in order to maintain system voltage as the load varies. In a typical situation, higher currents are required from the power supply as the electrical demand increases. This results in larger voltage drops through the distribution system and the available voltage decreases at the point of use.

Voltage regulation can be negative, zero, or positive. Positive voltage regulation occurs when the countervoltage at no load is greater than the terminal voltage at full load. This characteristic is typically available from separately excited shunt generators, the self-excited shunt generator, and the compound generator. Zero voltage regulation occurs when the countervoltage is stable or flat from no load to full load. This characteristic is typically available from compound generators. Negative voltage regulation occurs when the countervoltage at no load is less than the terminal voltage at full load. This type of regulation occurs only with the series generator.

A benefit of DC generators is their ability to increase the voltage under changing load conditions. Voltage regulation is provided by changing the generator speed and/or the field flux.

> **Definition**
>
> *Voltage regulation* is the ability of a source to vary the output voltage in order to maintain system voltage as the load varies.

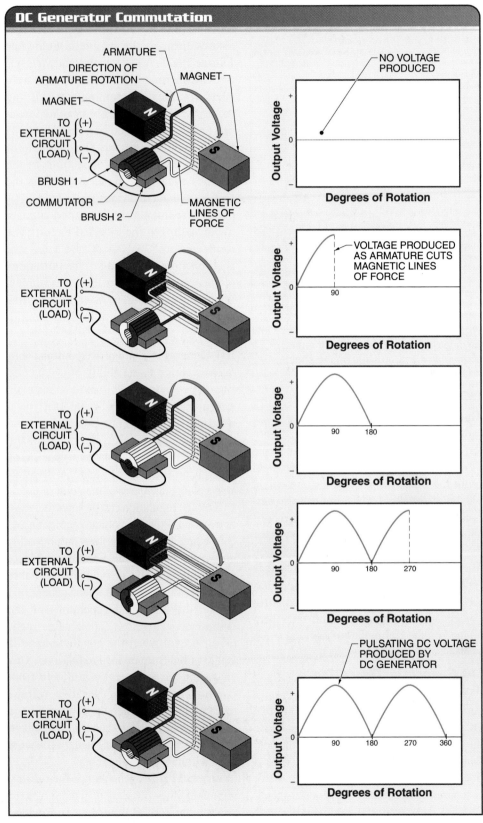

Figure 10-21. The armature is the movable coil of wire in a generator that rotates through the magnetic field.

DC GENERATOR TYPES

The three types of DC generators are series-wound, shunt-wound, and compound-wound generators. The difference between the types is based on the relationship of the field windings to the external circuit.

DC Series-Wound Generators

A *series-wound generator* is a generator that has its field windings connected in series with the armature and the external circuit (load). **See Figure 10-22.** In a series-wound generator, the field windings consist of a few turns of low-resistance wire because the large load current flows through them.

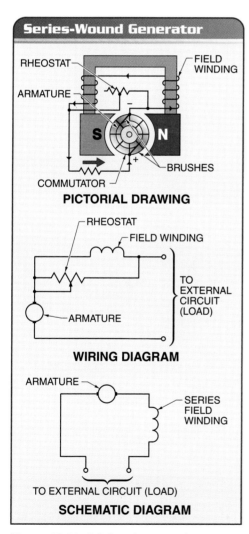

Figure 10-22. A DC series-wound generator has the field windings connected in series with the armature windings.

An external prime mover applies mechanical torque to the armature and series field circuit, which induces voltage and current in the generator. This provides terminal voltage and rated output power to the load. The generator provides a source of negative voltage regulation to the load it serves.

Series-wound generators have poor voltage regulation. Because of their poor voltage regulation, series-wound DC generators are not used frequently in industry. The output voltage of a series-wound generator may be controlled by a rheostat (variable resistor) connected in parallel with the field windings.

DC Shunt-Wound Generators

A *shunt-wound generator* is a generator that has its field windings connected as a shunt in parallel with the armature and the external circuit. **See Figure 10-23.** The armature winding, interpoles, and compensating windings, brushes and brush rigging, and commutator are connected in series with one another. The shunt field winding and shunt field resistance are connected in series with one another and in parallel with the armature circuit.

An external prime mover applies mechanical torque to the armature and series field circuit, which induces voltage and current in the generator. This provides terminal voltage and rated output power to the load.

A DC shunt-wound generator provides a source of positive and zero voltage regulation to the load it serves. Because the field windings are connected in parallel with the load, the current through the field windings is wasted as far as output is concerned. Therefore, the field windings consist of many turns of high-resistance wire to keep the current flow through them low. A shunt-wound generator is suitable if the load is constant. However, if the load fluctuates, the voltage also varies. The output voltage of a shunt-wound generator may be controlled by means of a rheostat connected in series with the shunt field.

Definition

A *series-wound generator* is a generator that has its field windings connected in series with the armature and the external circuit (load).

A *shunt-wound generator* is a generator that has its field windings connected as a shunt in parallel with the armature and the external circuit.

228 MOTORS

Definition

A *compound-wound generator* is a generator that includes series and shunt field windings.

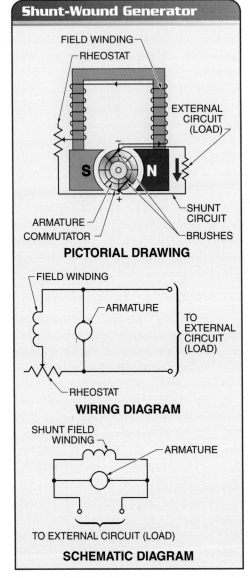

Figure 10-23. A DC shunt-wound generator has the field windings connected in parallel with the armature windings.

DC Compound-Wound Generators

A *compound-wound generator* is a generator that includes series and shunt field windings. In a compound-wound generator, the series field windings and shunt field windings are combined in a manner to take advantage of the characteristics of each. The shunt field is normally the stronger of the two. The series field is used only to compensate for effects that tend to decrease the output voltage. **See Figure 10-24.**

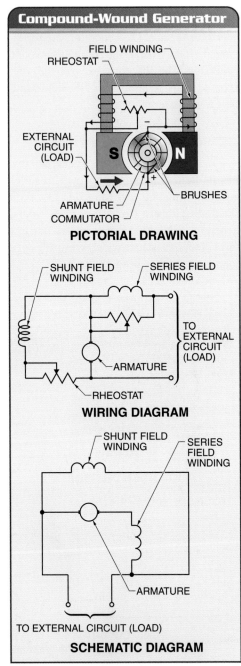

Figure 10-24. A DC compound-wound generator has the field windings connected in series and in parallel with the armature windings.

Tech Fact

DC motors that have wound poles are called shunt, series, or compound motors, depending on the connections. DC motors that use permanent magnets to magnetize their field poles are called PMDC motors.

DC Brushless Motors

DC brushless motors are very similar to AC synchronous motors. In a common design, brushless motors have permanent magnets mounted on the rotor and have AC power applied to the wound coils in the stator. This is reversed from a conventional brushed DC motor. There is no need for brushes and commutators because the permanent connections are on the fixed stator. Brushless motors use electronic switching instead of brushes and a commutator.

Power for brushless motors can come from square or sinusoidal waveforms generated in a motor drive. Square waveforms have more torque than sinusoidal waveforms, but have more torque ripple. Square waveforms are very easy to generate in a commutation encoder with a signal from an encoder or a Hall-effect sensor. Square waveforms are used for applications that are not affected by a small amount of torque ripple.

Sinusoidal waveforms are the same shape as the standard power waveforms in a circuit, but may have a different frequency. Sinusoidal waveforms have very little torque ripple and operate smoothly at low speed. Sinusoidal waveforms are normally used in finishing operations requiring a fine surface. Motors using sinusoidal waveforms are more expensive than an equivalent motor using square waveforms because of the extra electronics required to develop the sinusoidal waveform.

AC motor controllers feed power to all the legs simultaneously. DC brushless controllers operate similarly to AC motor controllers, but feed the maximum negative and positive current to two of the legs at a time.

Brushless motors are more efficient and quieter than conventional DC motors. They also require less maintenance, can operate at higher speeds, and have a longer lifetime because there are no brushes to wear. They are often used where high efficiency and small size are important.

Brushless motors are generally more expensive than similar brushed motors because of the complex electronics required. The electronics components in brushless motors use MOSFETs, insulated-gate switches, and integrated circuits for commutation, feedback, and modulation. As with most electronic components, the costs of the components used in brushless motors has been decreasing. These decreasing costs are leading to increasing market share for brushless motors.

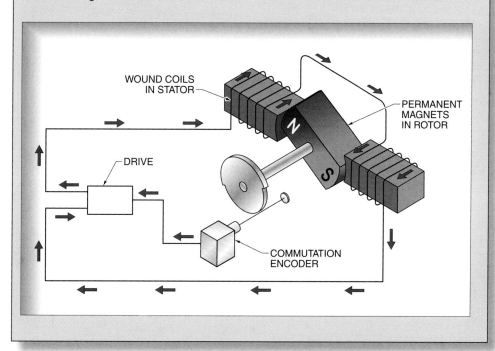

Supplemental Topic

DC Coreless Motors

There is nothing in the design of a DC motor that requires rotation of an iron core when the armature rotates. Torque is exerted only on the windings. A coreless DC motor uses the armature windings for structural support and eliminates the heavy iron armature core. As a result, the armature is hollow. Because of their low mass, coreless DC motors are prone to overheating and therefore are usually only used for very small loads. The motors are typically less than 1″ in size.

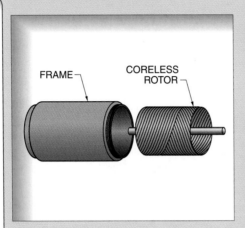

Some of the advantages of coreless motors include very low inertia, high efficiency, and the absence of magnetic fields acting on the laminations. This results in reduced torque ripple and the absence of iron eliminates cogging. A coreless motor operates smoothly, even at low speeds. Because the core is made without iron, the low armature mass allows more rapid acceleration and deceleration than any other class of DC motor. Coreless motors are often used for positioning applications with very small loads.

Commutator arcing in conventional DC motors is typically caused by the release of inductive energy stored in the armature upon commutation. Excessive arcing produces electrical noise and reduces the life of brushes. DC coreless motors have significantly less inductance and therefore have very little arcing at the commutator.

The two common rotor shapes in a DC coreless motor are cylindrical or disc shapes. The cylindrical shapes may be categorized as inside field or outside field. The disc shapes may be categorized as pancake, printed, or three-coil shapes.

The rotor is a hollow cylinder consisting of copper wire typically wound in a skewed honeycomb pattern. The cylinder is dipped in a varnish or epoxy to hold the wires together and provide structural strength. In outside-field designs, permanent magnets are mounted on the frame outside the cylinder. In inside-field designs, permanent magnets are mounted on the end and extend into the center of the hollow core. Rare-earth magnets are usually used because their magnetic fields are much stronger than the older ferritic magnets.

The commutators are typically made of precious metals, such as gold, silver, or platinum, for best conduction. Commutators are typically very small to minimize the use of the expensive precious metals and to minimize the commutator weight and inertia.

Application—DC Motor Reduced-Voltage Starting

DC motors smaller than one-half horsepower draw low current and are started by placing full-line voltage across the motor terminals. DC motors larger than one-half horsepower typically require reduced-voltage starting. In one method of starting large DC motors, a resistance unit is placed in series with the motor during starting. The moveable arm of the starting rheostat is connected directly to the positive side of the power supply and the motor is connected to the fixed side of the rheostat. The starter decreases the resistance as the motor accelerates.

The starting resistance unit is not needed when the motor is running because the motor generates a countervoltage. The amount of countervoltage depends on the speed of the motor. The faster the motor runs, the greater the countervoltage. For example, a DC motor that has a resistance of 4 Ω at standstill draws 57.5 A when connected to 230 V (230 ÷ 4 = 57.5).

When the motor accelerates to a speed that generates a countervoltage of 100 V, the total voltage in the motor is 130 V (230 − 100 = 130), and the current drawn is 32.5 A (130 ÷ 4 = 32.5).

When the motor accelerates to full speed and generates a countervoltage of 200 V, the total voltage in the motor is 30 V (230 − 200 = 30), and the current drawn is 7.5 A (30 ÷ 4 = 7.5).

Standard Wiring Procedures

	Series Motor	Shunt Motor	Compound Motor
Power supply	+ to A1 − to S2	+ to A1 + to F1 − to A2 − to F2	+ to A1 + to F1 − to F2 − to S2
Starting rheostat	Connect between positive side of power supply and motor with control arm on positive side of power supply	Connect between positive side of power supply and motor with control arm on positive side of power supply	Connect between positive side of power supply and motor with control arm on positive side of power supply
Speed-control rheostat	Connect between power supply and motor with control arm on positive side of power supply	Connect in series with field circuit with control arm on positive side of power supply	Connect in series with field circuit with control arm on positive side of power supply

Glossary...

A ***direct current motor,*** or ***DC motor,*** is a machine that uses DC connected to the field windings and armature to produce shaft rotation.

A ***direct current generator,*** or ***DC generator,*** is a power source that supplies DC when the armature is rotated.

A ***field frame*** is the stationary part in a DC motor or generator.

The ***field poles*** are metal pieces mounted to the field frame that are used as field windings.

The ***field windings*** are magnets or stationary windings used to produce the magnetic field in an alternator or motor.

Interpoles are auxiliary poles placed between the main field poles of the motor.

Compensating windings, or ***pole-face windings,*** are field windings placed in slots on the main poles.

An ***armature*** is the rotating part of a DC motor, consisting of the laminated core with slots for the coils, the main shaft, and the commutator and brushes.

A ***commutator*** is a ring made of insulated segments that keep the armature windings in the correct polarity to interact with the main fields.

A ***pigtail*** is an extended, flexible connection or a braided copper conductor.

The ***brush rigging*** is the entire assembly of the brush, brush holder, insulators, and any wiring included in the assembly.

A ***countervoltage,*** or ***counter EMF (CEMF),*** is a voltage induced in the windings that is opposite in polarity to that of the power supply.

Armature reaction is the distortion of the magnetic fields that happens when a current-carrying wire is placed within a fixed magnetic field.

The ***neutral plane*** is a line through the armature cross section that is perpendicular to the maximum amount of magnetic flux.

Brush neutral is the position of the brushes where commutation can occur with minimal induced voltage in the armature coils.

Commutation is the process where the armature current is periodically reversed in order to keep the motor torque in the same direction during the entire armature rotation.

A ***series motor*** is a DC motor that has the field winding connected in series with the armature.

A ***shunt motor*** is a DC motor that has the field wiring connected in parallel with the armature.

A ***self-excited shunt field*** is a shunt field connected to the same power supply as the armature.

A ***separately excited shunt field*** is a shunt field connected to a different power supply than the armature.

A ***stabilizing field winding*** is a small series field winding placed over the top of a shunt field winding that improves stability of the fields while running with reduced current in the field circuit.

A ***compound motor*** is a DC motor with the field connected in both series and shunt with the armature.

A ***DC cumulative-compounded motor,*** or ***DC overcompounded motor,*** is a motor where current flows in the same direction in the series and shunt coils and the flux surrounding the coils adds.

A ***DC differential-compounded motor,*** or ***DC undercompounded motor,*** is a motor where the current flows in the opposite direction in the series and shunt coils and the resulting net flux is the difference between the two fluxes.

A ***DC permanent-magnet motor*** is a motor that uses magnets, not a coil of wire, for the field windings.

A ***generator*** is a machine that converts mechanical energy into electrical energy by means of electromagnetic induction.

Voltage regulation is the ability of a source to vary the output voltage in order to maintain system voltage as the load varies.

...Glossary

A **series-wound generator** is a generator that has its field windings connected in series with the armature and the external circuit (load).

A **shunt-wound generator** is a generator that has its field windings connected as a shunt in parallel with the armature and the external circuit.

A **compound-wound generator** is a generator that includes series and shunt field windings.

Review

1. Describe the similarities in the construction of a DC motor and a DC generator.

2. Describe how a commutator and brushes work to develop torque.

3. Explain how armature reaction creates a distorted field that moves the neutral plane.

4. Describe the difference in winding connections between the four types of DC motors.

5. Describe the differences in the starting and running torque of the four types of DC motors.

6. List a typical application of each of the four types of DC motors.

Refer to the CD-ROM for Quick Quiz® questions related to chapter content.

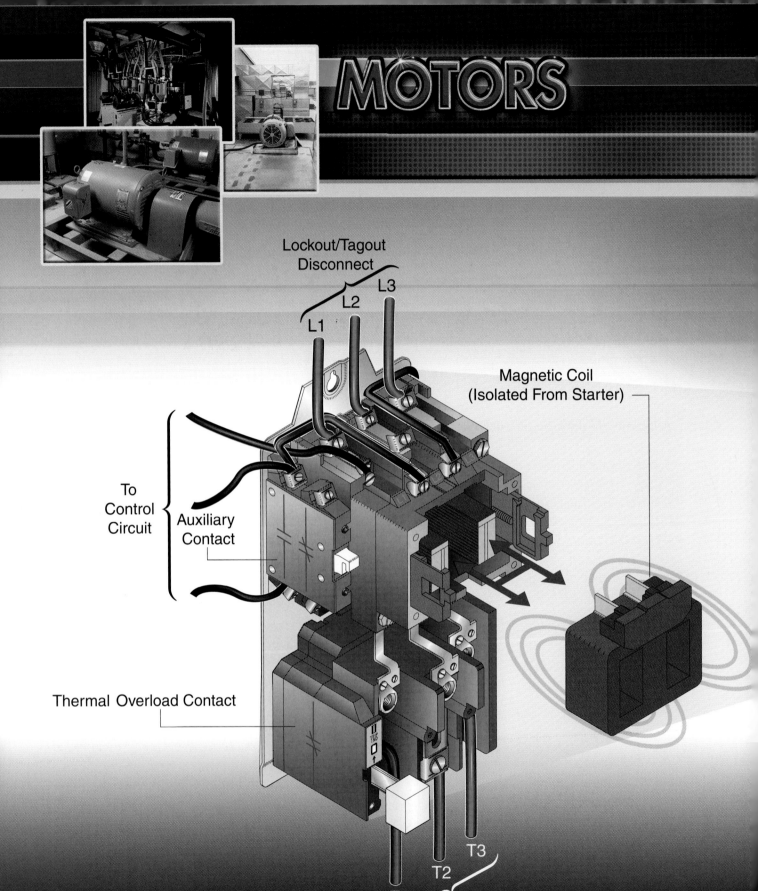

Starting

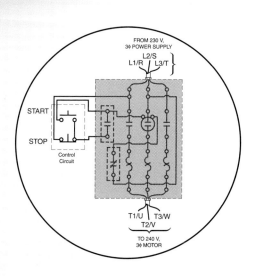

Motor Starting	238
Full-Voltage Starting	238
Reduced-Voltage Starting	239
Starting-Method Comparison	254
Supplemental Topic—NEMA and IEC Ratings	257
Supplemental Topic—Solid-State Switches	258
Application—Wye-Delta Starting Overload Protection	259
Summary	260
Glossary	261
Review	261

OBJECTIVES

- Explain why reduced-voltage starting is often used instead of full-voltage starting.
- Describe the difference between open-circuit and closed-circuit transitions.
- List the reduced-voltage starting methods and describe the method used to reduce the starting current.
- List the advantages and disadvantages of each starting method.

Full-voltage starting is the least expensive and most efficient means of starting a motor for applications involving small-horsepower motors. Many applications involve large-horsepower DC and AC motors that require reduced-voltage starting because full-voltage starting may create interference with other systems. Reduced-voltage starting reduces interference in the power source, the load, and the electrical environment surrounding the motor.

Definition

Full-voltage starting is a method of starting a motor with the full line voltage placed across the terminals.

A *manual contactor* is a control device that uses pushbuttons to energize or de-energize the load connected to it.

A *manual starter* is a contactor with an added overload protection device. Manual starters are used only in electrical motor circuits.

Locked rotor is a condition when a motor is loaded so heavily that the motor shaft cannot turn.

MOTOR STARTING

Motors typically draw starting current that is much larger than the current required to keep the motor running under load. Some motors and electrical systems are designed to withstand this large starting current. In this case, full-voltage starting is used. Some motors or electrical systems are not designed to withstand the large starting current. These motors use some type of reduced-voltage starting to reduce the starting current.

FULL-VOLTAGE STARTING

Full-voltage starting is a method of starting a motor with the full line voltage placed across the terminals. Full-voltage starting is also known as across-the-line starting and line-voltage starting. The simplest method for starting a motor is to apply the full line voltage across the stator terminals and establish a rotating magnetic field. This is the least expensive and most efficient means of starting a motor for applications involving small-horsepower motors.

Small motors and single-phase motors are almost universally started with full-voltage starting. The NEC® gives guidance on the current draw based on the code letter on the nameplate. Motors of several code letters are normally started on full voltage. In addition, the NEC® requires a suitable controller for all motors. For small portable motors at or below ⅓ HP, a plug and receptacle can be used as the controller to apply full-voltage starting. **See Figure 11-1.**

The two factors to consider when selecting a motor for full-voltage starting are whether it needs to develop high starting torque or high run efficiency. If a motor develops high starting torque and fairly low starting currents, then the rotor will usually have high resistance and poor efficiency at full speed. In order to induce enough current to run at full load, the rotor must slow down to increase the rotor frequency. This lowers the efficiency, as the rotor will run slower at load. A motor that develops high starting current and reasonable torque will run at a higher speed at load because of the low resistance of the rotor.

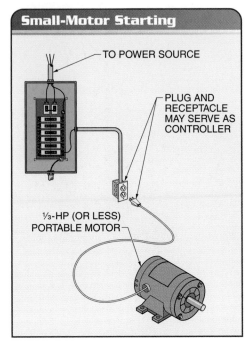

Figure 11-1. For small portable motors at or below ⅓ HP, a plug and receptacle can be used as the controller to apply full-voltage starting.

Manual Contactors and Starters

A *manual contactor* is a control device that uses pushbuttons to energize or de-energize the load connected to it. Manual contactors cannot be used to start and stop motors because they have no overload protection built into them. A *manual starter* is a contactor with an added overload protection device. Manual starters are used only in electrical motor circuits. The primary difference between a manual contactor and a manual starter is the addition of an overload protection device. **See Figure 11-2.**

The overload protection device must be added because the National Electrical Code® (NEC®) requires that a control device shall not only turn a motor ON and OFF, but shall also protect the motor from destroying itself under an overloaded situation, such as a locked rotor. *Locked rotor* is a condition when a motor is loaded so heavily that the motor shaft cannot turn. A motor with a locked rotor draws excessive current and burns up if not disconnected from the line voltage. To protect the motor, the overload device senses the excessive current and opens the circuit.

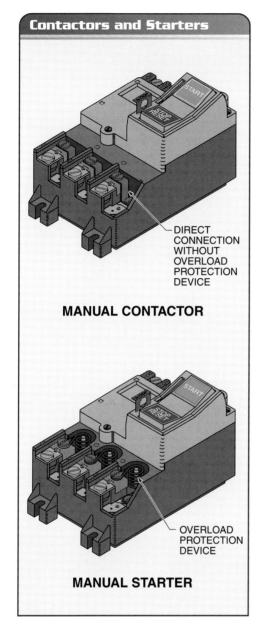

Figure 11-2. The primary difference between a manual contactor and a manual starter is the addition of an overload protection device.

Magnetic Starters

A *magnetic motor starter* is an electrically operated switch that includes motor overload protection. Magnetic motor starters include overload relays that detect excessive current passing through a motor and are used to switch all types and sizes of motors. Magnetic motor starters are available in sizes that can switch loads of a few amperes to several hundred amperes.

REDUCED-VOLTAGE STARTING

In those situations where a system is not designed to tolerate the starting current of large motors, some type of reduced-voltage starting is required. Reduced-voltage starting reduces the starting voltage, current, and torque at the expense of longer start times. These longer start times may affect the duty cycle. Common methods of reduced-voltage starting are primary resistor starting, autotransformer starting, part-winding starting, and wye-delta starting.

Open-Circuit and Closed-Circuit Transitions

Motors that are started at reduced voltage must be switched to line voltage before reaching full speed. The two methods used to switch motors from starting voltage to line voltage include open-circuit transition and closed-circuit transition. In open-circuit transition, a motor is temporarily disconnected from the voltage source when switching from a reduced starting voltage level to a running voltage level, before reaching full motor speed. **See Figure 11-3.** In closed-circuit transition, a motor remains connected to the voltage source when switching from a reduced starting voltage level to a running voltage level, before reaching full motor speed.

Closed-circuit transition is preferable to open-circuit transition because closed-circuit transition does not cause a high-current transition surge. However, closed-circuit transition is the more expensive circuit transition method. Open-circuit transition produces a higher-current surge than closed-circuit transition at the transition point because the motor is momentarily disconnected from the voltage source.

The high-current surge during open-circuit transition is based on the motor speed at the time of transition. Transfer from the low starting voltage to the high line voltage should occur as close to full motor speed as possible. If the transition occurs when the motor is at a low speed, a surge current even higher than the starting current can occur. **See Figure 11-4.**

Definition

*A **magnetic motor starter** is an electrically operated switch that includes motor overload protection.*

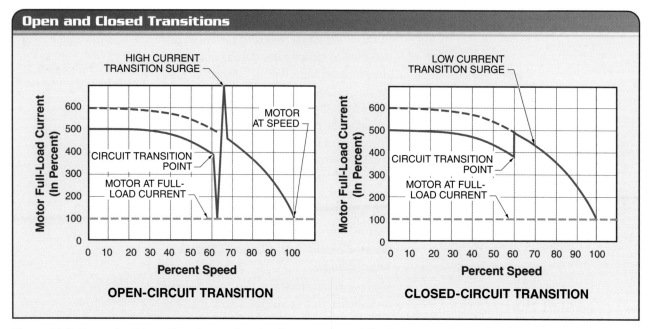

Figure 11-3. Open-circuit transitions from reduced-voltage starting to full-voltage operation can cause a high-current transient surge. Closed-circuit transitions minimize transients.

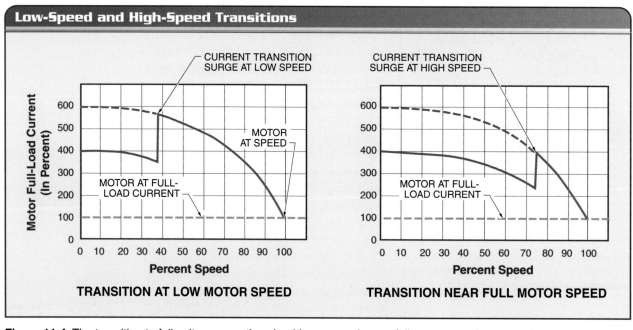

Figure 11-4. The transition to full-voltage operation should occur as close to full motor speed as possible.

Primary-Resistor Starting

Primary-resistor starting is a method of reduced-voltage starting that places resistors in series in the motor power circuit to reduce the voltage to the motor. **See Figure 11-5.** The resistors are removed from the circuit after the motor has had time to get up to speed and the current draw has decreased to the normal operating current.

Primary-resistor starters provide very smooth starting due to increasing voltage across the motor terminals as the motor accelerates. Because of the added resistance, the circuit is more resistive than with full-voltage starting. This improves the power factor over full-voltage starting. Primary-resistor starting is a slow method of starting, and the heat must dissipate from the resistors. The expensive, fixed resistors make it difficult to modify the starting torque of the motor for different operating conditions.

Standard primary-resistor starters provide two-point acceleration (one step of resistance) with approximately 60% to 70% of line voltage at the motor terminals. Multiple-step starting is possible by using additional contacts and resistors when extra smooth starting and acceleration are needed. This multiple-step starting may be required in paper or fabric applications where even a small jolt in starting may tear the paper or snap the fabric.

Definition

Primary-resistor starting is a method of reduced-voltage starting that places resistors in series in the motor power circuit to reduce the voltage to the motor.

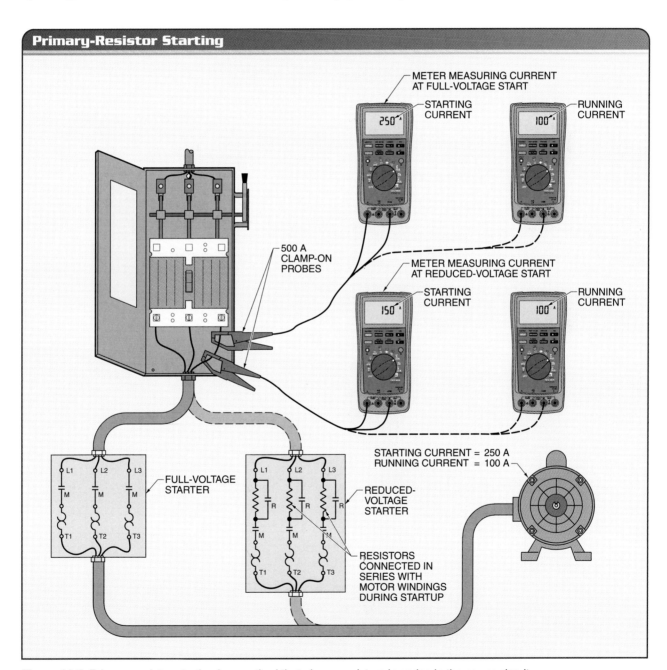

Figure 11-5. Primary-resistor starting is a method that places resistors in series in the power circuit.

Primary-Resistor Starting Circuits. In a primary-resistor starting circuit, external resistance is added to and taken away from the motor circuit. **See Figure 11-6.** The control circuit consists of the motor starter coil M, ON-delay timer TR1, and contactor coil C. Coil M controls the motor starter, which energizes the motor and provides overload protection. The timer provides a delay from the point where coil M energizes until contacts C close, shorting resistors R1, R2, and R3. Coil C energizes the contactor, which provides a short circuit across the resistors.

Pressing start pushbutton PB2 energizes motor starter coil M and the ON-delay timer coil TR1. Motor starter coil M closes contacts M to create memory. ON-delay timer coil TR1 causes contacts TR1 to remain open during reset, stay open during timing, and close after timing out. Once timed out, the contactor coil C energizes, causing contacts C to close and the resistors to short.

This circuit is a common reduced-voltage starting circuit. Changes are often made in the values of resistance and wattage to accommodate motors of different horsepower ratings.

DC Reduced-Voltage Starting

All DC motors are directly connected to the armature and field windings. During startup, current is limited by the resistance of the wire in the armature and the field windings. Larger motors have less resistance in the windings than smaller motors. Less resistance means more current during starting. In large DC motors, this starting current may be so high that it damages the motor. To prevent motor damage, reduced-voltage starting must be applied to DC motors larger than about 1 HP.

Tech Fact

The difference between a contactor and a starter is that contactors do not provide overload (OL) protection.

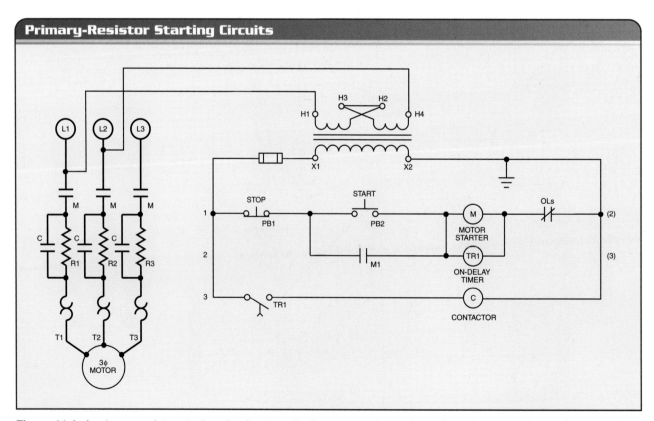

Figure 11-6. A primary-resistor starting circuit automatically removes the resistors from the power circuit after a predetermined time.

Reduced-voltage starting of DC motors reduces the amount of current during starting. As the motor accelerates, the reduced voltage may be removed because the current in the motor decreases with an increase in motor speed. This decrease in current results from the motor generating a countervoltage that is opposite to the applied voltage as it accelerates.

DC Reduced-Voltage Starting Circuits. A starting rheostat or a solid-state circuit is used when reduced-voltage starting is applied to DC motors. The starting rheostat is connected in series with the incoming power line (typically the positive DC line) and the motor. The rheostat reduces the voltage applied to the motor during starting by placing a high resistance in series with the motor. The resistance is decreased as the rheostat is moved to the run position. **See Figure 11-7.**

The starting rheostat is controlled manually, which means that the operator determines the exact starting time. Although a starting rheostat can also be used to control motor speed (the speed of a DC motor varies with the applied voltage), the purpose of a starting rheostat is to reduce the voltage (and thus current and torque) during starting. After the motor is started, a different circuit can be used to control motor speed.

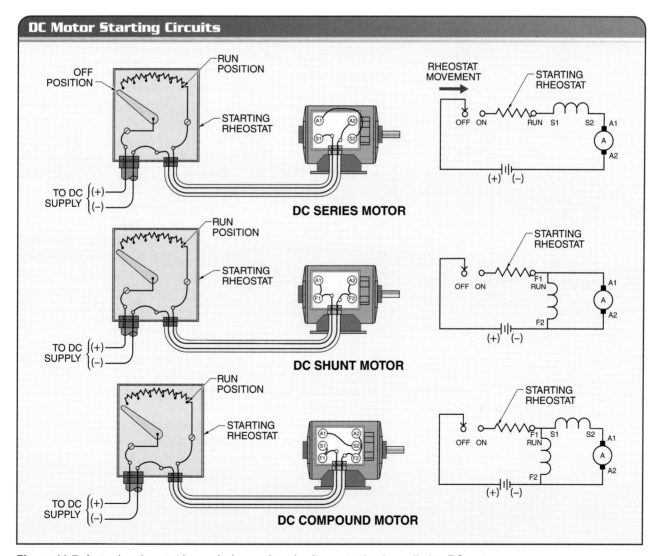

Figure 11-7. A starting rheostat is used when reduced-voltage starting is applied to DC motors.

Definition

Autotransformer starting is a method of reduced-voltage starting that uses a tapped three-phase autotransformer to provide reduced voltage for starting.

Autotransformer Starting

Autotransformer starting is a method of reduced-voltage starting that uses a tapped three-phase autotransformer to provide reduced voltage for starting. **See Figure 11-8.** After a predetermined time, timers actuate a circuit that connects the motor to full line voltage. Autotransformer starting is relatively expensive, but is preferred over primary resistor starting when the starting current drawn from the line must be held to a minimum value yet the maximum starting torque per line ampere is required.

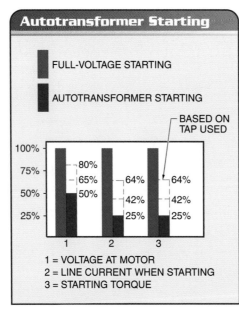

Figure 11-8. Autotransformer starting uses a tapped autotransformer to provide reduced voltage for starting.

A typical autotransformer may have a turns ratio of 1:0.8. Because of the transformer action, autotransformer starting has higher current in the coils than is drawn from the line. Because of the added inductance from the transformer coils, autotransformer starting reduces the power factor.

The electric utility commonly sets a limit of 400% current draw from the power line on the line side of the transformer. For example, a motor has a full-voltage starting torque of 120% and a full-voltage starting current of 600%. The power company has set a limitation of 400% current draw from the power line. This limitation is set only for the line side of the controller. Because the transformer has a step-down ratio, the motor current on the transformer secondary is larger than the line current. In this case, the current in the primary of the transformer must not exceed 400%.

In this example, with the line current limited to 400%, 80% voltage can be applied to the motor, generating 80% motor current. The motor draws only 64% line current ($0.8 \times 80 = 64\%$) due to the 1:0.8 turns ratio of the transformer. The advantage is that the starting torque is 77% (0.8×80 of 120%) instead of the 51% obtained in primary resistor starting. This additional percentage may be sufficient accelerating energy to start a load that may be difficult to start otherwise.

Autotransformer Starting Circuits. In an autotransformer starting circuit, the various windings of the transformer are added to and taken away from the motor circuit to provide reduced voltage when starting. **See Figure 11-9.**

The control circuit consists of an ON-delay timer TR1 and contactor coils C1, C2, and C3. Pressing start pushbutton PB2 energizes the timer, causing instantaneous contacts TR1 in line 2 and 3 of the line diagram to close. Closing the normally open (NO) timer contacts in line 2 provides memory for timer TR1, while closing NO timer contacts in line 3 completes an electrical path through line 4, energizing contactor coil C2. The energizing of coil C2 causes NO contacts C2 in line 5 to close, energizing contactor coil C3. The normally closed (NC) contacts in line 3 also provide electrical interlocking for coil C1 so that they may not be energized together. The NO contacts of contactor C2 close, connecting the ends of the autotransformers together when coil C2 energizes. When coil C3 energizes, the NO contacts of contactor C3 close and connect the motor through the transformer taps to the power line, starting the motor at reduced inrush current and starting torque. Memory is also provided to coil C3 by contacts C3 in line 6.

After a predetermined time, the ON-delay timer times out and the NC timer contacts TR1 open in line 4, de-energizing contactor coil C2, and NO timer contacts TR1 close in line 3, energizing coil C1. In addition, NC contacts C1 provide electrical interlock in line 4, and NC contacts C2 in line 3 return to their NC position. The net result of de-energizing C2 and energizing C1 is the connecting of the motor to full line voltage.

Note that during the transition from starting to full line voltage, the motor was not disconnected from the circuit, indicating closed-circuit transition. As long as the motor is running in the full-voltage condition, timer TR1 and contactor C1 remain energized. Only an overload or pressing the stop pushbutton stops the motor and resets the circuit. Overload protection is provided by a separate overload block.

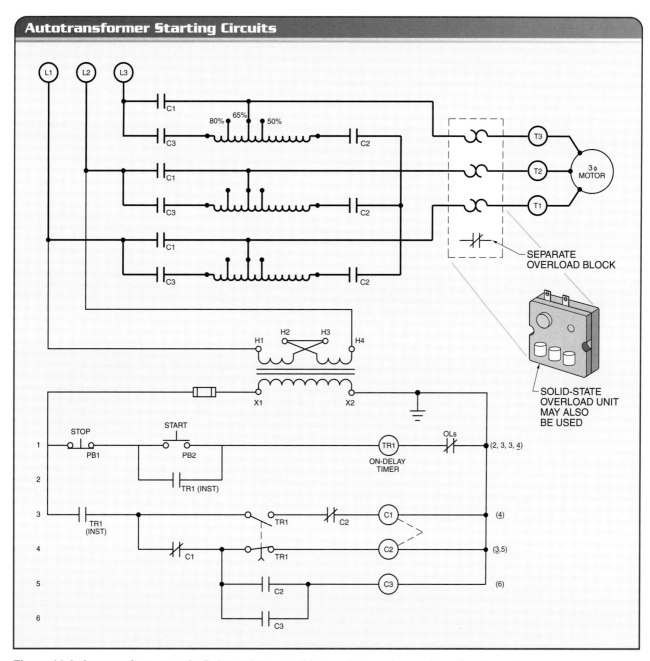

Figure 11-9. Autotransformers typically have three possible taps that can be used to adjust the voltage.

In this circuit, pushbuttons are used to control the motor. However, any NO and/or NC device may be used to control the motor. Thus, in an air conditioning system, the pushbuttons would be replaced with a temperature switch, and the circuit would be connected for two-wire control.

Impedance Reduced-Voltage Starting. A starter using reactors instead of autotransformers looks almost identical to an autotransformer starter. **See Figure 11-10.** The difference between the two methods is that an autotransformer is connected across the line and the reactor has one lead to the line and one lead to the stator lead. Rather than selecting different voltages with the taps on the autotransformer, impedance is added to the circuit to lower the voltage to the stator. The voltage drop depends on which tap is selected. Inductors in series act just as resistors in series do, with the source voltage dropped across the inductors (stator and reactors) in series. When sufficient time to accelerate the motor is monitored by a time-delay relay, the reactors are shunted by the contacts in the starter and the motor sees the source in its entirety.

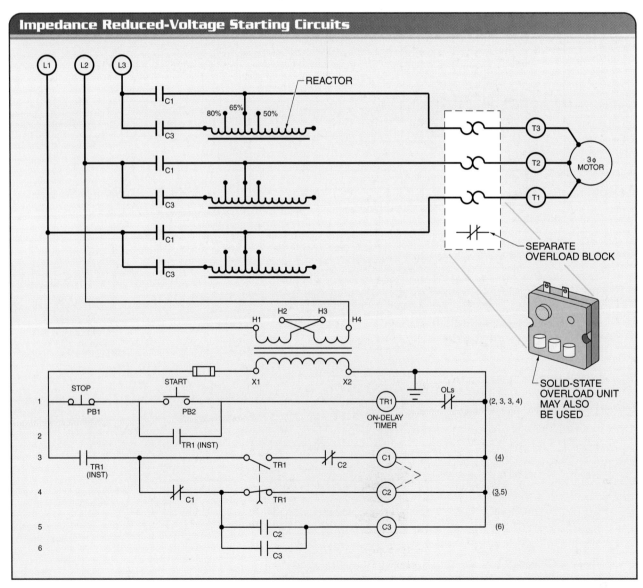

Figure 11-10. A starter using reactors instead of autotransformers looks almost identical to an autotransformer starter.

Part-Winding Starting

Part-winding starting is a method of reduced-voltage starting that applies voltage to only part of the motor coil windings for starting and then applies power to the remaining coil windings for normal running. The motor stator windings must be divided into two or more equal parts for a motor to be started using part-winding starting. Each equal part must have its termination available for external connections. In most applications, a wye-connected motor is used.

Part-winding starting is the least expensive reduced-voltage starting method. Part-winding starting has poor starting torque because the starting torque is fixed. It is unsuitable for heavy loads or long starts because it is subject to overheating until the entire winding is in the circuit.

Not all motors should be part-winding started. Consult the manufacturer specifications before applying part-winding starting to a motor. Some motors are wound sectionally with part-winding starting in mind. Indiscriminate application of part-winding starting to any dual-voltage motor can lead to excessive noise and vibration during starting, overheating, and extremely high transient currents on switching.

Part-winding starting is not truly a reduced-voltage starting method. Part-winding starting is usually classified as reduced-voltage starting because of the resulting reduced current and torque.

Part-Winding Starting Circuits. Part-winding reduced-voltage starting is less expensive than other starting methods and produces less starting torque. **See Figure 11-11.**

The control circuit consists of motor starter M1, ON-delay timer TR1, and motor starter M2. Pressing start pushbutton PB2 energizes starter M1 and timer TR1. M1 energizes the motor, and closes contacts M1 in line 2 to provide memory. With the motor starter M1 energized, L1 is connected to T1, L2 to T2, and L3 to T3, starting the motor at reduced current and torque through one-half of the wye windings.

The ON-delay NO contacts of ON-delay timer TR1 in line 2 remain open during timing and close after timing out, energizing coil M2. When M2 energizes, L1 is connected to T7, L2 to T8, and L3 to T9, applying voltage to the second set of wye windings. The motor now has both sets of windings connected to the supply voltage for full current and torque. The motor may normally be stopped by pressing stop pushbutton PB1 or by an overload in any line. Each magnetic motor starter need be only half-size because each one controls only one-half of the winding. Overloads must be sized accordingly.

Definition

Part-winding starting is a method of reduced-voltage starting that applies voltage to only part of the motor coil windings for starting and then applies power to the remaining coil windings for normal running.

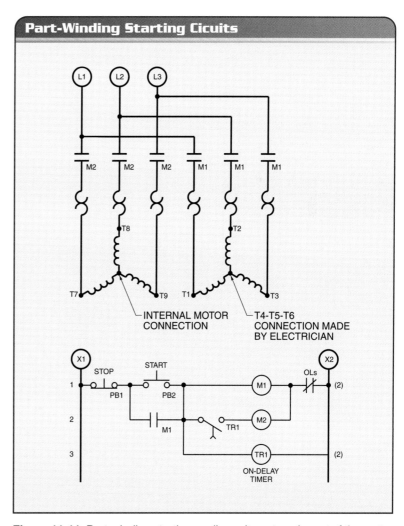

Figure 11-11. Part-winding starting applies voltage to only part of the motor coil windings for starting and then applies power to the remaining coil windings for normal running.

Wye-Connected Motors. A motor suitable for part-winding starting has two sets of identical windings in parallel. These windings produce reduced-starting current and reduced starting torque when energized in sequence. Most dual-voltage 230/460-V motors are suitable for part-winding starting at 230 V. **See Figure 11-12.**

Part-winding starters are typically available in either two- or three-step construction. The more common two-step starter is designed so that when the control circuit is energized, one winding of the motor is connected directly to the line. This winding draws about 65% of normal locked-rotor current and develops approximately 45% of normal motor torque. After a short time, the second winding is connected in parallel with the first winding in such a way that the motor is electrically complete across the line and develops its normal torque.

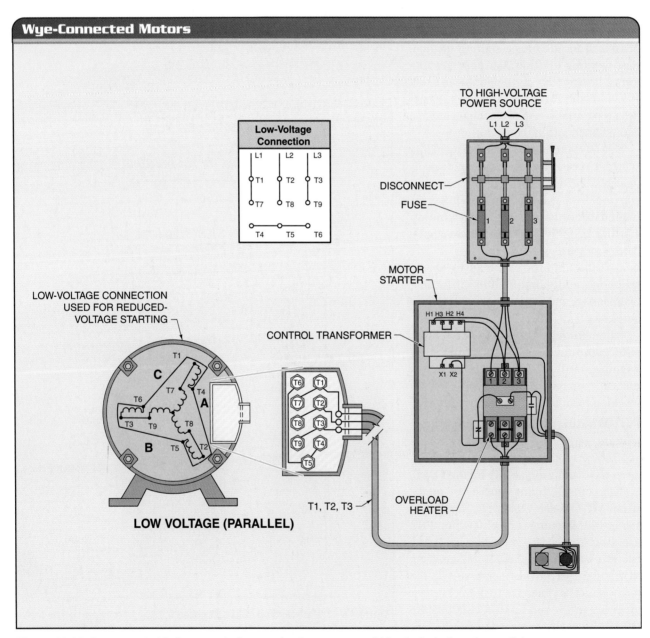

Figure 11-12. A motor suitable for part-winding starting has two sets of identical windings in parallel.

Wye-Delta Starting

Wye-delta starting accomplishes reduced-voltage starting by first connecting the motor leads in a wye configuration for starting. A motor started in the wye configuration receives approximately 58% of the normal voltage and develops approximately 33% of the normal torque. Wye-delta starting works well with high-inertia, long-starting loads because the voltage and current are normally well within the winding ratings.

Single-voltage, wye-delta motors are specially wound with six leads extending from the motor to enable the windings to be connected in either a wye or delta configuration. **See Figure 11-13.** When a wye-delta starter is energized, two contactors close. One contactor connects the windings in a wye configuration and the second contactor connects the motor to line voltage. After a time delay, the wye contactor opens, momentarily de-energizing the motor, and the third contactor closes to reconnect the motor to the power lines with the windings connected in a delta configuration. A wye-delta starter is inherently an open-transition system because the leads of the motor are disconnected and then reconnected to the power supply.

This starting method does not require any accessory voltage-reducing equipment such as resistors and transformers. Wye-delta starting gives a higher starting torque per line ampere than part-winding starting, with considerably less noise and vibration.

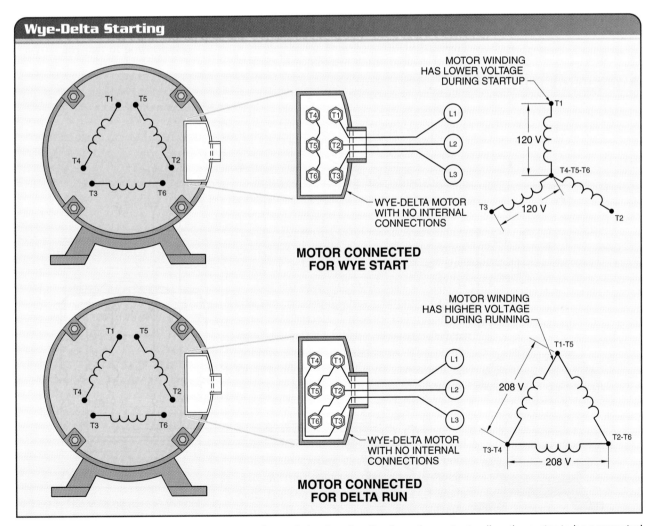

Figure 11-13. Single-voltage, wye-delta motors have six leads extending from the motor to allow the motor to be connected in a wye or delta configuration.

252 MOTORS

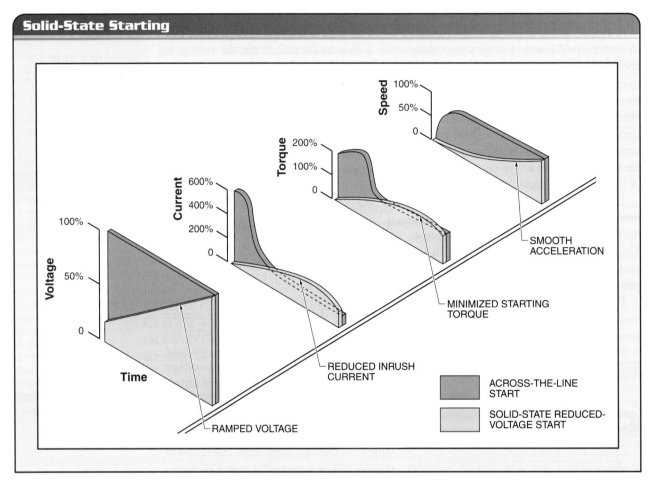

Figure 11-16. Solid-state starters ramp up the voltage as the motor accelerates.

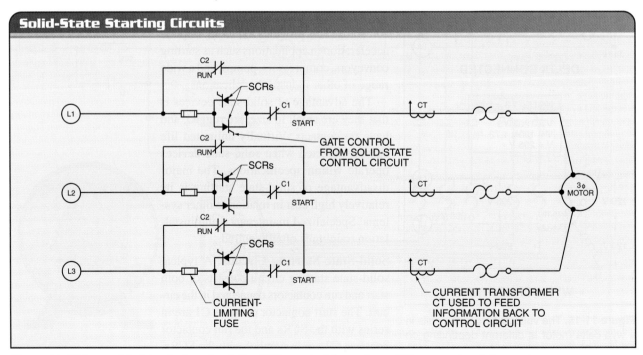

Figure 11-17. In a solid-state starting circuit, the SCRs control the motor until it approaches full speed.

The start contacts C1 close and the acceleration of the motor is controlled by triggering ON the SCRs when the starter is energized. The SCRs control the motor until it approaches full speed, at which time the run contacts C2 close, connecting the motor directly across the power line. At this point, the SCRs are turned OFF, and the motor runs with full power applied to the motor terminals.

Soft Starters. A *soft starter* is a device that provides a gradual voltage increase (ramp up) during AC motor starting. Most soft starters also provide soft stopping (ramp down) capabilities. Soft starters are part of solid-state starters used to control single-phase and three-phase motors.

Soft starting is achieved by increasing the motor voltage in accordance with the setting of the ramp-up control. A potentiometer is used to set the ramp-up time (normally 1 sec to 20 sec). Soft stopping is achieved by decreasing the motor voltage in accordance with the setting of the ramp-down control.

A second potentiometer is used to set the ramp-down time (normally 1 sec to 20 sec). A third potentiometer is in the circuit to adjust the starting level of motor voltage to a value at which the motor starts to rotate immediately when soft starting is applied. **See Figure 11-18.**

Like any solid-state switch, a soft starter produces heat that must be dissipated for proper operation. The heat dissipation requires large heat sinks and sometimes requires forced ventilation when high-current loads (motors) are controlled. For this reason, a contactor is often added in parallel with a soft starter. The soft starter is used to control the motor when the motor is starting or stopping. The contactor is used to short out or bypass the soft starter when the motor is running. This allows for soft starting and soft stopping without the need for large heat sinks during motor running. The soft starter includes an output signal that is used to control the time when the contactor is ON or OFF. **See Figure 11-19.**

> **Definition**
>
> A *soft starter* is a device that provides a gradual voltage increase (ramp up) during AC motor starting.

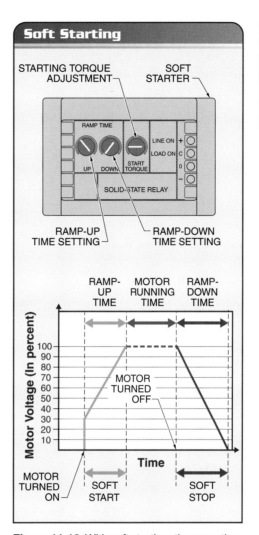

Figure 11-18. With soft starting, the ramp time and starting torque are adjustable.

Furnas Electric Co.
A magnetic motor starter is a contactor with added overload protection.

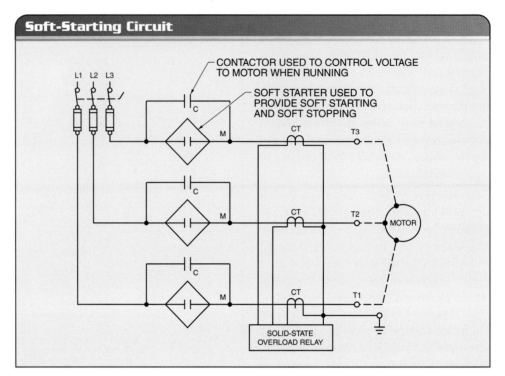

Figure 11-19. A soft starter includes an output signal that is used to control the time when the contactor is ON or OFF.

STARTING-METHOD COMPARISON

Several starting methods are available when an industrial application calls for using reduced-voltage starting. The amount of reduced current, the amount of reduced torque, and the cost of each starting method must be considered when selecting the appropriate starting method.

The selection is not simply a matter of selecting the starting method that reduces the current the most. The motor does not start and the motor overloads trip if the starting torque is reduced too much.

A general comparison can be made of the amount of reduced current for each type of starting method compared to across-the-line starting. **See Figure 11-20.** The amount of reduced current is adjustable when using solid-state or autotransformer starting. Autotransformer starting uses taps so the amount of reduced current is somewhat adjustable. Solid-state starting is adjustable throughout its range. Some primary resistor starters are adjustable, others are not. Part-winding and wye-delta starting are not adjustable.

A general comparison can be made of the amount of reduced torque for each type of starting method compared to across-the-line starting. The amount of reduced torque is adjustable when using the solid-state or autotransformer starting method. The autotransformer starting method has taps, so the amount of reduced torque is somewhat adjustable. Solid-state starting is adjustable throughout its range. The motor overloads trip if the load requires more torque than the motor can deliver. The torque requirements of the load must be taken into consideration when selecting a starting method.

A general comparison can also be made of the costs for each type of starting method compared to across-the-line starting. Although reducing the amount of starting current or starting torque in comparison to the load requirements is the primary consideration for selecting a starting method, cost may also have to be considered. The costs vary for each starting method. **See Figure 11-21.**

The primary resistor starting method is used when it is necessary to restrict inrush current to predetermined increments. Primary resistors can be built to meet almost any current inrush limitation. Primary resistors also provide smooth acceleration and can be used where it is necessary to control starting torque. Primary resistor starting may be used with any motor.

The autotransformer starting method provides the highest possible starting torque per ampere of line current and is the most effective means of motor starting for applications where the inrush current must be reduced with a minimum sacrifice of starting torque. Three taps are provided on an autotransformer, making it field adjustable. Cost must be considered because the autotransformer is the most expensive type of transformer. Autotransformer starting can be used with any motor.

The part-winding starting method is simple in construction and economical in cost. Part-winding starting provides a simple method of accelerating fans, blowers, and other loads involving low starting torque. The part-winding starting method requires a nine-lead wye motor. The cost is less than for other methods because no external resistors or transformers are required.

The wye-delta starting method is particularly suitable for applications involving long accelerating times or frequent starts. Wye-delta starting is commonly used for high-inertia loads such as centrifugal air conditioning units, although it can be used in applications where low starting torque is necessary or where low starting current and low starting torque are permissible. The wye-delta starting method requires a special six-lead motor.

The solid-state starting method provides smooth, stepless acceleration in applications such as starting conveyors, compressors, and pumps. Solid-state starting uses a solid-state controller, which uses SCRs to control motor voltage, current, and torque during acceleration. Although the solid-state starting method offers the most control over a wide range, it is also the most expensive.

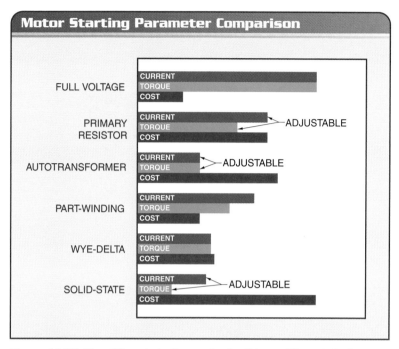

Figure 11-20. The different starting methods all have different starting currents, torques, and costs, allowing a starting method to be chosen for a specific application.

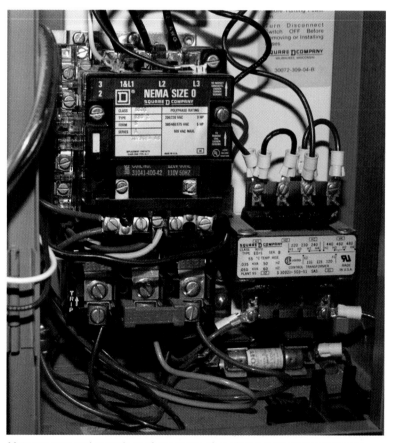

Motor starters and control transformers are often placed in the same enclosure.

Motor Starting Method Comparison

Starter Type	Starting Characteristics			Standard Motor	Transition	Extra Acceleration Steps Available	Installation Cost	Advantages	Disadvantages	Applications
	Volts at Motor	Line Current	Starting Torque							
Across-the-Line	100%	100%	100%	Yes	None	None	Lowest	Inexpensive, readily available, simple to maintain, maximum starting torque	High inrush, high starting torque	Many and various
Primary Resistor	65%	65%	42%	Yes	Closed	Yes	High	Smooth acceleration, high power factor during start, less expensive than autotransformer starter in low HPs, available with as many as 5 accelerating points	Low torque efficiency, resistors give off heat, starting time in excess of 5 sec, requires expensive resistors, difficult to change starting torque under varying conditions	Belt and gear drives, conveyors, textile machines
Autotransformer	80% 65% 50%	64% 42% 25%	64% 42% 25%	Yes	Closed	No	High	Provides highest torque per ampere of line current, 3 different starting torques available through autotransformer taps, suitable for relatively long starting periods, motor current is greater than line current during starting	Is most expensive design in lower HP ratings, low power factor, large physical size	Blowers, pumps, compressors, conveyors
Part-Winding	100%	65%	48%	*	Closed	Yes†	Low	Least expensive reduced-voltage starter, most dual-voltage motors can be started part-winding on lower voltage, small physical size	Unsuited for high-inertia, long-starting loads, requires special motor design for voltage higher than 230 V, motor does not start when torque demanded by load exceeds that developed by motor when first half of motor is energized, first step of acceleration must not exceed 5 sec or motor overheats	Reciprocating compressors, pumps, blowers, fans
Wye-Delta	100%	33%	33%	No	Open‡	No	Medium	Suitable for high-inertia, long-acceleration loads, high torque efficiency, ideal for especially stringent inrush restrictions, ideal for frequent starts	Requires special motor, low starting torque, momentary inrush occurs during open transition when delta contactor is closed	Centrifugal compressors, centrifuges
Solid-State	Adjust	Adjust	Adjust	Yes	Closed	Adjust	Highest	Energy-saving features available, voltage gradually applied during starting for a soft start condition, adjustable acceleration time, usually self-calibrating, adjustable built-in braking features included	High cost, requires specialized maintenance and installation, electrical transients can damage unit, requires good ventilation	Machine tools, hoists, packaging equipment, conveyor systems

* Standard dual-voltage 230/460 V motor can be used on 230 V systems
† Very uncommon
‡ Closed transition available for average of 30% more cost

Figure 11-21. All starting methods have advantages and disadvantages that influence which method is chosen for any particular application.

NEMA and IEC Ratings

Contactors and motor starters are rated according to the size (horsepower and/or current rating) and their voltage rating. The National Electrical Manufacturers Association (NEMA) and the International Electrotechnical Commission (IEC) are two primary organizations that rate contactors and motor starters.

NEMA contactors and motor starters are based on their continuous current (amperage) rating and voltage rating. NEMA ratings are listed as a size number, which ranges from size 00 to size 9.

In addition to the standard NEMA ratings, some motor starter manufacturers also provide motor-matched sizes (MM sizes). Motor-matched sizes fall between the standard NEMA sizes. Motor-matched sizes allow for a more closely matched size in applications in which the motor size is known and some cost savings can be gained by using a half-size rated starter.

Nema Contactor And Starter Ratings

NEMA Size	Continuous-Current Rating*	Horsepower 230 VAC	Horsepower 460 VAC
00	9	1	2
0	18	3	5
1	27	7	10
2	45	15	25
3	90	30	50
4	135	50	100
5	270	100	200
6	540	200	400
7	810	300	600
8	1215	450	900
9	2250	800	1600

* in A

Motor-Matched Size Ratings

MM Size	Continuous-Current Rating*	Horsepower 230 VAC	Horsepower 460 VAC
1¾	40	10	15
2½	60	20	31
3½	115	40	75
4½	210	75	150

* in A

IEC contactors and motor starters are based on their maximum operational current (amperage) rating. The IEC does not specify a size number. Instead, IEC contactors and motor starters state a utilization category rating that defines the typical duty of the IEC contactor or motor starter. Utilization categories AC-3 and AC-4 are used for most motor applications.

IEC Contactor and Starter Ratings

Utilization Category	IEC Category Description
AC-1	Used with noninductive or slightly inductive loads such as lamps and heating elements.
AC-2	Used with light inductive loads such as solenoids.
AC-3	Used with motors in which the motor is typically turned OFF only after the motor is operating at full speed.
AC-4	Used with motors in which the motor is used with rapid starting and stopping (jogging, inching, plugging, etc.).

Supplemental Topic

Solid-State Switches

Solid-state switches are electronic devices that have no moving parts (contacts). Solid-state switches can be used in most motor control applications. Advantages of solid-state switches include fast switching, no moving parts, long life, and the ability to be interfaced with electronic circuits (PLCs and PCs). However, solid-state switches must be properly selected and applied to prevent potential problems. Solid-state switches include transistors, silicon-controlled rectifiers (SCRs), insulated gate bipolar transistors (IGBTs), triacs, and alternistors.

Transistors

A *transistor* is a three-terminal device that controls current through the device depending on the amount of voltage applied to the base. Transistors may be NPN or PNP transistors. Transistors can be switched ON and OFF quickly. Transistors have a very high resistance when open and a very low resistance when closed. Transistors are used to switch low-level DC only. When transistors are used as switches, a diode can be mounted across the transistor to avoid damage from high-voltage spikes (transients).

Silicon-Controlled Rectifiers

A *silicon-controlled rectifier (SCR)* is a solid-state rectifier with the ability to rapidly switch heavy currents. SCRs are used as solid-state, low- and high-level DC switches. An SCR is either ON or OFF. The SCR is turned ON when voltage is applied to its gate. The SCR remains ON as long as current flows through the anode and cathode. The SCR is turned OFF when current flow is stopped. One SCR can be used to switch high-level DC. When controlling high-level AC, two SCRs can be mounted in an antiparallel configuration. Each SCR is used to control one-half of the AC sine wave. The advantage of using two separate SCRs (rather than one triac) is greater heat dissipation. One SCR in a diode bridge can be used when low-level current switching is required (often on printed circuit boards).

Insulated Gate Bipolar Transistors

An *insulated gate bipolar transistor (IGBT)* is a combination of several solid-state devices with high current and voltage capabilities and has a high switching speed and easy control. The widespread use of the IGBT as the switching element in inverters is a relatively recent development. The main benefits of an IGBT are the very fast switching time and high current-carrying capacity. This results in very efficient circuits.

Triacs

A *triac* is a three-terminal semiconductor thyristor that is triggered into conduction in either direction by a small current to its gate. Triacs are used as solid-state AC switches. Like an SCR, a triac is either ON or OFF. A triac is turned ON when voltage is applied to its gate. Once ON, the triac allows current to flow in both directions (AC). The triac is turned OFF when the gate voltage is removed. A triac that is used as a switch may have a snubber mounted across it to avoid damage from high-current transients.

Alternistors

An *alternistor* is two antiparallel thyristors and a triac mounted on the same chip. The alternistor was developed specially for industrial AC high-current switching applications. A combination of three alternistors is normally used in three-phase switching applications. An alternistor requires less space than antiparallel SCRs. The components are separated for increased heat dissipation.

Device/Application	Symbol
TRANSISTOR NPN or PNP DC switch, used to switch low-level DC	NPN / PNP (BASE, COLLECTOR, EMITTER)
SCR DC switching device, one SCR used to switch high-level currents	SCR (ANODE, CATHODE, GATE)
Two antiparallel SCRs used to switch AC, provides better cooling than triac alone	ANTIPARALLEL SCRs
One SCR in diode bridge used to switch low-level DC	SCR IN DIODE BRIDGE
IGBT DC switching device	IGBT (COLLECTOR, GATE, EMITTER)
TRIAC AC switching device	TRIAC
ALTERNISTOR AC switching device used to switch high-level AC, normally used when switching three-phase currents, provides better cooling than triac alone	ALTERNISTOR

Application—Wye-Delta Starting Overload Protection

A wye-delta motor is connected in a wye configuration during starting and then reconnected in a delta configuration during running. A wye-delta motor can be wound in single-voltage or dual-voltage configurations. A single-voltage, wye-delta motor needs 6 leads. A dual-voltage, wye-delta motor must have 12 leads.

For a delta-connected motor, the coil voltage is equal to the line voltage. For a wye-connected motor, the coil voltage is less than the line voltage. For a 100-A, 480-V motor, the coils are rated at 480 V. In a delta configuration, the voltage across the coils is 480 V. In a wye configuration, the voltage across the coils voltage is 58% of the delta voltage and is calculated as follows:

$$E_{coil} = \frac{E_{line}}{\sqrt{3}}$$

$$E_{coil} = \frac{480}{\sqrt{3}}$$

$$E_{coil} = \mathbf{277\,V}$$

For a wye-connected motor, the coil current is equal to the line current. For a delta-connected motor, the coil current is less than the line current. Since the motor nameplate current is 100 A, the coils are rated at 100 A. In a wye-configuration, the current through the coils is 100 A. In a delta configuration, the current through the coils is 58% of the wye current and is calculated as follows:

$$I_{coil} = \frac{I_{line}}{\sqrt{3}}$$

$$I_{coil} = \frac{100}{\sqrt{3}}$$

$$I_{coil} = \mathbf{58\,A}$$

Therefore, the current in a wye-connected motor is 58% of the current in a delta-connected motor. Since the voltage is also only 58% of the line voltage, the starting current of a wye-connected motor is 58% of 58%, or 33%, of the full-voltage starting current of a delta-connected motor.

The nameplate current is the line current at full load. The overload is placed in series with the stator coils to monitor the coil current, not the line current. Therefore, overloads in a wye-delta motor are sized for the stator coil current, not the line current. Since a wye-delta motor operates in a delta configuration, only 58% of the line current flows through each coil and the overloads are sized for 58% of the line current.

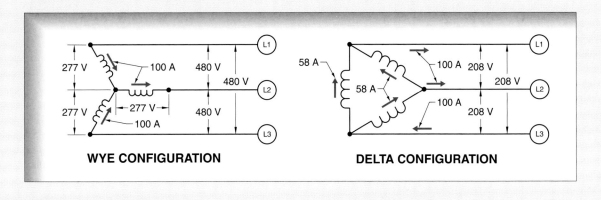

WYE CONFIGURATION   **DELTA CONFIGURATION**

Summary

- Full-voltage starting is the simplest and least expensive method of starting a motor.

- Full-voltage starting has high starting current and high starting torque.

- Primary resistor starting uses resistors in series in the power line to reduce the starting voltage and current to about 60% to 70% of full-voltage starting.

- Primary resistor starting has relatively slow starting, uses relatively expensive resistors, and makes it difficult to change the starting torque to match varying conditions.

- Autotransformer starting reduces the starting voltage to 80%, 65%, or 50% of line voltage, depending on the tap chosen.

- Autotransformer starting provides the highest torque per ampere of line current, with the motor current greater than line current during starting.

- Autotransformer starting is relatively expensive, especially in smaller motors, and has a low power factor.

- Impedance reduced-voltage starting uses a tapped reactor coil instead of an autotransformer to reduce the starting voltage.

- Part-winding starting uses full line voltage, but reduces starting current by using only some of the windings.

- Part-winding starting is the least expensive reduced-voltage starting method.

- Part-winding starting is unsuitable for high-inertia, long-starting loads because it may result in overheating with heavy loads or long starts.

- Wye-delta starting starts a motor in a wye configuration and switches to a delta configuration for full-load operation.

- Wye-delta starting has low starting current and works well with high-inertia, long-starting loads.

- Wye-delta starting requires a special motor.

- Solid-state starting uses solid-state components to control the voltage and current during starting.

- Solid-state starting has energy-saving features and allows soft starts and adjustable acceleration time.

- Solid-state starting has the highest cost of the starting methods and may require specialized maintenance and installation training.

- In an open-circuit transition, the motor is temporarily disconnected from the power circuit as the motor is switched from reduced-voltage starting to full-voltage running.

- In a closed-circuit transition, the motor remains connected to the power circuit as the motor is switched from reduced-voltage starting to full-voltage running.

Glossary

Full-voltage starting is a method of starting a motor with the full line voltage placed across the terminals.

A ***manual contactor*** is a control device that uses pushbuttons to energize or de-energize the load connected to it.

A ***manual starter*** is a contactor with an added overload protection device. Manual starters are used only in electrical motor circuits.

Locked rotor is a condition when a motor is loaded so heavily that the motor shaft cannot turn.

A ***magnetic motor starter*** is an electrically operated switch that includes motor overload protection.

Primary-resistor starting is a method of reduced-voltage starting that places resistors in series in the motor power circuit to reduce the voltage to the motor.

Autotransformer starting is a method of reduced-voltage starting that uses a tapped three-phase autotransformer to provide reduced voltage for starting.

Part-winding starting is a method of reduced-voltage starting that applies voltage to only part of the motor coil windings for starting and then applies power to the remaining coil windings for normal running.

A ***solid-state starter*** is a motor starter that uses a solid-state device, such as an insulated gate bipolar transistor (IGBT) or silicon-controlled rectifier (SCR), to control motor voltage, current, torque, and speed during acceleration.

A ***soft starter*** is a device that provides a gradual voltage increase (ramp up) during AC motor starting.

Review

1. Explain the difference between open-transition and closed-transition starting.

2. Explain why reduced-voltage starting is used instead of full-voltage starting.

3. List some advantages and disadvantages of full-voltage starting.

4. List some advantages and disadvantages of primary resistor starting.

5. List some advantages and disadvantages of autotransformer starting.

6. List some advantages and disadvantages of part-winding starting.

7. List some advantages and disadvantages of wye-delta starting.

8. List some advantages and disadvantages of solid-state starting.

9. Explain why an autotransformer starter has more torque than other starters for the same line current.

Refer to the CD-ROM for Quick Quiz® questions related to chapter content.

MOTORS

Plugging Switch

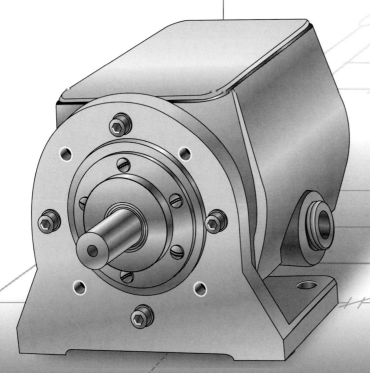

Friction Brake

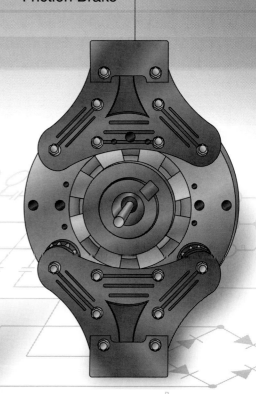

12

Braking

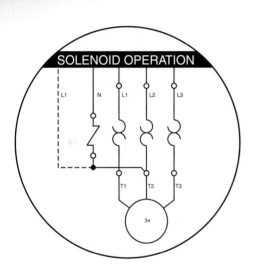

Braking	264
Friction Braking	264
Plugging	265
Electric Braking	269
Dynamic Braking	272
Braking Comparison	274
Supplemental Topic—Determining Brake Torque	274
Supplemental Topic—AC Motor Dynamic Brake Sizing	276
Application—Braking Solenoid Connections	278
Summary	278
Glossary	278
Review	279

OBJECTIVES

- Explain how a friction brake is used to stop a motor.
- Describe the operation of a plugging brake system.
- Describe how a DC field is used in electric braking.
- Explain how dynamic braking works to create generator action as a motor is turned OFF.
- Compare the different types of braking and understand the advantages and disadvantages of each type.

Braking is used when a motor must be stopped more quickly than coasting allows. The four most common methods of braking a motor are friction braking, plugging, electric braking, and dynamic braking. Each method has advantages and disadvantages.

BRAKING

A motor coasts to a stop when disconnected from the power supply. The time taken by the motor to come to rest depends on the inertia of the moving parts (motor and motor load) and friction. Braking is used when it is necessary to stop a motor more quickly than coasting allows. Hazard braking may be required to protect an operator even if braking is not part of the normal stopping method.

Braking is accomplished by different methods. The braking method used depends on the application, available power, circuit requirements, cost, and desired results. Braking applications vary greatly. For example, braking may be applied to a motor every time the motor is stopped, or it may be applied to a motor only in an emergency. In the first application, the braking action requires a method that is reliable with repeated use. In the second application, the method of stopping the motor may give little or no consideration to the damage braking may do to the motor or motor load.

FRICTION BRAKING

Friction brakes normally consist of two friction surfaces (shoes or pads) that come in contact with a wheel mounted on the motor shaft. Spring tension holds the shoes on the wheel and braking occurs as a result of the friction between the shoes and the wheel. **See Figure 12-1.** Friction brakes (magnetic or mechanical) are the oldest motor stopping method. Friction brakes are similar to the brakes on automobiles.

The advantages of using friction brakes are a lower initial cost and simplified maintenance. Friction brakes are less expensive to install than other braking methods because fewer expensive electrical components are required. Maintenance is simplified because it is easy to see whether the shoes are worn and if the brake is working. Friction brakes are available in both AC and DC designs to meet the requirements of almost any application. Friction brake applications include printing presses, cranes, overhead doors, hoisting equipment, and machine tool control.

Heidelberg Harris, Inc.

Figure 12-1. Friction brakes normally consist of two friction surfaces that come in contact with a wheel mounted on the motor shaft.

Solenoid Operation

Friction brakes are normally controlled by a solenoid that activates the brake shoes. The solenoid is energized when the motor is running. This keeps the brake shoes from touching the drum mounted on the motor shaft. The solenoid is de-energized and the brake shoes are applied through spring tension when the motor is turned OFF.

Two methods are used to connect the solenoid into the circuit so that it activates the brake whenever the motor is turned ON and OFF. **See Figure 12-2.** The first circuit is used if the solenoid has a voltage rating equal to the motor voltage rating. The second circuit is used if the solenoid has a voltage rating equal to the voltage between L1 and the neutral. The solenoid should always be connected directly into the motor circuit, not into the control circuit. This eliminates improper activation of the brake.

Tech Fact

Solenoids are characterized by their coil inrush current and by their sealed current. Manufacturers provide selection tables to help choose the correct solenoid for any application.

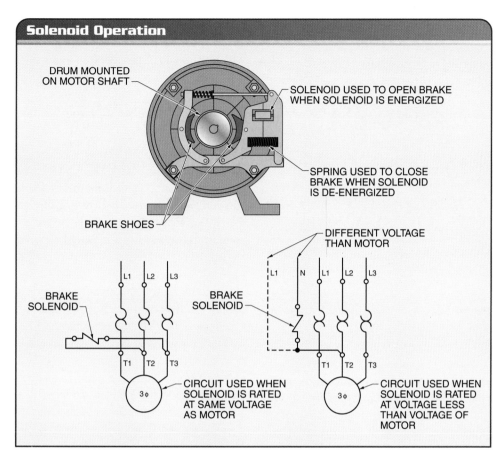

Figure 12-2. The solenoid should always be connected directly into the motor circuit, not into the control circuit. Another power source may be needed if the solenoid is not rated for line voltage.

Definition

Plugging is a method of motor braking in which the motor connections are reversed so that the motor develops a countertorque that acts as a braking force.

Brake Shoes

In friction braking, the braking action is applied to a wheel mounted on the shaft of the motor rather than directly to the shaft. The wheel provides a much larger braking surface than could be obtained from the shaft alone. This permits the use of large brake shoe linings and low shoe pressure. Low shoe pressure, equally distributed over a large area, results in even wear and braking torque. The braking torque developed is directly proportional to the surface area and spring pressure. The spring pressure is adjustable on nearly all friction brakes.

Limitation of Friction Braking

The disadvantage of friction brakes is that they require more maintenance than other braking methods. Maintenance consists of replacing the shoes. Shoe replacement depends on the number of times the motor is stopped. A motor that is stopped often needs more maintenance than a motor that is almost never stopped.

PLUGGING

Plugging is a method of motor braking in which the motor connections are reversed so that the motor develops a countertorque that acts as a braking force. The countertorque is accomplished by reversing the motor at full speed with the reversed motor torque opposing the forward inertia torque of the motor and its mechanical load. Plugging a motor allows for very rapid stopping. Although manual and electromechanical controls can be used to reverse the direction of a motor, a plugging switch is normally used in plugging applications. **See Figure 12-3.**

Figure 12-3. A plugging switch is normally used in plugging applications.

A plugging switch is connected mechanically to the shaft of the motor or driven machinery. The rotating motion of the motor is transmitted to the plugging switch contacts either by a centrifugal mechanism or by a magnetic induction arrangement. The contacts on the plugging switch are NO, NC, or both, and actuate at a given speed. The primary function of a plugging switch is to prevent the reversal of the load once the countertorque action of plugging has brought the load to a standstill. The motor and load would start to run in the opposite direction without stopping if the plugging switch were not present.

Plugging Switch Operation

Plugging switches are designed to open and close sets of contacts as the shaft speed on the switch varies. As the shaft speed increases, the contacts are set to change at a given revolution per minute (rpm). As the shaft speed decreases, the contacts return to their normal condition. As the shaft speed increases, the contact setpoint (point at which the contacts operate) reaches a higher revolution per minute than the point at which the contacts reset (return to their normal position) on decreasing speed. The difference in these contact operating values is the differential speed or revolutions per minute.

In plugging, the continuous running speed must be many times the speed at which the contacts are required to operate. This provides a high contact holding force and reduces possible contact chatter or false operation of the switch.

Continuous Plugging

A plugging switch may be used to plug a motor to a stop each time the motor is stopped. **See Figure 12-4.** In this circuit, the NO contacts of the plugging switch are connected to the reversing starter through an interlock contact. Pushing the start pushbutton energizes the forward starter, starting the motor in forward and adding memory to the control circuit. As the motor accelerates, the NO plugging contacts close.

The closing of the NO plugging contacts does not energize the reversing starter because of the interlocks. Pushing the stop pushbutton drops out the forward starter and interlocks. This allows the reverse starter to immediately energize through the plugging switch and the NC forward interlock. The motor is reversed and the motor brakes to a stop. After the motor is stopped, the plugging switch opens to disconnect the reversing starter before the motor is actually reversed.

Plugging for Emergency Stops

A plugging switch may be used in a circuit where plugging is required only in an emergency. **See Figure 12-5.** In this circuit, the motor is started in the forward direction by pushing the run pushbutton. This starts the motor and adds memory to the control circuit. As the motor accelerates, the NO plugging contacts close. Pushing the stop pushbutton de-energizes the forward starter but does not energize the reverse starter. This is because there is no path for the L1 power to reach the reverse starter, so the motor coasts to a stop.

Tech Fact
Plugging is accomplished by reversing the motor connections at full speed. This stops the motor very quickly.

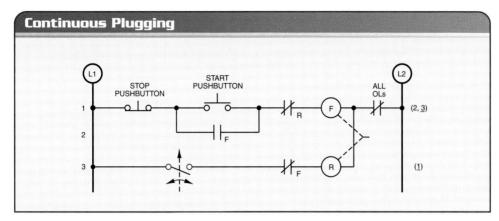

Figure 12-4. A plugging switch may be used to plug a motor to a stop each time the motor is stopped.

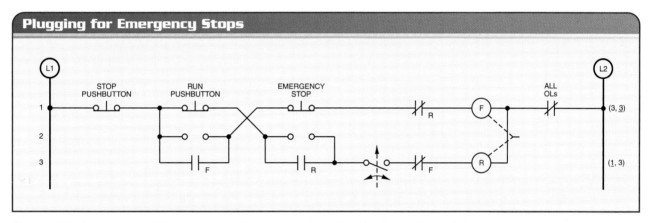

Figure 12-5. A plugging switch may be used in a circuit where plugging is required only in an emergency.

Pushing the emergency stop pushbutton de-energizes the forward starter and simultaneously energizes the reversing starter. Energizing the reversing starter adds memory in the control circuit and plugs the motor to a stop. When the motor is stopped, the plugging switch opens to disconnect the reversing starter before the motor is actually reversed. The de-energizing of the reversing starter also removes the memory from the circuit.

Plugging Using Timing Relays

Plugging can also be accomplished by using a timing relay. The advantage of using a timing relay is normally a lower cost since a timer is inexpensive and does not have to be connected mechanically to the motor shaft or driven machine. The disadvantage is that, unlike a plugging switch, the timer does not compensate for a change in the load condition once the timer is preset, which affects stopping time.

An OFF-delay timer may be used in applications where the time needed to decelerate the motor is constant and known. **See Figure 12-6.** In this circuit, the NO contacts of the timer are connected into the circuit in the same manner as a plugging switch. The coil of the timer is connected in parallel with the forward starter.

The motor is started and memory is added to the circuit when the start pushbutton is pressed. In addition to energizing the forward starter, the OFF-delay timer is also energized. The energizing of the OFF-delay timer immediately closes the NO timer contacts. The closing of these contacts does not energize the reverse contacts because of the interlocks.

The forward starter and timer coil are de-energized when the stop pushbutton is pressed. The NO timing contact remains held closed for the setting of the timer. The holding closed of the timing contact energizes the reversing starter for the period of time set on the timer. This plugs the motor to a stop. The timer's contact must reopen before the motor is actually reversed. The motor reverses direction if the time setting is too long.

An OFF-delay timer may also be used for plugging a motor to a stop during emergency stops. **See Figure 12-7.** In this circuit, the timer's contacts are connected in the same manner as the plugging switch. The motor is started and memory is added to the circuit when the start pushbutton is pressed. The forward starter and timer are de-energized if the stop pushbutton is pressed. Although the timer's NO contacts are held closed for the time period set on the timer, the reversing starter is not energized. This is because no power is applied to the reversing starter from L1.

If the emergency stop pushbutton is pressed, the forward starter and timer are de-energized and the reversing starter is energized. The energizing of the reversing starter adds memory to the circuit and stops the motor. The opening of the timing contacts de-energizes the reversing starter and removes the memory.

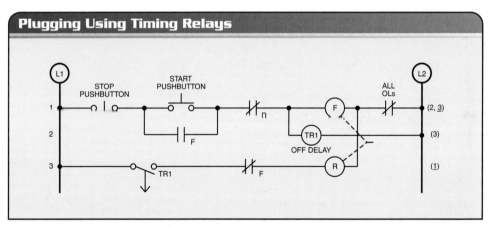

Figure 12-6. An OFF-delay timer may be used in applications where the time needed to decelerate the motor is constant and known.

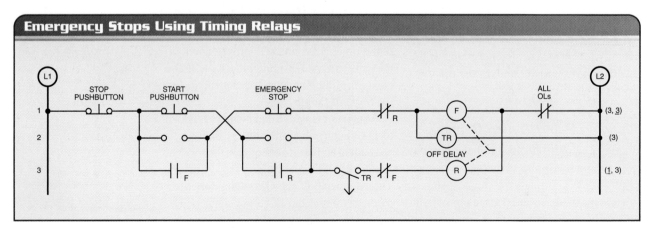

Figure 12-7. An OFF-delay timer may also be used for plugging a motor to a stop during emergency stops.

Limitations of Plugging

Plugging may not be applied to all motors and/or applications. Braking a motor to a stop using plugging requires that the motor can be reversed at full speed. Even if the motor can be reversed at full speed, the damage that plugging can do may outweigh its advantages.

Reversing. A motor cannot be used for plugging if it cannot be reversed at full speed. For example, a single-phase shaded-pole motor cannot be reversed at any speed. Thus, a single-phase shaded-pole motor cannot be used in a plugging circuit. Likewise, most single-phase split-phase and capacitor-start motors cannot be plugged because their centrifugal switches remove the starting windings when the motor accelerates. Without the starting winding in the circuit, the motor cannot be reversed.

Heat. All 3-phase motors and most single-phase and DC motors can be used for plugging. However, high current and heat result from plugging a motor to a stop. A motor is connected in reverse at full speed when plugging a motor. The current may be three or more times higher during plugging than during normal starting. For this reason, a motor designated for plugging or a motor with a high service factor should be used in all cases except emergency stops. The service factor (SF) should be 1.35 or more for plugging applications.

ELECTRIC BRAKING

Electric braking, or *DC injection braking*, is a method of braking in which a DC voltage is applied to the stator windings of a motor after the AC voltage is removed. **See Figure 12-8.** Electric braking is an efficient and effective method of braking most AC motors. Electric braking provides a quick and smooth braking action on all types of loads including high-speed and high-inertia loads. Maintenance is minimal because there are no parts that come in physical contact during braking.

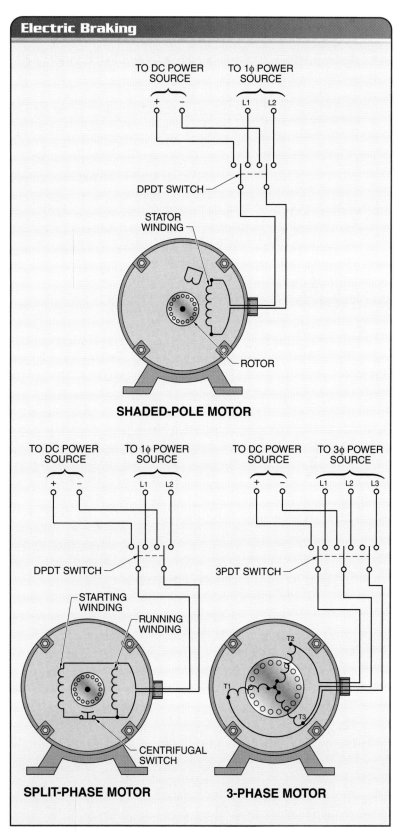

Figure 12-8. Electric braking applies a DC voltage to the stator windings of a motor.

Definition

Electric braking, or DC injection braking, is a method of braking in which a DC voltage is applied to the stator windings of a motor after the AC voltage is removed.

Electric Braking Operation

The principle that unlike magnetic poles attract each other and like magnetic poles repel each other explains why a motor shaft rotates. The method in which the magnetic fields are created changes from one type of motor to another. In AC induction motors, the opposing magnetic fields are induced from the stator windings into the rotor windings by transformer action. The motor continues to rotate as long as the AC voltage is applied. The motor coasts to a standstill when the AC voltage is removed because there is no induced field to keep it rotating. Electric braking can be used to provide an immediate stop if the coasting time is unacceptable, particularly in an emergency situation.

Electric braking is accomplished by applying a DC voltage to the stationary windings once the AC is removed. The DC voltage creates a constant magnetic field in the stator that does not change polarity. The rotor moving through the constant magnetic field in the stator induces current and a magnetic field in the rotor. Because the magnetic field of the stator does not change in polarity, it attempts to stop the rotor when the magnetic fields are aligned (N to S and S to N). **See Figure 12-9.** The only force that can keep the rotor from stopping with the first alignment is the rotational inertia of the load connected to the motor shaft. However, because the braking action of the stator is present at all times, the motor brakes quickly and smoothly to a standstill.

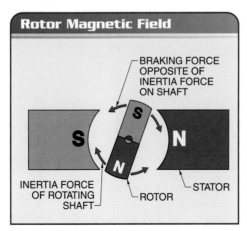

Figure 12-9. Electric braking uses the attraction between unlike poles to brake a motor.

DC Electric Braking Circuits

DC is applied after the AC is removed to bring the motor to a stop quickly. **See Figure 12-10.** This circuit, like most DC braking circuits, uses a bridge rectifier circuit to change the AC into DC. In this circuit, a 3-phase AC motor is connected to 3-phase power by a magnetic motor starter.

The magnetic motor starter is controlled by a standard stop/start pushbutton station with memory. An OFF-delay timer is connected in parallel with the magnetic motor starter. The OFF-delay timer controls a NO contact that is used to apply power to the braking contactor for a short period of time after the stop pushbutton is pressed. The timing contact is adjusted to remain closed until the motor comes to a stop.

The braking contactor connects two motor leads to the DC supply. A transformer with tapped windings is used to adjust the amount of braking torque applied to the motor. Current-limiting resistors could be used for the same purpose. This allows for a low- or high-braking action, depending on the application. The larger the applied DC voltage, the greater the braking force.

The interlock system in the control circuit prevents the motor starter and braking contactor from being energized at the same time. This is required because the AC and DC power supplies must never be connected to the motor simultaneously. Total interlocking should always be used on electrical braking circuits. Total interlocking is the use of mechanical, electrical, and pushbutton interlocking. A standard forward and reversing motor starter can be used in this circuit, as it can with most electric braking circuits.

Tech Fact

Electric braking is accomplished by applying a DC voltage to the stationary windings once the AC is removed. The DC voltage creates a constant-polarity magnetic field in the stator that quickly stops the rotor.

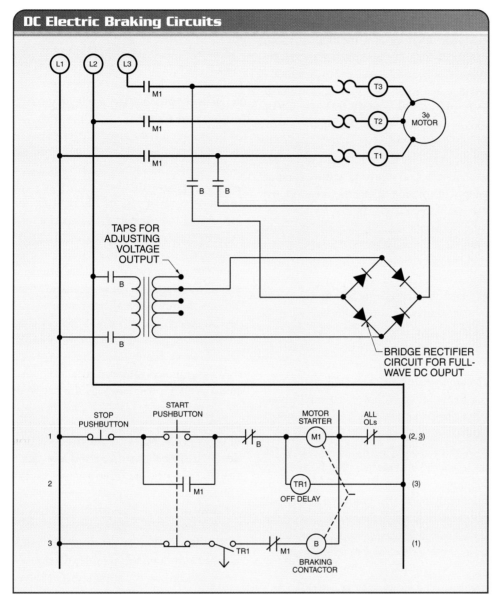

Figure 12-10. A bridge rectifier is used to rectify AC power to generate the DC required for electric braking.

Limitations of Electric Braking

Electric braking is relatively expensive because of the electronic components required to switch the DC field. Braking a motor one time with electric braking puts about the same amount of heat into the rotor as starting the motor at full voltage. Therefore, the time at which electric braking can be applied may be limited to prevent overheating of the rotor. Motion sensors can be used to determine when the motor has been stopped. At that point, the DC field is turned off to stop further heat input. The motor must be sized appropriately to account for this heating if the motor is to be braked regularly.

Electric braking requires a power source at all times. If the power source fails, the electric brake cannot stop the motor. In addition, electric braking cannot be used as a holding brake. For critical applications, some other type of external brake must be used to reduce the risk of personal injury or property damage.

Definition

Dynamic braking is a method of motor braking where the braking energy is dissipated as heat in a resistor as a motor is reconnected to act as a generator immediately after it is turned OFF.

Regenerative braking is a method of dynamic braking that reuses the braking energy to the AC source instead of dissipating the energy as heat.

DYNAMIC BRAKING

Dynamic braking is a method of motor braking where the braking energy is dissipated as heat in a resistor as a motor is reconnected to act as a generator immediately after it is turned OFF. Connecting the motor in this way makes the motor act as a loaded generator that develops a retarding torque, rapidly stopping the motor. The generator action converts the mechanical energy of rotation to electrical energy that can be dissipated as heat in a resistor.

Regenerative braking is a method of dynamic braking that reuses the braking energy to the AC source instead of dissipating the energy as heat. There are two common methods of reusing braking energy. The first method is to return the power to the utility bus, while the second method is to provide power to parallel motors on the same plant bus.

The need to stop a motor arises often in a situation where a mechanical brake is not a feasible solution. Although a dynamic brake cannot hold a load in position, it can bring the load to a stop in a very short time. A dynamic brake uses unlike magnetic poles that attract each other to provide stopping power to the rotating component. The unlike pole in the rotating component is produced as the result of generator action in the motor. The generator action can be used to stop the armature in a DC motor or the rotor in an AC motor.

DC Motor Dynamic Braking

Dynamic braking is easily applied to DC motors because there must be access to the rotor windings to reconnect the motor to act as a generator. This generator action in a DC motor is the countervoltage that provides the opposition to current in the armature. Access is accomplished through the brushes on DC motors. **See Figure 12-11.** Dynamic braking of a DC motor may be needed because DC motors are often used for lifting and moving heavy loads that may be difficult to stop.

When the source is removed from the armature and it continues to spin, the energized fields provide a stationary field through which the armature cuts as it coasts to a stop. As long as no path for current is provided, no stopping or decelerating action opposes the rotation. If a path is provided, current flows through the path. The fields that surround the armature conductors create an unlike pole and the two fields interact, stopping the motor.

In this circuit, the armature terminals of the DC motor are disconnected from the power supply and immediately connected across a resistor that acts as a load. The smaller the resistance of the resistor, the greater the rate of energy dissipation and the faster the motor comes to rest. The field windings of the DC motor are left connected to the power supply. The armature generates a countervoltage that causes current to flow through the resistor and armature.

The current causes heat to be dissipated in the resistor in the form of electrical watts. This removes energy from the system and slows the motor rotation. The generated countervoltage decreases as the speed of the motor decreases. As the motor speed approaches 0 rpm, the generated voltage also approaches 0 V. The braking action lessens as the speed of the motor decreases. As a result, a motor cannot be braked to a complete stop using dynamic braking. Dynamic braking also cannot hold a load once it is stopped because there is no braking action.

The direction of the armature conductor as it spins through the stationary field provides a current in the armature. The direction of the current through the conductor surrounds that conductor with a field that is unlike the stationary one. These two unlike poles attract one another, bringing the rotating member to a stop.

Electromechanical friction brakes are often used along with dynamic braking in applications that require the load to be held. A combination of dynamic braking and friction braking can also be used in applications where a large, heavy load is to be stopped. In these applications, the force of the load wears the friction brake shoes excessively. Therefore, dynamic braking can be used to slow the load before the friction brakes are applied.

Dynamic Braking

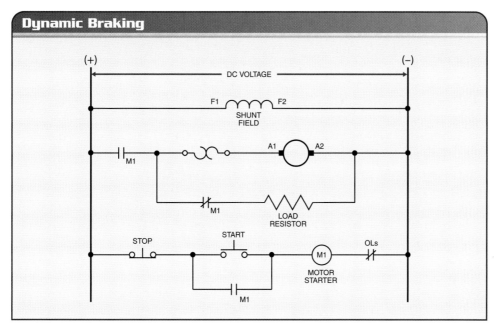

Figure 12-11. The generator action in a DC motor is the countervoltage that provides the opposition to current in the armature. The energy is dissipated in the shunt resistor.

AC Motor Dynamic Braking

Dynamic braking can be used in an AC motor if a stationary field surrounds the stator conductors. A stationary field can be created when a DC source is connected to the stator. With the stator providing a stationary field, the rotor conductors cut through this field and induce an unlike pole in the rotor. The two unlike poles attract each other, and the dynamic action stops the rotating component.

The DC power can be supplied by a small transformer, such as a control transformer, through a bridge rectifier. For a 3-phase motor, the DC source is connected across two arms of a wye-connected stator or across one arm of a delta-connected stator. For a single-phase motor, a dynamic brake needs to be connected to the run winding. The time that the DC current flows in the winding should be controlled by a time delay relay because the winding can overheat if left ON.

Electric Motor Drive Dynamic Braking

A motor drive can control the time it takes for a motor to stop. The stopping time is programmed by setting the deceleration parameter. The deceleration parameter can be set between 1 sec (or less) to several minutes. However, for fast stops (especially with high-inertia loads), a braking resistor is added. **See Figure 12-12.**

The braking resistor also helps control motor torque during repeated ON/OFF motor cycling. The size (wattage rating) and resistance (in ohms) of the braking resistor are typically selected using the drive manufacturer braking resistor chart. The resistor wattage must be high enough to dissipate the heat delivered at the resistor. The resistor resistance value must be low enough to allow current flow at a high enough level to convert the current to heat but not so low as to cause excessive current flow. The braking resistor(s) can be housed in an enclosure such as a NEMA 1 enclosure.

Limitations of Dynamic Braking

Dynamic braking uses power resistors to dissipate the heat of braking. Depending on the design of the resistor bank, dynamic braking may put limits on the number of allowed stops per hour. Dynamic braking is relatively expensive because of the electronic components required to reconnect the motor as a generator. The braking action is the strongest as the leads are reversed, and the breaking action decreases as the motor slows because the generator action decreases as the motor slows. Therefore, dynamic braking cannot brake a motor to a complete stop and cannot be used as a holding brake.

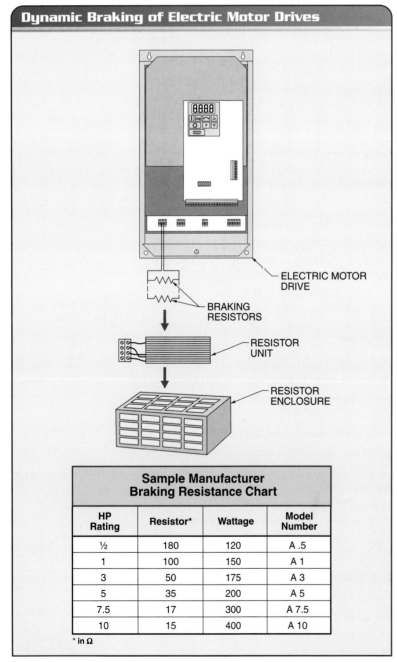

Figure 12-12. The stopping time of a motor drive can be programmed into the drive. Braking resistors are added for fast stops.

BRAKING COMPARISON

The four most common methods of braking a motor are friction braking, plugging, electric braking, and dynamic braking. Friction braking has the advantage of being relatively inexpensive with simple maintenance. Few expensive electrical components are needed. Friction braking has the disadvantage of requiring frequent maintenance to inspect and replace the brake shoes. Frequent stopping increases wear on the shoes and increases the amount of heat generated in the brake.

Plugging is more expensive than friction braking but less expensive than other braking methods. It allows for rapid stopping and can be used as an emergency stop. Plugging can only be used with motors that can be reversed at full speed. Plugging draws high current and generates considerable heat, so a motor with a high service factor should be used.

Electric braking provides a quick and smooth braking action on all types of loads. Maintenance is minimal because there are no parts that wear from physical contact. However, electric braking is relatively expensive because of the electronic components required to switch the DC field. Electric braking requires a power source at all times. It cannot stop a motor in the event of a power failure, and it cannot be used as a holding brake.

Dynamic braking provides a quick braking action and is often used to stop high-inertia loads. Dynamic braking is used most often with DC motors, but can be adapted for AC motors. Dynamic braking is relatively expensive because of the electronic components required to reconnect the motor as a generator. The braking action is strongest as the leads are reversed. Dynamic braking cannot brake a motor to a complete stop because the generator action decreases as the motor slows. It cannot be used as a holding brake.

> **Tech Fact**
>
> *The four most common methods of braking a motor are friction braking, plugging, electric braking, and dynamic braking. All of the methods have advantages and limitations that restrict their application.*

Determining Brake Torque

Full-load motor torque is calculated in order to determine the required braking torque of a motor. To calculate full-load motor torque, apply the following formula:

$$T = \frac{P \times 5252}{\upsilon}$$

where

T = full-load motor torque (in lb-ft)

P = motor power (in HP)

υ = rotor speed (in rpm)

The full-load torque of a 60 HP, 240 V motor rotating at 1725 rpm is calculated as follows:

$$T = \frac{P \times 5252}{\upsilon}$$

$$T = \frac{60 \times 5252}{1725}$$

$$T = \frac{315,125}{1725}$$

$$T = \mathbf{183\ lb\text{-}ft}$$

The torque rating of the brake selected should be equal to or greater than the full-load motor torque. In this case, a brake must have a torque rating of at least 183 lb-ft. Manufacturers of electric brakes list the torque ratings (in pounds per foot) for their brakes.

Braking torque may also be determined using a horsepower-to-torque conversion chart. A line is drawn from the horsepower (HP) to the revolutions per minute (rpm) of the motor. The point at which the line crosses the torque values is the full-load and braking torque of the motor. For example, a 50 HP motor that rotates at 900 rpm requires a braking torque of 300 lb-ft.

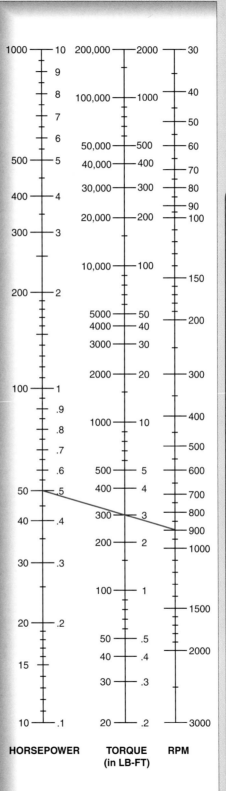

Supplemental Topic

AC Motor Dynamic Brake Sizing

With the dynamic braking of AC motors, DC power must be supplied to the stator windings to create the stationary field. The amount of power required depends on the amount of power being used to run the motor. The simplest method to determine the amount of DC power needed is to measure the winding resistance and then read the nameplate current. The voltage of the power supply can be calculated with Ohm's law.

If a motor has a winding resistance of 12.5 Ω with a nameplate current of 4 A, the power supply voltage required to drive that current is 50 V (12.5 × 4 = 50). A power supply that delivers 4 A at 50 V must be sized at a minimum of 200 VA (50 × 4 = 200). This means that 200 VA of DC power must be supplied to the stator windings.

If the power factor is 0.8, a 250 VA AC power supply would be needed. A 250 VA power supply would need to be placed in the control cabinet. If the control transformer is oversized by 250 VA, it can be used as the power source.

The AC power supply becomes the source for the bridge rectifier. The bridge rectifier must be sized and installed. The power resistors must be sized to dissipate the power. The power capabilities of the rectifier can be obtained from the manufacturer's literature.

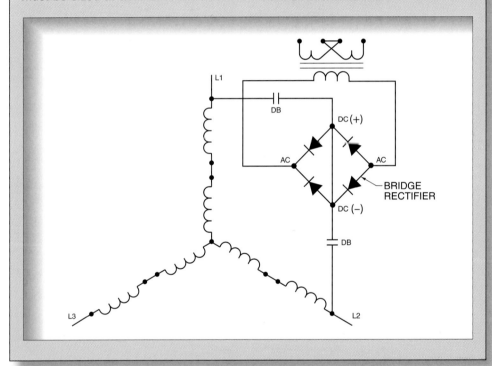

Application—Braking Solenoid Connections

Solenoid brake coils have a voltage rating that must be followed. A typical solenoid brake coil is rated for 240 VAC and can be used in both the low- and high-voltage connections. In a one-wye, dual-voltage motor wired for 480 V operation, the windings are wired in series with the terminals shorted together from T4 to T7, T5 to T8, and T6 to T9. In a two-wye, dual-voltage motor wired for 240 V operation, the windings are wired in parallel with the terminals shorted together from T1 to T7, T2 to T8, and T3 to T9.

In the motor wired for 480 V operation, the solenoid brake coil is wired from the short between two coils in each leg, such as from the T4-T7 junction to the T6-T9 junction, where the voltage drop is 240 V. In the motor wired for 240 V operation, the solenoid brake coil is wired from the line terminals, such as from the T1-T7 junction to the T3-T9 junction, where the voltage drop is 240 V.

In a one-delta, dual-voltage motor wired for 480 V operation, the windings are wired in series with the terminals shorted together from T4 to T7, T5 to T8, and T6 to T9. In a two-delta, dual-voltage motor wired for 240 V operation, the windings are wired in parallel with the terminals shorted together from T1 to T7, T2 to T8, and T3 to T9. In each case, the solenoid brake coil is wired across any one winding, where the voltage drop is 240 V.

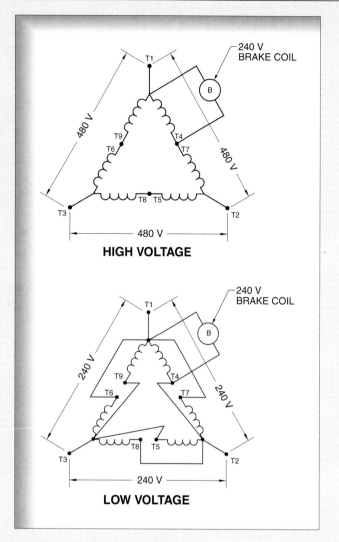

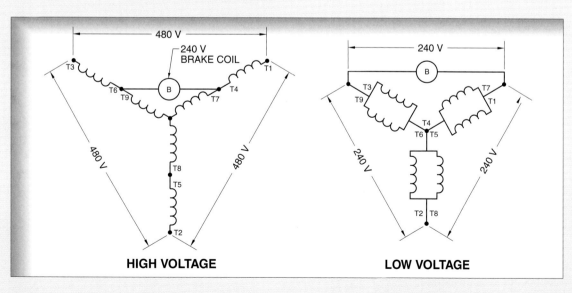

Summary

- Braking is used when a motor must be stopped more quickly than coasting allows.

- Friction brakes normally consist of two friction surfaces (shoes or pads) that come in contact with a wheel mounted on the motor shaft.

- The advantages of using friction brakes are a lower initial cost and simplified maintenance.

- The disadvantage of friction brakes is that they require more maintenance than other braking methods.

- Friction brakes are normally controlled by a solenoid that activates the brake shoes.

- With plugging, the motor is reversed at full speed to develop a counter torque.

- A plugging switch may be used to plug a motor to a stop each time the motor is stopped.

- Braking a motor to a stop using plugging requires that the motor can be reversed at full speed.

- Electric braking is accomplished by applying a DC voltage to the stationary windings once the AC is removed.

- Electric braking provides a quick and smooth braking action on all types of loads including high-speed and high-inertia loads.

- Electric braking requires a power source at all times and cannot be used as a holding brake.

- With dynamic braking, the motor acts as a loaded generator that develops a retarding torque, rapidly stopping the motor.

- Dynamic braking of a DC motor may be needed because DC motors are often used for lifting and moving heavy loads that may be difficult to stop.

- Dynamic braking uses power resistors to dissipate the heat of braking.

Glossary

Plugging is a method of motor braking in which the motor connections are reversed so that the motor develops a countertorque that acts as a braking force.

Electric braking, or **DC injection braking,** is a method of braking in which a DC voltage is applied to the stator windings of a motor after the AC voltage is removed.

Dynamic braking is a method of motor braking where the braking energy is dissipated as heat in a resistor as a motor is reconnected to act as a generator immediately after it is turned OFF.

Regenerative braking is a method of dynamic braking that reuses the braking energy to the AC source instead of dissipating the energy as heat.

Review

1. Describe the differences in braking requirements between typical braking and hazard braking.

2. Explain how friction braking operates to brake a motor.

3. List the advantages and disadvantages of friction braking.

4. Explain how plugging operates to brake a motor.

5. List the advantages and disadvantages of plugging.

6. Explain how electric braking operates to brake a motor.

7. List the advantages and disadvantages of electric braking.

8. Explain how dynamic braking operates to brake a motor.

9. List the advantages and disadvantages of dynamic braking.

Refer to the CD-ROM for Quick Quiz® questions related to chapter content.

MOTORS

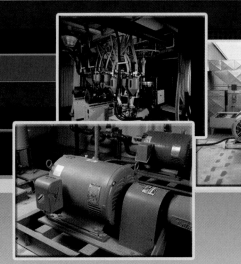
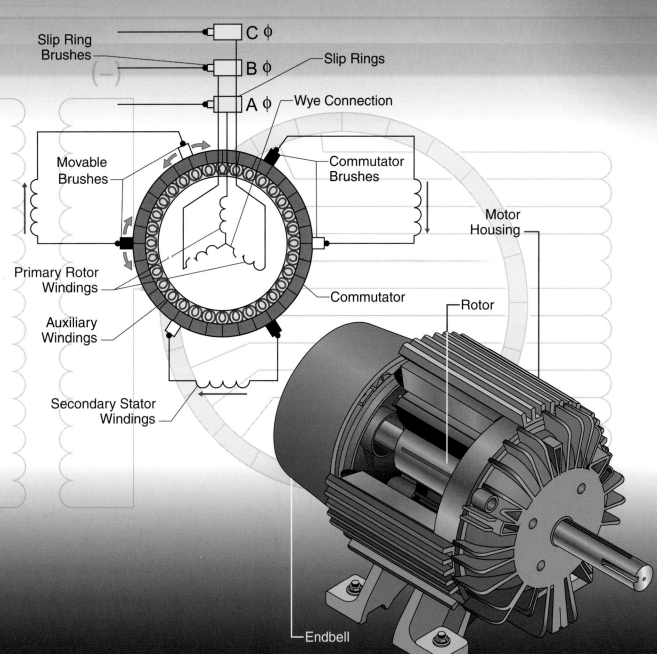

13

Multispeed Motors

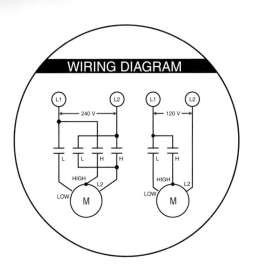

WIRING DIAGRAM

Multispeed Motor Types	282
Consequent-Pole Motor Circuits	285
Supplemental Topic—Schrage Motor Control Voltages	289
Application—Two-Speed Separate-Winding Motors	290
Application—Two-Speed Consequent-Pole Motors	291
Summary	292
Glossary	292
Review	293

OBJECTIVES

- Describe the operation of a two-speed consequent-pole motor.
- Explain how moving the brushes on a Schrage motor changes the motor speed.
- Describe the difference between compelling, accelerating, and decelerating circuit logic.

Some multispeed AC motors are designed to be operated at a constant line frequency. These motors have stator windings that can be connected in different ways to change the number of poles and thus to change the motor speed. These multispeed motors are available in two or more fixed speeds, which are determined by the connections made to the motor.

Some older multispeed AC motors use a commutator and slip rings, along with adjustable brushes, to vary the speed of the motor. These motors are rarely seen anymore. They have mostly been replaced by AC adjustable-speed drives.

Definition

A consequent-pole motor is a motor with stator windings that can be connected in two or more different ways so that the number of stator poles can be changed.

MULTISPEED MOTOR TYPES

In consequent-pole multispeed AC motors, the different speeds are determined by connecting the external winding leads to a multispeed starter. A magnetic motor starter is the most common means used for AC motor speed control, although multispeed manual starters are available.

Two-winding, consequent-pole motors have two separate windings in the same slots. Because the windings are separate and independent, they can be wound for any other pole combinations so that other speed ratios can be obtained. However, speed ratios greater than about 4:1 are rare because of cost, size, and weight concerns.

In Schrage AC motors, the rotor has a commutator, slip rings, and adjustable brushes. The relative position of the brushes is changed to change the speed of the motor.

Consequent-Pole Motors

AC motors can be designed to operate at multiple speeds. A *consequent-pole motor* is a motor with stator windings that can be connected in two or more different ways so that the number of stator poles can be changed. Since the speed of an induction motor depends on the number of poles in the stator, most consequent-pole motors have two speeds, one of which is half the other. Less commonly, two stator windings can be connected in more than two ways to give more than two speeds.

A motor can run at only one speed at a time. One starter is required for each speed of the motor. Each starter must be interlocked using mechanical, auxiliary contact, or pushbutton interlocking to prevent more than one starter from being ON at the same time. For two-speed motors, a standard forward/reverse starter is often used because it provides mechanical interlocking.

Consequent-Pole Motor Operation. Consequent-pole motors have an adjustable speed because the leads for the stator windings are brought out of the motor and connected in different ways to create different numbers of poles. The current can be made to travel through windings connected in series or parallel with each other. **See Figure 13-1.** A series connection of two windings has one pole. The current flows through both coils in the same direction, so both coils have the same pole. The poles merge since they are adjacent. The same windings connected in parallel have two poles. The current flows through the coils in opposite directions, so the coils have opposite poles. The poles are separate since they are opposite. For example, a motor with four poles when wired in series has eight poles when wired in parallel.

Starter contactors typically have several contacts that can be used to switch the stator windings from series to parallel operation. Typical synchronous speeds for 60 Hz consequent-pole motors are 3600 rpm and 1800 rpm with 2 and 4 poles; 1800 rpm and 900 rpm with 4 and 8 poles; and 1200 rpm and 600 rpm with 6 and 12 poles.

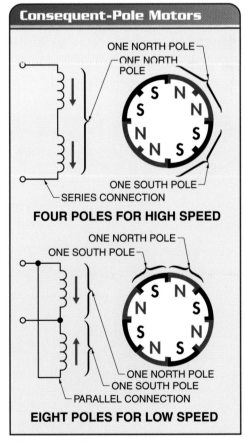

Figure 13-1. The current can be made to travel through windings connected in series or in parallel with other.

Consequent-Pole Motor Construction. Since each motor winding is divided into two equal parts, consequent-pole motors have one stator winding that may be connected in two ways to provide two speeds, one of which is half the speed of the other. A 3-phase motor can be wired so that six leads are brought out for connection. **See Figure 13-2.** The nameplate gives the required connections for the two speeds.

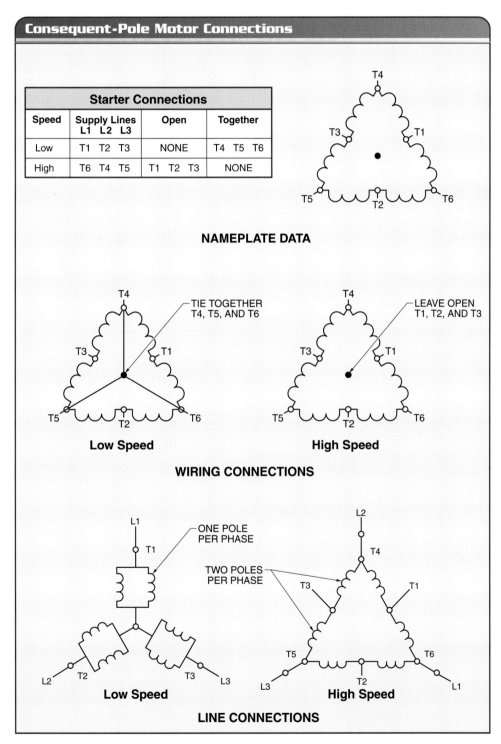

Figure 13-2. A 3-phase motor can be wired so that six leads are brought out for connection.

> **Definition**
>
> A *Schrage motor* is a 3-phase AC motor with a rotor fed by a commutator and a set of brushes connected to an external circuit.

Some consequent-pole motors have two separate stator windings. For a three-speed, consequent-pole motor, one winding is fixed with a specified number of poles. The other winding can be reconnected like the winding in a standard consequent-pole motor to create a varying number of poles. Three more leads are brought out, labeled T11, T12, and T13. There are many different wiring arrangements to give constant horsepower, constant torque, and variable torque. **See Appendix.**

For a four-speed, consequent-pole motor, both windings can be reconnected like the winding in a standard consequent-pole motor to create a varying number of poles. Four more leads are brought out, labeled T14, T15, T16, and T17. There are many different wiring arrangements to give constant horsepower, constant torque, and variable torque.

Schrage Motors

A *Schrage motor* is a 3-phase AC motor with a rotor fed by a commutator and a set of brushes connected to an external circuit. A Schrage motor is also known as both an ACA motor and as a shunt-commutator motor. Line voltage is applied to the rotor windings through the slip rings. Therefore, the rotor winding is the primary and always has current at line frequency. The stator winding is the secondary and the current frequency depends on the amount of slip. Schrage motors are seldom used anymore. They have been mostly replaced by adjustable-speed motor drives.

Schrage Motor Construction. Schrage motor construction is unique in that the rotor has slip rings and a commutator. **See Figure 13-3.** The rotor has two windings. The primary winding is located in the rotor and is excited through the slip rings. A control winding is placed in the same rotor slots as the primary windings. The control winding is also known as a tertiary winding, an auxiliary winding, and an armature winding. The secondary winding is in the stator and is excited through brushes on the commutator.

The coils in the control winding are connected in series. The number of control coils matches the number of commutator segments because each end of the control coil is connected to adjacent commutator segments. The rotor windings are wound in a wye, with the three leads connected to the slip rings. The brushes on the slip rings are connected to the 3-phase source.

In Figure 13-3, the commutator brushes are shown in two colors. The black brushes are all mounted in one brush rigging. The white brushes are all mounted in another brush rigging. The two brush riggings can be moved relative to each other so that each dark brush can be rotated relative to the corresponding white brush. In other words, the two brushes of each secondary stator winding can be moved relative to each other.

Schrage Motor Operation. A Schrage motor contains adjustable brush riggings that are rotated relative to one another by a handwheel or by a small 3-phase motor with a reversing contactor. The advantage of the motorized control is that the speed can be adjusted from many different locations by placing the buttons in strategic locations around the motor.

When the circuit is energized, the rotor is fed by the source through the slip rings and a voltage is induced in the control winding. The speed is controlled by changing the position of the brush at one end of the primary stator winding relative to the brush at the other end of the winding. As the brush riggings are rotated, the brushes move to different commutator segments.

The coils in the auxiliary winding have an induced voltage that is connected in series. There is a neutral point with the brush riggings where the motor operates like a standard induction motor. This neutral point is where both brushes of each secondary stator winding are on the same commutator segment. As the brushes are moved relative to one another, the brushes are placed on different commutator segments and the number of coils connected in series across the stator changes. Since voltage in series is additive, changing the number of coils changes the voltage across the primary stator windings.

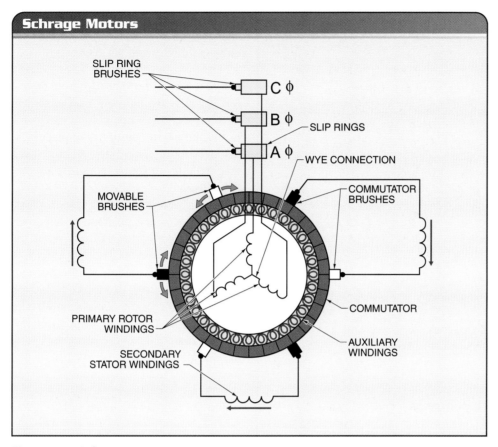

Figure 13-3. A Schrage motor has a rotor with slip rings and a commutator.

The voltage may be increased or decreased, depending on the direction of rotation of the brushes relative to one another. As the brushes are rotated in one direction from neutral, the polarity of the coils in the control winding shorted across the primary changes in one direction and the induced voltage in the control winding subtracts from the secondary winding. This causes reduced current and torque and the motor slows down. As the brushes are rotated in the other direction, the polarity of the coils in the control winding changes in the other direction, and the induced voltage in the control winding adds to the secondary winding. This causes increased current and torque, and the motor speeds up.

CONSEQUENT-POLE MOTOR CIRCUITS

When a motor runs at different speeds, the motor's torque or horsepower characteristics change with a change in speed. The motor chosen depends on the application in which the motor is used. Once this selection is made, the motor is connected into the circuit. Common motor connection arrangements conforming to NEMA standards are used when wiring motors in a circuit. **See Appendix.**

Control Circuit Logic

Several starting control circuits can be developed to control a multispeed motor, depending on the requirements of the circuit. In a simple two-speed motor control circuit, the motor can be started in the low or high speed. **See Figure 13-4.** In this circuit, the operator may start the motor from rest at either speed. In any control circuit, the pushbuttons may be replaced with any control device such as a pressure switch, photoelectric switch, etc. without changing the circuit logic.

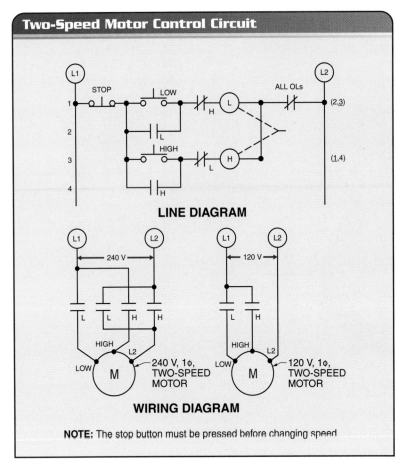

Figure 13-4. The control circuit for a two-speed consequent-pole motor can be wired so that the motor must be stopped before changing speed.

Definition

Compelling circuit logic is a control function that requires the operator to start and operate a motor in a predetermined order.

In this simple control circuit, pressing the low-speed pushbutton energizes the low-speed starter coil and starts the motor in low speed. The NC interlock contact from the low-speed starter prevents the high-speed starter coil from being energized. The motor can also be started by pressing the high-speed pushbutton. This energizes the high-speed starter coil and starts the motor in high speed. In this case, the NC contact from the high-speed starter prevents the low-speed starter coil from being energized. No matter how the motor is started, the stop pushbutton must be pressed before changing from low to high speed or from high to low speed.

In a modified control circuit, the motor can be changed from low speed to high speed without first stopping the motor. **See Figure 13-5.** In this circuit, the operator can start the motor from rest at either speed or change from low speed to high speed without pressing the stop pushbutton.

The stop pushbutton must be pressed before it is possible to change from high to low speed. This high-to-low arrangement prevents excessive line current and shock to the motor. In addition, the machinery driven by the motor is protected from shock that could result from connecting a motor at high speed to low speed.

In this modified control circuit, pressing the low-speed pushbutton energizes the low-speed starter coil and starts the motor in low speed. The NC interlock contacts open on the low starter, preventing the motor from being placed in high speed. The motor may be placed directly into high speed when it is running in low speed. Pressing the high-speed pushbutton de-energizes the low-speed circuit and energizes the high-speed circuit. However, the circuit is designed so that the motor cannot be directly switched to low speed when it is running in high speed. The stop pushbutton must first be pressed before the low starter coil can be energized when the motor is running in high speed. This prevents damage to the equipment the motor is driving.

Compelling Circuit Logic. In many speed control applications, a motor must always be started at low speed before it can be changed to high speed. *Compelling circuit logic* is a control function that requires the operator to start and operate a motor in a predetermined order. **See Figure 13-6.**

This circuit does not allow the operator to start the motor at high speed. The circuit compels the operator to first start the motor at low speed before changing to high speed. This arrangement prevents the motor and driven machinery from starting at high speed. The motor and driven machinery are allowed to accelerate to low speed before accelerating to high speed. The circuit also compels the operator to press the stop pushbutton before changing speed from high to low.

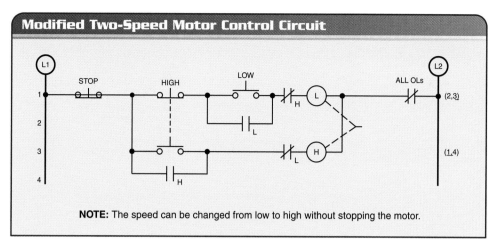

Figure 13-5. The control circuit for a two-speed consequent-pole motor can be wired so that the motor can be changed from low speed to high speed without first stopping the motor.

Definition

Accelerating circuit logic is a control function that permits the operator to select a motor speed so that the control circuit automatically accelerates the motor to that speed.

Decelerating circuit logic is a control function that permits the operator to select a low motor speed so that the control circuit automatically decelerates the motor to that speed.

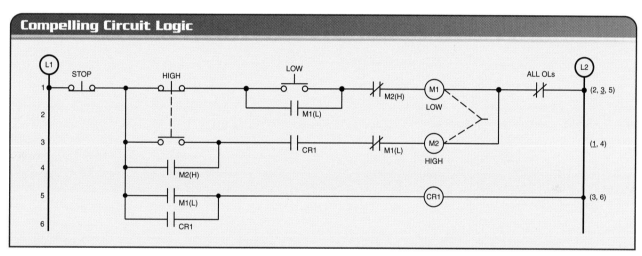

Figure 13-6. Compelling circuit logic requires the operator to start and operate a motor in a predetermined order.

Accelerating Circuit Logic. In many applications, a motor must be automatically accelerated from low to high speed even if the high pushbutton is pressed first. *Accelerating circuit logic* is a control function that permits the operator to select a motor speed so that the control circuit automatically accelerates the motor to that speed. **See Figure 13-7.**

A circuit with accelerating circuit logic allows the operator to select the desired speed by pressing either the low or high pushbutton. If the operator presses the low pushbutton, the motor starts and runs at low speed. If the operator presses the high pushbutton, the motor starts at low speed and runs at high speed only after the predetermined time set on the timer in the control circuit. This arrangement gives the motor and driven machinery a definite time period to accelerate from low to high speed. The circuit also requires that the operator press the stop pushbutton before changing speed from high to low.

Decelerating Circuit Logic. In some applications, a motor or load cannot take the stress of changing from a high to a low speed without damage. In these applications, the motor must be allowed to decelerate by coasting or braking before being changed to a low speed. *Decelerating circuit logic* is a control function that permits the operator to select a low motor speed so that the control circuit automatically decelerates the motor to that speed. **See Figure 13-8.**

This circuit allows the operator to select the desired speed by pressing either the low or high pushbutton. If the low pushbutton is pressed, the motor starts and runs at low speed. If the high pushbutton is pressed, the motor starts and runs at high speed.

If the operator changes from high to low speed, the motor changes to low speed only after a predetermined time. This gives the motor and driven machinery a period of time to decelerate from a high speed to a low speed.

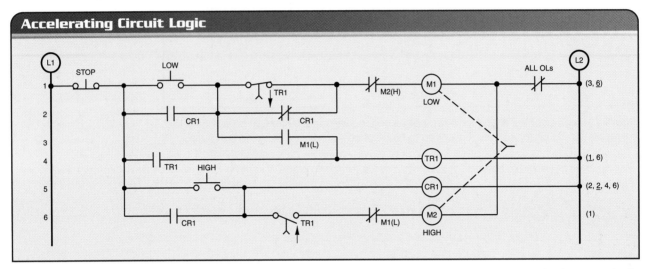

Figure 13-7. Accelerating circuit logic permits the operator to select a motor speed so that the control circuit automatically accelerates the motor to that speed.

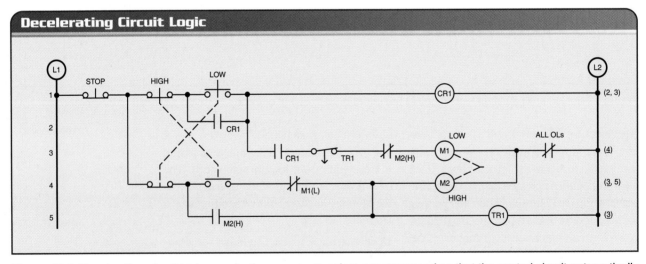

Figure 13-8. Decelerating circuit logic permits the operator to select a motor speed so that the control circuit automatically decelerates the motor to that speed.

Schrage Motor Control Voltages

In a Schrage motor, the commutator and brushes act as a rotating slide-wire voltage divider for tapping portions of the induced voltage in the adjusting winding. Each set of brushes is connected to the ends of each secondary stator winding. These brushes apply the control voltage from the commutator to the stator. The commutator also acts as a frequency-matching device, giving an adjusting voltage of the same frequency as the induced secondary voltage.

At the neutral point, the brushes of both ends of a given phase of the secondary winding are on the same segment of the commutator. When the brushes share the same segment, they are shorted together, and no adjusting voltage is connected to the secondary. The motor runs at its synchronous speed and operates like a squirrel-cage. The stator voltage depends on the motor design parameters.

When the brushes are rotated and placed on different commutator segments, the control winding voltage changes the net secondary voltage. When the brushes are moved apart in one direction, the control-adjusting voltage is added in series with the stator winding. When the brushes are moved apart in the other direction, the control-adjusting voltage is subtracted from the stator winding. In this example, the motor is designed so that the maximum control voltage is 50 V.

The brushes can be moved from the synchronous position in many small increments, allowing for fine speed control. The polarities of the windings determine whether the voltage is added or subtracted. A positive to negative polarity means that the voltages add and the net stator voltage is 150 V. A negative to negative polarity means that the voltages subtract and the net stator voltage is 50 V.

The stator frequency is up to twice that of the source. Because the rotor is driven against the direction of the rotating field, the source frequency and the slip frequency add together. The stator sees the flux from the rotating field in the rotor and the flux of the rotor poles as they spin by rotation. As the stator voltage and frequency increase, the force of repulsion between like stator and rotor poles also increases, speeding the rotor. As the stator voltage and frequency decrease, the force of repulsion between like stator and rotor poles also decreases, slowing the rotor.

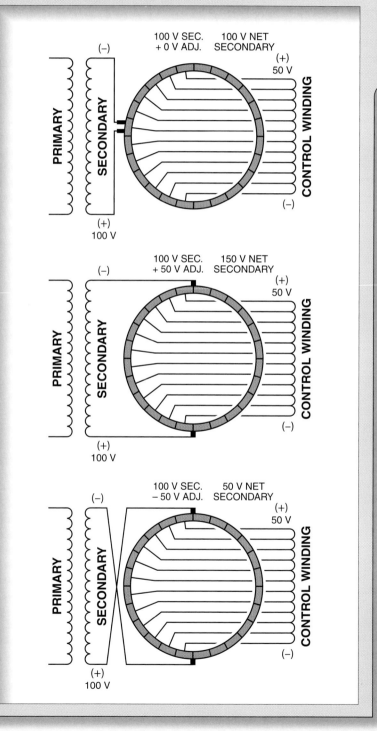

Application—Two-Speed Separate-Winding Motors

The speed of an AC motor is determined by the frequency of the power supply or the number of individual poles of the motor. The individual poles are determined by how the motor windings are connected. The speed of a motor decreases as the number of poles increases, while the speed of a motor increases as the number of poles decreases. To change the speed of a motor, the motor must have separate windings. Each winding has a different number of individual poles. When power is applied to the different windings, the motor speed changes. In a two-speed motor circuit, the motor can be started in either high or low speed.

Two-speed motors are wired the same basic way as any other motor. Basic steps are followed regardless of whether the motor is single-voltage, dual-voltage, reversible, two-speed, three-speed, etc. When wiring any motor, the nameplate of the motor gives written information and a wiring diagram. The written information states the motor's required voltage and operating speed. The wiring diagram shows how the motor is wired for a low voltage and high voltage on dual-voltage motors.

The wiring diagram also shows how the motor is wired for low speed and high speed on dual-speed motors. How the motor is reversed is also listed as part of the information included with the wiring diagram (usually below the wiring diagram).

With all power OFF, the motor wiring diagram is used to connect the motor to the motor side of the starter. Next, with all power OFF, the power lines are connected to the power side of the starter. After the circuit is connected and power is applied, the motor is tested at low speed. Only if the motor operates properly at low speed should the motor be tested at high speed.

Rotor Speed

	Synchronous Speed*	4% Slip*	Actual Speed*
 2	3600	144	3456
 4	1800	72	1728
 6	1200	48	1152

* in rpm

Application—Two-Speed Consequent-Pole Motors

Motor speed can be changed with consequent-pole motors. The speed of the motor changes as the number of poles on the winding changes. The number of poles changes because of the way the current from the power lines (L1, L2, or L3) travels through the motor windings. Since each motor winding is divided into two equal parts, the current can be made to travel through the windings connected in series or parallel with each other. A set of pushbuttons can be wired into the control circuit to allow the motor to be started in low speed or high speed.

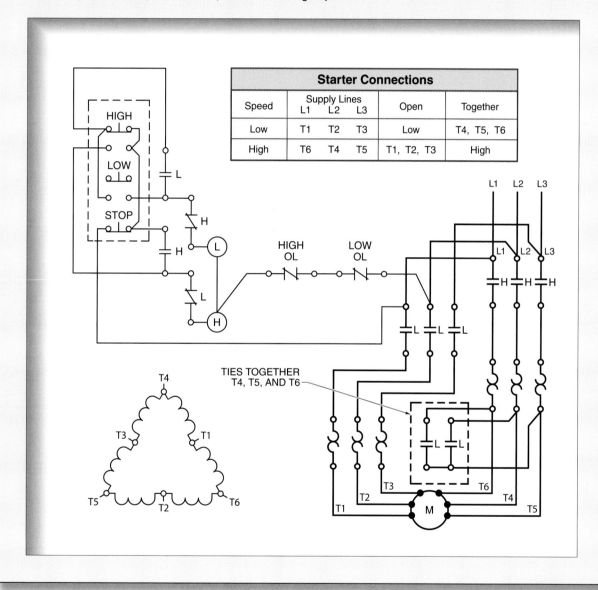

Starter Connections				
Speed	Supply Lines			
	L1	L2	L3	Open	Together
Low	T1	T2	T3	Low	T4, T5, T6
High	T6	T4	T5	T1, T2, T3	High

Summary

- A consequent-pole motor is a motor with stator windings that can be connected in two or more different ways so that the number of stator poles can be changed.

- In a consequent-pole motor, the current travels through the windings in series for one speed and in parallel for the other speed.

- A series connection of two windings has one pole. The current flows through both coils in the same direction, so both coils have the same pole. The poles merge since they are adjacent.

- A parallel connection of two windings has two poles. The current flows through the coils in opposite directions, so the coils have opposite poles. The poles are separate since they are opposite.

- A two-speed consequent-pole motor has six leads brought out for connection.

- In a Schrage motor, the rotor is the primary and the stator is the secondary. The rotor is energized through the slip rings.

- The rotor in a Schrage motor contains a control winding energized by the commutator.

- A Schrage motor has adjustable brushes on a commutator to provide a control voltage to the stator.

- A consequent-pole motor starting circuit usually allows the motor to be started at low or at high speed.

- Some consequent-pole motor circuits allow the speed to be changed while running.

- Compelling circuit logic requires a consequent-pole motor to be started at low speed.

- Accelerating and decelerating logic are methods of controlling the acceleration and deceleration of a motor and load.

Glossary

A **consequent-pole motor** is a motor with stator windings that can be connected in two or more different ways so that the number of stator poles can be changed.

A **Schrage motor** is a 3-phase AC motor with a rotor fed by a commutator and a set of brushes connected to an external circuit.

Compelling circuit logic is a control function that requires the operator to start and operate a motor in a predetermined order.

Accelerating circuit logic is a control function that permits the operator to select a motor speed so that the control circuit automatically accelerates the motor to that speed.

Decelerating circuit logic is a control function that permits the operator to select a low motor speed so that the control circuit automatically decelerates the motor to that speed.

Review

1. Explain why a parallel winding in a consequent-pole motor has two poles while a series winding has one pole.

2. Describe how line voltage is fed to a Schrage motor. Specifically, describe which part of the motor is the primary and which part is the secondary.

3. Explain how interlocks can be used to ensure that a motor must be stopped before changing the speed.

4. Explain why it is often necessary to place interlocks to prevent the motor speed from being changed from high to low.

5. Describe the behavior of compelling circuit logic during motor starting.

Refer to the CD-ROM for Quick Quiz® questions related to chapter content.

MOTORS

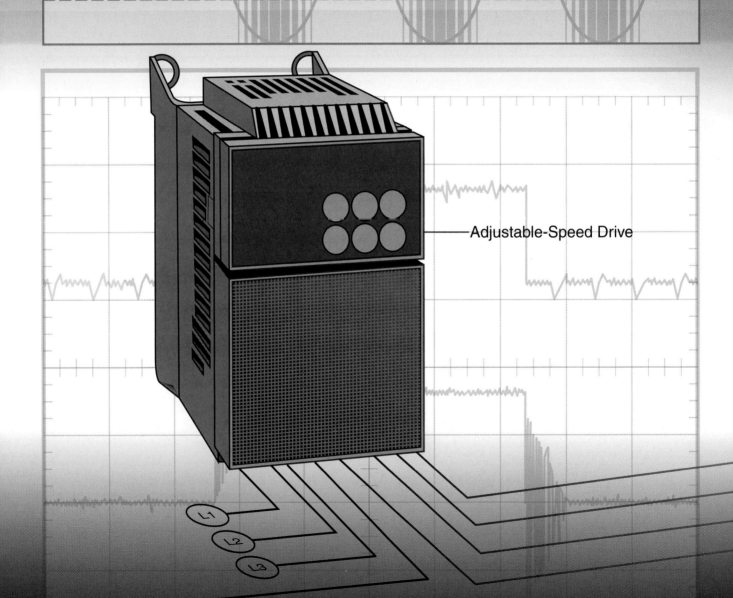

Adjustable-Speed Drive

14

Adjustable-Speed Drives

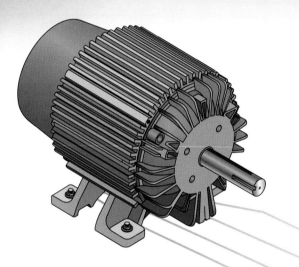

Adjustable-Speed Drive Components	296
AC Motor Speed Control	299
DC Motor Speed Control	310
Supplemental Topic—Shunt Motor Drive with Compound-Wound Motor	314
Supplemental Topic—DC Drive Classification	316
Supplemental Topic—Soft Start and Soft Stop	317
Application—Volts-per-Hertz Ratio	318
Application—Motor Lead Lengths	319
Summary	320
Glossary	321
Review	321

OBJECTIVES

- Describe the three main components of an adjustable-speed drive.
- Explain how the variable frequency applied to an AC motor is created.
- Describe the operation of inverter drives and vector drives.
- Explain how the variable voltage applied to a DC motor is created.
- Describe the operation of DC field-current control, SCR armature-voltage control, and chopper armature-voltage control drives.

Motor speed control is essential in many residential, commercial, and industrial applications. Many applications do not require precise speed regulation, so manual drives or controllers without feedback are sufficient. It may be desirable to vary the speed of bulk material conveyors, pumps, fans, or similar loads, but exacting speed control probably is not required. Any type of process in which material is fed from a roll at one end of a machine and collected on another roll at the other end of the machine probably requires very precise speed regulation, such as with newspaper printing and textile manufacturing.

ADJUSTABLE-SPEED DRIVE COMPONENTS

The three main sections of adjustable-speed drives are the converter section, DC bus section, and inverter section. The converter section includes a rectifier that receives the incoming AC voltage and changes it to DC. When the AC input voltage is different than the AC output voltage required by a motor, the converter section must first step up or step down the AC voltage to the proper level. For example, an adjustable-speed drive supplied with 115 VAC that must deliver 230 VAC to a motor requires a transformer to step up the input voltage. **See Figure 14-1.**

The DC bus section filters the voltage and maintains the proper DC voltage level. The DC link is part of the DC bus section. The DC bus section delivers the DC voltage to the inverter section for conversion back to AC voltage. The inverter section controls the speed and torque of a motor by controlling voltage, frequency, and current. A controller applies logic to the variable-voltage inverter section.

Converter Section

The converter section of an adjustable-speed drive is used to convert the AC source power into DC power that can be inverted back to variable-frequency voltage in the inverter section. Converter sections of adjustable-speed drives are single-phase full-wave rectifiers, single-phase bridge rectifiers, or 3-phase full-wave rectifiers. Small adjustable-speed drives supplied with single-phase power use single-phase full-wave or bridge rectifiers. Most adjustable-speed drives are supplied with 3-phase power. This requires 3-phase full-wave rectifiers. **See Figure 14-2.**

Power Supply Requirements. In order for a converter section to deliver the proper DC voltage to the DC bus section of an adjustable-speed drive, the rectifier section must be connected to the proper power supply. Adjustable-speed drives operate satisfactorily only when connected to the proper power supply. The proper power supply must not only be at the correct voltage level and frequency, but also provide enough current to operate an adjustable-speed drive at full power.

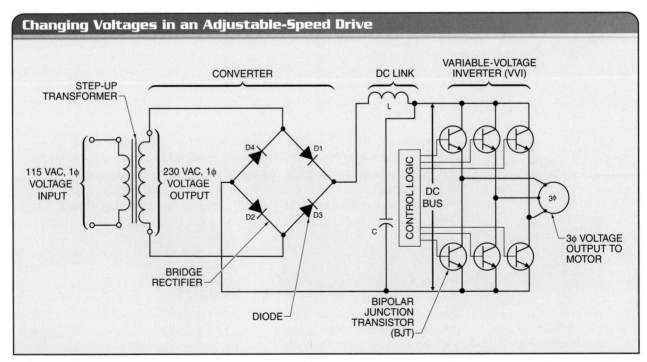

Figure 14-1. An adjustable-speed drive supplied with 115 VAC that must deliver 230 VAC to a motor requires a transformer to step up the input voltage.

Supply voltage at an adjustable-speed drive must be checked when installing a drive, servicing a drive, or adding additional loads or drives to a system. To determine if an adjustable-speed drive is underpowered, the voltage into the drive is measured under no-load and then under full-load operating conditions. **See Figure 14-3.** A voltage difference greater than 3% between no-load and full-load conditions indicates that the adjustable-speed drive is underpowered and/or overloaded.

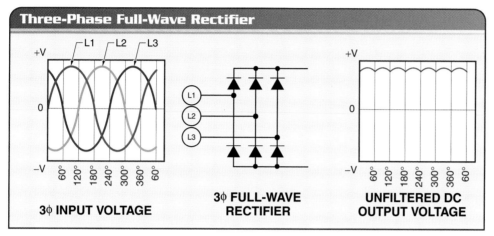

Figure 14-2. Most adjustable-speed drives are supplied with 3-phase power. This requires 3-phase full-wave rectifiers.

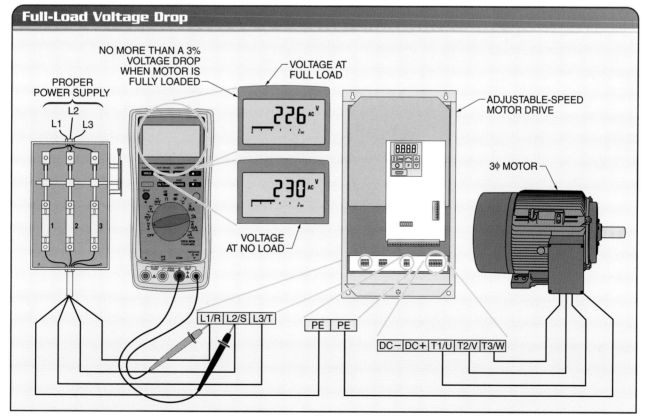

Figure 14-3. A voltage difference greater than 3% between no-load and full-load conditions indicates that the adjustable-speed drive is underpowered and/or overloaded.

DC Bus Section

The DC bus voltage is typically about 1.4 times the AC supply voltage to an adjustable-speed drive. The DC bus section, or DC link, includes DC voltage supplied by the converter section and the DC filter components. Capacitors and inductors in the DC bus section filter and maintain the proper DC bus voltage level. **See Figure 14-4.**

Capacitors. Capacitance is the ability of a component or circuit to store energy in the form of an electrical charge. Capacitors in a DC bus section are charged from the rectified DC voltage produced by the converter section. Capacitors oppose a change in voltage by holding a voltage charge that is discharged back into the circuit any time the circuit voltage decreases. For example, when the DC bus voltage starts to drop, capacitors discharge the electrical charge back into the system to counter the drop in voltage.

Inductors. An inductor stores energy in the form of a magnetic field. As current flows through a coil, a magnetic field is produced. The magnetic field remains at maximum potential until the current in the circuit is reduced. As the circuit current is reduced, the collapsing magnetic field around the coil sends current back into the circuit. The coil smoothes or filters the power by building a magnetic field as current is applied and by adding current back into the circuit as the magnetic field collapses. Capacitors and inductors (coils) are used together in DC filter circuits. Working together, capacitors and coils maintain a smooth waveform because capacitors oppose a change in voltage, and coils oppose a change in current.

Resistors. A resistor is used in the DC bus section to limit the charging current, to discharge capacitors, and to absorb unwanted voltages. Current-limiting resistors prevent capacitors from drawing too much current. Resistors are also used to discharge capacitors when power is removed from an adjustable-speed drive. Braking resistors dissipate power when a motor becomes a generator after a stop button is pushed.

Inverter Section

The inverter section of an adjustable-speed drive controls the voltage level, voltage frequency, and amount of current that a motor receives. The inverter sections of AC drives change the DC bus voltage to variable-frequency AC. The inverter sections of DC drives change the DC bus voltage to variable-voltage DC. There are several methods of accomplishing these changes. Inverter sections have undergone significant changes in recent years, while the rectifier section and DC bus have not changed as much. Adjustable-speed drive manufacturers are continuously developing inverter sections that can control motor speed and torque with the fewest problems.

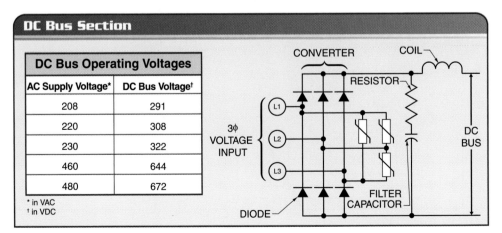

Figure 14-4. Capacitors and inductors in the DC bus section filter and maintain the proper DC bus voltage level.

Installation

Adjustable-speed drives must be protected from the environment by appropriate enclosures. Drives produce a fair amount of heat, so grouping several in a single cabinet can be a problem if forced cooling cannot be made available. Proper installation procedures must be followed when installing communications cable in order to minimize induced electrical noise. For example, power and low-voltage wiring must be run in separate conduits and should be separated in control cabinets and enclosures by metal barriers. Where separation cannot be maintained, power and signal wiring should cross at right angles. Care must also be taken to ensure that shields are continuous over the entire signal cable length and that they are grounded only at one point, preferably at the drive end.

AC MOTOR SPEED CONTROL

The supply frequency and the number of poles are the main variables that determine the speed of an AC motor. In North America, the supply frequency is 60 Hz. In Europe and many other parts of the world, the supply frequency is 50 Hz. AC adjustable-speed drives vary the AC frequency to the motor in order to adjust the motor speed. The number of poles can be changed with a consequent-pole motor. However, changing the number of poles is not really a method of adjustable-speed control. It is a method of changing the rated speed of the motor.

AC adjustable-speed drives are used to change the speed of AC motors by changing the frequency of the voltage applied to the motor. In addition to controlling motor speed, adjustable-speed drives can control motor acceleration time, deceleration time, motor torque, and motor braking. An AC adjustable-speed drive changes the frequency of the voltage applied to a motor by taking the incoming AC voltage, converting it to a DC voltage, and inverting it back to variable-frequency AC. **See Figure 14-5.**

The speed of an AC motor should not be changed by varying the applied voltage unless the motor is specifically rated for the varying voltage. Damage may occur to an AC motor if the supply voltage is varied more than 10% above or below the rated nameplate voltage. Lower voltage results in lower torque. This is because in an induction motor, the starting torque and breakdown torque vary as the square of the applied voltage. For example, with 90% of rated voltage, the torque is only 81% ($0.9^2 = 0.81$ or 81%) of its rated torque. Higher voltage can result in insulation damage and shortened motor life.

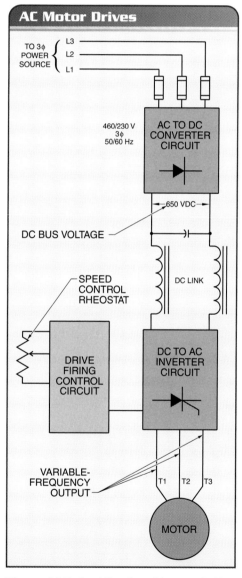

Figure 14-5. An AC adjustable-speed drive changes the frequency of the voltage applied to a motor by taking the incoming AC voltage, converting it to a DC voltage, and inverting it back to variable-frequency AC.

Motor Drive Names

An adjustable-speed drive is one piece of electrical equipment that is often called by many names, depending upon the type of drive, type of motor to be controlled, or the type of incoming supply voltage to the drive. Adjustable-speed drives are manufactured as variable-frequency drives, adjustable-frequency drives, electric motor drives, inverter drives, vector drives, direct torque-control drives, closed-loop drives, and regenerative drives. Commonly, the word drive is all that is used to indicate an adjustable-speed drive. The supply voltage to an adjustable-speed drive is either AC voltage or DC voltage. Regardless of the nomenclature (naming system) and the differences between the types of drives, the primary function of a drive is to convert the incoming supply power to an altered voltage level and frequency that can safely control the motor connected to the drive and the load connected to the motor. **See Figure 14-6.**

Inverter Duty Motors

The motor is one of the main considerations when choosing the speed control method for an application. Some motors offer excellent speed control through a wide range of speeds, while others may offer only two or three different speeds. Other motors offer only one speed that cannot be changed except by external means such as gears, pulley drives, or changes of power source frequency.

AC adjustable-speed drives can control the speed of a motor and the speed of the load the motor is driving. However, AC adjustable-speed drives create problems for standard adjustable-speed motors because adjustable-speed drives produce voltage spikes. The voltage spikes create heat that damages the insulation and shortens the life expectancy of standard electric motors. Because of this heat, inverter duty motors are used.

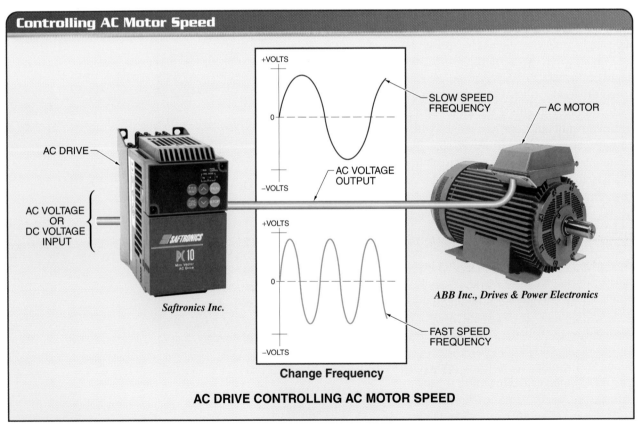

Figure 14-6. The primary function of a drive is to convert the incoming supply power to an altered voltage level and frequency that can safely control the motor connected to the drive and the load connected to the motor.

An *inverter duty motor* is an electric motor specifically designed to work with AC adjustable-speed drives. Inverter duty motors are designed to withstand the heat created by voltage spikes. These motors are made with special voltage-spike-resistant wire and improved insulation. Although a drive can be used to control a standard motor, it is best to use motors specifically designed for adjustable-speed drive use.

Control of Applied Frequency

Motors produce work to drive a load by rotating a shaft. The amount of work produced depends on the amount of torque produced by the motor shaft and the speed of the shaft. The primary function of all adjustable-speed drives is to control the speed and torque of a motor.

To safely control a motor, an adjustable-speed drive must monitor electrical characteristics such as motor current, motor voltage, drive temperature, and other operating conditions. All adjustable-speed drives are designed to remove power when there is a problem, and some drives allow conditions and faults to be monitored and displayed. In addition to controlling motor speed and torque, an adjustable-speed drive can include additional specialty functions that are built in, programmed into the drive, or sent to the drive through on-board communication with a PC or PLC.

Controlling the frequency to an AC motor controls the speed of the motor. AC drives control the frequency applied to a motor over the range 0 Hz to several hundred hertz. AC drives are programmed for a minimum operating speed and a maximum operating speed to prevent damage to the motor or driven load. Damage occurs when a motor is driven faster than its rated nameplate speed. AC motors should not be driven at a speed higher than 10% above the motor's rated nameplate speed unless a licensed engineer is consulted.

Controlling the volts-per-hertz ratio (V/Hz) applied to an AC motor controls motor torque. An AC motor develops rated torque when the volts-per-hertz ratio is maintained.

During acceleration over any speed between 1 Hz and 60 Hz, the motor shaft delivers constant torque because the voltage is increased at the same rate as the increase in frequency. Once an adjustable-speed drive is delivering full motor voltage, increasing the frequency does not increase torque on the motor shaft because voltage cannot be increased any more to maintain the volts-per-hertz ratio. **See Figure 14-7.**

> **Definition**
>
> An *inverter duty motor* is an electric motor specifically designed to work with AC adjustable-speed drives.

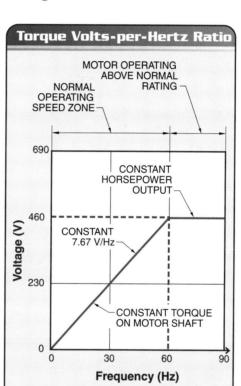

Figure 14-7. An AC motor develops rated torque when the volts-per-hertz ratio is maintained.

The simplest way to adjust the drive output is to set an internal reference voltage. The reference voltage is used to set the output frequency and voltage of an adjustable-speed drive. This approach can provide speed control within about 3% to 5% of the desired speed and can be used when precise speed control is not required.

Many drives produced today have internal circuitry that can monitor motor voltages or currents that can be used to improve speed control within about 1% of the desired speed. Drives with analog or digital feedback signals can be used where extremely precise speed

Definition

Pulse-width modulation (PWM) is a method of controlling the voltage sent to a motor by controlling the amount of time a transistor is ON and conducting current.

regulation is required. These drives can control speed within about 0.01% of the desired speed. The transducer used for this feedback signal can monitor either the speed of the motor itself or some part of the process being controlled.

Pulse-Width Modulation. *Pulse-width modulation (PWM)* is a method of controlling the voltage sent to a motor by controlling the amount of time a transistor is ON and conducting current. This creates voltage pulses that average out as a sine wave to represent line power, but at any desired frequency.

PWM uses transistors called insulated gate bipolar transistors (IGBTs) to create fixed-value pulses. The fixed-value pulses are produced by switching the transistors ON and OFF at high speed. By varying the time the transistor is ON, the width of each pulse varies and the average voltage can be varied. The wider the individual pulses, the higher the average voltage output. The PWM of a DC voltage is used to reproduce AC sine waves. **See Figure 14-8.**

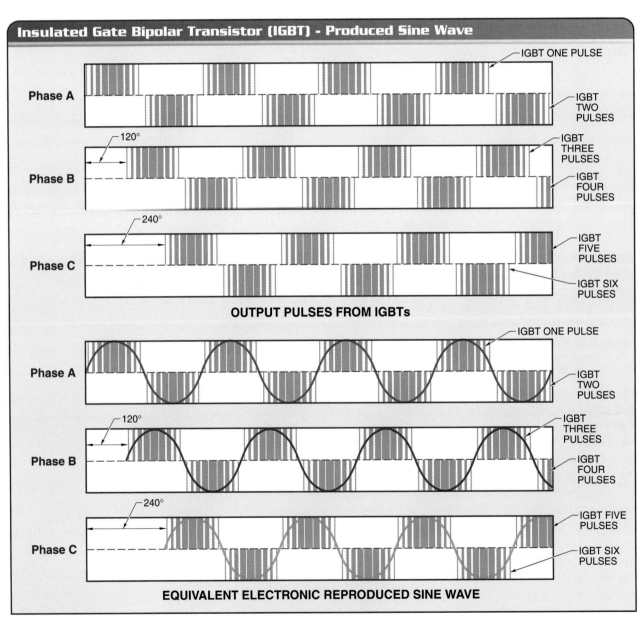

Figure 14-8. By varying the time the transistor is ON, the width of each pulse varies and the average voltage can be varied.

When PWM is used with AC voltage, two IGBTs are used for each phase. One IGBT is used to produce the positive pulses and another IGBT is used to produce the negative pulses of the sine wave. Because AC drives are typically used to control 3-phase motors, six IGBTs, two per phase, are used. The higher the switching frequency of the IGBTs, the closer the simulated AC sine wave is to a real sine wave and the lower the amount of heat produced by the motor.

Carrier Frequency. The *carrier frequency* is the frequency that controls the number of times the solid-state switches in the inverter section of a PWM adjustable-speed drive turn ON and OFF. The higher the carrier frequency, the more individual pulses there are to reproduce the fundamental frequency. **See Figure 14-9.** The *fundamental frequency* is the frequency of the voltage used to control motor speed. The carrier frequency pulses per fundamental frequency are found by applying the formula:

$$P = \frac{F_{CARR}}{F_{FUND}}$$

where
P = pulses per Hz
F_{CARR} = carrier frequency (in Hz)
F_{FUND} = fundamental frequency (in Hz)

For example, the number of pulses per fundamental frequency when a carrier frequency of 6 kHz (6000 Hz) is used to produce a 60 Hz fundamental frequency is calculated as follows:

$$P = \frac{F_{CARR}}{F_{FUND}}$$

$$P = \frac{6000}{60}$$

$P = \textbf{100 pulses per Hz}$

Definition

*The **carrier frequency** is the frequency that controls the number of times the solid-state switches in the inverter section of a PWM adjustable-speed drive turn ON and OFF.*

*The **fundamental frequency** is the frequency of the voltage used to control motor speed.*

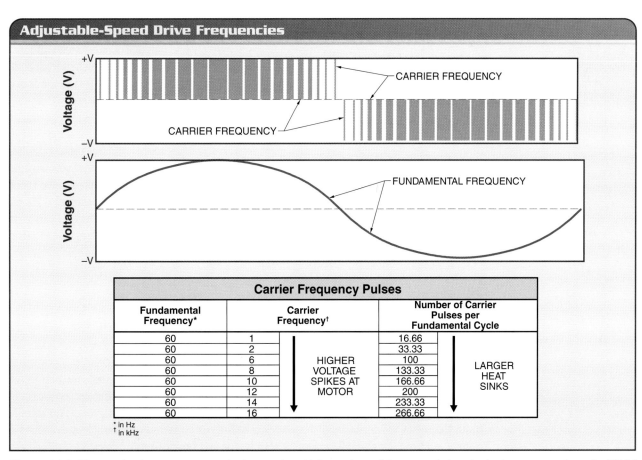

Figure 14-9. The carrier frequency controls the number of times the solid-state switches in the inverter section of a PWM adjustable-speed drive turn ON and OFF.

Definition

A volts-per-hertz ratio (V/Hz) is the ratio of voltage to frequency in a motor.

This means that 100 pulses of the bus voltage are used to create one pulse of 60 Hz voltage applied to the motor. If the desired frequency at the motor is 30 Hz, 50 pulses of the bus voltage are used to create one pulse of the 30 Hz voltage.

Fundamental frequency is the frequency of the voltage a motor uses, but the carrier frequency actually delivers the fundamental frequency voltage to the motor. The carrier frequency of most adjustable-speed drives can range from 1 kHz to about 16 kHz. The higher the carrier frequency, the closer the output sine wave is to a pure fundamental frequency sine wave.

Increasing the frequency to a motor above the standard 60 Hz also increases the noise produced by the motor. Noise is noticeable in the 1 kHz to 2 kHz range because it is within the range of human hearing and is amplified by the motor. A motor connected to an adjustable-speed drive delivering a 60 Hz fundamental frequency with a carrier frequency of 2 kHz is about three times louder than the same motor connected directly to a pure 60 Hz sine wave with a magnetic motor starter. Motor noise is a problem in adjustable-speed drive applications, such as HVAC systems, in which the noise can carry throughout an entire building.

Manufacturers have raised the carrier frequency beyond the range of human hearing to solve the noise problem. High carrier frequencies above about 8 kHz cause greater power losses in an adjustable-speed drive. Adjustable-speed drives must be derated or the size of heat sinks increased because of the increased power losses. Derating an adjustable-speed drive decreases the power rating of a drive, and increasing heat sink size adds additional cost to a drive. **See Figure 14-10.**

Carrier frequency can be changed at an adjustable-speed drive to meet particular requirements. The factory default value is usually the highest frequency, and changing to a lower frequency is done through a parameter change, such as changing 12 kHz to 2.2 kHz. One effect of high carrier frequencies is that the fast switching of the inverter produces larger voltage spikes that damage motor insulation. The voltage spikes become more of a problem as the cable length between an adjustable-speed drive and motor increases.

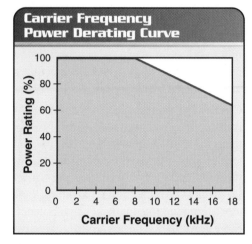

Figure 14-10. Derating an adjustable-speed drive decreases the power rating of the drive.

Voltage and Frequency. The motor heats excessively and damage occurs to the windings if the voltage is not reduced when frequency is reduced. The voltage applied to the stator of an AC motor must be decreased by the same amount as the frequency. The motor does not produce its rated torque if the voltage is reduced more than required. The ratio between the voltage applied to the stator and the frequency of the voltage applied to the stator must be kept constant when a motor is slowed. This ratio is referred to as the volts-per-hertz (V/Hz) ratio. A *volts-per-hertz ratio (V/Hz)* is the ratio of voltage to frequency in a motor. The motor develops rated torque if this ratio is kept constant.

Motor Lead Length

In any electrical system, the distance between components affects its operation. The primary limit to the distance between a magnetic motor starter and the motor is the voltage drop of the conductors. The voltage drop of conductors should not exceed 3% for any type of motor circuit.

When an adjustable-speed drive is used to control a motor, the distance between the drive and motor may be limited by other factors besides the voltage drop of the conductors. Conductors between an adjustable-speed drive and motor have line-to-line (phase-to-phase) capacitance and line-to-ground (phase-to-ground) capacitance. Longer conductors produce higher capacitance that causes high voltage spikes in the voltage to a motor. Since voltage spikes are reflected into the system, the voltage spikes are often called reflective wave spikes, or reflected waves. As the length of the conductors increases and/or an adjustable-speed drive's output carrier frequency increases, the voltage spikes become larger. **See Figure 14-11.**

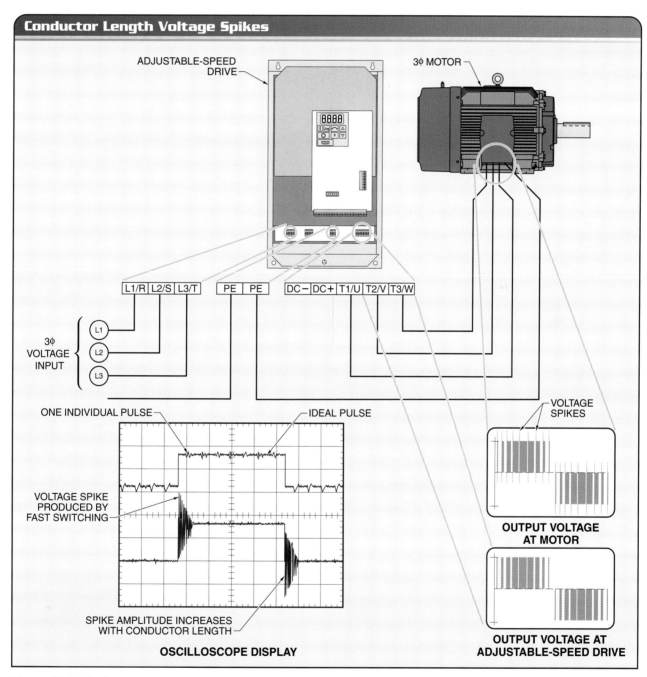

Figure 14-11. As the length of conductors increases and/or an adjustable-speed drive's output carrier frequency increases, the voltage spikes become larger.

> **Definition**
>
> An *inverter drive*, or *scalar drive*, is a standard adjustable-speed drive that uses pulse-width modulation to control the speed and torque.

Voltage spikes are a problem because spikes stress motor insulation. When voltage spikes become a problem, the lead length should be reduced and/or filters that suppress voltage spikes should be added. Drive manufacturers often provide tables that give the maximum lead length at specified frequencies. Smaller horsepower motors and multiple motors connected to one adjustable-speed drive can be more susceptible to voltage spikes. Reducing the carrier frequency also reduces reflected wave voltage spikes. Inverter rated motors that have spike-resistant insulation reduce the damage caused by voltage spikes.

Control of Applied Voltage

The speed of standard AC squirrel-cage induction motors is normally varied by changing the number of poles or the applied frequency. However, it is possible in some applications to control the speed of a load by varying the voltage applied to the motor. This method is not a standard method of speed control, and caution must be taken in applying it. By varying the voltage applied to the motor, the torque that the motor can deliver to the load is varied. The torque of a squirrel-cage induction motor varies with the square of the applied voltage.

The greater the torque, the faster the acceleration time. If the torque of a motor is reduced, the speed at which the motor performs the work is also reduced. Although it is possible to reduce the speed of a large motor by reducing the applied voltage, this method could damage the motor from excess heat buildup in the motor. This type of speed control is limited to applications of soft-start, light loads, such as fans and blowers. Shaded-pole or permanent-magnet motors are normally used in these applications. Changing the applied voltage should not be considered a standard method except in applications that are specifically designed for this type of speed control.

AC Drive Types

AC drives control motor speed and torque by converting incoming AC voltage to DC voltage and then converting the DC voltage to a variable-frequency AC voltage. Single-phase 208 V or 240 V AC drives are available for operating 3-phase motors up to about 3 HP. However, most AC drives operate on one of the common 3-phase systems. Medium-voltage electronic drives for use on 2400 V and higher-voltage systems are also available.

Three-phase motors are reversed by simply interchanging any two motor leads. This is normally much easier to do at the starter than at the motor terminal box. Many drive units are mounted in enclosures so that it is easy to swap two line leads. However, the incoming phase relationships are lost in the rectifier section, so interchanging two line leads does not reverse the motor. Therefore, any leads to be interchanged must be at the motor or at the drive output terminals.

Inverter Drives. An *inverter drive,* or *scalar drive,* is a standard adjustable-speed drive that uses pulse-width modulation to control the speed and torque. Inverter drives include variable-voltage inverters (VVI), current-source inverters (CSI), and pulse-width-modulated inverters (PWM). Inverter drives are the oldest type of AC drive. Older six-step inverter drives used SCRs in the inverter section of the AC drive to produce the AC sine wave. PWM inverters use transistors in the inverter section to produce the AC sine wave. **See Figure 14-12.** Transistors operate at much faster speeds than SCRs, thus allowing higher switching frequencies, to produce electronically reproduced sine waves that closely resemble a pure sine wave. The improved sine wave produces less heating in a motor than six-step inverters.

> **Tech Fact**
>
> AC drives often have higher power loss than DC drives because DC drives have one energy conversion (AC to DC), and AC drives have two energy conversions (AC to DC and DC to AC). New electric motor drives are more energy efficient than drives that are only a few years old.

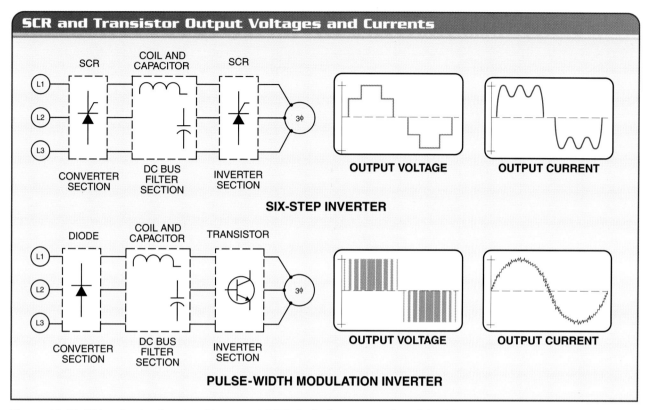

Figure 14-12. Older six-step inverter drives used SCRs in the inverter section of the AC drive to produce the AC sine wave. PWM inverters use transistors in the inverter section to produce the AC sine wave.

Vector Drives. Inverter drives that control motor speed by setting motor voltage and frequency work well for most applications. Historically, applications that required better motor torque control at different speeds used DC motors. Vector drives were developed to overcome the problem of using high-maintenance DC motors. A *vector drive* is a variable-speed drive that uses a microprocessor and PWM, usually with feedback, to calculate the precise vector between voltage and frequency that is needed to provide better control of speed and torque. Vector drives are rapidly replacing inverter drives in almost all applications as the cost of microprocessors decreases.

Closed-Loop Vector Drives. A *closed-loop vector drive* is a vector drive that uses shaft-mounted sensors to determine the rotor position and speed of a motor and send the information back to the adjustable-speed drive. Feedback from sensors, like encoders and tachometers, allows an adjustable-speed drive to automatically make adjustments to better meet the needs of the motor.

Vector drives, also called flux vectors or field orientation drives, are designed to operate AC motors with the same performance as DC motors. Vector drives achieve this performance by measuring the current drawn by a motor at a known speed and comparing the current drawn to the applied voltage. Induction motors produce torque at the rotor and shaft when there is slip. Induction motor torque is directly proportional to the amount of slip. When motor speed is known from encoder feedback, slip can be calculated and torque on the motor shaft can be controlled. **See Figure 14-13.**

Vector drives are capable of performing the same functions as inverter drives and more. Vector drives are a good choice for many applications, but inverter drives are more economical and are better for applications such as centrifugal pumps, fans, and blowers, or any applications that do not require full torque control.

Definition

A *vector drive* is a variable-speed drive that uses a microprocessor and PWM, usually with feedback, to calculate the precise vector between voltage and frequency that is needed to provide better control of speed and torque.

A *closed-loop vector drive* is a vector drive that uses shaft-mounted sensors to determine the rotor position and speed of a motor and send the information back to the adjustable-speed drive.

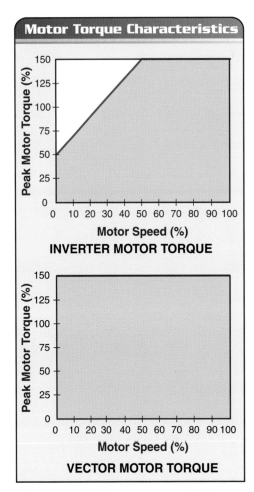

Figure 14-13. When motor speed is known from the encoder feedback, slip can be calculated and torque on the motor shaft can be controlled.

Vector drives have many advantages over inverter drives. For example, full rated torque can be delivered at zero speed. Inverter drives cannot maintain full rated torque at zero speed. Motor speed can be accurately controlled to within about 0.01% of the set speed of the drive when operating in a closed-loop condition. Inverter drives can typically control motor speed within 1% of the set speed of the drive. Motors run cooler with vector drives than with inverter drives.

There are also some disadvantages with vector drives. Vector drives are more expensive than inverter drives. They cannot be used to operate multiple motors from the same drive. Inverter drives can operate multiple motors from the same drive when individual motor protection is used with each motor. Vector drives require an encoder or tachometer. Mounting and maintaining the encoder adds cost to the system.

Open-Loop Vector Drives. An *open-loop vector drive,* or *sensorless vector drive,* is a vector drive that has no feedback from the motor. An open-loop vector drive uses an internal model of the motor and load to control the speed. For example, if the current is different than the model predicts, the vector drive compares it to the model and makes necessary adjustments. Open-loop vector drives are often called sensorless vector drives. Open-loop vector drives are programmed with motor and application data to anticipate how best to control the motor and load.

Open-loop vector drives are used where speed regulation is important, but not important enough to require encoder feedback. Open-loop vector drives typically can control speed within 0.1% of the drive setting.

Variable-Voltage-Inverter and Current-Source-Inverter Drives. AC adjustable-speed drives are classified as either variable-voltage-inverter (VVI) drives or as current-source-inverter (CSI) drives depending on how the DC bus filtering is accomplished. The filter sections of electronic drives can be either capacitive or inductive. The relative magnitude of the inductance and capacitance determines the type of inverter output. Large capacitive filters are used in VVI drives, which apply a voltage to the motor. Large inductive filters are used in CSI drives, which act as a constant current source for the motor. VVI drives are used for armature speed controls, while CSI drives are used for field controls.

Some VVI and CSI drives produce a square wave instead of a sine wave because of the slow SCRs or bipolar-junction transistors used in the inverters. Square waves produce high torque pulsations at the motor that cause a motor to have jerky movements at low speeds and operate at high temperatures. To overcome the square wave problem, better adjustable-speed drives use IGBTs and pulse-width modulation to produce a more accurate sine wave that is less susceptible to torque pulsations. **See Figure 14-14.**

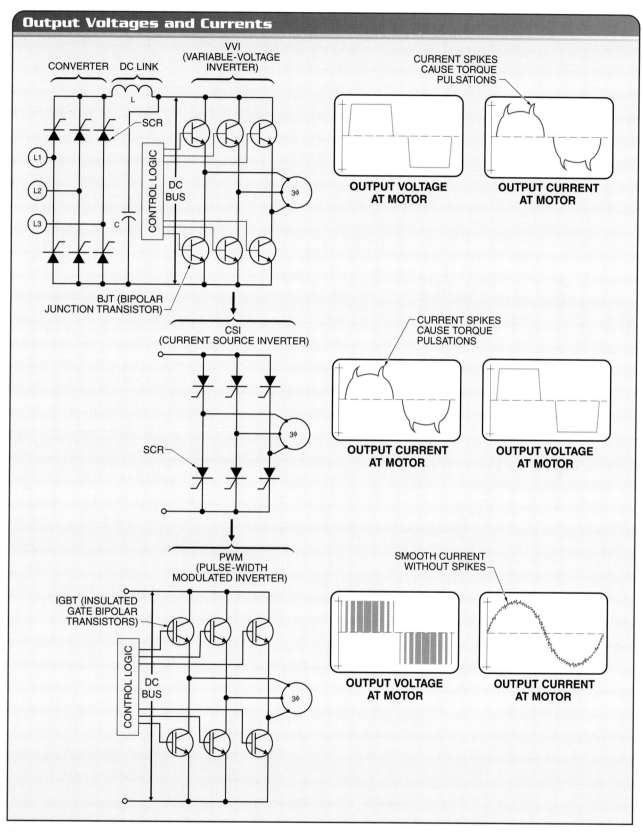

Figure 14-14. Better adjustable-speed drives use IGBTs and pulse-width modulation to produce a more accurate sine wave that is less susceptible to torque pulsations.

VVI drives can be used to control most types of induction and synchronous motors, and several motors can be operated from the same drive. VVI drives are inherently immune to damage from open circuits. Because of the capacitive filtering, voltage-source drives can tolerate voltage dips on the power lines. The main disadvantage of voltage-source drives is that deceleration control and braking cannot be accomplished without external braking resistors or relatively expensive regenerative braking controls.

CSI drives are specific to a particular type of motor, so drives must be purchased for each type of motor to be controlled. CSI drives can normally be used to control only a single motor because they require feedback signals. Because they are current devices, simple short-circuit protection is adequate. However, high open-circuit voltages can damage the drive. One major advantage is that current-source inverters are inherently capable of forward and reverse operation with braking capabilities in both directions, so external braking circuits are not required. The presence of large filtering reactors makes current-source drives much larger than their voltage-source counterparts.

Chapter 430 of the NEC® requires that disconnects be provided for all motors. CSI drives cannot tolerate open-circuit conditions. If a remote disconnect is opened while a current-source drive is operating, it can be destroyed almost instantly. A VVI drive is not affected by opening a disconnect, but cannot tolerate high motor-starting currents. However, the drive may be damaged when the disconnect is reclosed.

Regardless of any acceleration-rate parameter setting, all VVI drives must limit motor currents during startup. If the drive is already ON when the motor is reconnected, the current limit protection may not operate fast enough to prevent damage. These problems can be alleviated easily by installing auxiliary contacts in the remote disconnects. Virtually all AC drives have a set of terminals for this purpose.

DC MOTOR SPEED CONTROL

The primary difference between DC and AC drives is in the power output sections. With AC motors, the frequency is varied to adjust the speed of the motor. For many DC motor applications, the speed is controlled by varying the voltage supplied by the drive to the armature. The higher the applied voltage, the faster a DC motor rotates. For other DC motor applications, the speed is controlled by varying the current supplied by the drive to the shunt field.

The difference between these applications is the actual motor relative to the base speed. Base speed is the motor speed at full-line voltage and is listed on the motor nameplate. **See Figure 14-15.** At base speed, the motor has full armature current and full shunt current. The speed is controlled from 0 rpm up to the base speed by varying the voltage to the armature. If the speed must be controlled above the base speed, the armature voltage is held constant and the field current varied. When armature voltage is controlled, the motor delivers constant torque. When field voltage is controlled, the motor delivers constant horsepower.

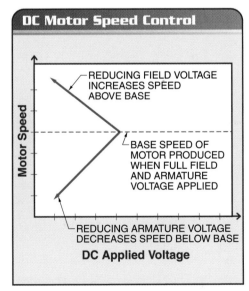

Figure 14-15. Base speed is the motor speed at full-line voltage and is listed on the motor nameplate.

DC motor torque is controlled by varying the amount of current in the armature. DC drives are designed to control the amount of voltage and current applied to the armature of DC motors to produce the desired torque and speed. The ideal operating condition is to deliver current to a motor to produce enough torque to operate the load without overloading the motor, adjustable-speed drive, or electrical distribution system.

DC Drives

A wide variety of solid-state, variable-speed drives for DC motors are available. Selection of the proper drive depends on the type of DC motor and the requirements of the load. The degree of speed regulation required affects the selection of both the drive and appropriate feedback components. Additional considerations include reversibility and braking requirements. Regenerative braking features may be beneficial when large motors with high inertial loads are to be controlled.

Regardless of the type of motor being driven, all DC drives include the same basic circuit components as AC drives. Most drives available today combine the field- and armature-control circuits in a single drive.

DC drives normally control the voltage applied to a motor over the range 0 V to the maximum nameplate voltage rating of the motor. **See Figure 14-16.** If an adjustable-speed drive can deliver more voltage than the rating of the motor, the drive should be programmed to limit the output voltage in order to prevent motor damage. DC motors are subject to the same reduced cooling and increased centrifugal-force effects as AC motors. However, since most DC motors are used for reduced-speed operations, they can be furnished with externally driven cooling systems.

Some modern DC drives include a field-voltage economizer function. The economizer function is used to apply a small voltage to the shunt field windings when the motor is OFF. This helps keep the motor warm and dry. Economizer voltage levels are often about 120 V, which can be hazardous for an electrician if the drive is not completely powered down during maintenance.

The three basic types of variable-speed DC motor controls are field-current control, SCR armature-voltage control, and chopper armature-voltage control. DC shunt motor and DC compound motor speed can be controlled by either using field-current control to vary the shunt field current or by varying the armature voltage. Armature-voltage control is the only method applicable to DC series motors because the same current flows through both the armature and the field coils.

Drives for small permanent-magnet motors are the simplest, while drives for compound-wound motors are the most complex. Drives for shunt-wound motors are probably the most common. Small DC drives are available for use on 12 V, 24 V, or 240 V single-phase power, while most larger DC drives usually operate on 208 V, 240 V, or 480 V 3-phase power.

DC Field-Current Control. DC motor speed is controlled by varying the voltage applied to the motor. *DC field-current control* is a method of controlling the voltage applied to a shunt field. The controller adjusts the current in the shunt field, and therefore controls the strength of the shunt field. DC field-current speed control is generally used only where higher speeds are required and reduced torque output can be tolerated. The controllers are usually included as a part of a complete drive system but can be stand-alone drives. Both types of field-current control drives include the same basic components. **See Figure 14-17.**

The feedback section receives and scales an input from a speed-sensing device, such as a tachometer generator. This conditioned input is then inverted and added to the speed-control signal set by the operator. The preamplifier section modifies the feedback signal so that it is compatible with the sawtooth generator voltage output. The output from the preamplifier is called the speed-reference voltage. The sawtooth generator section uses an oscillator to produce a 3 kHz sawtooth-shaped waveform. **See Figure 14-18.**

> **Definition**
>
> *DC field-current control is a method of controlling the voltage applied to a shunt field.*

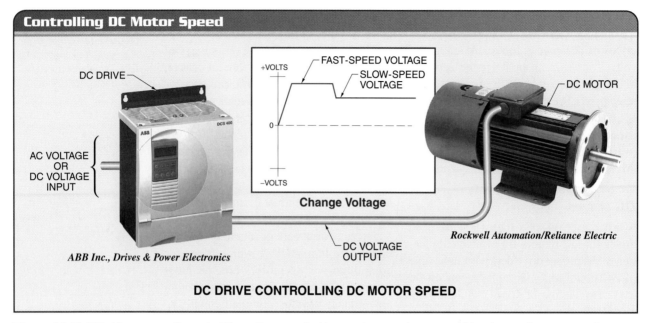

Figure 14-16. DC drives normally control the voltage applied to a motor over the range 0 V to the maximum nameplate voltage rating of the motor.

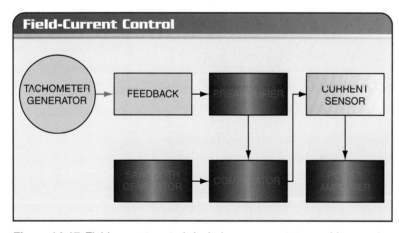

Figure 14-17. Field-current controls include components to provide a varying voltage to the shunt field.

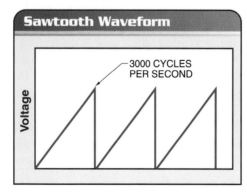

Figure 14-18. The sawtooth generator section uses an oscillator to produce a 3 kHz sawtooth-shaped waveform.

The comparator section compares the speed-reference voltage to the sawtooth voltage and acts as a trigger. If the sawtooth voltage is higher than the speed-reference voltage, the transistors in the power amplifier are turned ON and the DC bus voltage is applied to the DC motor. Since the sawtooth voltage is oscillating at high frequency, the sawtooth voltage quickly falls below the speed-reference voltage, and the DC bus voltage is removed from the DC motor. This is similar in operation to the pulse-width modulation used in AC drives.

The current sensor is a safety circuit that monitors current flow to the field coils. If current flow drops below a preprogrammed level, the drive is shut down to prevent damage.

If the motor speed decreases because of an increase in load, the speed-reference voltage increases. The trigger turns the voltage ON when the sawtooth voltage is higher than the speed-reference voltage. Therefore, a higher speed-reference voltage means the voltage is ON less time in each cycle and the average voltage is lower. This results in less current flowing to the field. The weakened field then causes the speed of the motor to increase. If the motor speed increases because of a reduction in load, the speed-reference voltage decreases. This results in more current flowing to the field with a corresponding decrease in speed.

While some drives are capable of feedback control of the shunt field, this method has limited application. Most often, field-current levels are entered as fixed parameters, and actual speed control is accomplished by silicon controlled rectifier (SCR) armature voltage control.

SCR Armature-Voltage Control. *SCR armature-voltage control* is a method of using SCR bridge rectifiers to control the voltage applied to the armature of a DC motor. SCR armature-voltage control drives include many conditioning circuits to precisely control the bridge output applied to the armature. **See Figure 14-19.**

A DC shunt or compound motor normally has full rated current applied to shunt field coils when armature voltage control is used. With rated current flowing in them, the shunt fields produce a nearly constant magnetic field strength. Therefore, armature-voltage control yields a constant torque output from very low speeds up to rated speed.

The control power supply is a separate regulated DC power supply that is optically isolated from the sawtooth generator. The pulse driver produces a highly conditioned pulsating output that is used to fire the SCRs precisely.

Feedback control may or may not be used. Many of the armature voltage controllers available today incorporate current and/or voltage-sensing circuits that can improve speed regulation when feedback is not used. These circuits monitor changes in motor speed by detecting changes in the voltage drop across and/or current flow through the armature.

Chopper Armature-Voltage Control. *Chopper armature-voltage control* is a method of using a high-speed chopper circuit to control the voltage applied to the armature of a DC motor. There are two primary differences between SCR and chopper armature-voltage controllers. SCR armature-voltage control occurs at the point where the power is being rectified by the SCRs. All chopper drive-control functions occur after power rectification has occurred.

SCR armature-voltage control uses relatively slow bipolar junction transistors to switch the power. Chopper circuits use faster MOSFETs or IGBT transistors. Higher switching speeds provide smoother motor operation and more rapid response to speed changes.

Chopper-based controllers may use diode or SCR rectifier bridges to produce the required DC power. Chopper controllers with SCR rectifiers can be used to drive motors with different voltage ratings because the DC bus voltage can be controlled by adjusting the firing angle of the SCRs. The voltage actually applied to the motor is always determined by the duty cycle of the chopper switches.

The choice of SCR or chopper armature-voltage control depends primarily on the type of power available. Facilities having a DC distribution bus can use chopper drives designed for battery operation. These drives are made without a rectifier section. Because chopper circuits operate at much higher frequencies than SCR circuits, these drives can provide better speed regulation.

> **Definition**
>
> *SCR armature-voltage control* is a method of using SCR bridge rectifiers to control the voltage applied to the armature of a DC motor.
>
> *Chopper armature-voltage control* is a method of using a high-speed chopper circuit to control the voltage applied to the armature of a DC motor.

Figure 14-19. SCR armature-voltage control drives include many conditioning circuits to precisely control the bridge output applied to the armature.

Supplemental Topic

Shunt Motor Drive with Compound-Wound Motor

A simple circuit can be constructed that will safely allow the use of a shunt motor drive with a compound-wound motor. A shunt motor drive has terminals labeled F1 and F2 for connection to the shunt field and terminals labeled A1 and A2 for connection to the armature. The series field has terminals labeled S1 and S2 that are connected in series with the armature. A typical compound-wound motor is a cumulative-compounded motor, where the magnetic fields in the shunt and the series winding are in the same direction, and the fields add together. A single-phase bridge rectifier can be used with the drive to ensure that the current flows in the correct direction through the series field at all times.

The A2 drive terminal is connected to one of the AC terminals of the rectifier and the A2 motor lead is connected to the other AC terminal. Proper polarity of the series field is maintained by connecting S1 to the positive bridge terminal and S2 to the negative bridge terminal. This ensures that the current through the series field is always from S2 to S1, even when the polarity of A1 and A2 is reversed. When A1 at the drive is negative, current flows from A1 at the drive to A1 at the armature, from A2 at the armature through the bridge and out the negative terminal, through the series field from S2 to S1, back through the bridge positive terminal, and back to the drive at terminal A2.

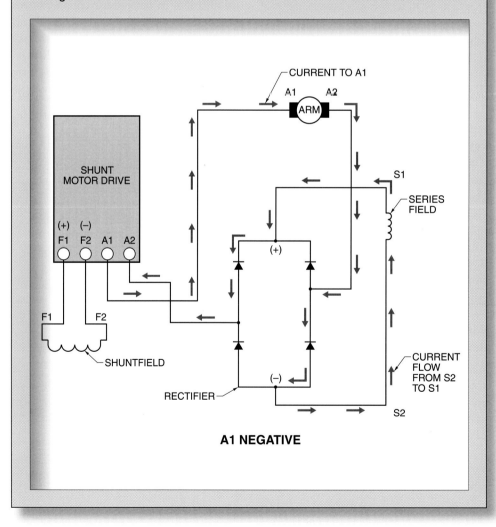

A1 NEGATIVE

When A2 at the drive is negative, current flows from A2 at the drive through the bridge and out the negative terminal, through the series field from S2 to S1, back through the bridge positive terminal to A2 at the armature, and from A1 at the armature back to the drive at terminal A1. In those rare instances where the application requires differential compounding, the S1 and S2 connections to the rectifier need to be reversed.

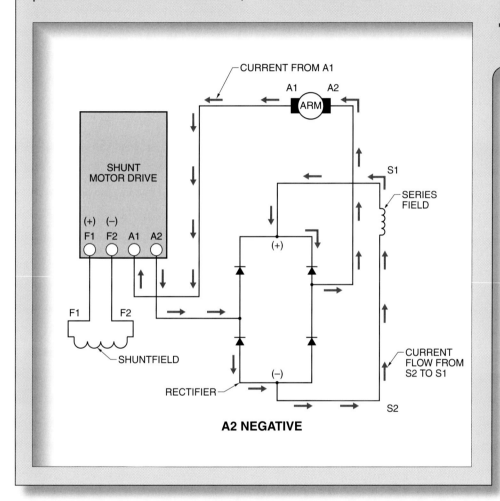

A2 NEGATIVE

Supplemental Topic

DC Drive Classification

DC drives can be classified by their operational capabilities. Form classifications are sometimes used to describe the operational characteristics of electronic drives. Positive torque means that torque is applied in the direction of the normal forward direction. Positive torque is used to accelerate a motor that is running in the forward direction or to brake a motor that is running in the reverse direction. Positive rotation means a motor is running in the normal forward direction. Negative rotation means a motor is running in the reverse direction.

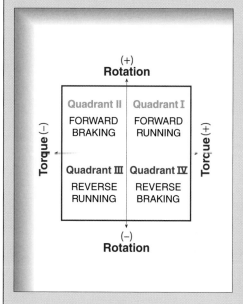

Form A drives are unidirectional, without braking capabilities. Their operation is limited to Quadrant I. Form A drives can be used to operate a motor in either direction by simply changing connections to the motor. An antiplugging switch or time delay may also be required to allow the machine to come to a complete stop before switching into reverse.

Form A drives are the simplest and least expensive. They are used in applications in which speed control is required but controlled stopping is not required. Both SCR and chopper drives are available. Stand-alone field-current controllers are Form A by definition because braking is not possible and reversing is not advisable. Form A drives can also be made reversing if the appropriate reversing contactors and safety interlocks are added. Very few drives can withstand plugging loads, and only very small DC motors can be started across the line. Therefore, reversing contactor design must include these features.

Form B drives are bidirectional and can operate in Quadrants I and III. Form B drives include the required reversing circuitry. This is accomplished either by incorporating two rectifier sections with opposite polarity or by using two switching paths to the armature output terminals. Since the switching circuitry is contained within the drive, the manufacturer can design an appropriate scheme to limit armature currents during reversal.

Form C drives are unidirectional with braking capabilities, and they operate in Quadrants I and II. Form C drives can be used to operate a motor in either direction by simply changing connections to the motor. Form C drives are simply Form A drives with the added feature of a dynamic braking circuit. Forms C drives can also be made reversing if the appropriate reversing contactors and safety interlocks are added. Some manufacturers offer basic Form A drives that include the diodes required for braking and then offer optional plug-in control boards for dynamic or regenerative braking.

Form D drives are capable of operating in all four quadrants. Form D drives are essentially expanded Form B drives that incorporate dynamic or regenerative braking control circuits. The most complex drives are Quadrant IV operation, which are capable of bidirectional operation with braking capabilities in both directions.

Soft Start and Soft Stop

A *soft starter* is a motor control device that provides a gradual voltage increase (ramp up) during AC motor starting and a gradual voltage decrease (ramp down) during stopping. Soft starters are used to control single-phase and 3-phase motors. The capabilities and advantages of soft starters fall between a magnetic motor starter and an adjustable-speed drive. A soft starter has advantages over a magnetic motor starter in that soft starters start and stop a motor gradually, causing less strain on the motor. A soft starter is the simplest solid-state starter available but provides fewer functions than basic variable-speed drives.

Soft starting is achieved by increasing the voltage of a motor gradually in accordance with the setting of the ramp-up control. A potentiometer is used to set the ramp-up time (typically 1 sec to 20 sec). Soft stopping is achieved with a second potentiometer by decreasing the motor voltage gradually in accordance with the setting of the ramp-down time (typically 1 sec to 20 sec). A third potentiometer is used to adjust the starting level of the voltage to a motor with a value at which the motor starts to rotate immediately when soft starting is applied.

Similar to any electrical device or solid-state switch, soft starters produce heat that must be dissipated for proper operation. Large heat sinks are required to dissipate the heat when high current loads (motors) are controlled. Typically, contactors are added in parallel with soft starters. The soft starter is used to control a motor when the motor is starting or stopping and the contactor is used to short out the soft starter when the motor is in operation. Placing contactors and soft starters in parallel allows for soft starting and soft stopping of motors without the need for large heat sinks during motor operation. Soft starters include an output signal that is used to control the time that a contactor is ON or OFF.

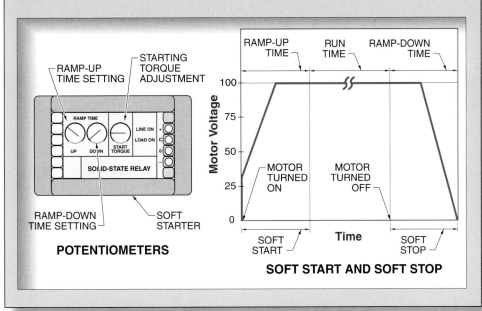

POTENTIOMETERS

SOFT START AND SOFT STOP

Application—Volts-per-Hertz Ratio

The volts-per-hertz ratio for an induction motor is found by dividing the rated nameplate voltage by the rated nameplate frequency. To find the volts-per-hertz ratio for an AC induction motor, apply the following formula:

$$V/Hz = \frac{V}{f}$$

where

V/Hz = volts-per-hertz ratio

V = rated nameplate voltage (in V)

f = rated nameplate frequency (in Hz)

For example, the volts-per-hertz ratio of an AC motor rated for 230 VAC, 60 Hz operation is calculated as follows:

$$V/Hz = \frac{V}{f}$$

$$V/Hz = \frac{230}{60}$$

$$V/Hz = 3.83$$

Above approximately 15 Hz, the amount of voltage needed to keep the volts-per-hertz ratio linear is a constant value. Below 15 Hz, the voltage applied to the motor stator may need to be boosted to compensate for the large power loss AC motors have at low speed. The amount of voltage boost depends on the motor.

A motor drive can be programmed to apply a voltage boost at low motor speeds to compensate for the power loss at low speeds. The voltage boost gives the motor additional rotor torque at very low speeds. The amount of torque boost depends on the voltage boost programmed into the motor drive. The higher the voltage boost, the greater the motor torque.

Motor drives can also be programmed to change the standard linear volts-per-hertz ratio to a nonlinear ratio. A nonlinear ratio produces a customized motor torque pattern that is required by the load operating characteristics. For example, a motor drive can be programmed for two nonlinear ratios that can be applied to fan or pump motors. Fans and pumps are normally classified as variable torque/variable horsepower loads. Variable torque/variable horsepower loads require varying torque and horsepower at different speeds.

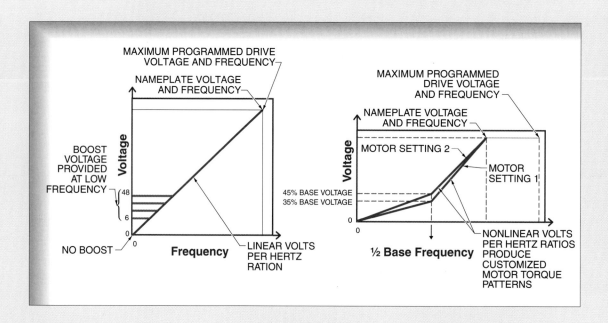

Application—Motor Lead Lengths

Drive manufacturers usually provide the recommended maximum cable length between the drive and motor. The length depends on the carrier frequency and the motor design used in the drive. Voltage spikes are created when high-frequency carrier frequencies are used. The voltage spikes are worse when the leads are longer. Some motor designs are able to tolerate the voltage spikes better than other designs because of improved insulation. Depending on the manufacturer, the maximum cable lengths may be given in feet or in meters.

Circuit 1 uses drive part number 01 and operates at 16 kHz. From the table, the maximum cable length is 10 m (32.8′). Circuit 2 uses drive part number 04 and operates at 4 kHz. From the table, the maximum cable length is 40 m (131′).

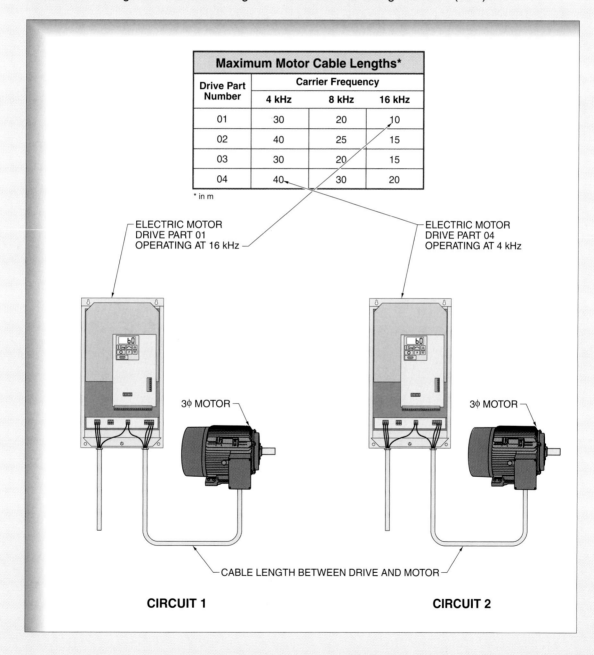

Maximum Motor Cable Lengths*			
Drive Part Number	Carrier Frequency		
	4 kHz	8 kHz	16 kHz
01	30	20	10
02	40	25	15
03	30	20	15
04	40	30	20

* in m

CIRCUIT 1   **CIRCUIT 2**

Summary

- The three main components of an adjustable-speed drive are the converter section, the DC bus section, and the inverter section.

- The converter section contains a rectifier that converts the AC line power to DC.

- The DC bus contains capacitors, inductors, and resistors that filter the DC.

- The inverter section uses the power from the DC bus to create variable-frequency AC.

- The speed of an AC motor is varied by varying the frequency of the applied voltage.

- The torque of an AC motor is kept constant at all frequencies by varying the volts-per-hertz ratio as the frequency is changed.

- Pulse-width modulation controls the amount of time a transistor is ON and conducting current.

- The carrier frequency controls the number of times the transistors turn ON and OFF to simulate an AC waveform. The carrier frequency is typically in the range of 1 kHz to 20 kHz.

- The fundamental frequency is the frequency of the voltage applied to the motor. The fundamental frequency is typically in the range of 0 Hz to 100 Hz.

- An inverter drive uses PWM without feedback to control AC motor speed.

- A vector drive uses a microprocessor and PWM, usually with feedback, to control AC motor speed.

- DC field-current control varies the voltage applied to the shunt coil of a shunt motor or a compound motor.

- SCR armature-voltage control varies the voltage applied to the armature of a DC motor. The DC bus voltage is controlled at the rectifier to control the voltage applied to the motor.

- Chopper armature-voltage control varies the voltage applied to the armature of a DC motor. The average voltage is determined by the duty cycle of the chopper switches.

Glossary

An **inverter duty motor** is an electric motor specifically designed to work with AC adjustable-speed drives.

Pulse-width modulation (PWM) is a method of controlling the voltage sent to a motor by controlling the amount of time a transistor is ON and conducting current.

The **carrier frequency** is the frequency that controls the number of times the solid-state switches in the inverter section of a PWM adjustable-speed drive turn ON and OFF.

The **fundamental frequency** is the frequency of the voltage used to control motor speed.

A **volts-per-hertz ratio (V/Hz)** is the ratio of voltage to frequency in a motor.

An **inverter drive,** or **scalar drive,** is a standard adjustable-speed drive that uses pulse-width modulation to control the speed and torque.

A **vector drive** is a variable-speed drive that uses a microprocessor and PWM, usually with feedback, to calculate the precise vector between voltage and frequency that is needed to provide better control of speed and torque.

A **closed-loop vector drive** is a vector drive that uses shaft-mounted sensors to determine the rotor position and speed of a motor and send the information back to the adjustable-speed drive.

An **open-loop vector drive,** or **sensorless vector drive,** is a vector drive that has no feedback from the motor.

DC field-current control is a method of controlling the voltage applied to a shunt field.

SCR armature-voltage control is a method of using SCR bridge rectifiers to control the voltage applied to the armature of a DC motor.

Chopper armature-voltage control is a method of using a high-speed chopper circuit to control the voltage applied to the armature of a DC motor.

Review

1. Describe how the converter section of an adjustable-speed drive operates.

2. Describe how the DC bus section of an adjustable-speed drive operates.

3. Describe how the inverter section of an adjustable-speed drive operates.

4. Explain how pulse-width modulation is used to create a variable-frequency source.

5. Explain why the length of the leads connecting a drive to a motor is important.

6. Explain why vector drives give better speed control than inverter drives.

7. Describe the difference between closed-loop vector drives and open-loop vector drives.

8. Describe the differences between DC field-current control, SCR armature-voltage control, and chopper armature-voltage control.

Refer to the CD-ROM for Quick Quiz® questions related to chapter content.

MOTORS

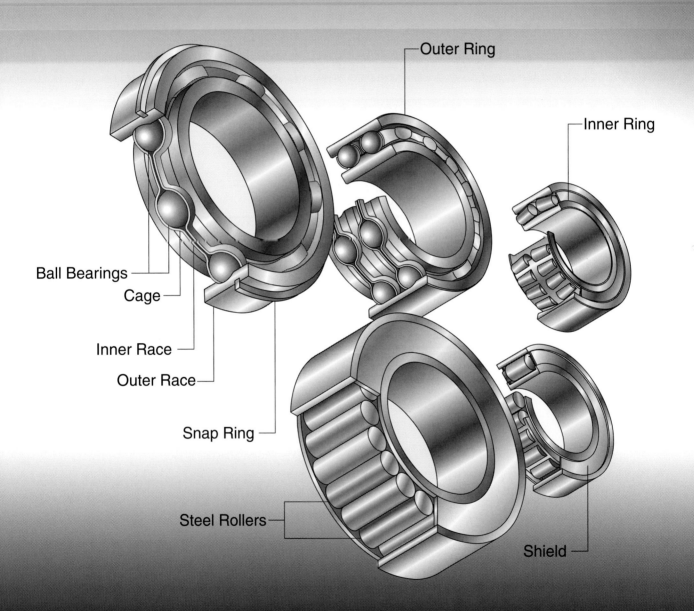

15

Bearings

Babbitt coating

Porous Bronze

Graphite inserts

Bearing and Loads	324
Bearing Installation	329
Bearing Operation	332
Bearing Removal	337
Supplemental Topic—Bearing Currents	342
Summary	344
Glossary	346
Review	347

OBJECTIVES

- Understand the difference between rolling-contact and friction bearings.
- Explain how to mount a bearing. Define the difference between press-fit and push-fit installation.
- Describe several methods for lubricating bearings.
- List various ways bearings can become damaged and describe how the type of damage can be identified through bearing failure analysis.

Bearings guide and position moving parts to reduce friction, vibration, and temperature. Motor efficiency and accuracy depends on proper bearing selection, installation and handling, and maintenance procedures. Bearings are classified as rolling-contact bearings or friction bearings. Rolling-contact bearings include ball, roller, and needle bearings. Successful bearing installation requires cleanliness and correct bearing selection, mounting methods, tool use, and tolerance specifications.

BEARINGS AND LOADS

A *bearing* is a machine component used to reduce friction and maintain clearance between stationary and moving parts. Motor bearings are mounted in the endbells at each end of the motor. Bearings guide and position moving parts to reduce friction, vibration, and temperature. The length of time a machine retains proper operating efficiency and accuracy depends on proper bearing selection, installation and handling, and maintenance procedures. Bearings are available with many special features, but all incorporate the same basic parts.

Bearing Loads

Bearings may be subjected to radial, axial (thrust), or a combination of radial and axial loads. A *radial load* is a load applied perpendicular to the rotating shaft, straight through the ball toward the center of the shaft. For example, a rotating shaft resting horizontally on, or being supported by, a bearing surface at each end has a radial load due to the weight of the shaft itself. **See Figure 15-1.** In a radial load, the shaft should have negligible end-to-end movement.

An *axial load,* or *thrust load,* is a load applied parallel to the rotating shaft. A rotating vertical shaft has an axial load due to the weight of the shaft itself. Combination radial and axial loads occur when both types of loads are present. The shaft of a fan blade is supported horizontally (radial load) and is pulled or pushed (axial load) by the fan blade. These loads can be transferred to the motor shaft through the coupling.

Rolling-Contact Bearings

A *rolling-contact bearing,* or *antifriction bearing,* is a bearing that contains rolling elements that provide a low-friction support surface for rotating or sliding surfaces. Rolling-contact bearings are normally manufactured with balls or rollers. **See Figure 15-2.** The rollers may be cylindrical, tapered, spherical, or needle.

Definition

A *bearing* is a machine component used to reduce friction and maintain clearance between stationary and moving parts.

A *radial load* is a load applied perpendicular to the rotating shaft, straight through the ball toward the center of the shaft.

An *axial load,* or *thrust load,* is a load applied parallel to the rotating shaft.

A *rolling-contact bearing,* or *antifriction bearing,* is a bearing that contains rolling elements that provide a low-friction support surface for rotating or sliding surfaces.

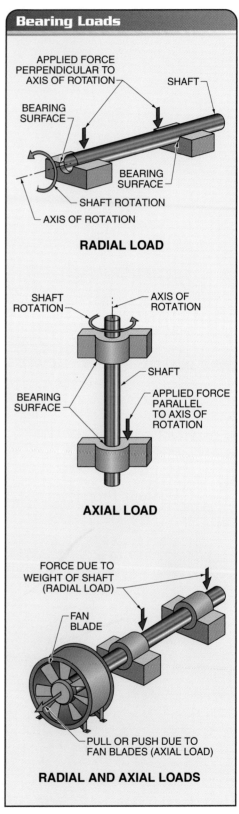

Figure 15-1. Bearings may be subjected to radial, axial (thrust), or a combination of radial and axial loads.

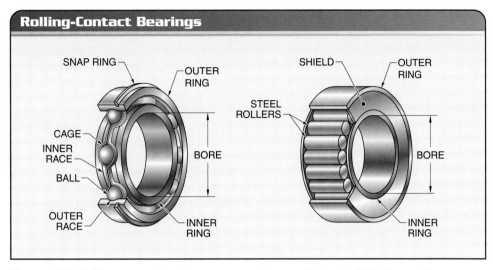

Figure 15-2. Rolling-contact bearings are normally manufactured with balls or rollers.

Definition

*A **ball bearing** is a rolling-contact bearing that permits free motion between a moving part and a fixed part by means of balls confined between inner and outer rings.*

*A **roller bearing** is a rolling-contact bearing that has parallel or tapered steel rollers confined between inner and outer rings.*

*A **needle bearing** is an rolling-contact roller-type bearing with long rollers of small diameter.*

*A **race** is the bearing surface of a rolling-contact bearing that supports the rolling elements.*

Rolling-contact bearings include ball, roller, needle, and tapered bearings. A *ball bearing* is a rolling-contact bearing that permits free motion between a moving part and a fixed part by means of balls confined between inner and outer rings. A *roller bearing* is a rolling-contact bearing that has parallel or tapered steel rollers confined between inner and outer rings. A *needle bearing* is a rolling-contact roller-type bearing with long rollers of small diameter. The rolling-contact bearing categories may be further divided into more specific designs or configurations.

Rolling-contact bearings are designed to roll on a film of lubricant that separates the metal components. Rolling-contact bearings reduce lubrication requirements and decrease starting and operating friction. Reduced friction results in less power required to rotate mechanical components and an increase in overall efficiency of the motor.

Rolling-Contact Bearing Construction. Rolling-contact bearings are constructed of an outer ring (cup), balls or rollers, and an inner ring (cone). The inner ring is generally pressed on a shaft with a tighter fit than the outer ring. A *race* is the bearing surface of a rolling-contact bearing that supports the rolling elements. Rolling-contact bearings are available separately or pre-installed in a housing. Common housings include pillow blocks, flanges, adjustable flanges, rubber cartridges, take-up units, and hanger units.

Needle bearings contain an outer ring (cup) and rollers. The rollers are retained in a cage and bear directly on the rotating shaft. Bearing precision and cost is determined by the smoothness of the ground surfaces (grade of finish) and the quality of tolerances. Better finishes produce less friction, lower temperatures, smoother movements, and longer bearing life, but usually cost more.

Additional bearing components include cages, separators, and snap rings. A cage is used to hold the balls or rollers in place. Bearing designs are open for lubrication injection, or sealed to hold lubricant for the life of the bearing. Seals are used to retain lubrication as well as prevent contamination from dust, dirt, or other solids. Snap rings allow the bearing to be inserted into a housing and held at a certain depth.

Ball Bearings. A ball bearing is a rolling-contact bearing that permits free motion between a moving part and a fixed part by means of hardened balls confined between inner and outer rings. Ball bearings are selected based on the application of the bearing. Ball bearings may be designed for light or heavy loads, radial or axial loads (or combination of each), and harsh or clean environments. General-use ball bearings

Definition

A *Conrad bearing* is a single-row ball bearing that has races that are deeper than normal.

A *roller bearing* is a rolling-contact bearing that has cylinder-shaped or tapered steel rollers confined between an outer ring (cup) and an inner ring (cone).

A *cylindrical roller bearing* is a roller bearing having cylinder-shaped rollers.

A *tapered roller bearing* is a roller bearing having tapered rollers.

are designed as single-row radial, single-row angular-contact (axial), or double-row radial or axial based on the direction of applied force. **See Figure 15-3.** A *Conrad bearing* is a single-row ball bearing that has races that are deeper than normal. Conrad bearings allow for axial and radial loads. Installing Conrad bearings in the wrong direction results in immediate damage.

Double-row bearings, also known as duplex bearings, are matched pairs of angular-contact bearings. They can be used with heavy radial and thrust loads in both directions. Double-row bearings are designed as matched sets and are identified based on their configuration, such as back-to-back or face-to-face. Two single-row bearings should never be used when replacement requires the use of double-row bearings. Pairing unmatched single-row bearings adversely affects shaft rotation.

Roller Bearings. A *roller bearing* is a rolling-contact bearing that has cylinder-shaped or tapered steel rollers confined between an outer ring (cup) and an inner ring (cone). **See Figure 15-4.** Roller bearings are designed for loads and applications similar to those of ball bearings. Roller bearings are designed for heavy radial and axial loads. Cylindrical rollers are used for radial loads and tapered rollers are used for radial and axial loads. Roller bearings are precision devices and must be kept clean and handled with care.

A *cylindrical roller bearing* is a roller bearing having cylinder-shaped rollers. Cylindrical roller bearings are also known as radial roller bearings. Cylindrical roller bearings are used in high-speed, high-load applications and may contain as many as four rows of rollers. These bearings have a high radial load capacity but are not designed for axial loads and cannot tolerate misalignment.

A *tapered roller bearing* is a roller bearing having tapered rollers. Tapered roller bearings are normally used for heavy radial and adjustable axial loads. Tapered roller bearings are available in more than 20 configurations from many different manufacturers. The different tapered roller bearings are generally interchangeable and may be cross-referenced.

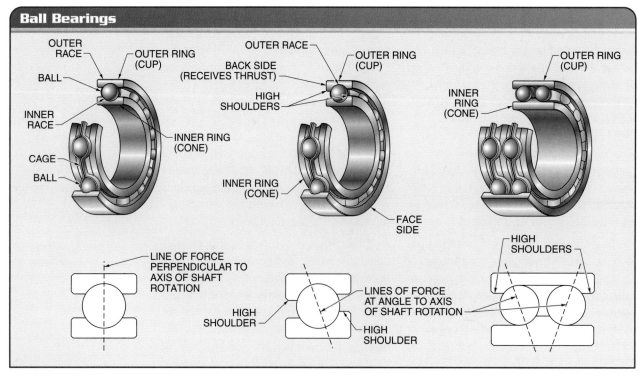

Figure 15-3. General-use ball bearings are designed as single-row radial, single-row angular-contact (axial), or double-row radial or axial based on the direction of applied force.

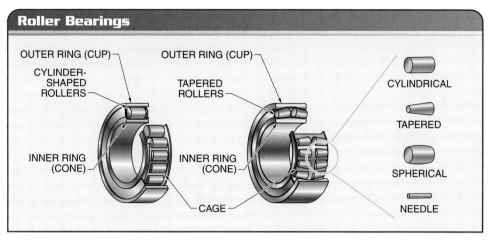

Figure 15-4. Roller bearings have steel rollers confined between an outer ring (cup) and an inner ring (cone).

Definition

A *needle bearing* is a roller-contact roller-type bearing with long rollers of small diameter.

A *friction bearing* is a bearing consisting of a stationary bearing surface, such as machined metal or pressed-in bushings, that provides a low-friction support surface for rotating or sliding surfaces.

Needle Bearings. A *needle bearing* is a roller-contact roller-type bearing with long rollers of small diameter. Needle bearings, similar to cylindrical roller bearings, are chosen for relatively high radial loads. Needle bearings are characterized by their rollers being of small diameter compared to their length. The ratio of length to diameter may be as much as 10:1. Needle bearings are often used in applications with limited space. Needle bearings normally have tightly packed rollers without separators or inner rings. Special needle bearings that are used for oscillating motion in aircraft elements may have separators with inner rings. Needle bearing cases may have machined surfaces or be drawn and formed for roller retention. Needle bearings are generally press fit. A firm, even, and square press during installation prevents damage to the bearing case.

Friction Bearings

A *friction bearing* is a bearing consisting of a stationary bearing surface, such as machined metal or pressed-in bushings, that provides a low-friction support surface for rotating or sliding surfaces. **See Figure 15-5.** Friction bearings are also known as plain bearings or sleeve bearings. Friction bearings commonly use lubricating oil to separate the moving component from the stationary bearing surface. Friction bearing surfaces normally consist of a material that is softer than the supported component. This allows foreign matter to embed in the bearing material. This prevents the spreading of the foreign matter throughout the lubrication system. Friction bearings can conform to slightly irregular mating surfaces. Friction bearings may be integrally machined, one-piece sleeve, or split sleeve.

Large motors can be equipped with sleeve bearings. Large sleeve bearings are split along horizontal lines running parallel with the shaft. After installation they are coated with a dye called bluing. The high spots in the bearings are machined out so that the entire bearing surface contacts the shaft. An oil reservoir supplies lubricant. Even though the shaft is lubricated, friction opposes the rotation and causes energy losses.

Friction bearings are used in areas of heavy loads where space is limited. Friction bearings are quieter, less costly, and, if kept lubricated, have little metal fatigue compared to other bearings. Friction bearings may support radial and axial (thrust) loads. In addition, friction bearings can conform to the part in contact with the bearing because of the sliding rather than rolling action. This allows the friction bearing material to yield to any abnormal operating condition rather than distort or damage the shaft or journal. A *journal* is the part of a shaft that moves in a friction bearing. The sliding motion of a shaft or journal is generally against a softer, lower-friction bearing material.

Friction Bearings

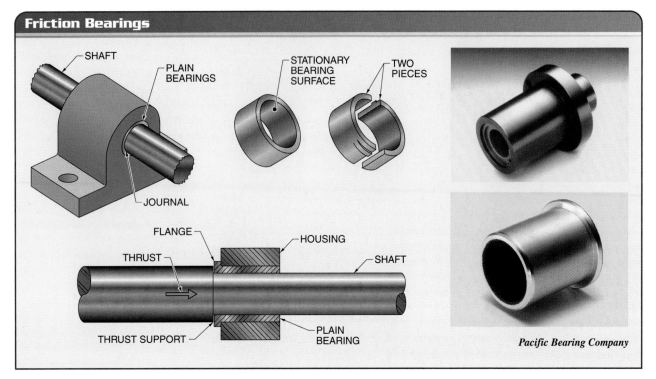

Figure 15-5. Friction bearings provide a low-friction support surface for rotating surfaces.

Definition

*A **babbitt metal** is an alloy of soft metals such as copper, tin, and lead, and a hardening material such as antimony.*

The service life of a friction bearing depends on the surface condition of the shaft. A shaft with nicks, gouges, scratches, or rough machine marks wears a friction bearing rapidly. In addition, a shaft that is ground too fine does not allow lubricant retention and wears once the lubricant is squeezed out.

Friction Bearing Materials. Special materials, or combinations of materials, must be selected for friction bearings because of the momentary metal-to-metal contact that occurs during shaft stopping and starting. Friction bearing material must be corrosion- and fatigue-resistant, able to handle running loads and thermal activity, and compatible with other materials used. **See Figure 15-6.**

Babbitt metals are usually the best metals for friction bearing loads. A *babbitt metal* is an alloy of soft metals such as copper, tin, and lead, and a hardening material such as antimony. Copper-leads, bronze, and aluminum base metals are used for friction bearings requiring increased load-carrying capacities. Babbitt metals are used in a thin layer over a steel support for heavy commercial applications, such as armature bearings used in hand drills.

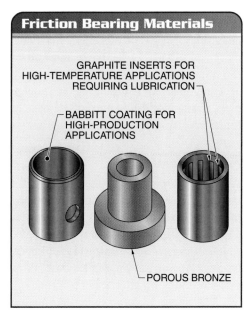

Figure 15-6. Special materials must be used for friction bearings because of the momentary metal-to-metal contact that occurs.

Bearing hardness must also be considered in addition to operating conditions. Normal wear and scoring must take place on the less costly bearing surface, not on the surface of the shaft or journal. For this to occur, friction bearings must be at least 100 Brinell points softer than the shaft or journal. A Brinell hardness test measures the hardness of a metal or alloy by hydraulically pressing a hardened steel ball into the metal to be tested and then measuring the area of indentation. The Brinell hardness number is found by measuring the diameter of the indentation and finding the corresponding hardness number on a calibrated chart.

Bearing Selection

No universal bearing exists that can do all of the functions and applications required in industry. **See Figure 15-7.** In many cases, a review of the machine function and its bearing requirements may indicate if proper bearings are being used. Bearings are often selected without the use of manufacturer's specifications. Certain factors other than dimensions must be observed when a replacement bearing is chosen by comparison of a removed bearing instead of from an equipment manual or parts book. Factors to be considered include the exact replacement part number, the type and position of any seal, the direction of force and positioning of a required high shoulder, and whether a retaining ring is required.

Provisions made for thermal expansion within a machine are generally published by the machine manufacturer and are listed as space tolerances between housing, bearing components, and shaft. Greater space tolerances are allowed for friction bearings than for rolling-contact bearings because friction bearings are more susceptible to damage from higher temperatures.

BEARING INSTALLATION

Successful bearing installation requires cleanliness, correct bearing selection, mounting methods, tool use, and tolerance specifications. Proper bearing assembly is required for proper bearing performance, durability, and reliability. More bearings fail because of poor installation practices than from malfunction during their useful life. Bearings can easily be damaged by force, dirt, or misalignment.

Parts Preparation

When a bearing has been removed and taken apart for maintenance, the parts should be cleaned and spread out on a clean surface for inspection. Cleaning is accomplished by dipping or washing the housing, shaft, bearing, spacers, and other parts in a clean, nonflammable cleaning solvent. All traces of dirt, grease, oil, rust, or any other foreign matter must be removed.

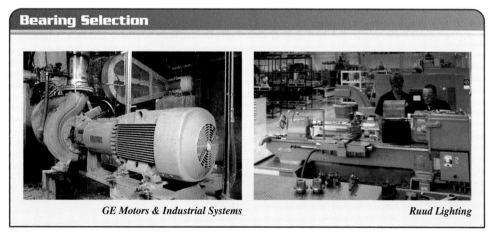

Figure 15-7. The requirements for bearings are different, depending on the environment. Some bearings must be able to withstand harsh industrial use, while others operate in clean environments.

Definition

Press fit *is a bearing installation where the bore of the inner rotating ring is smaller than the diameter of the shaft and considerable force must be used to press the bearing onto the shaft.*

Push fit *is a bearing installation where the diameter of the outer fixed ring is smaller than the diameter of the bearing housing and the ring can be pushed in by hand.*

Caution should be taken when using part-cleaning solutions. Seals, O-rings, and other soft materials may deteriorate due to incompatibility. The cleaning solution label should be checked before use. If in doubt, a lightweight, warm mineral oil is a good cleaning and flushing fluid. Care should be taken to clean housing and shaft bearing seats, corners, and keyways.

All parts should be wiped with a clean towel soaked in lightweight oil and then wrapped or covered to protect them from dust and dirt. All parts need to be inspected for nicks, burrs, or corrosion on shaft seats, shoulders, or faces. All bearing components should be inspected for indication of abnormality or obvious defects, such as cracks and breaks.

Any nicks, corrosion, rust, and scuffs on shaft or housing surfaces should be removed. All corners and bearing seats should be square and all diameters must be round, in tolerance, and without runout. Any worn spacers, shafts, bearings, or housings must be replaced. The housing shoulder must be square enough to clear the bearing corner.

Bearing Mounting

Bearing mounting procedures affect the performance, durability, and reliability of the motor. Precautions should be taken to allow a bearing to perform without excessive temperature rise, noise from misalignment or vibration, and shaft movements. Bearings are mounted with one rotating ring installed as a press fit over the shaft and the other rotating ring installed as a push fit in the bearing housing. *Press fit* is a bearing installation where the bore of the inner rotating ring is smaller than the diameter of the shaft and considerable force must be used to press the bearing onto the shaft. *Push fit* is a bearing installation where the diameter of the outer fixed ring is smaller than the diameter of the bearing housing and the ring can be pushed in by hand.

During a press-fit installation, force must be applied uniformly on the face or ring that is to be press fit. This can be accomplished by using a piece of tubing, steel plate, and a hammer or an arbor press. **See Figure 15-8.** Wood should not be used because of the possibility of contaminating a bearing with wood splinters or fibers. A push fit allows the outer ring to be slid into the bearing housing by hand. With some heavy-duty cylindrical roller bearings, the extra loads require that both rings be press fit.

Bearings that are designed for thrust loads must be installed in the correct direction to prevent the load from separating the bearing components. These bearings have a face and back side for ease in identifying the thrust direction. The back side receives the thrust and is marked with the bearing number, tolerance, manufacturer, and, in some cases, the word "thrust." Rolling-contact bearings must be firmly mounted so endplay and shaft expansion and contraction due to temperature changes is minimized.

Heating and Cooling. Bearings can be mounted by heating (expanding) or cooling (shrinking) the bearing, depending on the application. Heating the inner ring of a bearing increases the size of the bearing, allowing it to slip over a shaft. **See Figure 15-9.** Cooling the outer ring of a bearing reduces the size of the bearing, allowing it to fit inside a housing.

Heating bearings may be accomplished using a light bulb, an oven, clean hot oil that has a high flash point, a hot plate, or induction heat. The light bulb, oven, and induction heating methods are reliable because their temperature is easy to control. A hot oil bath is the best heating method because it provides even and controlled heating of the bearing.

The light bulb heating method is easy and economical. This method uses a light bulb placed in the bore of a bearing to provide heat to increase the diameter of the bearing. The temperature is controlled by the length of time the bulb is placed in the bearing. This method works well because the inner ring is the only component to be brought to high heat, allowing handling and assembly using the outer ring.

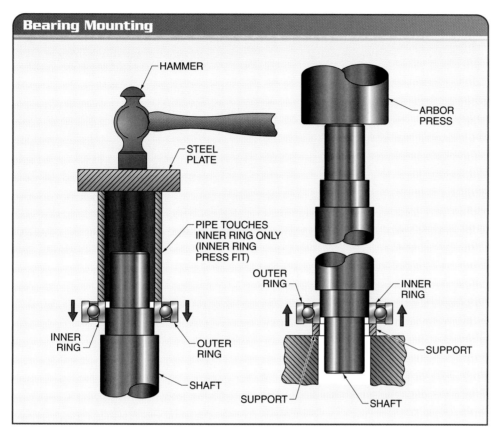

Figure 15-8. Uniform force must be applied on the face that is to be press fit with a piece of tubing, steel plate, and hammer, or with an arbor press.

Temperatures must be even throughout the inner bearing ring and controlled up to about 200°F. Torches heat unevenly, distort diameters, and must not be used. Temperatures over 250°F may reduce the hardness of bearing metals, resulting in early failure. Bearing temperatures may be determined using thermal crayons, which melt at specific temperatures. Induction heaters, operating electrically and rapidly, leave metals magnetized and may cause problems. Small metal particles are drawn to the assembly if bearings are not demagnetized.

Freezing may be required to reduce the outside diameter of a bearing to allow installation into a housing when heating methods are not possible. Prelubricated bearings must not be heated for installation. Freezing causes shaft sizes to be reduced to allow installation of prelubricated bearings. Liquid nitrogen or a mixture of dry ice and alcohol can be used to lower the temperature of a bearing or shaft. Condensation forms on bearings in areas where ambient conditions are humid. Corrosion is prevented if the condensate is wiped or blown off after assembly, followed by thorough lubrication. Corrosion should not be a problem if components were lightly oiled before freezing.

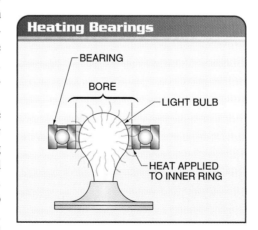

Figure 15-9. Heating the inner ring of a bearing increases the size of the bearing, allowing it to slip over a shaft.

Definition

A *grease fitting* is a hollow tubular fitting used to direct grease to bearing components.

A *pressure cup* is a pressurized grease reservoir that provides constant lubrication to a bearing.

Precautions During Mounting. Bearings must be installed properly to ensure proper life and service. Precautions that must be taken when mounting bearings include the following:

- Know the bearing function in a machine.
- Keep all bearings wrapped or in the original sealed container until ready to use. Reusable bearings should be treated as new.
- Maintain clean tools, hands, and work surface, and work in a clean environment.
- Use clean, lint-free cloths when wiping bearings.
- Never attempt to remove the rust preventive compound used by the manufacturer unless specifically recommended.
- Use the best bearings available within reason. The life and reliability of a bearing is generally related to its cost.
- Always follow the instructions of the heating equipment manufacturer when bearings are to be heated for assembly.
- Use rings, sleeves, or adaptors that provide uniform, square, and even movements.
- Prevent cocking during bearing installation by starting races evenly on the shaft without pressure devices.
- Never strike the bearing with a wooden mallet or wooden block.
- Never apply pressure on the outer ring if the inner ring is press fit.
- Never apply pressure on the inner ring if the outer ring is press fit.
- Be careful not to abuse, strike, force, press on, scratch, or nick bearing seals or shields.

BEARING OPERATION

After a bearing is installed on a motor, the bearing must be lubricated and tested. Many smaller bearings are factory sealed and should not be lubricated. Motors are normally lubricated at the factory to provide long operation under normal service conditions without relubrication. Excessive and frequent lubrication can damage a motor. The time period between lubrications depends upon the motor's service conditions, its ambient temperature, and its environment.

Lubrication

The lubrication instructions provided with the motor should always be followed. These instructions are usually listed on the nameplate or terminal box cover. Alternately, there may be separate instructions furnished with the motor. If lubrication instructions are not available, relubricate sleeve bearings and ball bearings according to a set schedule. **See Figure 15-10.**

Friction and rolling-contact bearings normally require periodic lubrication to prevent premature failure. Sealed bearings do not require lubrication. Many bearings are fitted with a shield that helps to contain the grease inside the bearing. Shielded bearings require regular lubrication, and dirt can enter the bearing by going around the shield. Devices used for lubricating bearings include grease fittings, pressure cups, oil cups, and oil wicks. **See Figure 15-11.**

A *grease fitting* is a hollow tubular fitting used to direct grease to bearing components. The head of the grease fitting is designed to open when a grease gun is attached and close when it is removed. Grease fittings are available with different head designs and configurations to provide easy access during lubrication. Grease fittings are attached to a bearing housing by a pipe thread. The pipe thread provides a sealed connection between the bearing housing and the grease fitting.

A *pressure cup* is a pressurized grease reservoir that provides constant lubrication to a bearing. A pressure cup consists of a grease reservoir with an internal spring and pressure plate. As grease is pumped into the reservoir, the spring is compressed and the pressure plate is pushed toward the top of the cup. During bearing movement, grease is forced into the bearing components by pressure from the compressed spring.

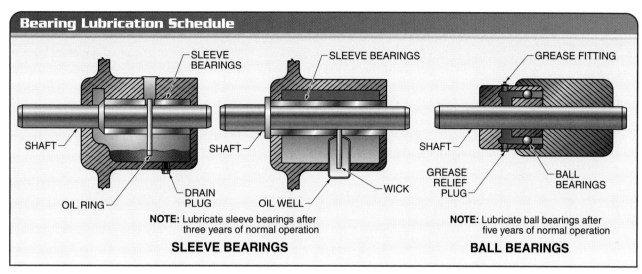

Figure 15-10. Bearings should be lubricated according to a set schedule.

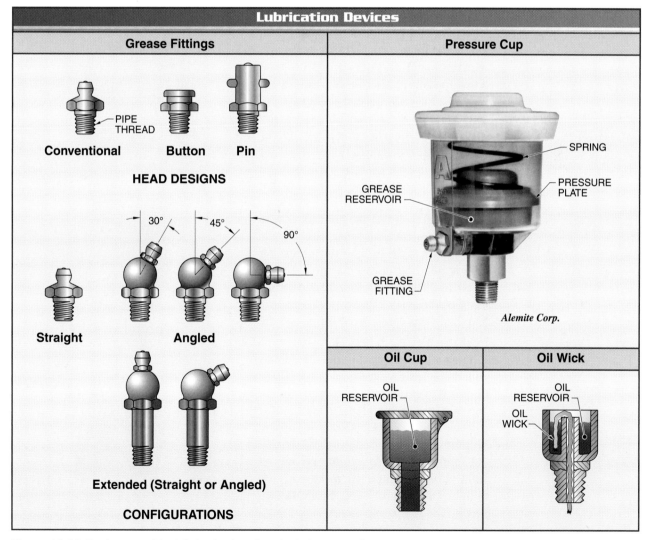

Figure 15-11. Devices used for lubricating bearings include grease fittings, pressure cups, oil cups, and oil wicks.

Definition

*An **oil cup** is an oil reservoir located on a bearing housing to provide lubrication to a bearing.*

An *oil cup* is an oil reservoir located on a bearing housing to provide lubrication to a bearing. Oil cups rely on gravity to feed the lubricant through a direct passage to the bearing. Lubrication continues as long as there is a supply of oil in the reservoir. Some oil cups have an oil wick that serves as a conduit for oil from the oil cup to the bearing surface. The flow of oil from the cup is regulated by the maximum flow allowed through the oil wick material. This provides a controlled flow of lubricant to bearing components over an extended period of time. To prevent contaminants from entering the lubrication system, the tools and grease fittings or oil cups should be cleaned before adding the lubricant.

Friction bearings usually require a light film of oil to operate properly. The endbell contains a lubrication tube and wicking material. Lubricant is dropped down the lubrication tube to saturate a wick. The wick is located in the bearing sleeve so that it can transfer a light film of oil to the bearing and shaft. Other types of sleeve bearings are made of porous metal that absorbs oil to be used to create a film between the bearing and the shaft.

It is important that the lubricating tube points up to make sure that gravity pulls the oil down to the wick. If the lubricating tube is pointing down, the oil will flow away from the wick. When the wick dries out, the lubricant will not reach the bearing and shaft. This causes overheating and the shaft can seize to the bearing.

Oil Seals. An *oil seal* is a device used to contain oil inside a housing. Oil seals are designed to close spaces between moving and stationary components in mechanical systems. Oil seals extend bearing life by keeping the oil in the bearing and sealing out foreign matter, providing protection from abrasive materials, moisture, lack of lubrication, and other harmful elements. Oil seal components include the outer case, inner case, lip, and spring. **See Figure 15-12.**

The outer case of an oil seal is made of metal and provides structural support for the oil seal. The diameter of the outer case is slightly larger than the bore into which the oil seal fits. This size difference provides a tight fit in the housing. The inner case is used with larger-diameter seals to provide extra rigidity to the oil seal, but provides no sealing capabilities. An oil seal lip contacts the moving part of the equipment to prevent material from passing by the oil seal. The lip provides the actual seal to the shaft. Some seals use a spring to provide extra contact to the shaft.

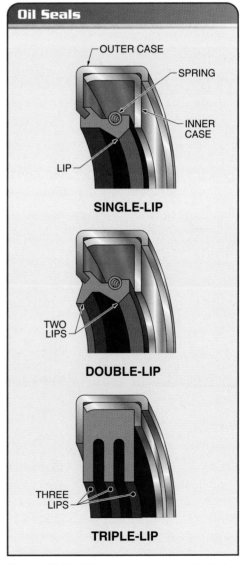

Figure 15-12. Oil seal components include the outer case, inner case, lip, and spring.

Oil seal lip material includes synthetic materials, leather, and felt. Most general-purpose seals have a synthetic lip molded directly to the metal case on the inside or outside diameter. Synthetic general-purpose seals may have up to three sealing lips, depending on the application. The flexible synthetic lip may be spring-loaded or springless. The lip is designed to work like a squeegee, wiping oil from the shaft and preventing it from escaping. The lip also prevents contaminants from entering the bearing. The springless single-lip seal is considered a grease seal because it retains thick lubricants but leaks if used with thin lubricants and high pressures. Oil seals located in dirty environments require double or triple lips to provide extra protection for the bearing.

Oil Seal Installation. Oil seals are commonly installed next to a bearing on the outside of a housing. **See Figure 15-13.** The oil seal outside diameter is usually larger than the bearing outside diameter, allowing the seal to seat against a recess when it is pressed into the housing. The recess provides a press fit for the seal and also keeps the seal from touching the bearing.

Correct installation of an oil seal ensures proper performance. The lip of an oil seal must be lubricated before installing the seal to allow it to slide during installation. The same oil used in the motor bearing is used to lubricate the lip to prevent mixing of oils. The seal outside diameter and housing bore must remain clean and dry, providing a leakproof fit. The oil seal outside diameter should also be coated with a bore sealant before installation.

Hammering an oil seal normally results in a leaky seal. The proper installation tools should be used when installing oil seals. **See Figure 15-14.** Oil seal installation tools apply force evenly and at the correct location on the seal. A variety of installation tools are available for specific applications, such as proper positioning and installation over a shaft. A piece of steel or heavywall plastic tubing may also be used as an oil seal installation tool.

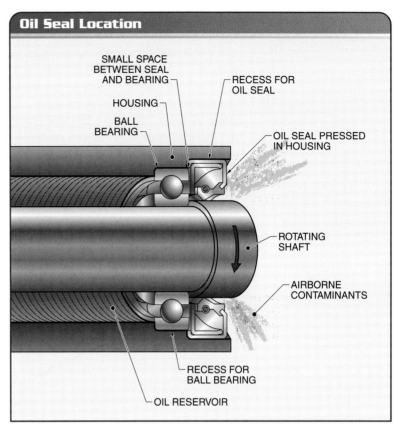

Figure 15-13. Oil seals are commonly installed next to a bearing on the outside of a housing.

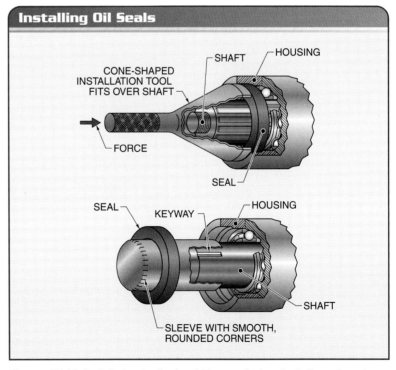

Figure 15-14. Installation tools should be used when installing oil seals.

An oil seal lip must also be protected when being installed over a shaft. Sharp edges on a shaft damage the lip and must be removed. A cone-shaped installation tool, a sleeve with smooth rounded corners, or piece of shim stock rolled in the shape of a cone can be used when installing a seal over the end of a shaft. Keyways on shafts can also damage oil seals. A sleeve is commonly used to cover the keyed area protecting the seal as it is slid on the shaft.

Improper Lubrication. Improper lubrication is a major cause of bearing failure. Improper lubrication includes underlubrication, overlubrication, lubricant contamination, and mixing lubricants. A bearing that experiences starvation experiences wear immediately. If the bearing is underlubricated, the metal surfaces touch, causing rapid failure.

Bearings must be prelubed to ensure lubricant is present during initial startup. New bearings are generally ⅓ to ½ filled with grease. This is their normal operating level. A common mistake when greasing a bearing is to add too much grease. This causes overheating and damages the bearing. The overheating is caused when the rollers must force their way through the excess grease that is packed into an overlubricated bearing.

Contamination of the lubricant can also reduce the effectiveness of the lubricant. Common lubricant contaminants include water and dirt. Lubricant contaminated with dirt subjects moving components to a constant flow of abrasives. Water gives the oil a milky appearance and causes bearing components to rust, increasing friction and causing eventual bearing failure. Sources of water may be condensation and environments with high humidity. Periodic oil changes are necessary to ensure that the bearing has pure lubricant.

Contamination

Contamination causes many bearing failures. Dirt or other impurities can enter a bearing during operation, lubrication, and/or assembly. Dirt is very abrasive to bearing components and causes premature failure of the bearing. Dirt is also a poor heat conductor, which reduces the ability of a lubricant to conduct heat away from the bearing. Bearings located inside a gearbox may have dirt and metal chips introduced during manufacturing or rebuilding. During operation, the bearing may be subjected to dirt from the surrounding environment. Poor bearing seals allow dirt and other contaminants to enter the bearing components. Bearings located inside a crankcase may also be contaminated during operation due to leaky gaskets. Replace leaking gaskets or seals immediately.

Dirt can contaminate a bearing during lubrication because oily or greasy areas such as oil cups and grease fittings attract and gather dirt and dust. Dirt large enough to damage a bearing can be invisible to the naked eye. Bearing life is extended by using a lint-free rag and keeping visible dirt away from lubricants. This dirt must be removed from the fitting before the bearing is lubricated because the dirt mixes with the lubricant and enters the bearing. Lubrication tools and fill caps located on gearboxes must also be cleaned before any lubricant is added or checked. For example, the tip of a grease gun collects dirt when not in use and should be cleaned before it is placed on a grease fitting.

Dirt may also enter a bearing during assembly. Care should be taken to not contaminate the bearing, seals, or gearbox during assembly or rebuild operations. Bearing journals must also be free of dirt before the bearing is put into position. Dirt lodged behind a bearing can cause misalignment, an isolated pressure point, and incorrect location. During assembly, hands, bearings, seals, and gearboxes must be clean and work must be performed in a clean environment. Bearing surfaces should not be touched with bare hands because hands can contain enough oil and dirt to cause damage.

Machine Run-In

A machine run-in check should be made after bearing assembly is complete. A run-in check starts with a hand check of the torque of the shaft. For safety reasons, the power must be locked out when manually rotating a shaft. Unusually high torques normally indicate a problem with a tight fit, misalignment, or improper assembly of machine parts. Restore machine power and listen for unusual noises. High noise levels may indicate excessive loading or cocked or damaged bearings. The problem must be corrected before continuing.

Final checks are accomplished by measuring machine temperatures. High initial temperatures are common because bearings are packed with grease, which can produce excess friction when the motor is first started. Run-in temperatures should decrease to within recommended ranges. Any machine with temperatures that continue to run high should be corrected before proceeding. Continued high temperatures are generally a sign of tight fit, misalignment, or improper assembly. After a machine is placed back in operation, record motor amperage readings and bearing temperature for proactive and predictive maintenance procedures.

BEARING REMOVAL

Proper tools and maintenance procedures are required when removing bearings. Many bearing failures are due to contaminants that have worked their way into or around a bearing before it has been placed in operation. Work benches, tools, clothing, wiping cloths, and hands must be clean and free from dust, dirt, and other contaminants. Internal abrasive particles permanently indent balls, rollers, and raceways. This alters the shape of the surface and begins bearing erosion. Bearing tolerances are such that a solid particle of a few thousands of an inch (0.001″ to 0.003″) lodged between the housing and the outer ring can distort raceways enough to reduce critical clearances.

Bearing removal is more difficult than bearing installation. A firm, solid contact must be made for bearing removal. Bearings should always be removed from a shaft with even pressure against the ring that was press fit. Bearings are removed from shafts using bearing pullers, gear pullers, arbor presses, or manual impact. These methods enable easy bearing removal and reduce the damage to the bearing. **See Figure 15-15.**

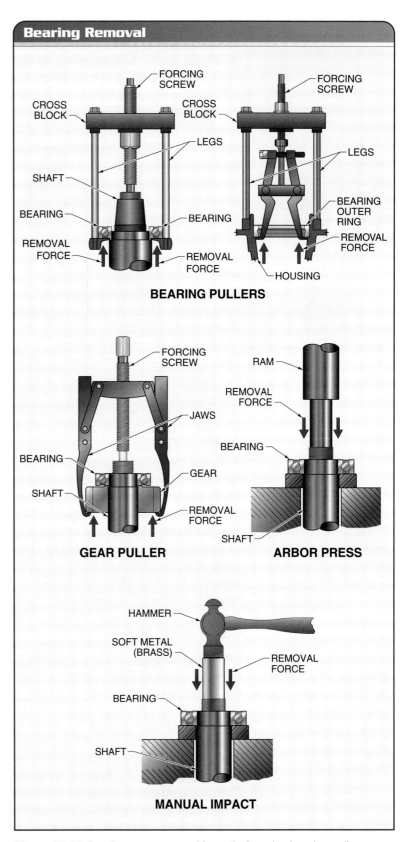

Figure 15-15. Bearings are removed from shafts using bearing pullers, gear pullers, arbor presses, or manual impact.

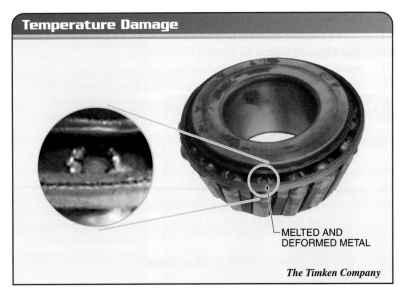

Figure 15-19. Temperature damage can cause melted and deformed metal.

Definition

Fretting corrosion *is the rusty appearance that results when two metals in contact are vibrated, rubbing loose minute metal particles that become oxidized.*

Misalignment wear *is bearing damage that occurs when the two bearing rings are not aligned with one another and the rolling-contact points cause eccentric wear.*

Thrust damage *is bearing damage due to axial force.*

Galling, *or* ***adhesive wear,*** *is a bonding, shearing, and tearing away of material from two contacting, sliding metals.*

Fretting Corrosion. *Fretting corrosion* is the rusty appearance that results when two metals in contact are vibrated, rubbing loose minute metal particles that become oxidized. In many cases, fretting is a normal condition that appears as discoloration on the outer surface of the outer ring between the outer ring and the housing. This happens as moisture from the air settles between the two contacting and unprotected metal surfaces. Fretting corrosion becomes harmful when the oxidation breaks down supporting wall surfaces, creating looseness. Fretting corrosion is also harmful as its oxidation particles (oxides) mix with and break down the bearing lubricant. **See Figure 15-20.**

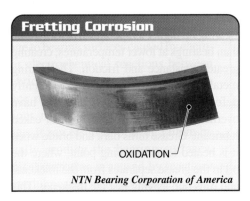

Figure 15-20. Fretting corrosion appears as discoloration on the outer surface of the outer ring between the outer ring and the housing.

Misalignment Wear. *Misalignment wear* is bearing damage that occurs when the two bearing rings are not aligned with one another and the rolling-contact points cause eccentric wear. Bearing surfaces that are misaligned appear as worn surfaces on one side or opposing sides of a bearing. Rollers in roller bearings can leave wear marks on one side of the bearing inner race. The roller may also show high and low trails on the inside of the outer race. Misalignment wear may also appear as lack of fretting on two sides of the outer surface of the outer ring. **See Figure 15-21.**

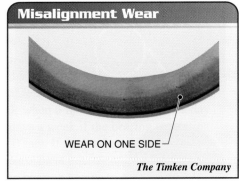

Figure 15-21. Bearing surfaces that are misaligned appear as worn surfaces on one side or opposing sides of a bearing.

Thrust Damage. *Thrust damage* is bearing damage due to axial force. Thrust damage on ball bearings appears as marks on the shoulder or upper portion of the inner and outer race. The appearance can be vary from a slight discoloration to heavy galling. *Galling,* or *adhesive wear,* is a bonding, shearing, and tearing away of material from two contacting, sliding metals. The amount of galling is proportional to the applied load forces. Thrust damage on friction bearings appears as heavy wear at the bearing ends. **See Figure 15-22.**

Tech Fact

The load capacity of friction bearings can greatly exceed the load capacity of rolling-contact bearings because of the increased surface area in contact with the shaft.

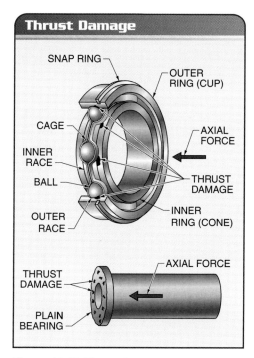

Figure 15-22. Thrust damage on friction bearings appears as heavy wear at the bearing ends.

Electrical Pitting and Fluting. *Electrical pitting* is bearing damage in the form of pits formed on the balls or race caused by electrical discharge through the bearing. This happens as current passes from its introduction, such as electrical system feedback or welding currents, to a grounded connection. **See Figure 15-23.** When the current passes through bearings, the current can etch or pit bearing surfaces. Mild electrical currents may not etch the metal, but can create high enough temperatures as current transfers through the bearing to burn and break down lubricants.

Welding current damage is observed as short, pitted lines on balls or rollers that were stationary when the current was present. The race has corresponding damage, but this is not normally observed unless the bearing is destroyed. Electrical feedback created by certain forces throughout plant electrical usage, faulty wiring, and static electricity can be prevented from flowing through a machine if extra grounding is provided. Extra grounding of a machine can be as simple as running a wire from the machine to a pneumatic or water line. **WARNING:** Never ground a machine by connecting a wire from the machine to a pipe containing flammable materials, such as gas or oil.

Fluting is observed in roller bearings that were rotating while welding currents passed through them. *Fluting* is the elongated and rounded grooves or tracks left by the etching of each roller on the rings of an improperly grounded roller bearing when current passes through the bearing. Fluting marks do not occur at the same distance apart as the ball bearings, as happens with Brinnell damage. In roller bearings, fluting is caused by electrical arcing and pitting the length of each roller in the bearing. Damage from welding can be prevented by attaching the welding ground clamp in a location where no bearings are between the ground and the weld.

Definition

Electrical pitting is bearing damage in the form of pits formed on the balls or race caused by electrical discharge through the bearing.

Fluting is the elongated and rounded grooves or tracks left by the etching of each roller on the rings of an improperly grounded roller bearing when current passes through the bearing.

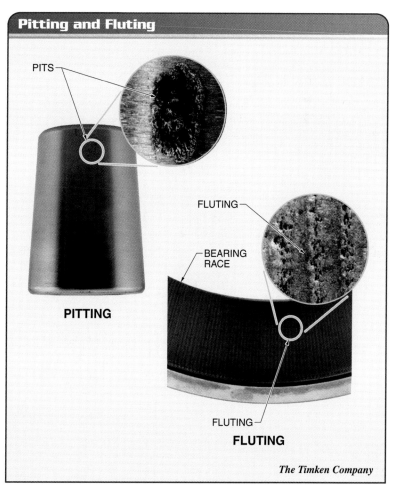

Figure 15-23. Electrical pitting happens as current passes through a bearing.

Supplemental Topic

Bearing Currents

Bearing current is unintended current flowing through a motor bearing that causes motor bearings to fail because of electrical pitting and fluting. Bearing currents are caused by voltage induced into the bearing, housing, or shaft. A common source of the induced voltage is high-frequency adjustable-speed motor drives.

Modern adjustable-speed drives switch at carrier frequencies up to about 20 kHz. The generated pulses have a high-voltage spike at the leading edge of the pulse. The spike represents a very high rate of change of the voltage, which induces a voltage potential on the rotor shaft relative to ground. One possible path to ground from the rotor shaft is through the bearing and grounded stator. For example, bearing currents are common in ungrounded HVAC fan applications because the bearings are the only path for the voltage potential to discharge to ground.

Bearing lubricants are insulators and therefore form a capacitor, separating the potential on the rotor shaft from the potential on the grounded stator. If the shaft voltage becomes large enough, an electrical discharge across the lubrication gap goes through the lubricant and discharges the rotor voltage. After the discharge, the rotor shaft charges again and the cycle repeats.

The discharge current through the bearings causes an electrostatic discharge machining (EDM) effect, eventually causing bearing damage. This EDM causes pitting that develops into a grooved pattern called fluting. The fluting eventually causes spalling, which causes small flakes of metal to break loose and get trapped within the lubricant. This causes even more wear of the bearing. One of the first observable symptoms is increased noise as the bearing wears. In addition, the bearing current can burn or char the oil or grease, which causes further damage to the bearing from the loss of lubrication.

Solutions to Bearing Current Problems

There are several solutions to bearing current problems. Bearing currents are caused by a voltage difference between the rotor shaft and ground. Solutions include ways to reduce the induced voltage on the shaft and ways to bypass or break the conductive path through the bearing.

The length of the leads from the drive to the motor affects the size of the voltage spikes. Longer leads cause larger spikes from line-to-line capacitance between the power lines. During motor installation, the shortest leads possible should be used. Shielded cable, such as low-impedance, corrugated, aluminum-sheathed cable, can minimize the voltage spikes. If long leads cannot be avoided, inductive filters can often reduce the size of the voltage spikes.

Electric motor manufacturers have developed grounding systems for motor shafts that reduce or eliminate the bearing current. A grounding system of this type connects the rotor to ground through a brush or other device touching the shaft. This provides a low-impedance path to ground. However, this adds to maintenance costs because the grounding system brush needs to be inspected regularly.

Another method to reduce or eliminate bearing currents is to lower the carrier frequency of the adjustable-speed drive. Modern adjustable-speed drives switch at a carrier frequency up to about 20 kHz to simulate the best possible sine wave and minimize motor heating. In most drives, the carrier frequency is programmable and can be adjusted to about 1 kHz to 2 kHz to reduce the induced rotor voltage, which reduces the damage caused by overheating.

Some motor designs use insulated bearings to break the conduction path from rotor to ground. Some bearing surfaces are coated with ceramic insulation that withstands the rotor voltage. However, this does not eliminate the induced rotor voltage. The voltage potential is still present and may take another path to ground, such as through the load.

A modified motor design can eliminate the induced voltage entirely. An electrostatic shielded induction motor (ESIM) places a Faraday shield in the air gap between the rotor and the stator. The shield eliminates the rotor voltage entirely by blocking the capacitive coupling between the rotor and stator. These motors are expensive because of the precise construction required during motor manufacture.

All of these solutions cost money, and bearing currents are somewhat uncommon. Careful analysis of the motor application is needed to determine whether any of these solutions are required. In many cases, one or more of these solutions is applied after damage has already occurred. Preventive maintenance, such a vibration analysis, may indicate that damage is occurring and a solution is needed. Fluting shows up in vibration analysis as a grouping of peaks in the range of 2 kHz to 4 kHz.

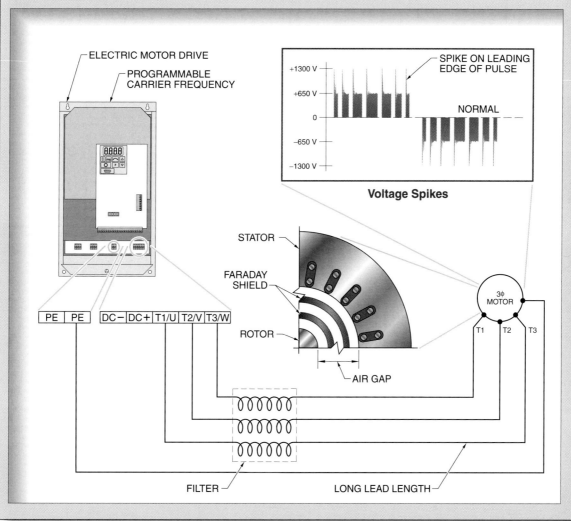

Voltage Spikes

Summary...

- A bearing is a machine component used to reduce friction and maintain clearance between stationary and moving parts.

- A radial load is a load applied perpendicular to the rotating shaft, straight through the ball toward the center of the shaft.

- An axial load is a load applied parallel to the rotating shaft.

- A rolling-contact bearing is a bearing that contains rolling elements that provide a low-friction support surface for rotating or sliding surfaces.

- Rolling-contact bearings are constructed of an outer ring (cup), balls or rollers, and an inner ring (cone).

- A friction bearing is a bearing consisting of a stationary bearing surface, such as machined metal or pressed-in bushings, that provides a low-friction support surface for rotating or sliding surfaces.

- Press fit is a bearing installation where the bore of the inner rotating ring is smaller than the diameter of the shaft and considerable force must be used to press the bearing onto the shaft.

- Push fit is a bearing installation where the diameter of the outer fixed ring is smaller than the diameter of the bearing housing and the ring can be pushed in by hand.

- An arbor press, or hammer and tube can be used to apply force uniformly on the face or ring that is to be press fit.

- Bearings can be mounted by heating (expanding) or cooling (shrinking) the bearing.

- A grease fitting is used to direct grease to bearing components. The head opens when a grease gun is attached and closes when it is removed.

...Summary

- A pressure cup is a reservoir that provides constant lubrication to a bearing and consists of a grease reservoir with an internal spring and pressure plate.

- An oil cup is a reservoir located on a bearing housing to provide lubrication to a bearing and rely on gravity to feed the lubricant through a direct passage to the bearing.

- Friction bearings usually require a light film of oil, provided from lubrication tube and wicking material.

- An oil seal is used to contain oil inside a housing and seal out foreign matter.

- Spalling can be caused by overloading a bearing until microscopic flakes break off the surface.

- Brinell damage happens where applied force exceeds the yield strength of the surface. The balls are pressed into the surface to cause indentations that are spaced the same as the distance between the balls.

- False Brinell damage is caused by vibration that moves one ring relative to another and causes axial elliptical indentations at ball positions when the bearing is not rotating.

- Temperature damage happens when a bearing overheats. It is visible as discoloration of the metal.

- Fretting corrosion is the rusty appearance that results when two metals in contact are vibrated, rubbing loose minute metal particles that become oxidized.

- Bearing surfaces that are misaligned appear as worn surfaces on one side or opposing sides of a bearing.

- Thrust damage on ball bearings appears as marks on the shoulder or upper portion of the inner and outer race. The appearance can vary from a slight discoloration to heavy galling.

- Electrical pitting and fluting occurs when the current passes through bearings. The current can etch or pit bearing surfaces.

Glossary...

A **bearing** is a machine component used to reduce friction and maintain clearance between stationary and moving parts.

A **radial load** is a load applied perpendicular to the rotating shaft, straight through the ball toward the center of the shaft.

An **axial load,** or **thrust load,** is a load applied parallel to the rotating shaft.

A **rolling-contact bearing,** or **antifriction bearing,** is a bearing that contains rolling elements that provide a low-friction support surface for rotating or sliding surfaces.

A **ball bearing** is a rolling-contact bearing that permits free motion between a moving part and a fixed part by means of balls confined between inner and outer rings.

A **roller bearing** is a rolling-contact bearing that has parallel or tapered steel rollers confined between inner and outer rings.

A **needle bearing** is an rolling-contact roller-type bearing with long rollers of small diameter.

A **race** is the bearing surface of a rolling-contact bearing that supports the rolling elements.

A **Conrad bearing** is a single-row ball bearing that has races that are deeper than normal.

A **cylindrical roller bearing** is a roller bearing having cylinder-shaped rollers.

A **tapered roller bearing** is a roller bearing having tapered rollers.

A **friction bearing** is a bearing consisting of a stationary bearing surface, such as machined metal or pressed-in bushings, that provides a low-friction support surface for rotating or sliding surfaces.

A **babbitt metal** is an alloy of soft metals such as copper, tin, and lead, and a hardening material such as antimony.

Press fit is a bearing installation where the bore of the inner rotating ring is smaller than the diameter of the shaft and considerable force must be used to press the bearing onto the shaft.

Push fit is a bearing installation where the diameter of the outer fixed ring is smaller than the diameter of the bearing housing and the ring can be pushed in by hand.

A **grease fitting** is a hollow tubular fitting used to direct grease to bearing components.

A **pressure cup** is a pressurized grease reservoir that provides constant lubrication to a bearing.

An **oil cup** is an oil reservoir located on a bearing housing to provide lubrication to a bearing.

Surface reaction is damage to bearing surfaces caused by chemical or electrochemical reactions between the lubricant and the metal of the bearing.

Spalling is general wear of rolling contacts where metal pieces flake away the surfaces in contact, leaving a roughened surface.

Brinell damage is bearing damage where applied force exceeds the yield strength of the surface and presses the balls into the surface to cause indentations.

False Brinell damage is bearing damage caused by vibration or other forces that move one ring relative to another and cause axial elliptical indentations at ball positions when the bearing is not rotating.

Temperature damage is bearing damage caused by high temperatures in the bearing.

Fretting corrosion is the rusty appearance that results when two metals in contact are vibrated, rubbing loose minute metal particles that become oxidized.

Misalignment wear is bearing damage that occurs when the two bearing rings are not aligned with one another and the rolling-contact points cause eccentric wear.

Thrust damage is bearing damage due to axial force.

...Glossary

Galling, or **adhesive wear**, is a bonding, shearing, and tearing away of material from two contacting, sliding metals.

Electrical pitting is bearing damage in the form of pits formed on the balls or race caused by electrical discharge through the bearing.

Fluting is the elongated and rounded grooves or tracks left by the etching of each roller on the rings of an improperly grounded roller bearing when current passes through the bearing.

Review

1. Describe the difference between a radial load and an axial load.

2. Describe the difference between a rolling-contact bearing and a friction bearing.

3. Describe the difference between a press fit and a push fit.

4. Explain how to mount a rolling-contact bearing on a shaft.

5. List several methods used to lubricate bearings.

6. Explain how to tell the difference between spalling damage and temperature damage.

7. Explain how to tell the difference between Brinell damage and fluting damage.

Refer to the CD-ROM for Quick Quiz® questions related to chapter content.

MOTORS

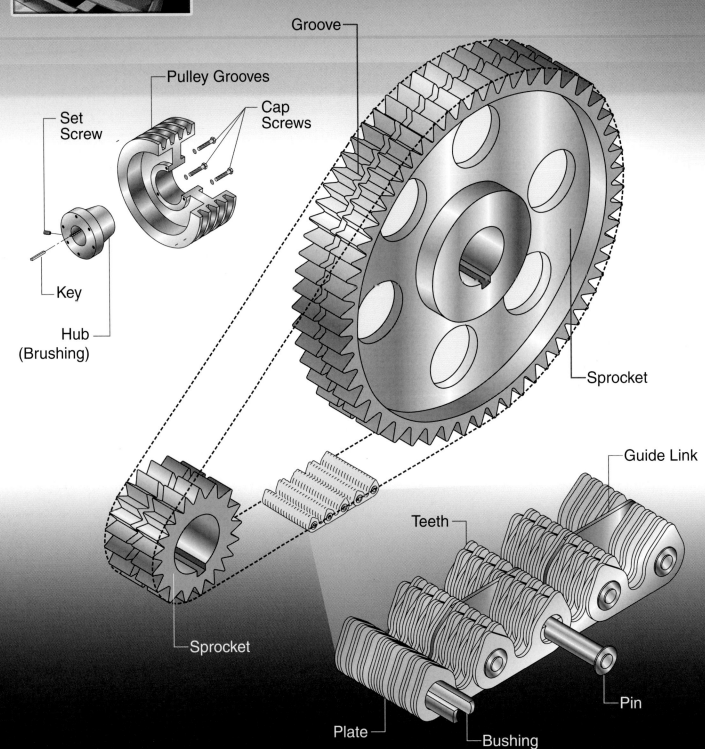

16

Drive Systems and Clutches

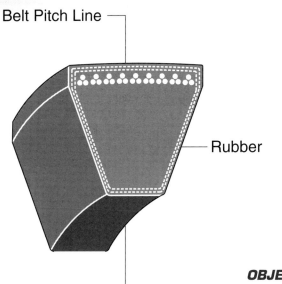

Belt Pitch Line
Rubber
Cloth Wrap

Flexible Drives 350
Mechanical Drives 355
Clutches 360
Supplemental Topic—Gear Reducers 363
Supplemental Topic—Calculating V-Belt Length 364
Application—Belt Tensioning 365
Application—Selecting Motor Couplings 366
Summary 367
Glossary 368
Review 369

OBJECTIVES

- Explain how flexible belts are used to transfer torque and change speed between a motor and a load.
- Explain how chains and gears are used to transfer torque and change speed between a motor and a load.
- Describe the different types of gears and explain how they operate.
- Explain how friction clutches and eddy-current clutches are able to connect or disconnect a load to a motor while the motor is running.

A motor must be coupled to a load in order to produce useful work. Flexible drives use belts and pulleys. Mechanical drives use chains or gears. Clutches are used to connect or disconnect a load to a motor while the motor is running. Drives and clutches can also be used to operate a load at a different speed than the motor.

Definition

A *flexible drive* is a system in which a resilient flexible belt is used to drive one or more shafts.

FLEXIBLE DRIVES

A *flexible drive* is a system in which a resilient flexible belt is used to drive one or more shafts. Flexible drives may also change the magnitude, direction, and speed of the applied force. Flexible drives are nonsynchronous drive systems because they do not provide positive engagement between the drive and driven sides of the system. Nonsynchronous drive systems slip if the resistance to movement exceeds the friction between the belt and the pulley.

Belt Drives

Belt drives are one of the most common drive systems used in industry. Belts are attached to a motor shaft and a load shaft. As the motor shaft turns, friction between the belt and a pulley provides the torque needed to turn the other shaft. Belt drives are relatively inexpensive, quiet, easy to maintain, and provide a wide range of speed and torque. The material used for the belt is normally selected for the application. Belts normally have tensile members that run the length of the belt to provide tensile strength for the belt. Belts commonly used in industry include flat belts, V-belts, and timing belts. **See Figure 16-1.**

Characteristics of belt drives include belt tension, creep, and slip. Belt tension includes tight side, slack side, and centrifugal. Tight-side tension is the tension on a belt when it is approaching the drive pulley. Slack-side tension is the tension on a belt when it is approaching the driven pulley. Centrifugal tension is the tension needed to offset the centrifugal force on the belt as it engages the pulley. Centrifugal force pushes the belt away from the pulley, reducing friction. **See Figure 16-2.**

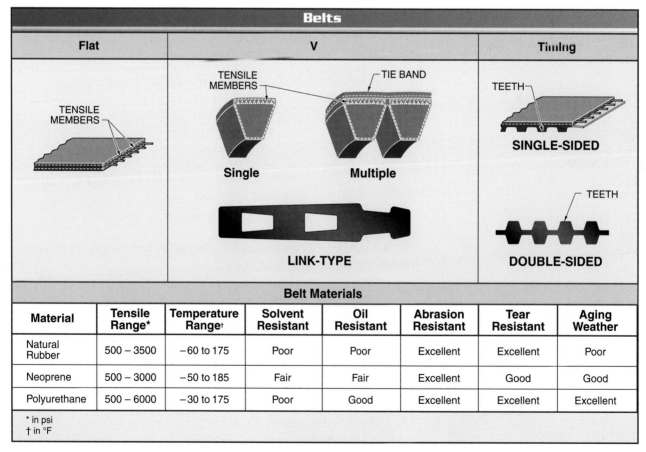

Figure 16-1. Belts used in industry include flat, V-, and timing belts.

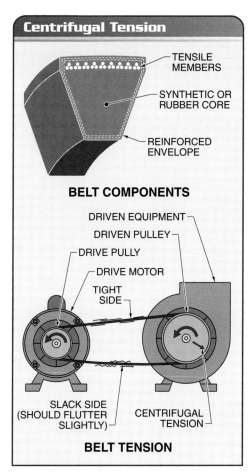

Figure 16-2. Belts are pulled tight to prevent centrifugal force from causing a belt to fly off a pulley. Tensile members in the belt keep it from stretching too much.

Belt creep is the natural movement of the belt on the face of the pulley when it is subjected to changes in tension. Belt creep occurs at startup or when the load is increased or decreased. Belt slip is the movement of a belt on the face of the pulley when belt tension is too loose and the belt slides on the pulley. Belt slip reduces pulley speed and causes premature wear on the belt and pulley.

Protecting the belts from the surrounding environment is a high priority. A protective cover should be designed to keep objects and foreign substances such as grease, oil, and dirt from contacting the belt or pulleys. Foreign material on a belt causes glazing. Glazing is a slick polished surface caused by dirt and other debris being rubbed on the surface of a belt. A glazed belt has reduced friction with the pulleys, resulting in belt slippage and a loss of power transmission. Glazed belts should be replaced and the pulleys inspected for possible damage. High humidity and temperatures may cause premature belt failure.

V-Belts. A *V-belt* is a flexible drive belt that has a cross-section in the shape of a V. V-belts are the most frequently used belts for belt drive systems. Compared to flat belts, V-belts require less tension, can tolerate some minor misalignment, run with very little noise, and require less maintenance. The lower operating tension required for V-belts reduces bearing loads.

A *fractional-horsepower (FHP) V-belt* is a V-belt designed for light-duty applications. FHP V-belts are used on drive systems that range from 3 HP to 17 HP. FHP V-belts are made from lightweight materials, allowing them to bend over small pulleys. Some small FHP V-belts are notched part way through the cross section, allowing more flexibility in the belt.

A *standard V-belt* is a V-belt designed for moderate-duty applications. Standard V-belts are available in five different cross-sectional sizes, indicated by the letters A through E. **See Figure 16-3.** Standard belts range from 25″ to 660″ in length. For example, a standard V-belt may be labeled C105. The C indicates a cross-sectional measurement of ⅞″ and the 105 is a nominal size designation.

A *narrow V-belt* is a V-belt having a smaller cross section and a higher profile than a standard belt that is designed for heavy-duty applications. The shape of a narrow belt provides more surface contact with the pulleys, allowing the belt to transmit large amounts of power with a small cross section. Narrow V-belts are often referred to as heavy-duty belts. Narrow V-belts require high operating tension and are sized by the same system as FHP belts. Cross-sectional sizes include 3, 5, and 8. For example, a narrow V-belt may be labeled 3VX300. The 3 indicates a cross section of ⅜″, the V indicates a narrow V-belt, the X indicates a notched belt, and the 300 indicates an outside length of 30″.

Definition

*A **V-belt** is a flexible drive belt that has a cross-section in the shape of a V.*

*A **fractional-horsepower (FHP) V-belt** is a V-belt designed for light-duty applications.*

*A **standard V-belt** is a V-belt designed for moderate-duty applications.*

*A **narrow V-belt** is a V-belt having a smaller cross section and a higher profile than a standard belt that is designed for heavy-duty applications.*

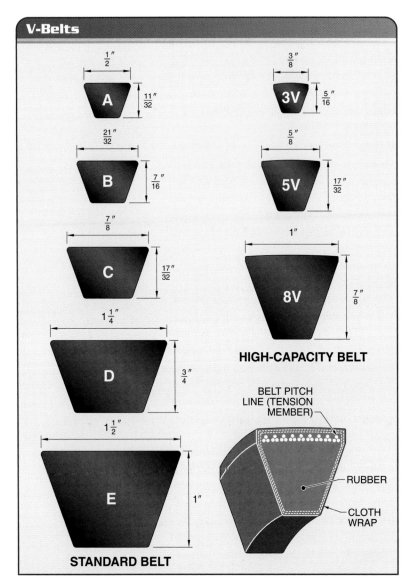

V-belts transmit torque to and from a pulley through the wedging action of the tapered sides of the belt in the pulley groove. **See Figure 16-4.** V-belts should not normally run in contact with the bottom of the pulley. V-belts are classified as fractional horsepower (FHP), standard, and narrow and are available in a variety of lengths. Proper V-belt selection is determined by pulley diameter, groove angle, motor frame number, pulley center distance, and belt cross section. Manufacturers provide tables that can be used to specify belt and pulley combinations.

Some industrial applications use a multiple belt drive system to transfer large amounts of horsepower. Multiple V-belts must be a matched set and have a consistent length and cross section. If the belts are different sizes, one belt carries the majority of the load, resulting in premature failure of the belt and an overall loss of power transfer. All belts should be replaced when any one belt in a multiple-belt drive is replaced. Used belts are no longer the same size as new belts because they have been stretched during use.

Tech Fact

An excessive number of belts, or belts that are too large, can severely stress motor or driven shafts. This can happen when load requirements are changed on a drive, but the belts are not redesigned accordingly. This can also happen when a drive is greatly overdesigned for the load and forces created from belt tensioning are too great for the shafts.

Figure 16-3. Standard V-belts are designated as A, B, C, D, or E. High-capacity V-belts are designated as 3V, 5V, or 8V.

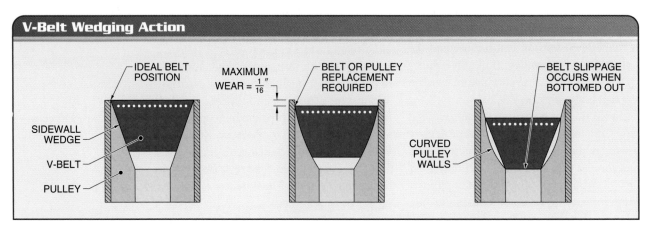

Figure 16-4. V-belts transfer power through the wedging action of the tapered sides of the belt in a pulley groove.

Flat Belts. A *flat belt* is a belt that has a rectangular cross section and relies on friction for proper operation. Proper friction requires tensioning a flat belt using a tension tester. Flat belts are commonly used in applications that require high speeds, low noise, and small pulley diameters. Some flat belts can operate at speeds of 140,000 rpm. Flat belts are used to drive a wide range of industrial machinery and vary in width and thickness depending on the torque requirements.

Timing Belts. A *timing belt* is a flat belt containing gear teeth that are used for synchronous drive systems. **See Figure 16-5.** Timing belt teeth are evenly spaced, providing timing and positive engagement with the mating pulley. A timing belt provides a drive system that has no slippage or creep, does not stretch, needs no lubricant, requires low belt tension, and has very little backlash. Timing belts can provide more efficiency than V-belts by combining the advantages of a flat belt drive system and a positive synchronous drive system. Timing belt selection is determined by horsepower and pulley size.

Pulleys. A *sheave* is a grooved wheel used to hold a V-belt. **See Figure 16-6.** A pulley consists of one or more sheaves and a frame or block to hold the sheaves. Pulleys are used to change the speed of a driven load relative to the motor speed. A pulley of one size is placed on the drive shaft and a pulley of another size is placed on the driven shaft. A V-belt connects the two pulleys to transfer the torque.

A pulley system depends on friction between the V-belt and the sheave. The best situation is where the V-belt wraps around a sheave and makes contact with an entire half of the sheave. However, when the two sheaves have different diameters, the V-belt makes contact with more than half of the larger sheave and less than half of the smaller sheave. This reduces the force available at the smaller sheave.

As a rule of thumb, the pulleys should be placed a distance apart equal to 3 times the diameter of the largest pulley. In addition, the sheaves should be placed as close as possible to the motor housing or shaft bearings. This minimizes the loading on the bearing.

There are many situations where a load must be rotated at a speed slower than the speed of an inexpensive 2-pole induction motor. The size and the cost of a motor that runs slower because it has more poles can be much higher than the standard motor. In addition, the low-speed motor often has a lower power factor and lower efficiency. Pulleys are often used to drive a load at a different speed than the motor speed.

> **Definition**
>
> A *flat belt* is a belt that has a rectangular cross section and relies on friction for proper operation.
>
> A *timing belt* is a flat belt containing gear teeth that are used for synchronous drive systems.

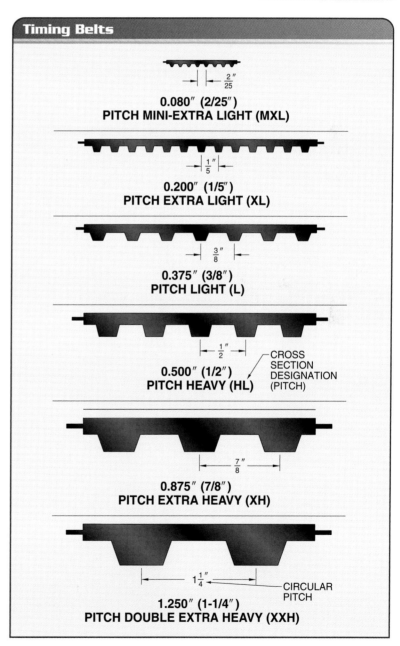

Figure 16-5. Timing belts are classified by their standard length, cross-section designation, and circular pitch.

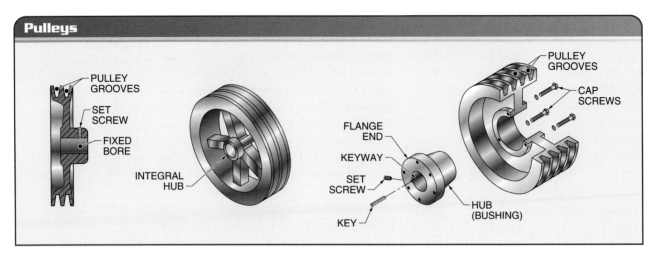

Figure 16-6. V-belt pulleys contain grooves to hold the belt.

Definition

*A **sheave** is a grooved wheel used to hold a V-belt.*

The speeds of the motor and the load and the diameters of the sheaves are related to each other. **See Figure 16-7.** The full-load speed of a motor can be found on the nameplate. The speed of the load is determined by the load requirement. The diameter of the sheaves can be measured or specified. The speeds and diameters are related as follows:

$$\frac{S_M}{S_L} = \frac{D_L}{D_M}$$

or

$$S_M \times D_M = S_L \times D_L$$

where
S_M = motor speed, in rpm
S_L = load speed, in rpm
D_L = load-sheave diameter, in in.
D_M = motor-sheave diameter, in in.

For example, the operating speed of a motor is 1745 rpm and the desired speed of the load is twice that, or 3490 rpm. The motor sheave diameter is 6″ in diameter. The required diameter of the load sheave is calculated as follows:

$$\frac{S_M}{S_L} = \frac{D_L}{D_M}$$

$$\frac{1745}{3490} = \frac{D_L}{6}$$

$$D_L = \frac{1745 \times 6}{3490}$$

$$D_L = \frac{10470}{3490}$$

$$D_L = 3''$$

Belt Speed. V-belts have a maximum speed rating. Different types of V-belts have different speed ratings, but a good rule of thumb is to use a top speed of about 6000 fpm to 7000 fpm. The centrifugal force of the rotating sheave applied force to the belt that tends to stretch the belt. If the belt stretches too much, contact between the belt and the pulleys will be lost.

Belt speed can be calculated when the motor speed and diameter of the motor sheave is known. **See Figure 16-8.** Alternatively, the load speed and load sheave can be used if those are known. The belt speed is the same at each sheave even though the speed of rotation is different. The speed of rotation is in revolutions per minute (rpm)

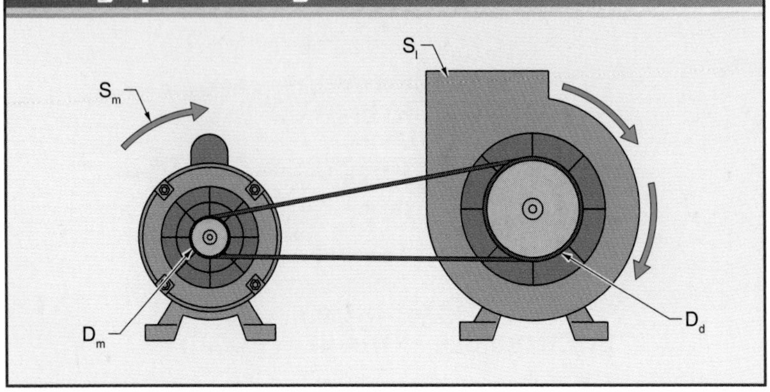

Figure 16-7. The diameter or speed of a driven or drive pulley may be determined by solving for any one value when the other three are known.

and the belt speed is in feet per minute (fpm). The sheave diameter is used to convert rpm to fpm as follows:

$$S_B = S_M \times \frac{\pi \times D}{12}$$

where
S_B = belt speed, in fpm
S_M = motor speed, in rpm
D = sheave diameter, in in.

For example, a motor operating at a speed of 1745 rpm is fitted with a sheave with a diameter of 4″. The load sheave has a diameter of 6″. Since the motor speed is known, the motor sheave is used to calculate the belt speed as follows:

$$S_B = S_M \times \frac{\pi \times D}{12}$$

$$S_B = 1745 \times \frac{\pi \times 4}{12}$$

$$S_B = 1745 \times 1.047$$

$$S_B = \mathbf{1825\ fpm}$$

MECHANICAL DRIVES

A *mechanical drive* is a combination of mechanical components that transfer torque from one location to another. Mechanical drives may also change the magnitude, direction, and speed of the applied force. Mechanical drives include chain drives and gear drives. Couplings provide a direct connection between a motor and a load.

Chain Drives

A *chain drive* is a synchronous mechanical drive system that uses a chain to transfer torque from one sprocket to another. A chain is a series of interconnected links that form a loop. A sprocket is a wheel with evenly spaced teeth located around the perimeter of the wheel. The chain meshes with the teeth on the sprocket, creating a synchronous drive system. Chain drives use roller and silent chain.

Roller Chain. *Roller chain* is a chain that contains roller, pin, and connecting master links. **See Figure 16-9.** A roller link consists of two bushings placed inside two rollers that are pressed into two side bars. The side bar has two precision holes used to connect two pins or two bushings. When the roller link is pressed together, the rollers are free to spin around the bushings, providing a pivot point. Roller chain must be lubricated because it wears quickly.

Definition

*A **mechanical drive** is a combination of mechanical components that transfer torque from one location to another.*

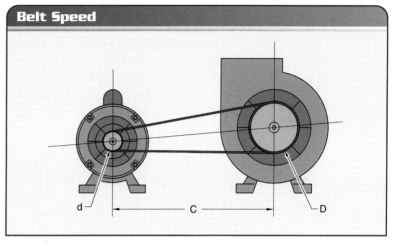

Figure 16-8. Belt length is calculated from the center-to-center distance between pulleys and the pulley diameters.

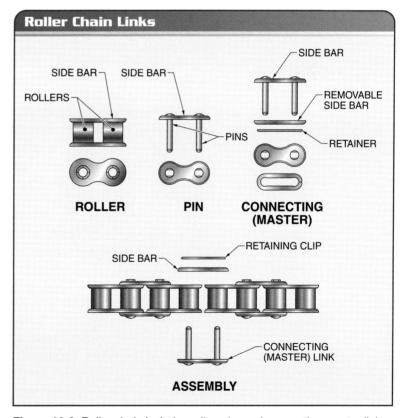

Figure 16-9. Roller chain includes roller, pin, and connecting master links.

Definition

A *chain drive* is a synchronous mechanical drive system that uses a chain to transfer torque from one sprocket to another.

Roller chain is a chain that contains roller, pin, and connecting master links.

Silent chain is a synchronous chain that consists of a series of links joined together with bushings and pins.

A pin link is used to join roller links and consists of two steel pins pressed into two side bars with matching holes. The width of a pin link is greater than the width of a roller link. This enables the pin link side bars to fit on the outside of the roller link side bars. Assembly of the roller chain requires sliding one pin of a pin link through one roller of a roller link and the other pin through one roller of another roller link. The link is made by pressing on the second side bar of the pin link. Pin and roller links are joined until the desired chain length is achieved.

A connecting master link has a removable side bar and retainer that snaps on the pin and outside the side bar to hold the side bar in place. Connecting links are used to connect two ends of a chain, making a complete loop. An offset link is used to shorten or lengthen a chain and to connect two ends of a chain. The length of an offset link is half the length of a pin, roller, or connecting master link. An offset link should only be used for short length adjustments because it is weaker than other chain links.

A sprocket has evenly spaced teeth located around the perimeter of the wheel. **See Figure 16-10.** The teeth of a sprocket provide positive engagement with a chain in order to transfer force from one sprocket to another. Sprockets are classified as A, B, and C sprockets. An A sprocket is a sprocket that has no hub and is usually mounted on a hub or flange with mechanical fasteners. A B sprocket has an integrated hub on one side of the sprocket. A C sprocket has an integrated hub on both sides of the sprocket. B and C sprockets are commonly attached to a shaft using a key and set screws.

Silent Chain. *Silent chain* is a synchronous chain that consists of a series of links joined together with bushings and pins. **See Figure 16-11.** The bushing and pin arrangement provides a pivot point for the silent chain. The links are flat on the top and have teeth on the bottom. Tooth width is determined by the number of plates. Silent chain sprockets also have teeth that mesh with the chain teeth, providing a positive drive system.

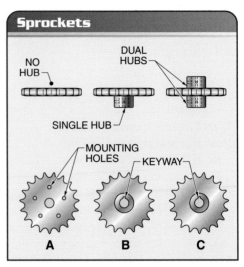

Figure 16-10. A sprocket provides positive engagement with a chain to transfer force from one sprocket to another.

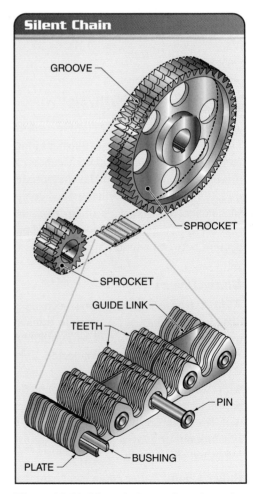

Figure 16-11. Silent chain consists of a series of links that are flat on the top and have teeth on the bottom.

A silent chain sprocket has a groove cut in the center of the teeth to prevent the silent chain from moving laterally on the sprocket by controlling the position of the guide link at the center of the chain. The guide link fits into a groove on the sprocket and keeps the chain properly aligned in the sprocket. Silent chain is more expensive than roller chain but has higher operating speeds, is more efficient, runs quieter and smoother, and provides longer life.

Gear Drives

A *gear drive* is a synchronous mechanical drive system that uses the meshing of two or more gears to transfer motion from one shaft to another. A gear is a toothed wheel used to transfer torque or motion from one gear to another or from one shaft to another using interlocking gear teeth. Gear drives provide positive contact between gears and the teeth prevent significant slippage. Gear-driving and gear-driven shafts should be located close to each other. Long distances between shafts are not economical for gear drive systems. Lubrication of a gear drive system is necessary for the drive system to function properly. The lubricant must provide a protective barrier between the mating gears to eliminate any metal-to-metal contact.

Gear Ratio. A *gear ratio* is the ratio between the diameter of the drive gear and the driven gear. Matching gears have teeth of the same size. The number of teeth on a gear determines its diameter. The gear ratio determines how fast the driven gear rotates in relation to the drive gear. Gears with equal diameters have a 1:1 ratio and rotate at the same rate. A gear ratio of 1:2 means that the driven gear has a diameter that is twice that of the drive gear. Therefore, the drive gear rotates more quickly and makes 2 revolutions for each revolution of the driven gear.

Any two gears that mesh rotate in opposite directions. For example, a drive gear rotating counterclockwise causes the second gear to rotate clockwise. An added third gear rotates in the same direction as the first gear. **See Figure 16-12.** The driven gear is forced to move by the drive gear. The driven gear has a load attached to it, while the drive gear is often attached to a motor. To find the gear ratio, the diameter of each gear must be known. The gear ratio is found by applying the formula:

$$R = \frac{D}{d}$$

where
R = gear ratio
D = drive-gear diameter (in inches)
d = driven-gear diameter (in inches)

For example, the gear ratio with a 3″ drive gear and a 1⅞″ (1.875″) driven gear is calculated as follows:

$$R = \frac{D}{d}$$
$$R = \frac{3}{1.875}$$
$$R = \textbf{1.6, or 1.6 : 1}$$

Definition

*A **gear drive** is a synchronous mechanical drive system that uses the meshing of two or more gears to transfer motion from one shaft to another.*

*A **gear ratio** is the ratio between the diameter of the drive gear and the driven gear.*

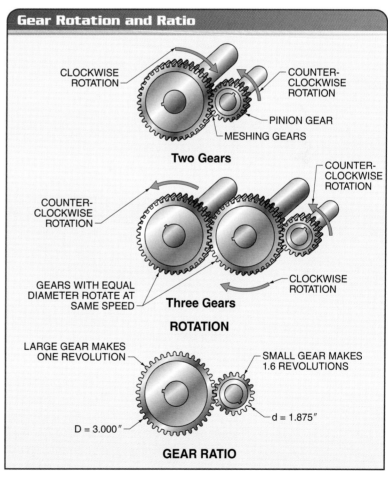

Figure 16-12. A gear drive uses the meshing of two or more gears to transfer motion from one shaft to another.

Definition

Backlash is the amount of movement, or play, between meshing gear teeth.

A *spur gear* is a gear with straight teeth cut parallel to the shaft axis.

A *helical gear* is a gear with teeth that are not parallel to the shaft axis.

A *herringbone gear* is a gear with two rows of helical teeth.

A *bevel gear* is a gear with straight tapered teeth used in applications where shaft axes intersect.

A *worm* is a screw thread that rotates the worm gear.

A *worm gear* is a spur gear that is driven by a worm.

Backlash. Gears are designed to have backlash between meshing teeth for maximum life and efficiency. *Backlash* is the amount of movement, or play, between meshing gear teeth. Backlash is required to prevent full contact on both flanks of the teeth. The space created enables the flow of lubricant between the teeth flanks. Inadequate backlash can cause resistance, resulting in overheating or jamming of the meshing gears. Excessive backlash can cause problems if the direction of rotation is reversed frequently. Each time the direction of rotation is reversed, the gear teeth are subjected to impact. Backlash is generally measured by holding one gear stationary while rocking the engaged gear back and forth. The movement is measured with a dial indicator placed at the pitch diameter. **See Figure 16-13.**

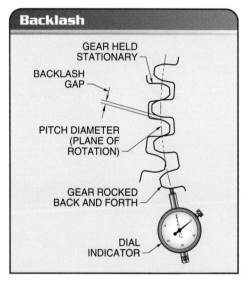

Figure 16-13. Backlash between two gears can be measured with a dial indicator placed on the pitch diameter.

Gears. A *spur gear* is a gear with straight teeth cut parallel to the shaft axis. **See Figure 16-14.** Spur gears are the most common gears used in industry and are used where gears are mounted on parallel shafts. Spur gears may be external, internal, or rack gears. External gears have teeth cut on their outside perimeter. Internal gears have an external gear that meshes on the inside circumference of a larger gear. This permits a large ratio of speed reduction in a small space. A pinion gear is the smaller of two meshing gears. A rack has teeth spaced along a straight line. A pinion gear is used with a rack to convert rotary motion into linear motion.

A *helical gear* is a gear with teeth that are not parallel to the shaft axis. Helical gears are manufactured in pairs so their helix axes match. Helical gears are commonly used in gearboxes because they provide smoother operation than spur gears. Helical gear teeth mesh with each other in a sliding motion. This results in more tooth contact at any given time. End thrust is produced because of the angle of the teeth.

A *herringbone gear* is a gear with two rows of helical teeth. Herringbone gears have parallel shafts and teeth cut at opposite angles. End thrust is avoided because the teeth are at opposite angles. Herringbone gears provide quiet and efficient operation.

A *bevel gear* is a gear with straight tapered teeth used in applications where shaft axes intersect. Bevel gears are conical in shape rather than cylindrical in shape. Bevel gears are manufactured in pairs to ensure matching tapers. Bevel gears are primarily used in gearboxes requiring shafts at right angles to each other. A spiral bevel gear has curved teeth, which provide smoother operation at high speeds. A miter gear is used at right angles to transmit torque at a 1:1 ratio. Miter gears operated together must have the same number of teeth.

Pairs of gears consisting of a worm and a worm gear are used for large speed reductions and smooth, quiet service. A *worm* is a screw thread that rotates the worm gear. A *worm gear* is a spur gear that is driven by a worm. Worm gear teeth are cut at an angle and in a concave shape to mate securely with the worm. A single-threaded worm is designed to advance the worm gear one tooth for every revolution of the worm.

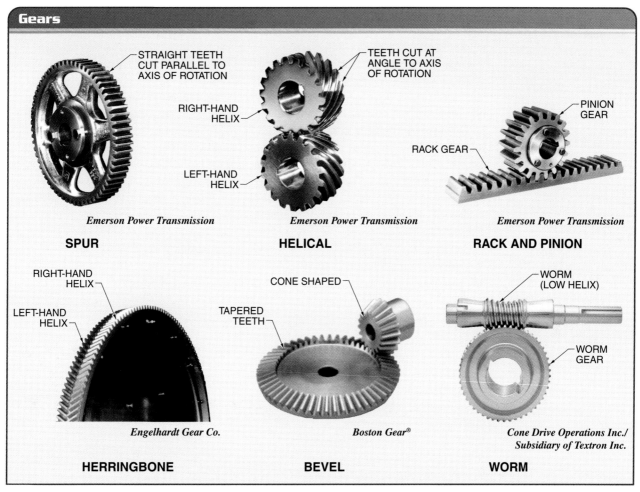

Figure 16-14. The gear used in a mechanical drive is determined by the required gear speed, load placed on the gear, angular requirements, and space constraints.

Gearboxes. Gears used in gear drive systems may be exposed or encased in a gearbox. A *gearbox* is a sealed container that has an input shaft and an output shaft and houses at least one set of mating gears. Gearboxes are used to change shaft rotation, speed, and/or direction of rotation, and to protect gears from the surrounding environment. **See Figure 16-15.**

An inline gearbox normally contains spur or helical gears and is primarily used to reduce shaft speed. Speed is controlled by the gear ratio. Inline gearboxes can also provide single-axis offset that can be useful for joining shafts that are on different axes. A right-angle gearbox normally contains bevel, miter, or worm gears and is used to change the direction of the output shaft in relation to the input shaft. Worm gears provide a large speed reduction.

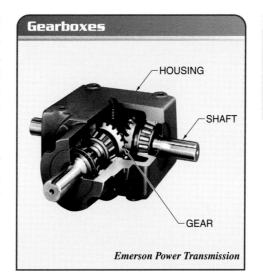

Figure 16-15. A set of gears is used in a gearbox to transfer torque from the motor to the load.

Definition

*A **gearbox** is a sealed container that has an input shaft and an output shaft and houses at least one set of mating gears.*

Definition

A *coupling* is a device that connects the ends of rotating shafts.

A *rigid coupling* is a device that joins two precisely aligned shafts within a common frame.

A *flexible coupling* is a coupling with a resilient center that flexes under temporary torque or misalignment due to thermal expansion.

A *clutch* is a coupling between a motor and a load that connects or disconnects the motor shaft to a drive shaft while the motor is running.

The lower portion of a gearbox is usually used as an oil reservoir for lubricating the internal gears. Proper gearbox maintenance includes changing the gearbox oil according to manufacturer specifications, using the proper gear oil, and visually inspecting the dirty oil for contaminants. Oil level is checked by looking through the sight glass or by removing a plug located on the gearbox. The oil should just begin dripping from the opening revealed by the plug. Oil that is removed may be analyzed to monitor the conditions inside the gearbox. Chemical breakdown, contamination levels, and particle type and amount indicate the level of gear wear inside the gearbox.

Couplings

A *coupling* is a device that connects the ends of rotating shafts. There are many applications where the load speed can be the same as the motor speed, such as many pumping and fan applications. Couplings are the most common and least expensive method of connecting two shafts. Couplings are classified as rigid or flexible. Couplings require accurate alignment of the mating shafts.

A *rigid coupling* is a device that joins two precisely aligned shafts within a common frame. **See Figure 16-16.** Rigid couplings are made of metal and are secured with bolts or setscrews. One advantage of rigid couplings is the ability to transmit more torque than flexible couplings because of their simple design and rigidity. Rigid couplings include flange and sleeve couplings.

A *flexible coupling* is a coupling with a resilient center that flexes under temporary torque or misalignment due to thermal expansion. Flexible couplings reduce the conduction of heat, sound, and electricity through the drive system, although they can allow enough vibration to cause excessive wear to seals and bearings. Where flexible couplings are used, shaft alignment should be as accurate as if solid couplings were used.

CLUTCHES

A *clutch* is a coupling between a motor and a load that connects or disconnects the motor shaft to a drive shaft while the motor is running. The input shaft and output shaft of a friction clutch rotate at the same speed. The output shaft of an eddy-current clutch rotates slower than the input shaft.

Tech Fact
The correct belt tension is the least amount of tension that enables the belt to run without slipping when a full load is applied.

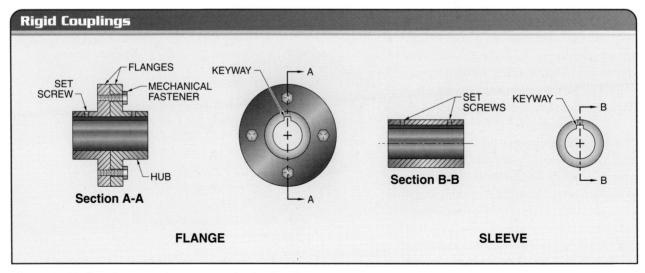

Figure 16-16. Couplings transfer rotary motion by directly connecting two shafts.

The advantage of using a clutch instead of a direct coupling is that the load is connected to the motor shaft while the motor is running at its no-load speed. This means that the motor can be started under no load and can deliver maximum horsepower without having a long startup. For this reason, the motor does not draw locked-rotor current during starting and does not stress the distribution system and overheat the motor. The load can be stopped and started many times without affecting the duty cycle of the motor.

Friction Clutches

A *friction clutch* uses the force of friction between two or more rotating disks, drums, or cones to engage or disengage the two shafts. **See Figure 16-17.** Engaging the clutch brings the clutch components together and starts torque transfer from the motor to the load. This brings the load up to operating speed. Disengaging the clutch separates the clutch components and stops torque transfer from the motor to the load. This allows the load to stop turning while the motor continues turning. Clutches are sometimes combined with brakes in one unit. Friction clutches are less expensive than most other types of clutches, but generally cannot control the torque over a wide operating range. They are well suited for hoists and cranes.

Kebco Power Transmission

Figure 16-17. A friction clutch brings the clutch components together where the force of friction transfers the torque from the motor to the load.

The term "magnetic clutch" is sometimes used in the field. It has several meanings and is not used consistently. It may mean a friction clutch that uses a magnetic field to engage the friction disks. It may also mean a clutch that uses a magnetic field to couple a motor to a load, as with an eddy-current clutch.

Eddy-Current Clutches

An *eddy-current clutch* uses a magnetic field to couple a motor to a load. The clutch contains a drum that is attached to the motor shaft or input shaft of the clutch. Inside the drum is a rotor with an output shaft. In a common design, the motor is attached to the input shaft and the load is attached to the output shaft. At startup, the drum rotates at the speed of the motor shaft and the rotor within the drum remains stationary. When the field coil is energized, a magnetic field couples the drum and the rotor, and the rotor is dragged around with the drum. When the coil is de-energized, springs push the drum and rotor apart and disengage the load.

As with an induction motor, an eddy-current clutch must operate with slip. Torque can be increased by increasing the slip or by increasing the field current in the rotor coils. Slip causes heat to be generated in the clutch. As the output shaft slows, slip increases and heating of the clutch can cause damage. This may require external cooling to ensure long drum life. Air-cooled and water-cooled clutches can run at a higher slip without damage to the drum.

When motor speed control is needed, a tachometer generator can be added to the clutch. The tachometer feedback is used as a signal back to a controller known as an eddy-current drive. The controller uses the feedback signal to adjust the field current. This allows the clutch to transmit just enough torque to operate at the desired speed.

For example, steel, paper, or film is wound onto large rolls in a mill. **See Figure 16-18.** The roll can be fairly large in diameter. At startup, the input speed of the clutch almost matches the output speed. Some slip is needed to generate eddy currents. As the cores on which the product is being wound

Definition

*A **friction clutch** uses the force of friction between two or more rotating disks, drums, or cones to engage or disengage the two shafts.*

*An **eddy-current clutch** uses a magnetic field to couple a motor to a load.*

start to rotate, the roll turns quickly to allow the product to wrap around the small core while maintaining constant tension. The output shaft must turn quickly to match. As the roll finishes off, the diameter is much larger and the roll turns slowly to allow the product to wrap the larger diameter while maintaining the same tension. The speed is controlled by varying the field current, which controls the slip. The same kind of situation exists where a roll is unrolled for use in a machine.

Eddy-current drives are generally less efficient than adjustable-speed drives. Power is proportional to speed. Since slip is required in an eddy-current drive, the output speed of the clutch is less than the input speed. Therefore, output power of an eddy-current clutch is less than the input power. The lost power shows up as heat in the clutch. Because of this power loss, many applications that formerly used eddy-current drives now use adjustable-speed AC drives.

Figure 16-18. An eddy-current clutch is used to control the speed when a roll is wound or unwound.

Gear Reducers

Gear reducers transmit torque from a motor to a driven machine. They change torque, speed, direction, and position. In most applications, gear reducers are used to amplify torque. When the output speed of the gear reducer is slower than the motor speed, the torque is increased proportionally to the reducer ratio. However, the final torque is reduced because of inefficiencies in torque transfer.

To find the output torque of a gear reducer, apply the formula as follows:

$$T_O = T_I \times R_R \times E_R$$

where

T_O = output torque (in lb-ft)

T_I = input torque (in lb-ft)

R_R = gear reducer ratio

E_R = reducer efficiency

For example, a motor with 8 lb-ft of torque is connected to a gear reducer with a ratio of 30:1 and an efficiency of 85%. The output torque of the gear reducer is calculated as follows:

$$T_O = T_I \times R_R \times E_R$$
$$T_O = 8 \times 30 \times 0.85$$
$$T_O = \mathbf{204 \text{ lb-ft}}$$

To find the output speed of a gear reducer, apply the formula as follows:

$$S_O = \frac{S_I}{R_R}$$

where

S_O = output speed (in rpm)

S_I = input speed (in rpm)

R_R = gear reducer ratio

The output speed of the gear reducer is proportional to the gear ratio. There is no slip. For example, a motor shaft that turns at 1740 rpm is connected to a gear reducer with a ratio of 30:1. The output speed of the gear reducer is calculated as follows:

$$S_O = \frac{S_I}{R_R}$$
$$S_O = \frac{1740}{30}$$
$$S_O = \mathbf{58 \text{ rpm}}$$

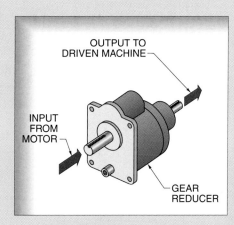

Supplemental Topic

Calculating V-Belt Length

The length of a V-belt of this machine can be found by adding the length of the belt that wraps around the sheaves to the length between the centers of the sheaves. For sheaves that are the same size, the length of the belt that wraps around each of the sheaves is the same. For sheaves that are different sizes, the length of the belt that wraps around each of the sheaves is different. The belt wraps more than halfway around the larger sheave and less than halfway around the smaller sheave. The belt length can be calculated as follows:

$$L = (D+d) \times \pi/2 + 2 \times CC + \left(\frac{(D-d)^2}{4 \times CC}\right)$$

where
L = belt length, in in.
D = diameter of larger sheave, in in.
d = diameter of smaller sheave, in in.
CC = center-to-center distance between sheaves, in in.

For example, a motor with a 24″ sheave is connected to a load with a 6″ sheave. The center-to-center distance between sheaves is 18″. Belt length is calculated as follows:

$$L = (D+d) \times \pi/2 + 2 \times CC + \left(\frac{(D-d)^2}{4 \times CC}\right)$$

$$L = (24+6) \times \pi/2 + 2 \times 18 + \left(\frac{(24-6)^2}{4 \times 18}\right)$$

$$L = 30 \times \pi/2 + 36 + \frac{18^2}{72}$$

$$L = 47.1 + 36 + \frac{324}{72}$$

$$L = \mathbf{87.6″}$$

A belt with a length of 87.6″ would not normally be available off the shelf. Therefore, a standard length longer than 87.6″ would be chosen and an idler roller would be used to adjust belt tension.

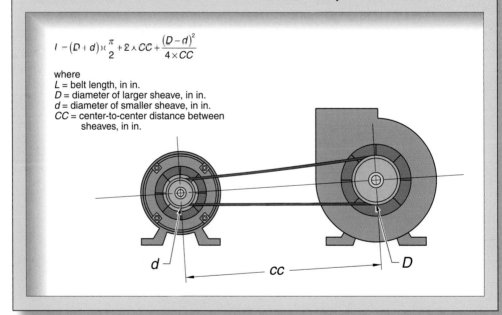

$$L = (D+d) \times \frac{\pi}{2} + 2 \times CC + \frac{(D-d)^2}{4 \times CC}$$

where
L = belt length, in in.
D = diameter of larger sheave, in in.
d = diameter of smaller sheave, in in.
CC = center-to-center distance between sheaves, in in.

Application—Belt Tensioning

All belts must operate at the proper tension. The belt slips if the tension is too low, causing rapid wear of the belt and pulley. The belt overheats if the tension is too high, resulting in rapid wear and increased load on the motor and bearings. The proper tension is just tight enough to prevent slippage at full load and is determined by the belt, size, and belt material. The proper belt tension should result in approximately 1/64″ of belt deflection for every inch of span between pulley centers. Belt manufacturers also specify belt tension requirements, which can be checked with a belt tension tester. Proper belt tension is determined with a tension tester by applying the following procedure:
1. Measure span length (t).
2. Slide the lower O-ring to the position equal to 1/64″ per inch of span length on the deflection distance scale. The scale is read at the bottom edge of the O-ring.
3. Position a straightedge even with the top of the belt.
4. Apply force with the tension tester at the center of the span with enough force to deflect the belt until the bottom edge of the lower O-ring is even with the straightedge.
5. Locate amount of deflection force on the deflection force scale. The upper O-ring slides down the scale as the tension tester is compressed and remains there when force is released. *Note:* Read scale at the top edge of the O-ring.
6. Compare the deflection force with the min/max range in the recommended belt deflection force table. Tighten the belt if force is less than the minimum recommended deflection force. Loosen the belt if force is greater than the maximum recommended deflection force.

Belt operation also provides clues to proper belt tension. For example, excessive fluttering of a belt between pulleys indicates that belt tension may be too low. Many electricians and maintenance technicians judge the tension of a belt by feel. The tight side of the belt should feel springy, not excessively tight or excessively loose.

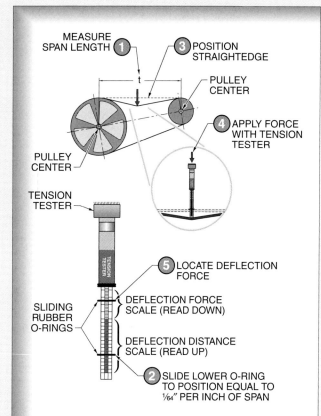

Recommended Belt Deflection Force*			
V-Belt Cross Section	Small Pulley Diameter†	Small Pulley Speed‡	Min/Max*
A	3.0 3.2 3.4 – 3.6 3.8 – 4.2 4.6 – 7.0	1750 – 3600	2.7 – 3.8 2.9 – 4.2 3.3 – 4.8 3.8 – 5.5 4.9 – 7.1
B	4.6 5.0 – 5.2 5.4 – 5.6 6.0 – 6.8 7.4 – 9.4	1160 – 1800	5.1 – 7.4 5.8 – 8.5 6.2 – 9.1 7.1 – 10 8.1 – 12
C	7.0 7.5 8.0 – 8.5 9.0 – 10.5 11.0 – 16.0	870 – 1800	9.1 – 13 9.7 – 14 11 – 16 12 – 18 14 – 21
D	12.0 – 13.0 13.5 – 15.5 16.0 – 22.0	690 – 1200	19 – 27 21 – 30 24 – 36

* in lb
† in in.
‡ in rpm

Application—Selecting Motor Couplings

The correct coupling for an application is selected by determining the nominal torque rating of the power source, determining the application service factor, calculating the coupling torque rating, selecting a coupling with an equal or greater torque rating, and ensuring that the coupling has the correct shaft size to fit the drive unit. To select the correct coupling for an application, apply the procedure:

1. Determine the nominal torque rating of the electric motor. The nominal torque rating is calculated or found on a conversion table. If the nominal torque rating is not available, it can be calculated as follows:

$$T_i = \frac{P \times 63,025}{S}$$

or

$$T_f = \frac{P \times 5252}{S}$$

where

T_i = torque, in lb-in.

T_f = torque, in lb-ft

P = power, in HP

S = speed, in rpm

A 2 HP motor operates at 1725 rpm to power a heavy-duty conveyor. The nominal torque rating in lb-in. is calculated as follows:

$$T_f = \frac{P \times 63,025}{S}$$

$$T_f = \frac{2 \times 63,025}{1725}$$

$$T_f = \mathbf{73.1 \text{ lb-in.}}$$

2. Determine the application service factor that adjusts for the operating conditions of the coupling. A larger multiplier is used when greater stress is placed on the coupling. By applying a multiplier, the size of the coupling is increased to adjust for severity of the load placed on the motor. From the table, the application service factor is 2.00.

3. Calculate the coupling torque rating by multiplying the nominal torque rating of the power source by the service factor of the application. The coupling torque rating is calculated as follows:

$$T_{CR} = T_{NR} \times SF$$

where

T_{CR} = coupling torque rating, in lb-in. or lb-ft

T_{NR} = nominal torque rating, in lb-in. or lb-ft

SF = application service factor

The nominal torque rating is 73.1 lb-in. The application service factor is 2.00. The coupling torque rating for the conveyor application is calculated as follows:

$$T_{CR} = T_{NR} \times SF$$

$$T_{CR} = 73.1 \times 2.00$$

$$T_{CR} = \mathbf{146 \text{ lb-in.}}$$

4. Select a coupling with a torque rating equal to or greater than the calculated coupling torque rating. From the table, the coupling with the next greater torque rating is coupling number 10-105A, rated at 176 lb-in.

5. Ensure that the coupling has the correct shaft size to fit the drive unit. The exact size of a motor shaft can be determined by the motor frame number, located on the nameplate. A 2 HP motor typically has a 145T frame. The shaft size of a 145T frame is ⅞". The coupling must have a bore size that accepts a ⅞" shaft.

Common Service Factors

Equipment	Service Factor
Blowers	
Centrifugal	1.00
Vane	1.25
Compressors	
Centrifugal	1.25
Vane	1.50
Conveyors	
Uniformly loaded or fed	1.50
Heavy-duty	2.00
Elevators	
Bucket	2.00
Freight	2.25
Extruders	
Plastic	2.00
Metal	2.50
Fans	
Light-duty	1.00
Centrifugal	1.50

Coupling Selections

Coupling Number	Rated torque (lb-in)	Maximum shock torque (lb-in)
10-101-A	16	45
10-102-A	36	100
10-103-A	80	220
10-104-A	132	360
10-105-A	176	480
10-106-A	240	660
10-107-A	325	900
10-108-A	525	1450
10-109-A	875	2450
10-110-A	1250	3500
10-111-A	1800	5040
10-112-A	2200	6160

Summary...

- Belt drives are relatively inexpensive, quiet, easy to maintain, and provide a wide range of speed and torque.

- Centrifugal tension is the tension needed to offset the centrifugal force on the belt as it engages the pulley.

- Belt creep is the natural movement of the belt on the face of the pulley when it is subjected to changes in tension.

- Belt slip is the movement of a belt on the face of the pulley when belt tension is too low and the belt slides on the pulley.

- V-belts transmit torque to and from a pulley through the wedging action of the tapered sides of the belt in the pulley groove.

- Flat belts are commonly used in applications that require high speeds, low noise, and small pulley diameters.

- Timing belts provide a drive system that has no slippage or creep, does not stretch, needs no lubricant, requires low belt tension, and has very little backlash.

- Pulleys are used to change the speed of a driven load relative to the motor speed.

- Mechanical drives are used to transfer torque as well as change the magnitude, direction, and speed of the applied force.

- A chain is used with sprockets to transfer torque from a motor to a load.

- A gear is a toothed wheel used to transfer torque or motion from one gear to another or from one shaft to another using interlocking gear teeth.

...Summary

- The gear ratio determines how fast the driven gear rotates in relation to the drive gear.

- A spur gear is a gear with straight teeth cut parallel to the shaft axis.

- A pinion gear is the smaller of two meshing gears. A rack has teeth spaced along a straight line. A pinion gear is used with a rack to convert rotary motion into linear motion.

- A helical gear is a gear with teeth that are not parallel to the shaft axis.

- A herringbone gear is a gear with two rows of helical teeth.

- A bevel gear is a gear with straight tapered teeth used in applications where shaft axes intersect.

- A worm is a screw thread that rotates the worm gear. A worm gear is a spur gear that is driven by a worm.

- A gearbox is a sealed container that has an input shaft and an output shaft and houses at least one set of mating gears.

- Couplings are the most common and least expensive method of connecting two shafts.

- A clutch can be used instead of a direct coupling so that the load is connected to the motor shaft while the motor is running at its no-load speed.

- An eddy-current clutch must operate with slip. Torque can be increased by increasing the slip or by increasing the field current in the rotor coils.

Glossary...

A **flexible drive** is a system in which a resilient flexible belt is used to drive one or more shafts.

A **V-belt** is a flexible drive belt that has a cross-section in the shape of a V.

A **fractional-horsepower (FHP) V-belt** is a V-belt designed for light-duty applications.

A **standard V-belt** is a V-belt designed for moderate-duty applications.

A **narrow V-belt** is a V-belt having a smaller cross section and a higher profile than a standard belt that is designed for heavy-duty applications.

A **flat belt** is a belt that has a rectangular cross section and relies on friction for proper operation.

A **timing belt** is a flat belt containing gear teeth that are used for synchronous drive systems.

A **sheave** is a grooved wheel used to hold a V-belt.

A **mechanical drive** is a combination of mechanical components that transfer torque from one location to another.

A **chain drive** is a synchronous mechanical drive system that uses a chain to transfer torque from one sprocket to another.

Roller chain is a chain that contains roller, pin, and connecting master links.

Silent chain is a synchronous chain that consists of a series of links joined together with bushings and pins.

...Glossary

A **gear drive** is a synchronous mechanical drive system that uses the meshing of two or more gears to transfer motion from one shaft to another.

A **gear ratio** is the ratio between the diameter of the drive gear and the driven gear.

Backlash is the amount of movement, or play, between meshing gear teeth.

A **spur gear** is a gear with straight teeth cut parallel to the shaft axis.

A **helical gear** is a gear with teeth that are not parallel to the shaft axis.

A **herringbone gear** is a gear with two rows of helical teeth.

A **bevel gear** is a gear with straight tapered teeth used in applications where shaft axes intersect.

A **worm** is a screw thread that rotates the worm gear.

A **worm gear** is a spur gear that is driven by a worm.

A **gearbox** is a sealed container that has an input shaft and an output shaft and houses at least one set of mating gears.

A **coupling** is a device that connects the ends of rotating shafts.

A **rigid coupling** is a device that joins two precisely aligned shafts within a common frame.

A **flexible coupling** is a coupling with a resilient center that flexes under temporary torque or misalignment due to thermal expansion.

A **clutch** is a coupling between a motor and a load that connects or disconnects the motor shaft to a drive shaft while the motor is running.

A **friction clutch** uses the force of friction between two or more rotating disks, drums, or cones to engage or disengage the two shafts.

An **eddy-current clutch** uses a magnetic field to couple a motor to a load.

Review

1. Describe how a belt and pulleys are used to transfer torque from a motor to a load.

2. Describe the difference between a fractional-horsepower V-belt, a standard V-belt, and a narrow V-belt.

3. Explain the advantage of using a timing belt instead of another type of belt.

4. Describe how a chain drive is used to transfer torque from a motor to a load.

5. Describe how a gear drive is used to transfer torque from a motor to a load.

6. List the common types of gears.

7. Describe the operation of a friction clutch.

8. Describe the operation of an eddy-current clutch.

Refer to the CD-ROM for Quick Quiz® questions related to chapter content.

MOTORS

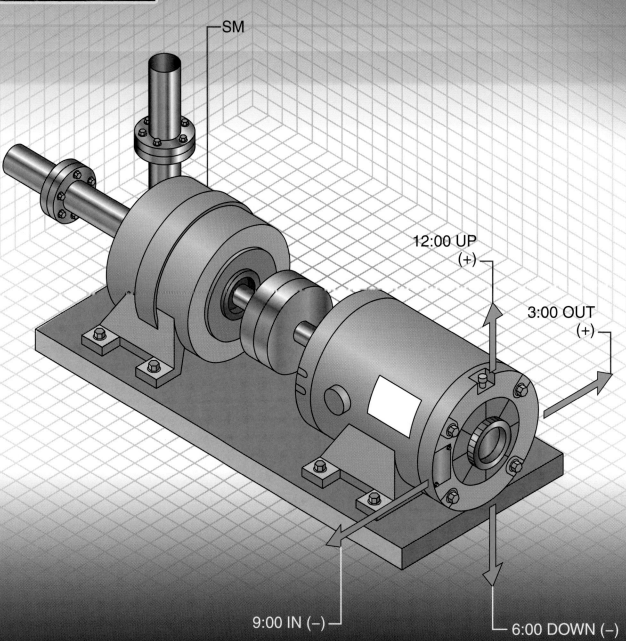

SM

12:00 UP (+)

3:00 OUT (+)

9:00 IN (−)

6:00 DOWN (−)

17

Motor Alignment

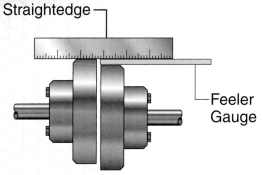

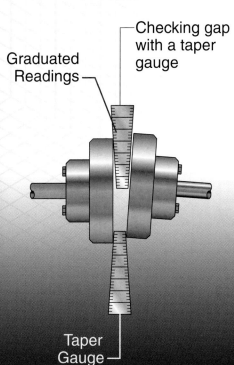

Shaft Alignment	372
Motor Placement	373
Dial Indicators	383
Alignment Methods	384
Application—Rim-and-Face Alignment	396
Summary	399
Glossary	400
Review	401

OBJECTIVES

- Describe the factors to consider when placing a motor.
- Describe soft foot and explain how it affects anchoring and alignment.
- Explain how the straightedge method is used to align a motor and load.
- Explain how the rim-and-face method is used to align a motor and load.
- Explain how the reverse dial method is used to align a motor and load.

A motor must be coupled to a load in order to produce useful work. Flexible drives use belts and pulleys. Mechanical drives use chains or gears. Clutches are used to connect or disconnect a load to a motor while the motor is running. Drives and clutches can also be used to operate a load at a different speed than the motor.

SHAFT ALIGNMENT

Alignment is the process where the centerlines of two machine shafts are placed within specified tolerances. The objective of proper alignment is to connect two shafts under operating conditions so all forces that cause damaging vibration between the two shafts and their bearings are minimized. The objective is to align the shafts, not the couplings. Couplings are imperfect. Aligning the couplings may result in misaligned shafts with aligned but irregularly shaped couplings.

Misalignment

Misalignment is the condition where the centerlines of two machine shafts are not aligned within specified tolerances. Properly aligned rotating shafts reduce vibration and add many years of service to equipment seals and bearings. As a rule of thumb, misalignment of a coupling by 0.004″ can shorten its life by 50%. **See Figure 17-1.**

Poor condition of equipment, such as worn bearings, bent shafts, stripped mounting bolts, bad gear teeth, or insufficient foundations or base plates, can create enough vibration to render any alignment effort useless. Once vibration starts, rapid wear of other components begins.

Misalignment exists when two shafts are not aligned within specific tolerances. *Offset misalignment* is a condition where two shafts are parallel but are not on the same axis. *Angular misalignment* is a condition where one shaft is at an angle to the other shaft. Shaft misalignment is usually a combination of offset and angular misalignment. **See Figure 17-2.** Offset and angular misalignment may be in the vertical or horizontal planes or both. Misalignment may be vertical angularity, vertical offset, horizontal angularity, or horizontal offset. Most misalignments are a combination of each type.

> **Definition**
>
> *Alignment* is the process where the centerlines of two machine shafts are placed within specified tolerances.
>
> *Misalignment* is the condition where the centerlines of two machine shafts are not aligned within specified tolerances.
>
> *Offset misalignment* is a condition where two shafts are parallel but are not on the same axis.
>
> *Angular misalignment* is a condition where one shaft is at an angle to the other shaft.

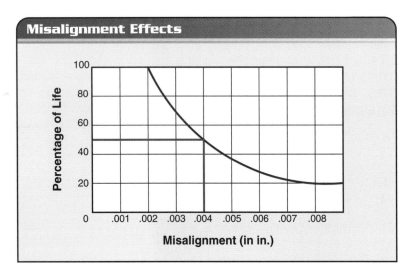

Figure 17-1. A coupling misalignment of 0.004″ can shorten the life of the coupling by 50%.

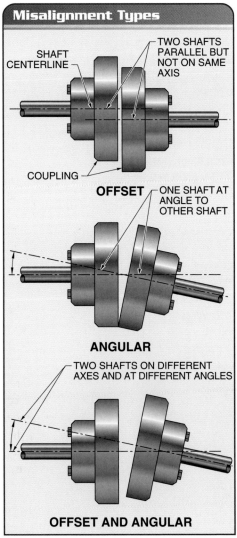

Figure 17-2. Misalignment can be offset or angular, but is generally a combination of the two.

MOTOR PLACEMENT

Each element related to placing a motor has a direct effect on the alignment forces on the equipment. Motor placement must consider piping and plumbing and the condition of anchoring components such as bolts and washers. Each component that directly or indirectly affects the proper alignment of machinery must be identified and considered before actual alignment begins. Considerations for good alignment include preparation of foundations and base plates; piping strain; anchoring; machinery adjustments during alignment; soft foot; and thermal expansion.

Foundations and Base Plates

Aligning any equipment begins with the foundation and base plate to which the equipment is anchored. A *foundation* is an underlying base or support. A *base plate* is a rigid steel support for firmly coupling and aligning two or more rotating devices. Foundations must be level and strong enough to provide support without movement. Base plates must be rigid enough to firmly support the equipment without stress and be securely anchored to the foundation. **See Figure 17-3.**

Originally, equipment base plates were made of thick cast iron that was strong enough to support the equipment in all operating conditions. The mounting surfaces were machined level. Currently, many equipment base plates are only sheet metal or plate metal welded or bolted to angle iron or I-beams. Flexing base plates must not be used. Good alignment is wasted when a flexing base plate is used.

The feet on machines such as motors, gearboxes, and pumps must be checked for cracks, breaks, rust, corrosion, or the presence of paint. The contacting surfaces between the motor, pump, gearbox, etc., and the base plate must be smooth, flat, and free of paint, rust, or foreign materials. This equipment should be bolted to a base plate, not anchored to concrete. The feet need to be inspected for burrs, rust, cracks, breaks, or any other damage. Finally, the couplings, shafts, and bearings need to be inspected for damage, contamination, or inaccurate sizes.

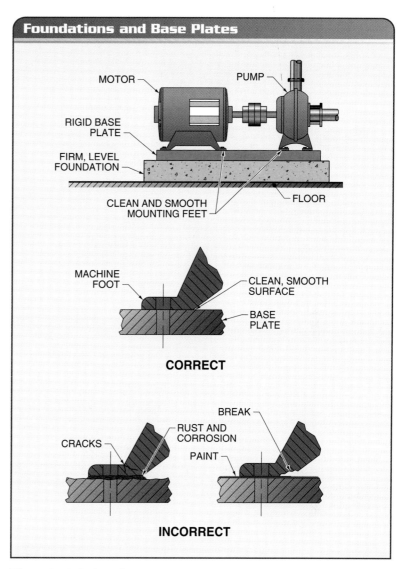

Figure 17-3. A clean, level base plate and foundation are required for proper alignment.

Piping Strain

Pipe and conduit connections, if improperly installed, and thermal expansion can produce enough force to affect machine alignment. To ensure that outside forces do not affect the proper motor alignment, motors should initially be aligned unattached from any piping if possible. Therefore, all plumbing must be properly aligned and have its own permanent support even when unattached. In some cases, flexible plumbing connections are necessary to separate stresses and vibrations between pump/motor and product lines. **See Figure 17-4.**

> **Definition**
>
> A *foundation* is an underlying base or support.
>
> A *base plate* is a rigid steel support for firmly coupling and aligning two or more rotating devices.

Piping Strain

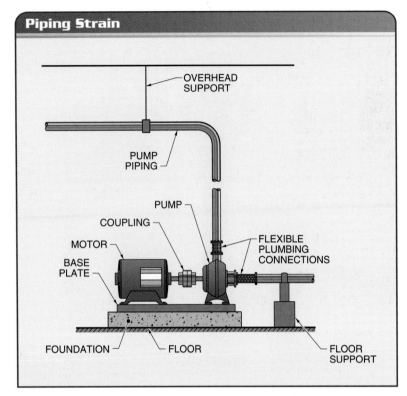

Figure 17-4. Piping must be independently supported to prevent external forces from working against bearing and alignment tolerances.

Definition

Anchoring is any means of fastening a mechanism securely to a base or foundation.

Bolt bound is the condition where the horizontal movement of a machine is restricted because the machine anchor bolts contact the sides of the machine anchor holes.

Anchoring

Anchoring is any means of fastening a mechanism securely to a base or foundation. Firm but adjustable anchoring of mechanisms on a base plate is accomplished using the proper mechanical fasteners (nuts, bolts, screws, etc.). Adverse anchoring includes bolt-bound machines, bolts or boltholes not of the proper size to allow for sufficient movement, and improper washers that create a dowel effect. Improper use, type, or fit of the anchoring bolts can make alignment of any machine impossible. **See Figure 17-5.**

Proper Anchoring. A base plate that is drilled and tapped to anchor a machine must be a minimum thickness of 1½ times the root diameter of the anchoring bolts. The threaded depth must be a minimum of 1½ times the root diameter of the bolt when a base plate is thicker than 1½ times the root diameter of the bolt. The correct grade and size of the anchor bolts must be used. If the base plate is less than 1½ times the bolt diameter, the mounting bolts require the use of washers, lock washers, and nuts. Mounting bolts require the use of nuts and lock washers if the base plate is less than 1½ times the bolt diameter.

Bolt Bound. *Bolt bound* is the condition where the horizontal movement of a machine is restricted because the machine anchor bolts contact the sides of the machine anchor holes. Bolt-bound bolts prevent horizontal movement of a machine in any needed horizontal direction. This condition should be checked before any alignment checks begin. In many cases, bolt-bound conditions may be checked using a straightedge placed along the machine shafts to determine if enough horizontal movement is available for proper alignment.

If sufficient movement is not available at the machine to be shimmed, all anchor bolts can be loosened and both machines shifted to try to align the machines. If repositioning both machines does not work, the anchor bolts must be turned down or the machine-mounting holes enlarged. Anchor bolts may be turned down without decreasing the tensile strength as long as the cut does not go beyond the root diameter (bottom portion) of the threads. Also, mounting holes may be enlarged, but this can lead to the dowel effect. In some cases, a combination of undercutting the bolt and enlarging the holes may be necessary.

Excess Bolt Body. Bolts come in many lengths. Some anchor bolts with the same root diameter have different lengths for the unthreaded portion of the bolts. The unthreaded portion of the bolt does not fit into the tapped bolt hole and the bolt cannot be properly tightened without damaging the tapped bolt hole. The correct anchor bolts must be used in order to ensure that the proper amount of the threads is engaged in the bolt hole to provide the required holding strength and to ensure that the bolt can be tightened down on the machine foot.

Bolt Bottoms Out. Base plates are drilled and tapped to create a bolt hole for the anchoring bolt. The bolt hole must be deep enough for the bolt and the threads must extend far enough into the bolt hole to take the entire bolt length. If the hole or threads are not deep enough, the bolt bottoms out and the bolt head cannot be tightened down on the machine foot. This situation should not be corrected by using a shorter bolt because that would reduce the holding strength. The bolt hole must be drilled and tapped to the correct length.

Dowel Effect. The *dowel effect* is a condition that exists when the bolt hole of a machine is so large that the bolt head forces the washer into the hole opening on an angle. Angled washers force the bolt to the center of the hole, making any horizontal movement impossible. The dowel effect is corrected by using machined washers two to three times thicker than the original washer. This prevents any deformation of the washer by bolt forces.

> **Definition**
>
> The *dowel effect* is a condition that exists when the bolt hole of a machine is so large that the bolt head forces the washer into the hole opening on an angle.

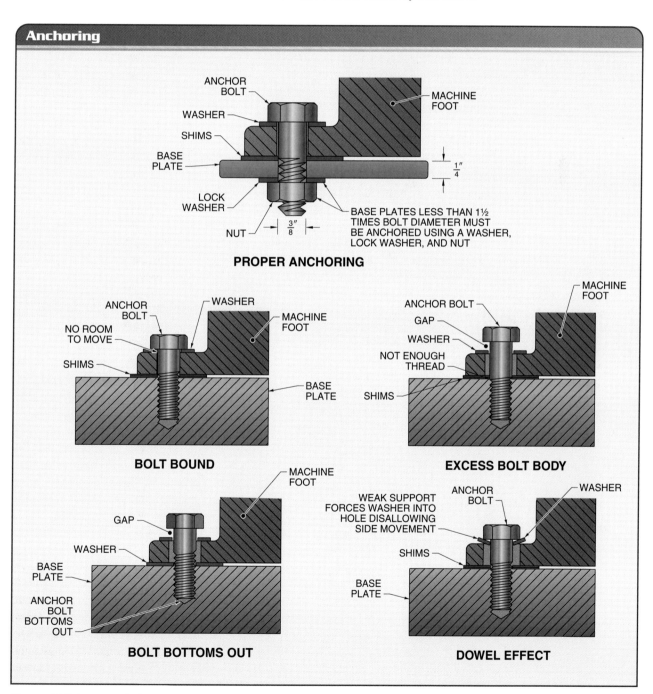

Figure 17-5. Improper bolt-hole diameter, excessive bolt body or length, and weak washers can make alignment impossible.

Definition

*A **jackscrew** is a screw inserted through a block attached to a machine base plate that is used to move a machine horizontally.*

Machine Adjustments

One machine, the motor or the load, is usually chosen as the stationary machine (SM) and the other machine is chosen as the machine to be shimmed (MTBS). The choice of which machine to move during alignment, the proper tools to use to prevent equipment damage, and the components used for precise movements must be known for an accurate, fast, and damage-free alignment. This knowledge includes the proper use and installation of jack screws and the proper use and selection of spacers and shim stock.

Either machine may be the moved and shimmed machine, but the heaviest machine or the machine attached to plumbing is generally the SM. The motor is generally the MTBS. Regardless of which machine is moved, the SM must initially be higher than the MTBS to allow for proper vertical alignment. A common practice is to initially install the SM using 0.125″ shims under each foot. This practice requires raising the MTBS, but prevents any vertical movement requirements of the SM. **See Figure 17-6.**

Jackscrews. A *jackscrew* is a screw inserted through a block attached to a machine base plate that is used to move a machine horizontally. Jackscrews should be installed on the base plate to easily and accurately control the horizontal movement of a machine. Typical jackscrew blocks are about 1½″ × 2½″ rectangular blocks made from ½″ or larger steel and are bolted or welded to the sides of the base plate with the jackscrews directly in-line with each mounting hole. Each jackscrew block is drilled and tapped to allow for a ½″ bolt assembly. **See Figure 17-7.**

Jackscrews are used for machine movement only. To prevent additional forces from being applied, jackscrew pressure must be backed each time the anchor bolts are tightened. Jackscrews are invaluable when a machine is to be moved just a few thousandths of an inch. A screwdriver or crowbar can be used for machine movement if it is not possible to install jackscrews or if they are not available. An easy, steady prying force is safer and less damaging than a blow from a hammer. A soft-blow hammer should be used if a hammer is necessary.

Shim Stock. Shims are used as spacers between machine feet and a base plate. The feet of a machine must be firmly anchored to the base plate without creating excessive forces or movement between mating shafts. It is rare for any machine to have all of its feet in contact with the base plate and also be within tolerance. Shims and spacers are used to adjust the height of a machine. Shim stock is used to make shims. Shim stock is manufactured in thicknesses ranging from 0.0005″ to 0.125″. Shim stock can be purchased as a sheet or roll or in precut shapes. A spacer is used for filling spaces ¼″ or greater. Good shim packs are laser cut with each size printed (not stamped) on the shim. When checking for accuracy, shims should be stacked with the printed size reversed on every other shim. This arrangement produces a true size and condition of the stack. **See Figure 17-8.**

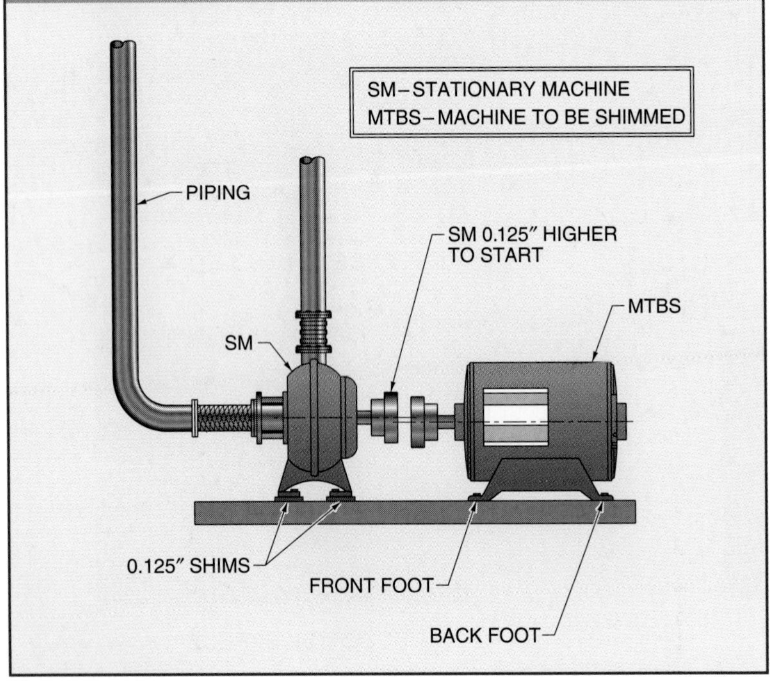

Figure 17-6. A machine is chosen as the MTBS because it is easier to move than the other machine.

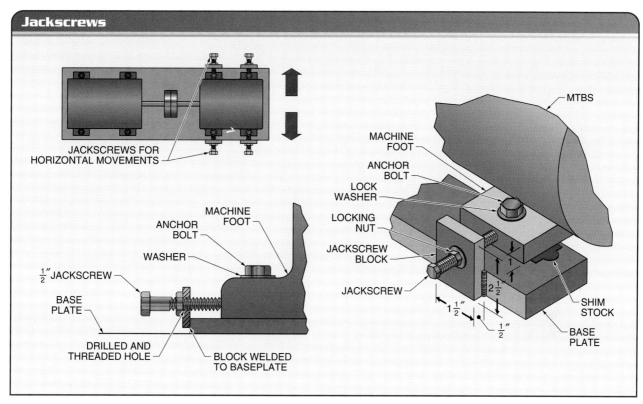

Figure 17-7. A jackscrew is attached to a block that is bolted or welded to a machine base to allow precise control of horizontal movement.

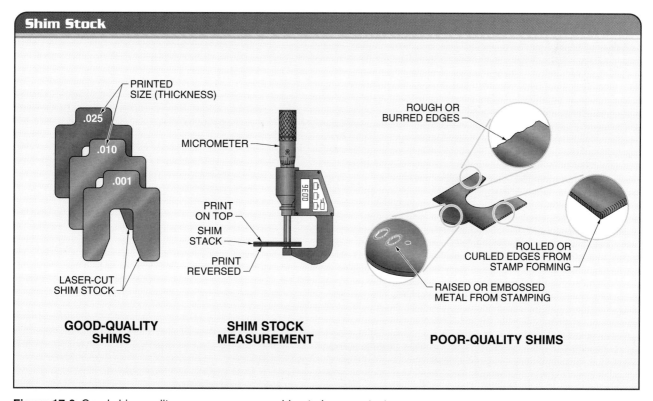

Figure 17-8. Good shim quality ensures proper machine-to-base contact.

Definition

Runout *is an out-of-round variation from a true circle.*

Soft foot *is a condition that occurs when one or more feet of a machine do not make complete contact with its base plate.*

Parallel soft foot *is a condition that exists when one or two machine feet are higher than the others and parallel to the base plate.*

Angular soft foot *is a condition that exists when one machine foot is bent and not on the same plane as the other feet.*

To prevent stacking inaccuracies, the number of shims or spacers on each foot should be limited to five or less. The combination that uses the fewest number of shims or spacers should be used because extra spring is added each time a shim is added to a stack. For example, a spacer and shim combination is required to raise a motor 0.683″. To keep the spacer to 1 piece and the shim stack to 5 or less, the spacer selected should be ⅝″ (0.625″) with the remaining 0.058″ taken up by shims. Shim stacks used are sometimes based on the sizes available in a shim pack set. Variations for the 0.058″ shim stack include one 0.050″ and two 0.004″ or two 0.025″, one 0.005″, one 0.002″, and one 0.001″.

Shaft Runout. *Runout* is an out-of-round variation from a true circle. Any shaft that runs eccentric to the true centerline of a machine by more than 0.002″ is very difficult to align and the runout should be corrected. An eccentric shaft produces high vibrations similar to those caused by a bent shaft. An eccentric shaft may be identified by the use of a dial indicator to measure the deviation from a true circular path. Shaft runout can be corrected by having the shaft recut. **See Figure 17-9.**

Soft Foot

Soft foot is a condition that occurs when one or more feet of a machine do not make complete contact with its base plate. A common reason for bearing failure is frame distortion due to soft foot. Distorted frames cause internal misalignment to bearing housings, shaft deflection, and distorted bearings, resulting in premature bearing and coupling failure. This condition also creates great difficulty in shaft alignment. Before aligning, any machine that has soft foot must be shimmed for equal and parallel support on all feet.

Soft foot should be checked on the MTBS and SM because soft foot can occur on any machine. Soft-foot tolerance must be within 0.002″ of shaft movement. Soft foot may be parallel, angular, springing, or induced. Each is independent of the others. All four conditions may exist on the same machine. **See Figure 17-10.**

Parallel. *Parallel soft foot* is a condition that exists when one or two machine feet are higher than the others and parallel to the base plate. Parallel soft foot occurs when a machine leg is longer or shorter than the others or when spacers of different thicknesses are used. Parallel soft foot is corrected by first rocking the machine from side to side to determine the highest foot. Feeler gauges are used to determine the air gap between the foot and base plate to within thousandths of an inch.

Shim stock equal to the thickness of the soft-foot gap is placed under the high foot and all four machine feet are rechecked. The shims should be moved from under the first foot to the other rocking foot if soft foot is noticed under a foot that was not shimmed. The shim stock should be divided between the feet that were rocking if soft foot is noticed under two opposing feet. Checking more than once for proper parallel spacing is done as an attempt to find true, or close to true, parallel spacing without creating angular soft foot.

Angular. *Angular soft foot* is a condition that exists when one machine foot is bent and not on the same plane as the other feet.

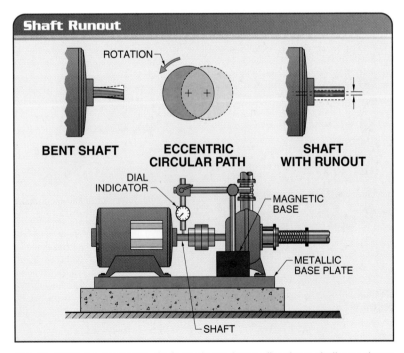

Figure 17-9. An eccentric shaft produces large vibrations similar to those caused by a bent shaft.

Generally, one corner of the angular foot is touching the base plate. Angular soft foot is usually caused by rough handling or by uneven mounting pads due to poor machining or welding. Angular soft foot may be identified when a 0.002″ feeler gauge can be placed under one side of a foot but not under the other side of the same foot.

Angular soft foot can be corrected by machining the foot to be on the same plane as the other feet or by step shimming the foot to fill the gap. Step shimming begins by determining the direction and amount of slope and filling the sloping opening with a series of 5 or 6 shims placed at an angle. This is done by measuring the largest portion of the gap and dividing by 5 or 6, giving the thickness of each step. Finally, each shim is placed by hand, in steps, to fill the gap without lifting the machine. Shaft movement can be checked with a dial indicator while the machine bolts are tightened.

Springing. *Springing soft foot* is a condition that exists when a dial indicator at the shaft shows soft foot but feeler gauges show no gaps. The machine acts as if it were mounted on springs due to each imperfection. This condition occurs from shims that are burred or bent, corroded bases or feet, dirt, grease, or rust between the feet, shims, and base plate, or too many shims.

Springing soft foot is prevented by using solid bases that are cleaned to the metal by removing all paint, grime, rust, and corrosion. The top and bottom of the machine feet must also be free of rust, paint, and grime. Shims must be flat, clean, and without stamping imperfections.

Induced. *Induced soft foot* is a condition that exists when external forces are applied to the machine. Forces such as coupling misalignment, piping strain, tight jackscrews, or improper structural bracing can cause induced soft foot. Coupling forces from vertical or horizontal misalignment are noticed when couplings are difficult to bolt up or a spring or snap is observed as couplings are disconnected.

Any external force in any direction to coupled pipe flanges on a pump strains the machine. This condition may be seen when checking for soft foot before, during, and after piping connections or when checking for structural bracing strain.

Definition

Springing soft foot is a condition that exists when a dial indicator at the shaft shows soft foot but feeler gauges show no gaps.

Induced soft foot is a condition that exists when external forces are applied to the machine.

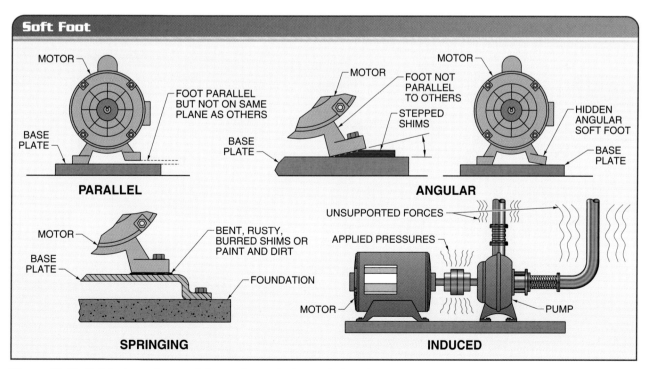

Figure 17-10. Soft foot may be parallel, angular, springing, or induced.

Measuring Soft Foot

The two methods of measuring soft foot are the shaft deflection method and the at-each-foot method. **See Figure 17-11.** Both methods use a dial indicator with a magnetic base to determine movement. More than one set of readings must be made in the same direction of movement to ensure that readings are constant. When checking for soft foot, the anchor bolt for one foot at a time is loosened while keeping the bolts for the other three feet tightened. Angular soft foot must be corrected before either shaft can be aligned. The shaft deflection method is easier to use than the at-each-foot method because the indicator does not interfere with loosening the anchor bolts.

Once soft foot conditions have been corrected, shaft alignment may begin. Shaft alignment is difficult or impossible if soft foot conditions have not been corrected to within tolerance.

> **Tech Fact**
> *The correct torque must be used when tightening anchor bolts. Excessive torque can cause bolt stretch or distortion in the base plate or machine frame.*

Shaft-Deflection Method. In the shaft-deflection method, the shaft is checked for deflection when anchor bolts are tightened or loosened. **See Figure 17-12.** If the shaft moves more than 0.002″, critical distortion has occurred and correction is necessary. This method is quicker and more accurate than the at-each-foot method. The shaft-deflection method is performed as follows:

1. Place a magnetic-base dial indicator on the base plate. Adjust it so the stem is above and perpendicular to the top of the shaft or coupling, whichever is the farthest from the MTBS. The farther the dial indicator is from the feet, the greater the dial movement and the greater the accuracy.
2. Ensure that all four feet are anchored firmly and zero the dial on the indicator.
3. Slowly loosen the bolt on the first foot. If the shaft moves more than 0.002″, place shim stock beneath that foot with a thickness equal to the amount of movement. Tighten the anchor bolt for that foot and recheck. Rezero the indicator and check and shim each foot individually until shaft movement is within the 0.002″ tolerance.
4. Recheck all feet. If one foot rises, check all feet for hidden angular soft foot.

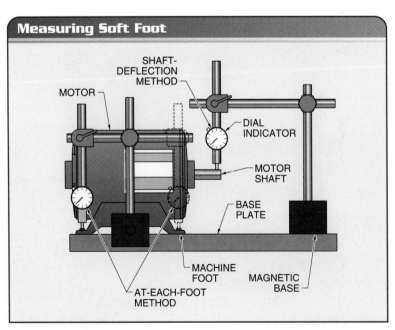

Figure 17-11. Soft foot may be measured at the shaft or at each foot of the machine.

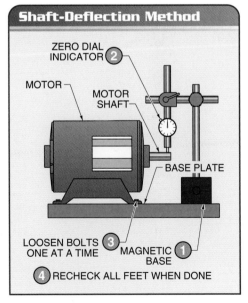

Figure 17-12. The shaft-deflection method measures the movement of the shaft as each foot is loosened and tightened.

At-Each-Foot Method. In the at-each-foot method, soft foot is measured at each foot of the machine. **See Figure 17-13.** The at-each-foot method is performed as follows:

1. Place a magnetic-base dial indicator on the base plate at the foot to be tested. The dial indicator is adjusted so the stem is above and perpendicular to the top of the foot.
2. Ensure that all four feet are anchored firmly and zero the dial on the indicator.
3. Slowly loosen the bolt on the first foot. If the foot moves more than 0.002″, place shim stock beneath that foot with a thickness equal to the amount of movement. Tighten the anchor bolt for that foot and recheck.
4. Move the indicator to another foot and rezero the indicator. Loosen the bolt and check the movement of the foot. If the foot moves more than 0.002″, place shim stock beneath that foot with a thickness equal to the amount of movement. Tighten the anchor bolt for that foot and recheck. Check and shim each foot individually until each foot movement is within the 0.002″ tolerance.
5. Recheck all feet. If one foot rises when rechecking, check all feet for hidden angular soft foot.

The at-each-foot method is seldom used because it is slower and less accurate than the shaft deflection method. The biggest problem with this method is that it does not measure the actual movement of the shaft. If the front and back feet are close together with a long shaft, the shaft moves a relatively large amount for very small adjustments to the feet. If the front and back feet are far apart with a short shaft, the shaft moves a relatively small amount for very large adjustments to the feet. If this method is used, extra calculations are needed to correlate the foot movement tolerance with a shaft movement tolerance.

Thermal Expansion

Thermal expansion is the dimensional change in a substance due to a change in temperature. For proper alignment, two coupled shafts must be on the same horizontal and vertical plane under operating conditions. However, there can be a significant change in physical dimensions when there is a temperature change from resting conditions to operating conditions and thermal expansion moves the load shaft. **See Figure 17-14.**

Definition

Thermal expansion is the dimensional change in a substance due to a change in temperature.

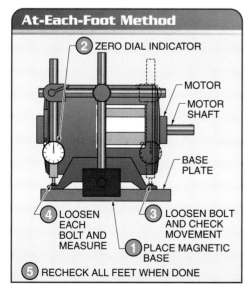

Figure 17-13. The at-each-foot method measures the movement of each foot as it is loosened and tightened.

Motors are used to drive many types of loads. The motor and load must be properly aligned to prevent damage and optimize useful life.

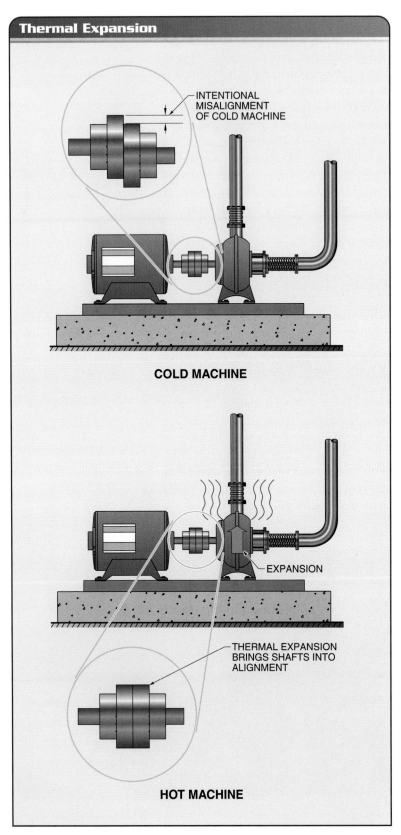

Figure 17-14. Cold machines can be intentionally misaligned to allow for thermal expansion.

Temperature changes may be caused by the temperature of product being pumped, excessive room or ambient temperature, or heat from a loaded motor. Pump manufacturers often provide the amount of thermal expansion for a given temperature change. Manufacturers of motors and other equipment rarely provide this information. The thermal expansion can easily be calculated when it is not known.

The change in material length for different materials can be calculated to determine specific alignments for thermal expansion. A material constant is used to calculate the amount of thermal expansion. Typical material constants are 0.0000063 for cast iron, 0.000009 for stainless steel, and 0.0001 for plastic. Thermal expansion is found as follows:

$$\Delta L = L \times \Delta T \times C$$

where
ΔL = change in length, in in.
L = original length, in in.
ΔT = change in temperature, in °F
C = material constant, in in./in.-°F

For example, a cast-iron pump used to move a hot fluid changes temperature as the fluid moves through the pump at startup. The motor used to turn the pump is lightly loaded and does not get very hot during operation. Since the pump gets hot, the distance from the bottom of the feet to the centerline of the shaft increases. When the operating temperature of the pump increases from 75°F to 140°F, the temperature change is 65°F (140 – 75 = 65). The pump measures 15″ from its base to the centerline of the shaft. The change in length is calculated as follows:

$$\Delta L = L \times \Delta T \times C$$
$$\Delta L = 15 \times 65 \times 0.0000063$$
$$\Delta L = \mathbf{0.006''}$$

The horizontal plane of the motor should be increased by 0.006″ because the pump shaft rises as it gets warmer while operating. A flexible coupling is normally required because it must work for a cold startup as well as normal hot operating conditions. If the motor were to get hot during operation, that expansion would also need to be calculated.

DIAL INDICATORS

A dial indicator is a precise instrument that must be treated with care if it is to be a useful tool. Dial indicators must be mounted perpendicular to the surface being measured. Blows from mallets or hammers on a machine into an indicator can damage the indicator or throw off readings. Slow, forceful movements should be used when adjusting toward an indicator.

Dial Indicator Use

Dial indicators are read using the total movement of the dial needle. Readings do not have to begin at zero to determine total movement. Total indicator readings (TIRs) are used as indicator readings regardless of whether the needle begins before, at, or after zero. Total indicator readings are found by subtracting one reading from the other reading and taking the absolute value of the difference. **See Figure 17-15.** It does not matter which reading is subtracted from the other as long as the negative signs are accounted for.

For example, readings on an indicator have an ending reading of +0.006″ and a beginning reading of +0.022″. The TIR is 0.016″ (0.006″ − 0.022″ = −0.016″). The absolute value is 0.016″. If the readings are subtracted in the opposite order, the result is still 0.016″ (0.022″ − 0.006″ = +0.016″).

The same method is used with dial readings that are negative. An indicator reading of −0.006″ is subtracted from a reading of −0.022″ to give a TIR of 0.016″ (−0.022″ − (−0.006″) = −0.016″). The absolute value is 0.016″. If the readings are subtracted in the opposite order, the result is still 0.016″ (−0.006″ − (−0.022″) = +0.016″).

The same method is also used with mixed dial readings that include negative and positive numbers. For example, an indicator reading of +0.022″ is subtracted from a reading of −0.006″ to give a TIR of 0.028″ (0.006″ − (+0.022″) = −0.028″). The absolute value is 0.028″. If the readings are subtracted in the opposite order, the result is still 0.028″ (+0.022″ − (−0.006″) = +0.028″).

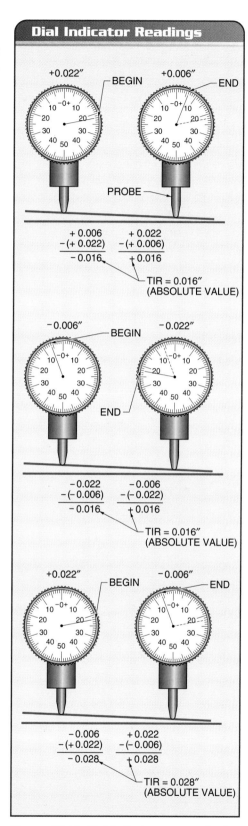

Figure 17-15. Dial indicators are read using the total movement of the dial needle and do not have to begin at zero.

Rod Sag

The indicating device that measures the alignment of one shaft must be clamped or strapped to the opposing shaft. The entire assembly generally consists of the clamp or strap, a riser rod, two 90° rod couplings, a spanning rod, and an indicating device. The indicating device may be a dial indicator or an electronic indicator. The greater the distance between the rod couplings, the more the weight of the indicating device creates a sag in the spanning rod. **See Figure 17-16.** This sag can create errors if it is not accounted for in the calculations. Modern indicator kits have very little sag because of stiffer spanning rods. Older indicator kits with steel spanning rods often have significant sag errors.

Electronic indicators calculate rod sag when measurements are keyed in. Rod sag from a dial indicator must be measured manually. All parts (rods, couplings, indicator) are assembled on a solid shaft or pipe using the established coupling distance. A solid shaft is helpful in determining rod sag because it is not misaligned. The dial indicator is adjusted to get a reading off the bar and then zeroed at the 12:00 position. The dial shows a negative reading at the 6:00 position equal to twice the actual amount of rod sag when the bar is rotated 180°. Record half of the total reading so it may be subtracted from the actual alignment readings. The actual sag is –0.005″ if the readings on the bar indicated total sag of 0.010″.

For example, an actual alignment reading indicates a rotational misalignment of 0.024″. This is divided by two, giving a vertical reading of 0.012″. The indicator rod sag of 0.005″ is then subtracted from the vertical reading, giving an actual vertical offset reading of 0.007″ (0.012″ – 0.005″ = 0.007″).

ALIGNMENT METHODS

The three common methods used to align machinery include straightedge, rim-and-face (indicator and laser), and reverse dial (indicator and electronic). **See Figure 17-17.** All alignment techniques require that adjustments be made in a specific order. The specific order of shaft alignment adjustments is as follows:

1. angular in the vertical plane (up-and-down angle)
2. parallel in the vertical plane (up-and-down offset)
3. angular in the horizontal plane (side-to-side angle)
4. parallel in the horizontal plane (side-to-side offset)

Once angular and parallel alignment in the vertical plane have been corrected, they generally are maintained when angular and parallel alignment in the horizontal plane are corrected. This step-by-step process is used regardless of the alignment method. Each corrective move should be double checked.

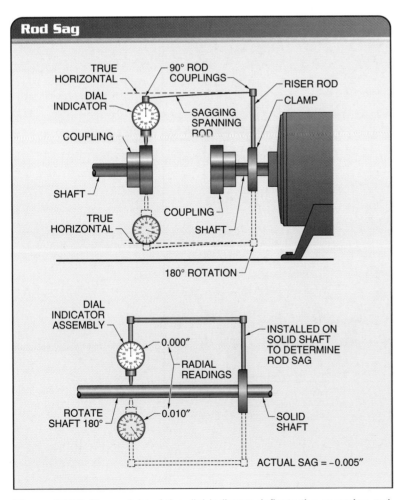

Figure 17-16. The weight of the dial indicator deflects the spanning rod, causing rod sag.

Accuracy Expectations

The choice of alignment method is based on cost, accuracy required, ease of use, and time required to perform the alignment. The accuracy of any alignment is based on the skill level of the individual doing the alignment and the alignment method used. For example, straightedge measurements are usually made without the knowledge of coupling irregularities and require the feel of thickness gauge measurements. Therefore, the accuracy of straightedge alignment generally is no better than about 1/64″.

Dial indicators and electronic measuring devices (except laser) measure in the thousandths of an inch, which allows for an alignment accuracy within 0.001″. Laser alignment methods are generally exact and quick with a possible accuracy of 0.0002″. Straightedge measurement equipment is the least expensive, dial indicators and electronic measurement equipment are more expensive, and laser measurement equipment is the most expensive. Laser equipment is generally connected to a computer or similar controller that performs all the calculations and speeds up the alignment process.

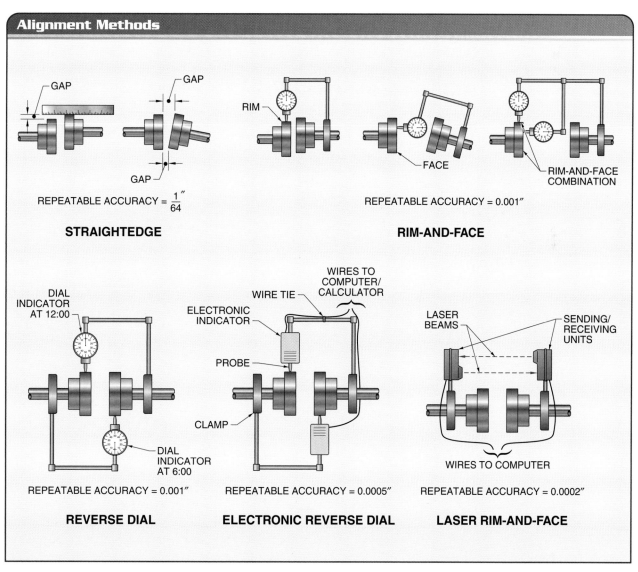

Figure 17-17. The three common methods used to align machinery include straightedge, rim-and-face (indicator and laser), and reverse dial (indicator and electronic).

Alignment Tolerance. Alignment tolerance requirements of two or more shafts are based on the speed of the motor or drive unit. At times, a manufacturer may indicate the alignment tolerance for its machine. A shaft alignment tolerances chart may be used if manufacturer tolerances are not available. **See Figure 17-18.** This chart provides approximate values for equipment typically seen by electricians. There are no universally accepted values for all types of machines under all circumstances.

For example, an electrician needs to align a pump and motor combination that has a 5″ coupling and operates at 1200 rpm. The horizontal offset measured by the electrician is 0.003″ and the vertical offset measured by the electrician is 0.005″. From the chart, the maximum acceptable offset at 1200 rpm is 0.004″. Therefore, the horizontal offset is within tolerance and the vertical offset is out of tolerance.

The angularity values in the table are given in mils/in. One mil is 0.001″. The value in the table must be multiplied by the size of the coupling. For a 1″ coupling, the values are multiplied by 1, meaning that the values can be used directly from the table. For a 5″ coupling, the value in the table must be multiplied by 5.

The measured horizontal angularity is 0.005″ and the measured vertical angularity is 0.006″. From the chart, the maximum acceptable angularity at 1200 rpm is 1.0 mil/in. Since a 5″ coupling is being used, the chart value is multiplied by 5 and the maximum acceptable angularity is 5.0 mil, or 0.005″. Therefore, the horizontal angularity is within tolerance and the vertical angularity is out of tolerance.

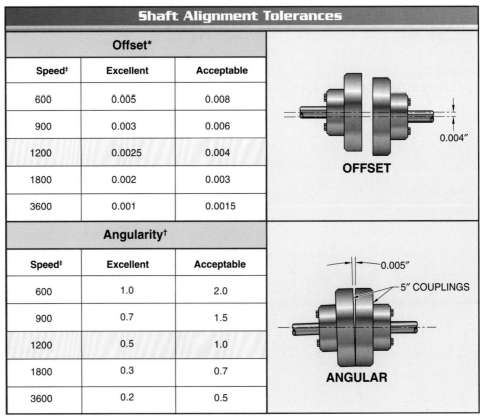

Shaft Alignment Tolerances

Offset*		
Speed‡	Excellent	Acceptable
600	0.005	0.008
900	0.003	0.006
1200	0.0025	0.004
1800	0.002	0.003
3600	0.001	0.0015

Angularity†		
Speed‡	Excellent	Acceptable
600	1.0	2.0
900	0.7	1.5
1200	0.5	1.0
1800	0.3	0.7
3600	0.2	0.5

* in in.
† in mils/in.
‡ in rpm

Figure 17-18. A shaft alignment tolerances chart gives suggested tolerances for different operating speeds.

Horizontal Movements

Angular and offset movements in the horizontal plane are normally made as jackscrews are screwed in and dial indicator movement is observed. To move the MTBS away 0.020″, a dial indicator is placed at the back edge of the machine base plate and directly in line with the adjusting jackscrew. The back jackscrews are backed out to allow the machine to move. The front jackscrew is turned in until the indicator registers a 0.020″ movement. For horizontal movements, front and back refer to the direction of movement, not the location on the machine. Front is the side pushed upon, and back is the opposite side. **See Figure 17-19.**

When only one indicator is used, a movement at one end of a machine changes the machine position at the other end, complicating accurate movement at both ends. To overcome possible confusion during horizontal movement, two indicators, one at each foot, are used to display exact front and back movement.

Angular Movements

Angular movements are generally made by adjustments at two machine feet. To prevent compounding errors, new readings must be taken after each adjustment to the MTBS. To reduce the chances for error, dial indicators should be rotated in one direction. The indicator should be reset to the zero setting if movement direction has been reversed. Clock-position movements (clockwise/counterclockwise) are those viewed from the MTBS toward the SM. **See Figure 17-20.** Vertical or horizontal angularity corrections using shim stock must be performed at one location only, such as at the MTBS front feet or back feet. The opposing feet are used as the angle pivot point. Offset adjustments may require that shims be placed under all the feet.

Angular measurements are easier to interpret when it is realized that each misalignment angle in the same plane (vertical or horizontal) is the same whether it is measured at the coupling face or the misalignment at the feet of the machine.

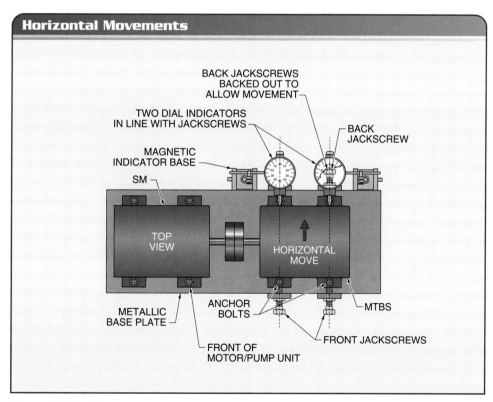

Figure 17-19. Jackscrews are used to make movements in the horizontal plane.

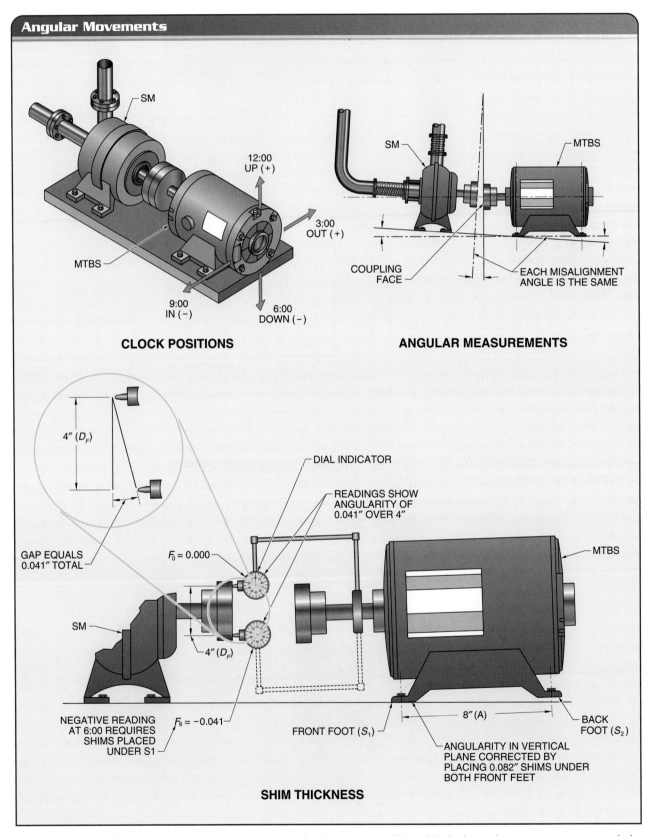

Figure 17-20. The horizontal and vertical planes are defined by clock positions. Vertical angular movements are made by placing shims under the feet.

Measurements of distance and gap may be used to determine angularity and gap at any other distance. This principle can be used to determine the shim thickness to eliminate any angle. The required shim thickness at the desired distance is found as follows:

$$S = g \times \frac{D}{d}$$

where
S = shim thickness (in in.)
g = measured gap at coupling (in in.)
d = distance between gap measurements (in in.)
D = distance between adjustment and pivot points (in in.)

For example, the mounting holes on the motor are 8″. The gap misalignment at 4″ from zero is 0.041″. The shim-stock thickness at the motor is calculated as follows:

$$S = g \times \frac{D}{d}$$
$$S = 0.041 \times \frac{8}{4}$$
$$S = 0.041 \times 2$$
$$S = \mathbf{0.082″}$$

The choice of which feet are adjusted and which act as the pivot is determined by the 6:00 reading. The front feet are shimmed if the 6:00 reading is positive (greater than the 12:00 reading). The back feet are shimmed if the 6:00 reading is negative. In this case, the measured gap at the face is –0.041″ over a distance of 4″. This shows that this measured gap can be used to calculate the required gap of 0.082″ over a distance of 8″. A shim or shims with a thickness of 0.082″ must be placed under the front feet to eliminate the gap.

Tech Fact

Sudden shock loads during startup and operation can damage couplings and cause misalignment. A coupling should operate quietly and smoothly. If a coupling vibrates or makes unusual sounds, the coupling and alignment should be rechecked.

Straightedge Alignment

Straightedge alignment is a method of coupling alignment in which a straightedge is used to align couplings and a taper gauge is used to measure the gap between the coupling halves. The reading on the taper gauge at the gap edge is the diameter of the gap at that point. A feeler gauge is used to measure offset. **See Figure 17-21.** A taper gauge is more reliable than a feeler gauge for measuring angular misalignment.

Straightedge alignment is the oldest method used for measuring misalignment at couplings. This method is easy to understand and perform but is not very accurate. Any system with tolerances of a few thousandths of an inch must be read in thousandths of an inch. The straightedge method is not accurate enough to align a system with that tolerance. With straightedge alignment, coupling hub face and OD runout must be checked and compensated for in making corrective calculations. Straightedge alignment is a good method for aligning a motor and load in slow movement applications with flexible couplings or for getting shafts close before using other alignment methods.

Definition

Straightedge alignment is a method of coupling alignment in which a straightedge is used to align couplings and a taper gauge is used to measure the gap between the coupling halves.

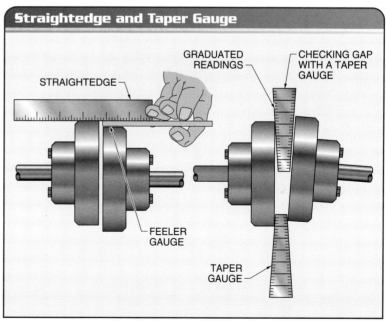

Figure 17-21. The straightedge alignment method uses a straightedge, feeler gauge, and taper gauge to align couplings.

Vertical Plane. To begin alignment, the SM must be firmly anchored and the MTBS must be free to move. The angular misalignment in the vertical plane (up-and-down angle) should be checked with a taper gauge. **See Figure 17-22.** This measurement is made at the 12:00 and 6:00 positions on the shafts or couplings.

The difference between the 12:00 and 6:00 readings is used to determine the thickness of the shim stock placed under the feet to eliminate the angle. Shim-stock thickness to eliminate angular misalignment in the vertical plane is based on the vertical angular gap (gap at 12:00 minus gap at 6:00), the diameter of the coupling, and the distance between front and back MTBS mounting holes. Shim-stock thickness to eliminate angular misalignment in the vertical plane is found as follows:

$$S = g \times \frac{D}{d}$$

where
S = shim thickness (in in.)
g = measured gap at coupling (in in.)
d = distance between gap measurements (in in.)
D = distance between adjustment and pivot points (in in.)

For example, a pump and motor assembly have a 5″ diameter coupling. The distance between mounting holes on the MTBS is 6.5″. The measured gap at the top of the coupling is 0.410″ and the measured gap at the bottom of the coupling is 0.360″. The misalignment is 0.050″ (0.410 − 0.360 = 0.050). The distance between the gap measurements is 5″. The required shim-stock thickness is calculated as follows:

$$S = g \times \frac{D}{d}$$
$$S = 0.050 \times \frac{6.5}{5}$$
$$S = 0.050 \times 1.3$$
$$S = \mathbf{0.065″}$$

The misalignment in the vertical plane is adjusted by placing shims with a thickness of 0.065″ under both back feet of the MTBS because the gap between the couplings is wider at 12:00 (0.410″) than at 6:00 (0.360″).

After the angular misalignment in the vertical plane has been corrected, the MTBS is checked for offset misalignment in the vertical plane (vertical offset). This is accomplished by laying a straightedge across the top of the shafts or couplings. With

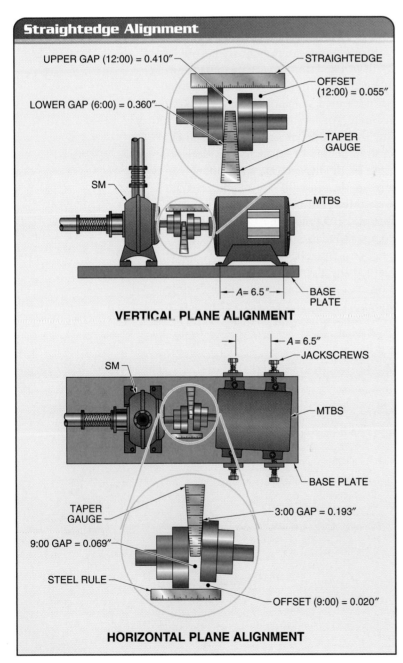

Figure 17-22. Straightedge alignment is generally used for rough alignment with a flexible coupling or prior to using a more precise method of alignment.

the straightedge held firmly and parallel against the highest shaft or coupling, feeler gauges are slid into the offset to determine the distance that the MTBS must be raised or lowered.

For example, feeler gauges are used to make an offset measurement of 0.055″ with a straightedge placed on top of a 5″ coupling. This means the MTBS must be raised 0.055″. After inserting shims equaling 0.055″ under all four feet and tightening the mounting bolts, the angular position in the vertical plane and the offset position in the vertical plane should be rechecked before making the horizontal alignments.

Horizontal Plane. The angular position in the horizontal plane (side-to-side angle) and the offset position in the horizontal plane (side-to-side offset) are checked and corrected similar to the vertical plane adjustments. These movements are accomplished with the use of jackscrews. The angular position in the horizontal plane is checked with the jackscrews fingertight against the feet of the MTBS and the mounting bolts tight. The angular gap is checked with the taper gauge at the 3:00 and the 9:00 positions.

The adjustment to eliminate angular misalignment in the horizontal plane is based on the horizontal angular gap (gap at 3:00 minus gap at 9:00, or gap at 9:00 minus gap at 3:00), the diameter of the coupling, and the distance between the front and back MTBS mounting holes. Horizontal adjustment to eliminate angular misalignment in the horizontal plane is found as follows:

$$S = g \times \frac{D}{d}$$

where
S = distance to be moved at adjustment point (in in.)
g = measured gap at coupling (in in.)
d = distance between gap measurements (in in.)
D = distance between adjustment and pivot points (in in.)

For example, a pump and motor assembly have a 5″-diameter coupling. The distance between mounting holes on the MTBS is 6.5″. The measured gap of the coupling at the 3:00 position is 0.193″ and the measured gap at the 9:00 position is 0.069″. The misalignment is 0.124″ (0.193 − 0.069 = 0.124). The distance between the gap measurements is 5″. The amount of movement to correct the angular misalignment is calculated as follows:

$$S = g \times \frac{D}{d}$$
$$S = 0.124 \times \frac{6.5}{5}$$
$$S = 0.124 \times 1.3$$
$$S = \mathbf{0.161''}$$

The horizontal angular adjustment of 0.161″ is at the back foot (S_2) because the gap is wider at the 3:00 position (0.193″) than at the 9:00 position (0.069″). The angular misalignment should be rechecked.

Offset misalignment in the horizontal plane is corrected similarly to correcting offset misalignment in the vertical plane. After the horizontal angular measurement is completed, a straightedge and feeler gauge is placed at the 9:00 position and the offset is measured.

For example, a straightedge placed at the 9:00 position of a 5″ coupling shows an offset of 0.020″. The gap being between the straightedge and the MTBS coupling half at the 9:00 position requires the motor to be moved to the front a distance of 0.020″ by the jackscrews. Recheck the angular position in the horizontal plane and the offset position in the horizontal plane after making any adjustments.

Straightedge alignment is not necessarily true to the shaft axis because straightedge alignment measures the condition of a coupling and shaft assembly along with machined surfaces of the coupling. Final results may easily be far from tolerance if this method is used as the total alignment method. Upon completion of alignment, release (unscrew) any pressure from jackscrews.

Definition

Offset misalignment is a condition where two shafts are parallel but are not on the same axis.

Pulley Alignment. When installing offset drives of pulleys and belts, the three types of alignments to be checked are offset alignment, nonparallel alignment, and angular alignment. Pulley alignment is normally done with a long straightedge placed against the side of both sheaves. **See Figure 17-23.**

Offset misalignment is a condition where two shafts are parallel but are not on the same axis. Offset misalignment may be corrected using a straightedge along the pulley faces. The straightedge may be solid or a string. Offset misalignment must be within 1/10″ per foot of drive center distance. *Nonparallel misalignment* is misalignment where two pulleys or shafts are not parallel. Nonparallel misalignment is also corrected using a string or straightedge. The device connected to the pulley that touches the straightedge at one point is rotated to bring it parallel with the other pulley so that the two pulleys touch the straightedge at four points. *Angular misalignment* is a condition where one shaft is at an angle to the other shaft. Angular misalignment is corrected using a level placed on top of the pulley parallel with the pulley shaft. Angular misalignment must not exceed 1/2°.

Rim-and-Face Alignment

Rim-and-face alignment is a method of coupling alignment in which the offset and angular misalignment of two shafts is determined by measuring at the rim and face of a coupling. **See Figure 17-24.** A dial indicator at the rim position measures offset directly under the indicator stem. Also, the difference in offset over the distance between two indicators is the angularity. Before shim size adjustments are determined, offset and angular gaps must be calculated and used to determine if a machine must be moved up, down, backward, or forward.

Rim-and-face alignment may be accomplished by using the individual rim-and-face method or the combination rim-and-face method. The combination method is considerably faster and more accurate than the individual method.

Individual Rim-and-Face Alignment. The individual rim-and-face alignment method uses an indicator that is used to measure the coupling face (angularity) and then repositioned to measure the coupling rim (offset). Rim and face readings must be taken with the coupling disconnected and both coupling halves rotated together. This is best accom-

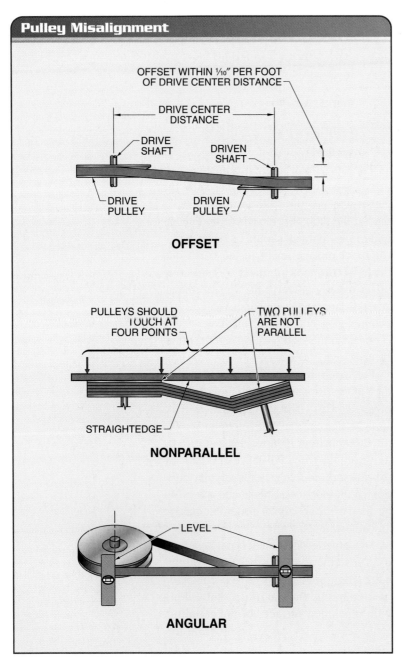

Figure 17-23. Pulley misalignment is corrected by placing a straightedge across the pulleys and adjusting the position of the equipment so the pulleys touch the straightedge at four points.

plished when a mark is made on both coupling halves and kept in-line as the couplings are rotated. By rotating both couplings, shaft centerlines are measured, whereas rotating only one shaft measures one shaft in relation to the opposing coupling face or diameter.

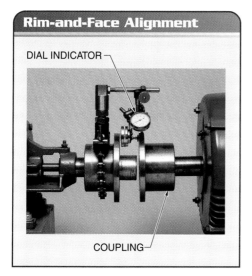

Figure 17-24. Dial indicators must be mounted perpendicular to the contact surface when used for rim-and-face alignment.

Combination Rim-and-Face Alignment. The combination rim-and-face alignment method uses two dial indicators. One indicator reads the rim offset and the other reads the face to measure the angularity. Both indicators are assembled using rods and couplings placed at the same end of the spanning rod. The two indicators measure the vertical and horizontal planes of the same shaft simultaneously. One dial indicator can be clamped on one coupling and rotated around with the foot of the indicator riding on the face of the other half to measure the angular misalignment. The other indicator is attached to the same spanning rod so that the foot of the indicator rides on the outer diameter of the coupling to measure the offset.

Laser Rim-and-Face Alignment. *Laser rim-and-face alignment* is a method of coupling alignment in which laser devices are placed opposite each other to measure alignment. The laser rim-and-face method is used when accuracy and fast alignment are required. The initial cost of laser equipment is higher than that of other methods, but laser alignment is more accurate.

Laser alignment devices operate using the rim-and-face method, with the dial indicator replaced with a laser beam. **See Figure 17-25.** The beam is directed to a 90° prism reflector, which is directed back to the sending unit where a receiving transducer (photo position detector) accepts the signal and converts it into an electrical signal. A benefit of using a laser beam is that there is no rod sag. The beam sensor is able to detect up and down movement. Offsets and angles are read and measured accurately when the return beam is detected at one position (12:00) and then detected at another position (6:00). A computer or calculator measures offset directly by the location of the reflected beam and determines the angularity by measuring angles.

Definition

Nonparallel misalignment is misalignment where two pulleys or shafts are not parallel.

Angular misalignment is a condition where one shaft is at an angle to the other shaft.

Rim-and-face alignment is a method of coupling alignment in which the offset and angular misalignment of two shafts is determined by measuring at the rim and face of a coupling.

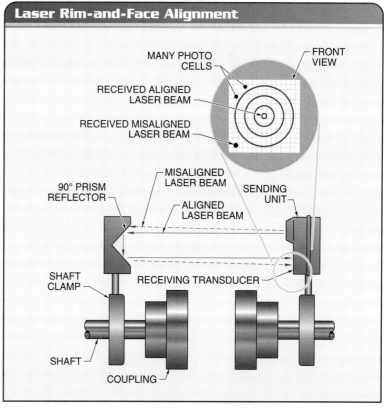

Figure 17-25. Laser accuracy is based on being able to send a light beam, reflect it, and measure the location of the return beam.

Definition

Laser rim-and-face alignment is a method of coupling alignment in which laser devices are placed opposite each other to measure alignment.

Reverse dial alignment is a method of coupling alignment that uses two dial indicators to take readings of opposing sides of coupling rims, giving two sets of shaft runout readings.

Once set up, laser alignment devices check soft foot, compensate for thermal expansion, indicate the movement of a machine during alignment, rapidly couple machines with multiple couplings, and determine shim placement. A graphic display presents the condition of each foot when checking for soft foot or angular soft foot. Finally, all rechecking is completed and corrected in a matter of seconds.

Reverse Dial Alignment

Reverse dial alignment is a method of coupling alignment that uses two dial indicators to take readings of opposing sides of coupling rims, giving two sets of shaft runout readings. Each indicator shows both angle and offset. Reverse dial indicator readings can be illustrated by a plotted layout. **See Figure 17-26.** Electronic reverse dial alignment has mostly replaced manual reverse dial alignment because of the complexity of the plotting and calculations.

The plot shows the relative position of shaft centerlines and the indicator reading dimensions. Each square represents a horizontal and vertical measurement. In this plot, the horizontal squares are 1″ per three-square division, providing a representative view of the overall machine dimensions. The vertical squares are 0.005″ per division, providing a representative view of total indicator readings (TIRs).

Plotting begins by drawing a horizontal centerline and placing a mark on the centerline that represents an indicator stem point. The appropriate number of squares is counted, and a mark is placed at the second indicator stem point. This distance represents the distance between the indicator stem points on the two shafts, shown as D_1 on the plot.

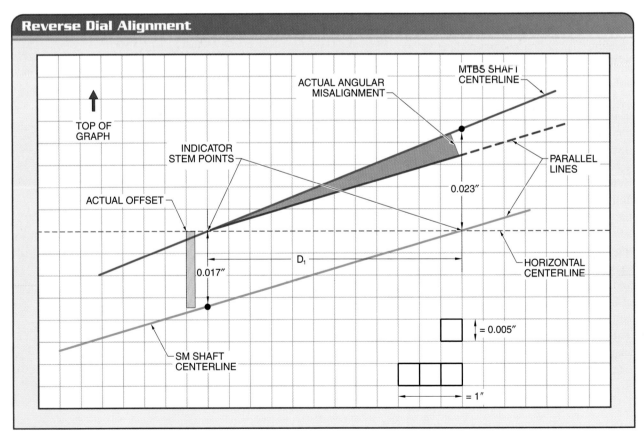

Figure 17-26. The reverse dial method uses a plot of the location of one shaft relative to the other to determine the adjustments needed.

To plot the MTBS shaft centerline, the MTBS reading is divided by two and plotted toward the top of the graph from one indicator stem point. For example, if the total MTBS reading is 0.046″, a mark is made 0.023″ toward the top of the graph from one of the indicator stem points (0.023″ = 4.6 squares). A line is drawn from this mark through the opposing indicator stem point.

To plot the SM shaft centerline, the SM reading is divided by two and plotted toward the bottom of the graph from the other indicator stem point. A line is drawn from this mark through the opposing indicator stem point. For example, if the total SM reading is 0.034″, a mark is made 0.017″ toward the bottom of the graph from that indicator stem point. These lines give offset and angular positions. The alignment objective is to end up with both lines parallel. The difference between these lines should be adjusted for proper alignment. Plotting provides a graphic illustration as well as an indication of the movements required for alignment. In this case, any corrective move must be made by raising the SM or lowering the MTBS.

Electronic Reverse Dial Alignment

Electronic reverse dial alignment replaces dial indicators with electromechanical sensing devices. **See Figure 17-27.** The electronic reverse dial method is supported by computer-aided electronic instrumentation. The sensing devices detect physical movement and convert the movement to an electrical signal, which is sent to a calculator. Physical dimensions must be entered by the electrician before the calculator can process an adjustment response. The calculator computes MTBS movements as its response.

The electronic reverse dial method requires that two sensing devices be used, each attached to a coupling or shaft. Installation is similar to using dial indicators where one sensor is opposite the other. Sensor wires must be secured with slack to prevent unwanted tugging forces. Some manufacturers of alignment kits recommend that the coupling be disconnected while others suggest that they be connected. Manufacturer's recommendations must be followed. However, any force, including coupling force, can create enough resistance to produce adverse responses.

The shafts are rotated together and readings are entered with the press of a button at the 12:00, 3:00, 6:00, and 9:00 positions. Some manufacturers recommend entering readings at 12:00, then forward (clockwise) to 3:00, back (counterclockwise) to 9:00, forward (clockwise) to 6:00, and back (counterclockwise) to 12:00. Any thermal expansion compensation information can be entered into the calculator.

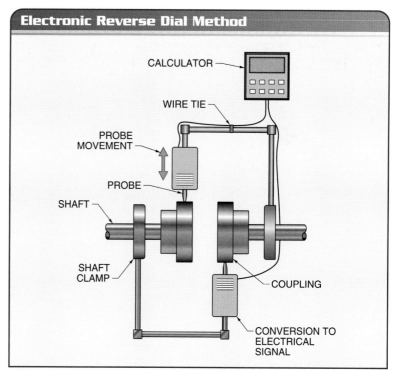

Figure 17-27. In the electronic reverse dial method, electromechanical devices convert a mechanical movement into an electrical signal.

Application—Rim-and-Face Alignment

Angular gap and offset information must be obtained before shim thickness and location can be determined. Angular gap and offset information used for shim placement is found by checking angular misalignment in the vertical plane (up and down), offset misalignment in the vertical plane (vertical offset), angular misalignment in the horizontal plane (side to side), and offset misalignment in the horizontal plane (side-to-side offset). The misalignment values are found by measuring with dial indicators. Combination rim-and-face alignment is performed by applying the procedure as follows:

1. Check for angular misalignment in the vertical plane (up-and-down angle).

Angular misalignment in the vertical plane is measured at the face of the coupling at the 12:00 and 6:00 positions. The angular misalignment in the vertical plane is the absolute value of the difference between the face readings at the 12:00 and 6:00 positions.

The shim-stock thickness to adjust for angular misalignment in the vertical plane can be found once the vertical angular gap is determined. The shim-stock thickness is based on the distance between gap measurements, the measured gap at the coupling, and the distance between front and back MTBS mounting holes (distance between adjustment and pivot points). Shim-stock thickness in the vertical plane is found by applying the formula:

$$S = g \times \frac{D}{d}$$

where

S = shim thickness (in in.)

g = measured gap at coupling (in in.)

d = distance between gap measurements (in in.)

D = distance between adjustment and pivot points (in in.)

From the table, the face reading at 12:00 is 0.000″ and at 6:00 is –0.016″. The measured gap at the coupling is the absolute value of the difference between the readings, or 0.016″ (0.016 – 0.000 = 0.016). From the illustration, the distance between the gap measurements is 5″. The distance between the adjustment and pivot points is 9″. The shim-stock thickness is calculated as follows:

$$S = g \times \frac{D}{d}$$

$$S = 0.016 \times \frac{9}{5}$$

$$S = 0.016 \times 1.8$$

$$S = 0.029″$$

Shim stock with a thickness of 0.029″ is needed to correct the vertical misalignment. The shim stock is placed under the front feet because the 6:00 reading is negative.

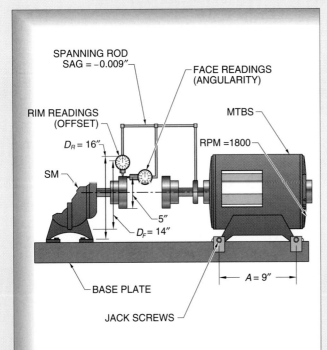

Dial Indicator Data		
Reading Position	Face Reading	Rim Reading
12:00	0.000	0.000
3:00	–0.006	–0.022
6:00	–0.016	–0.021
9:00	–0.010	+0.001

2. Check for offset misalignment in the vertical plane (up-and-down offset).

Offset misalignment in the vertical plane is checked by measuring the rim of the coupling at the 12:00 and 6:00 positions. The shim-stock thickness to adjust for offset misalignment in the vertical plane can be found once the vertical offset is determined. The shim-stock thickness is based on the difference between the 12:00 and 6:00 rim readings, adjusted for the rod sag. Shim-stock thickness to eliminate offset misalignment in the vertical plane is found as follows:

$$O_V = \frac{R_0 - R_6 - RS}{2}$$

where
O_V = offset in vertical plane (in in.)
R_0 = reading of rim at 12:00 (in in.)
R_6 = reading of rim at 6:00 (in in.)
RS = rod sag (in in.)

From the table, the rim reading at 12:00 is 0.000″ and the rim reading at 6:00 is –0.021″. The rod sag is 0.009″. The offset in the vertical plane is calculated as follows:

$$O_V = \frac{R_0 - R_6 - RS}{2}$$

$$O_V = \frac{0.000 - (-0.021) - 0.009}{2}$$

$$O_V = \frac{0.000 + 0.021 - 0.009}{2}$$

$$O_V = \frac{0.012}{2}$$

$$O_V = \mathbf{0.006''}$$

Shims with a thickness of 0.006″ are placed under all the feet of the motor to correct the vertical offset misalignment.

3. Check for angular misalignment in the horizontal plane (side-to-side angle).

Angular misalignment in the horizontal plane is checked by measuring the face of the coupling at the 3:00 and 9:00 positions. The horizontal adjustment to eliminate angular misalignment in the horizontal plane is based on the horizontal angular gap, the diameter traveled by the face indicator tip, and the distance between the front and back MTBS mounting holes. Horizontal adjustment to eliminate angular misalignment in the horizontal plane is found as follows:

$$S = g \times \frac{D}{d}$$

where
S = shim thickness (in in.)
g = measured gap at coupling (in in.)
d = distance between gap measurements (in in.)
D = distance between adjustment and pivot points (in in.)

From the table, the face reading at 3:00 is –0.006″ and the face reading at 9:00 is –0.010″. The measured gap at the coupling is the absolute value of the difference between the readings, or 0.004″ (–0.006 – (–0.010) = 0.004). From the illustration, the distance between the gap measurements is 5″. The distance between the adjustment and pivot points is 9″. The horizontal angular misalignment is calculated as follows:

$$S = g \times \frac{D}{d}$$

$$S = 0.004 \times \frac{9}{5}$$

$$S = 0.004 \times 1.8$$

$$S = \mathbf{0.007''}$$

Jackscrews are used to move the machine 0.007″ to correct the horizontal misalignment.

Application—Motor Alignment

4. Check for offset misalignment in the horizontal plane (side-to-side offset).

Offset misalignment in the horizontal plane is checked by measuring the rim of the coupling at the 3:00 and 9:00 positions. The shim-stock thickness to adjust for offset misalignment in the horizontal plane can be found once the horizontal offset is determined. The shim-stock thickness is based on the difference between the 3:00 and 9:00 rim readings. Rod sag deviation has no effect in the horizontal plane. Shim-stock thickness to eliminate offset misalignment in the horizontal plane is found as follows:

$$O_H = \frac{R_3 - R_9}{2}$$

where
O_H = offset in horizontal plane (in in.)
R_3 = reading of the rim at 3:00 (in in.)
R_9 = reading of the rim at 9:00 (in in.)

From the table, the face reading at 3:00 is –0.022″ and the face reading at 9:00 is 0.001″. From the illustration, the distance between the gap measurements is 5″. The distance between the adjustment and pivot points is 9″. The horizontal offset is calculated as follows:

$$O_H = \frac{R_3 - R_9}{2}$$

$$O_H = \frac{-0.022 - 0.001}{2}$$

$$O_H = \frac{-0.023}{2}$$

$$O_H = -\mathbf{0.011''}$$

Jackscrews are used to move the machine 0.011″ to correct the horizontal misalignment. After all four adjustments have been made, all measurements should be taken again to verify that the changes resulted in the correctly aligned motor and pump.

Summary

- The objective of proper alignment is to connect two shafts under operating conditions so all forces that cause damaging vibration between the two shafts and their bearings are minimized.

- Misalignment may be vertical angularity, vertical offset, horizontal angularity, or horizontal offset.

- Considerations for good alignment include preparation of foundations and base plates; piping strain; anchoring; machinery movement during alignment; and proper use of alignment methods or procedures.

- The contacting surfaces between the motor, pump, gearbox, etc., and the base plate must be smooth, flat, and free of paint, rust, or foreign materials.

- Pipe and conduit connections, if improperly installed, can produce enough force to affect machine alignment.

- Improper use, type, or fit of the anchoring bolts can make alignment of any machine impossible.

- One machine, the motor or the load, is usually chosen as the stationary machine (SM) and the other machine is chosen as the machine to be shimmed (MTBS).

- Jackscrews are used for horizontal machine movement. Shims are used for vertical machine movement.

- Soft foot occurs when one or more feet do not make contact with the base plate. Soft foot may be parallel, angular, springing, or induced.

- The shaft-deflection method of measuring soft foot measures the deflection of the shaft as each foot is loosened. The at-each-foot method measures the movement of each foot as it is loosened.

- Thermal expansion can cause misalignment, as the dimensions of a machine change when the temperature changes.

- With dial indicators, total indicator readings (TIRs) are found by subtracting one reading from the other reading and taking the absolute value of the difference.

- The spanning rod of an indicator assemble can deflect under the weight of the indicator, causing rod sag.

- The three common methods used to align machinery include straightedge, rim-and-face (indicator and laser), and reverse dial (indicator and electronic).

- With straightedge alignment, the reading on a taper gauge at the edges of coupling is the coupling angular gap. A feeler gauge is used to measure offset.

- The three types of pulley alignments to be checked are offset alignment, nonparallel alignment, and angular alignment.

- With rim-and-face alignment, a dial indicator at the rim position measures offset directly and the difference in offset over the distance between two indicators is the angularity misalignment.

- Reverse dial alignment uses two dial indicators to take readings of opposing sides of coupling rims.

Glossary

Alignment is the process where the centerlines of two machine shafts are placed within specified tolerances.

Misalignment is the condition where the centerlines of two machine shafts are not aligned within specified tolerances.

Offset misalignment is a condition where two shafts are parallel but are not on the same axis.

Angular misalignment is a condition where one shaft is at an angle to the other shaft.

A ***foundation*** is an underlying base or support.

A ***base plate*** is a rigid steel support for firmly coupling and aligning two or more rotating devices.

Anchoring is any means of fastening a mechanism securely to a base or foundation.

Bolt bound is the condition where the horizontal movement of a machine is restricted because the machine anchor bolts contact the sides of the machine anchor holes.

The ***dowel effect*** is a condition that exists when the bolt hole of a machine is so large that the bolt head forces the washer into the hole opening on an angle.

A ***jackscrew*** is a screw inserted through a block attached to a machine base plate that is used to move a machine horizontally.

Runout is an out-of-round variation from a true circle.

Soft foot is a condition that occurs when one or more feet of a machine do not make complete contact with its base plate.

Parallel soft foot is a condition that exists when one or two machine feet are higher than the others and parallel to the base plate.

Angular soft foot is a condition that exists when one machine foot is bent and not on the same plane as the other feet.

Springing soft foot is a condition that exists when a dial indicator at the shaft shows soft foot but feeler gauges show no gaps.

Induced soft foot is a condition that exists when external forces are applied to the machine.

Thermal expansion is the dimensional change in a substance due to a change in temperature.

Straightedge alignment is a method of coupling alignment in which a straightedge is used to align couplings and a taper gauge is used to measure the gap between the coupling halves.

Nonparallel misalignment is misalignment where two pulleys or shafts are not parallel.

Rim-and-face alignment is a method of coupling alignment in which the offset and angular misalignment of two shafts is determined by measuring at the rim and face of a coupling.

Laser rim-and-face alignment is a method of coupling alignment in which laser devices are placed opposite each other to measure alignment.

Reverse dial alignment is a method of coupling alignment that uses two dial indicators to take readings of opposing sides of coupling rims, giving two sets of shaft runout readings.

Review

1. List the main factors to be considered when placing a motor.

2. List and describe anchoring problems that can occur when placing a motor.

3. Describe jackscrews and shims and explain when they are used.

4. List and describe the four types of soft foot.

5. Describe how the shaft deflection method is used to measure soft foot.

6. Describe thermal expansion and explain how to account for thermal expansion during alignment.

7. Explain how to obtain a total indicator reading (TIR) from two readings of a dial indicator.

8. List the order of adjustments that should be followed when adjusting alignment.

9. Demonstrate how to calculate the shim thickness needed to adjust vertical angular misalignment when the measured gap at the coupling is 0.030″, the distance between mounting holes is 10″, and the distance between gap measurements is 4″.

10. Demonstrate how to calculate the shim thickness required for vertical offset alignment when the reading at 12:00 is 0.002″, the reading at 6:00 is 0.024″, and the rod sag is 0.005″.

Refer to the CD-ROM for Quick Quiz® questions related to chapter content.

MOTORS

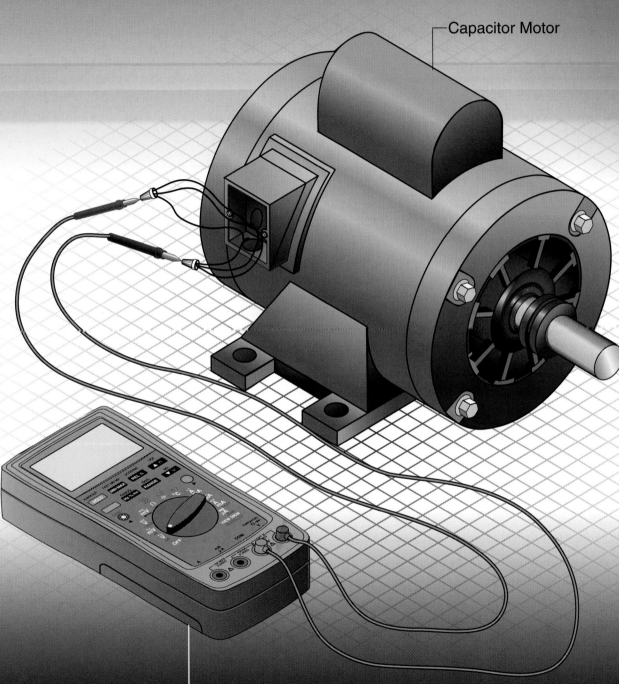

Capacitor Motor

Digital Multimeter (turned off)

18

Troubleshooting Motors

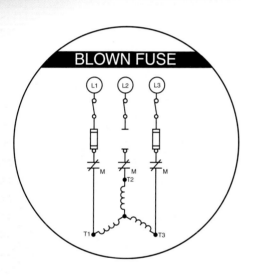

BLOWN FUSE

Motor Failure	404
AC Motors	412
DC Motors	417
Motor Controls	422
Application—Electric Motor Drive Troubleshooting	425
Motor Lead Identification	432
Test Tools	438
Application—Troubleshooting Guide	443
Summary	448
Glossary	449
Review	449

OBJECTIVES

- Describe the major causes of motor failure.
- Explain how to troubleshoot AC motors.
- Explain how to troubleshoot DC motors.
- Explain how to troubleshoot motor controls and circuits.
- Explain how to identify unmarked motor leads.

Motor failure occurs for a variety of reasons. More motors fail due to overloading than for any other reason. Troubleshooting procedures vary for DC motors, single-phase AC motors, and 3-phase AC motors. DC motors require the most troubleshooting because of their brushes. Single-phase AC motors require more troubleshooting than 3-phase motors because of their centrifugal switches, capacitors, starting windings, and built-in thermal switches.

MOTOR FAILURE

Electric motors are highly effective and reliable drivers of machinery and other loads. However, motors do fail at times. Each motor type has its advantages and disadvantages. For any given application, one type of motor generally offers the best performance. If a motor has a recurring problem, the motor and application may be mismatched. A change of motor type may be required for better service.

Motor failure commonly occurs due to overheating, improper power supply, improper application, improper maintenance, and motor damage. Motor manufacturers have found that overloading is one of the most common causes of motor failure. A motor is overloaded anytime it is required to deliver more power than it was designed for. The most common type of overloading is placing a load on the motor that is too large for the motor design. **See Figure 18-1.** In addition, a motor that is turned OFF remains warm for some time. This warmth can attract insects, snakes, and rodents. These pests can damage a motor by restricting ventilation or corroding the insulation. In high-pest areas, a special zip-seal jacket can be placed around the motor.

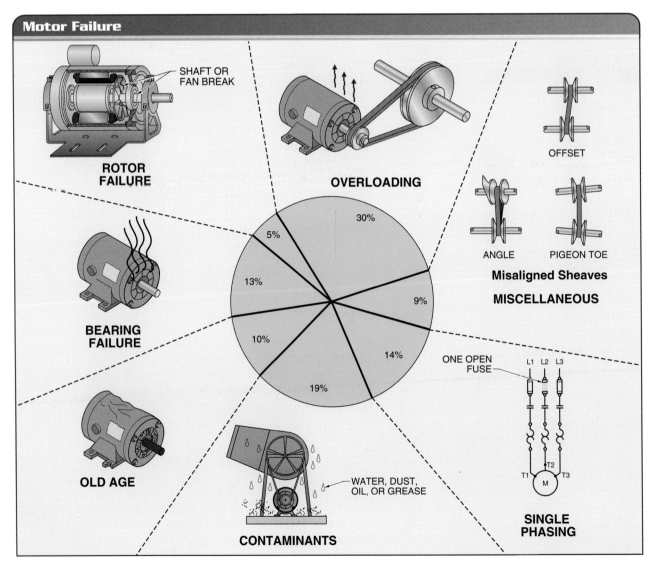

Figure 18-1. The most common type of overloading is placing a load on the motor that is too large for the motor design.

Overheating

Overheating is a major cause of motor failure and a sign of other motor problems. As the temperature in a motor increases beyond the temperature rating of the insulation, the life of the insulation is shortened. The higher the temperature, the sooner the insulation will fail. When motor insulation is damaged, the windings short and the motor is no longer functional.

The temperature rating of motor insulation is listed as the insulation class. The insulation class is given in Celsius (°C) and/or Fahrenheit (°F). The motor nameplate typically lists the insulation class of the motor. **See Figure 18-2.** Heat buildup in a motor can be caused by several factors, such as:

- Incorrect motor type or size for the application
- Improper ventilation, usually from dirt buildup
- Excessive load, usually from improper use
- Excessive friction, usually from misalignment or vibration
- Electrical problems, typically voltage unbalance, phase loss, or surge voltages

Improper Ventilation. All motors produce heat as they convert electrical energy to mechanical energy. This heat must be removed or it will destroy the motor insulation, and consequently the motor. Motors are designed with air passages that permit a free flow of air over and through the motor. This airflow removes the heat from the motor. Anything that restricts airflow through the motor causes the motor to operate at a higher temperature than it is designed for. Airflow may be restricted by the accumulation of dirt, dust, lint, grass, pests, rust, etc. If the motor becomes coated with oil from leaking seals or from excess lubrication, airflow is restricted much faster. **See Figure 18-3.**

Overheating can also occur if the motor is placed in an enclosed area. When a motor is installed in a location that does not permit the heated air to escape, the motor will overheat due to the recirculation of the heated air. Vents can be added at the top and bottom of the enclosed area to allow a natural flow of heated air.

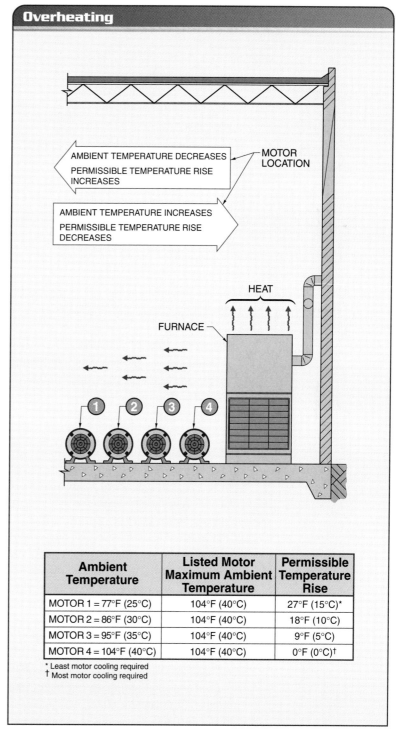

Figure 18-2. The temperature rating of motor insulation is listed as the insulation class.

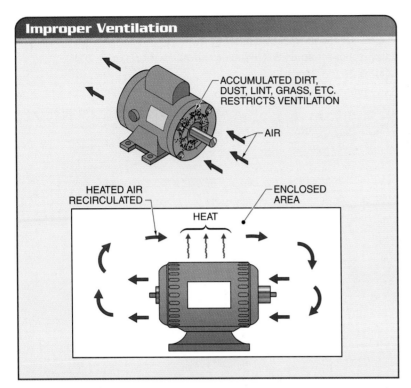

Figure 18-3. Anything that restricts airflow through the motor causes the motor to operate at a higher temperature than it is designed for.

Definition

Phase unbalance is the unbalance that occurs when lines are out of phase.

Voltage unbalance is the unbalance that occurs when the voltages at the motor terminals are not equal.

Improper Power Supply

There are many situations where problems with the power source show up as problems at a motor. Common types of power problems are phase unbalance, voltage unbalance, single phasing, and surge voltages.

Phase Unbalance. *Phase unbalance* is the unbalance that occurs when lines are out of phase. Phase unbalance of a 3-phase power system occurs when single-phase loads are applied, causing one or two of the lines to carry more or less of the load. The loads of 3-phase power systems are balanced by the electricians during installation. However, as additional single-phase loads are added to the system, an unbalance begins to occur. This unbalance causes the 3-phase lines to move out of phase, and are consequently no longer 120 electrical degrees apart.

Phase unbalance causes 3-phase motors to run at temperatures higher than their listed ratings. The greater the phase unbalance, the greater the temperature rise. These high temperatures produce insulation breakdown and other related problems. **See Figure 18-4.**

Three-phase motors cannot deliver their rated horsepower when a system is unbalanced. For example, a phase unbalance of 3% can cause a motor to work at only 90% of its rated power, requiring the motor to be derated for any given application.

Voltage Unbalance. *Voltage unbalance* is the unbalance that occurs when the voltages at the motor terminals are not equal. This voltage unbalance can range from as little as a few millivolts to full voltage loss on one line. The voltage unbalance results in a current unbalance. If the voltage is not balanced, one winding will overheat, causing thermal deterioration of that winding.

Voltage should be checked for voltage unbalance periodically and during all service calls. Whenever more than 2% voltage unbalance is measured, the surrounding power system should be checked for excessive loads connected to one line. If the voltage unbalance cannot be corrected, the load or motor rating should be reduced.

Voltage unbalance is measured by taking a voltage reading between each of the incoming power lines. **See Figure 18-5.** The readings are taken from L1 to L2, L1 to L3, and L2 to L3. The average voltage is calculated and then the maximum voltage deviation is found by subtracting the voltage average from each voltage and choosing the largest deviation from the average. The voltage unbalance is found as follows:

$$V_u = \frac{V_d}{V_a} \times 100$$

where
V_u = voltage unbalance (in %)
V_d = voltage deviation (in V)
V_a = voltage average (in V)

For example, the voltage unbalance of a feeder system with readings of L1 to L2 = 442 V, L1 to L3 = 474 V, L2 to L3 = 455 V is calculated as follows:

1. Turn disconnect OFF and measure the voltage between incoming power lines. The readings are taken from L1 to L2, L1 to L3, and L2 to L3.

2. Calculate the average voltage, V_a:

$$V_a = \frac{442 + 474 + 455}{3}$$

$V_a = \mathbf{457\,V}$

3. Calculate the maximum voltage deviation, V_d:

$V_d = 442 - 457 = -15$
$V_d = 474 - 457 = 17$
$V_d = 455 - 457 = -2$

V_d is the largest deviation from the average. In this case, $V_d = 17$ V.

4. Calculate the voltage unbalance, V_u:

$$V_u = \frac{V_d}{V_a} \times 100$$

$$V_u = \frac{17}{457} \times 100$$

$V_u = 0.0372 \times 100$

$V_u = \mathbf{3.72\%}$

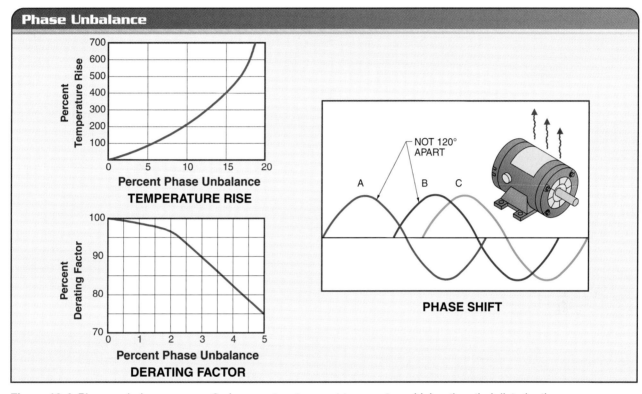

Figure 18-4. Phase unbalance causes 3-phase motors to run at temperatures higher than their listed ratings.

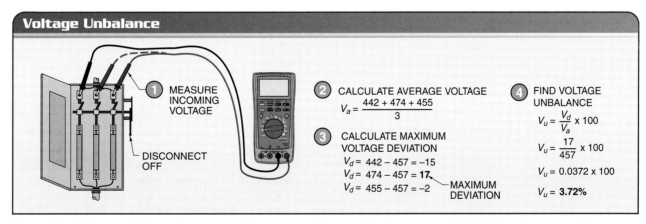

Figure 18-5. Voltage unbalance is measured by taking a voltage reading between each of the incoming power lines.

Definition

Single phasing is the operation of a motor designed to operate on three phases operating on only two phases because one phase is lost.

A *voltage surge* is any higher-than-normal voltage that temporarily exists on one or more of the power lines.

A troubleshooter can observe the blackening of one or two of the stator windings, which occurs when a motor has failed due to voltage unbalance. If a large unbalance exists on one winding, it is darkened the most. If the unbalance is divided over two windings, they are both darkened. The winding that had the largest unbalance is the darkest. Single phasing is the maximum condition of voltage unbalance.

Single Phasing. When one of the 3-phase lines leading to a 3-phase motor no longer delivers voltage to the motor, the motor will single phase. *Single phasing* is the operation of a motor designed to operate on three phases operating on only two phases because one phase is lost. Single phasing occurs when one phase opens on either the primary or secondary power distribution system. This can happen when one fuse blows, there is a mechanical failure within the switching equipment, or lightning takes out one of the lines. **See Figure 18-6.**

A 3-phase motor running on 2-phase will continue to run in many applications. Therefore, single phasing can go undetected on most systems for a long enough time to burn out a motor. When single phasing, the motor will draw all its current from two of the lines.

Measuring the voltage at the motor will not usually detect a single phasing condition. The open winding in the motor generates a voltage almost equal to the phase voltage that was lost. In this case, the open winding acts as the secondary of a transformer, and the two windings connected to power act as the primary.

Single phasing can be reduced by using the proper-size dual-element fuse and by using the correct heater sizes. In motor circuits, or other types of circuits in which a single-phasing condition cannot be allowed to exist for even a short period, an electronic phase-loss monitor is used to detect phase loss. When a phase loss is detected, the monitor activates a set of contacts to drop out the starter coil.

A troubleshooter can observe the blackening of one of the 3-phase windings, which occurs when a motor has failed due to single phasing. The coil that experienced the voltage loss will indicate obvious and fast damage, which includes the blowing out of the insulation on the one winding.

Single phasing is distinguished from voltage unbalance by the severity of the damage. Voltage unbalance causes less blackening (but usually over more coils) and little or no distortion. Single phasing causes severe burning and distortion to one phase coil.

Voltage Surges. A *voltage surge* is any higher-than-normal voltage that temporarily exists on one or more of the power lines. Lightning is a major cause of large surge voltages. A lightning surge on the power lines comes from a direct hit or induced voltage. The lightning energy moves in both directions on the power lines, much like a rapidly moving wave. This traveling surge causes a large voltage rise in an extremely short period. The large voltage is impressed on the first few turns of the motor windings, destroying the insulation and burning out the motor.

A troubleshooter can observe the burning and opening of the first few turns of the windings, which occur when a motor has failed due to a voltage surge. **See Figure 18-7.** The rest of the windings appear normal, with little or no damage.

Lighting arresters with the proper voltage rating and connection to an excellent ground assure maximum protection. Surge protectors are also available. These are placed on the equipment or throughout the distribution system.

Voltage surges can also occur from normal switching of higher-rated power circuits. These are of much less magnitude than lightning strikes and usually do not cause any problems in motors. A surge protector should be used on circuits with computer equipment to protect sensitive electronic components.

Tech Fact

Voltage surges can reach thousands of volts. They are often caused by lightning or when large, high-current loads are switched ON or OFF.

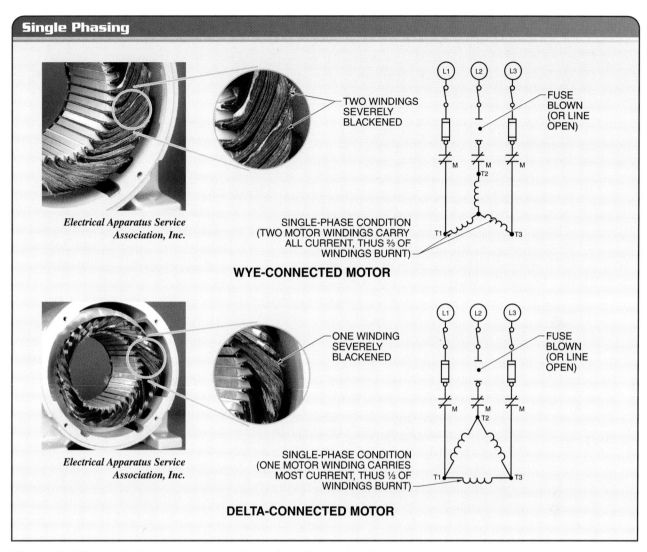

Figure 18-6. Single phasing causes overloading in the coils carrying the current.

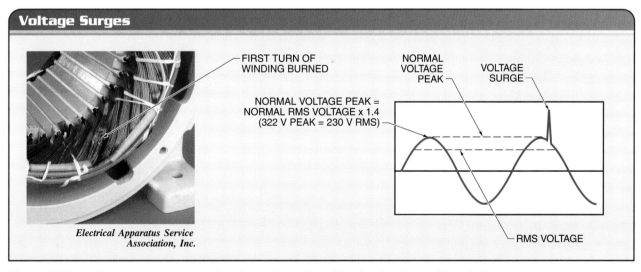

Figure 18-7. A voltage surge can cause burning and opening of the first few turns of the windings.

Definition

*An **overload** is the application of too much load to a motor.*

***Overcycling** is the process of turning a motor ON and OFF more often than the motor design allows.*

Improper Application

Many motor failures occur because of application failures where the application does not match the motor. Common causes of application failures include overloads, overcycling, and exposure to moisture.

Motor Overloading. An *overload* is the application of too much load to a motor. Motors attempt to drive the connected load when the power is ON. All motors have a limit to the load they can drive. A larger load requires more power and more current than a smaller load. An overload draws excess current that overheats a motor.

Motor overloads should not harm a properly protected motor. Any overload that is present longer than the built-in time delay of the protection device will be detected and removed. Properly sized heaters in the motor starter assure that an overload is removed before any damage is done.

An overload can occur when overloads are improperly sized or applied. A troubleshooter can observe the even blackening of all motor windings, which occurs when a motor has failed due to overloading. **See Figure 18-8.** The even blackening is caused by the slow destruction of the motor over a long period of time. There is no obvious damage to the insulation, and there are no isolated areas of damage.

Current readings are taken at the motor to identify an overload problem. **See Figure 18-9.** If the motor is drawing rated current, the motor is working to its maximum. If the motor is drawing more than rated current, the motor is overloaded. If overloads become a problem, the motor size may be increased or the load on the motor decreased.

Motor Overcycling. *Overcycling* is the process of turning a motor ON and OFF more often than the motor design allows. **See Figure 18-10.** Motor starting current is usually five to six times the full-load running current of the motor. Overcycling occurs when a motor is at its maximum operating temperature and still cycles ON and OFF. This will further increase the temperature of the motor, destroying the motor insulation. Totally enclosed motors can better withstand overcycling than open motors, because they can withstand heat longer.

Several things can be done when overcycling a motor is necessary. First, a motor with more overtemperature protection can be used. This can be a motor with an allowable 50°C temperature rise instead of the standard 40°C temperature rise. It can also be a motor with a higher service factor, such as 1.25 instead of 1.00. The second option is to provide additional cooling by blowing air over the motor.

Moisture. Moisture causes metal parts to rust and motor coil insulation to lose some of its insulating properties. A motor cools when it is turned OFF. This causes air (with its moisture) to be sucked into the motor. Motors that operate every day will heat enough to remove any moisture. Moisture is usually a problem for a motor that is seldom operated, or is shut down for a period of time.

Any motor that is not operated on a regular basis should contain a heating element to keep the motor dry. If adding a heating element is not practical, a maintenance schedule calling for short operation of motors that are seldom used should be developed. This schedule should also consider new motor installations, since in some plants, motors may be installed some time in advance of the plant startup.

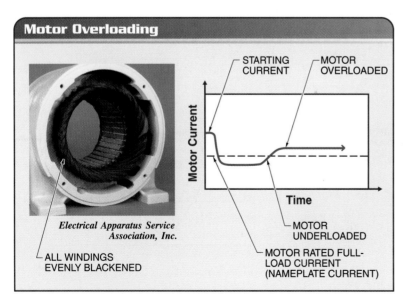

Figure 18-8. Overloading causes even blackening of all motor windings.

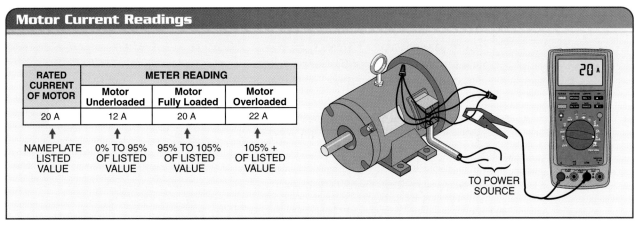

Figure 18-9. Current readings are taken at the motor to identify an overload problem.

Improper Maintenance

Motor failures can occur because of improper maintenance. Common causes of motor failure from improper maintenance include improper belt tension and misalignment and vibration.

Belt Tension. Belt drives provide a quiet, compact, and durable form of power transmission and are widely used in industrial applications. A belt must be tight enough not to slip, but not so tight as to overload the motor bearings.

Belt tension is usually checked by placing a straightedge from sheave to sheave and measuring the amount of deflection at the midpoint, or by using a tension tester. Belt deflection should equal 1/64″ per inch of span. For example, if the span between the center of the drive pulley and the center of the driven pulley is 16″, the belt deflection is 1/4″ (16 × 1/64 = 1/4). If the belt tension requires adjustment, it is usually accomplished by moving the drive component away from or closer to the driven component. This reduces or increases deflection. **See Figure 18-11.**

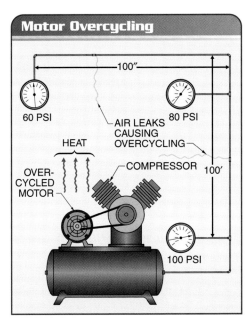

Figure 18-10. Overcycling increases the temperature of the motor, destroying the motor insulation.

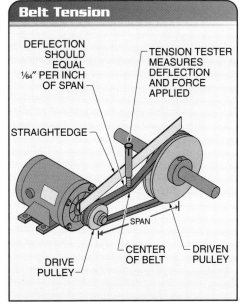

Figure 18-11. A belt must be tight enough not to slip, but not so tight as to overload the motor bearings.

Misalignment and Vibration. Misalignment of the motor and driven load is a major cause of motor failure. If the motor and driven load are misaligned, premature failure of the motor bearings, load, or both may occur. Equipment shafts should be properly aligned on all new installations and checked during periodic maintenance inspections. Misalignment is usually corrected by placing shims under the feet of the motor or driven equipment.

All motors produce vibration as they rotate. This vibration can loosen mechanical and electrical connections. Loose mechanical connections generally cause noise and can be easily detected. Loose electrical connections do not cause noise, but do cause a voltage drop to the motor and excess heat. Always check mechanical and electrical connections when troubleshooting a motor.

Motor Damage

Motor damage is any damage that occurs to a properly manufactured motor. The damage may occur before or during installation and during operation. A sound maintenance schedule and proper operation of a motor minimize the occurrence of motor damage.

As with any machine, a motor can fail due to a motor defect or motor damage. A motor defect is an imperfection created during the manufacture of the motor that impairs its use. If it impairs initial motor operation, the defect is usually caught by the manufacturer. If the defect manifests itself after the motor has been in operation for some time, a troubleshooter determines that the problem is a defect in the motor. Motors with defects should be replaced, and the manufacturer should be notified.

A troubleshooter can observe the effect when a motor has failed due to a defect, which is usually confined to a small area of the motor. Typical defects that may occur in a motor include windings grounded in the slot, windings grounded at the edge of the slot, windings shorted phase-to-phase, and shorted connections. **See Figure 18-12.**

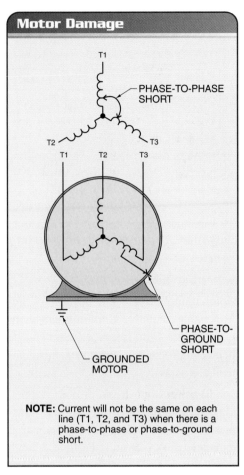

Figure 18-12. A motor can be damaged from a phase-to-ground short.

AC MOTORS

Troubleshooting is the systematic elimination of the various parts of a system, circuit, or process to locate a malfunctioning part. The first step of any troubleshooting procedure is to identify the symptoms of motor failure. The motor may have failed catastrophically and burned up, or it may be running hot. The motor may not start, or it may be vibrating severely.

Troubleshooting 3-Phase Motors

Three-phase motors have fewer components that may malfunction than other motor types. Therefore, 3-phase motors usually operate for many years without any problems. If a 3-phase motor is the problem, the motor is serviced or replaced. Servicing often requires that the motor be sent to a motor repair shop for rewinding. If the motor is less than 1 HP and more than five years old, it is normally replaced. If the motor is more than 1 HP, but less than 5 HP, it may be serviced or replaced. If the motor is more than 5 HP, it is usually serviced.

The extent of troubleshooting performed on a 3-phase motor depends on the motor's application. If the motor is used in an application that is critical to the operation or production, testing is usually limited to checking the voltage at the motor. If the voltage is present and correct, the motor is assumed to be the problem. Unless it is very large, the motor is usually replaced at this time so production can be resumed. If time is not a critical factor, further tests can be performed in order to determine the exact problem. **See Figure 18-13.** To troubleshoot a 3-phase motor, apply the following procedure:

1. Using a voltmeter, measure the voltage at the motor terminals. If the voltage is present and at the correct level on all three phases, the motor must be checked. If the voltage is not present on all three phases, the incoming power supply must be checked.

2. If voltage is present but the motor is not operating, turn the handle of the safety switch or combination starter OFF. Lock out and tag the starting mechanism per company policy.

3. Disconnect the motor from the load.

4. After the load is disconnected, turn power ON to try restarting the motor. If the motor starts, check the load.

5. If the motor does not start, turn it OFF and lock out the power.

6. With an ohmmeter, check for open or shorted windings. Take a resistance reading of the T1-T4 coil. Since the coil winding is made of wire only, the resistance is low. However, there is resistance on a good coil winding. This coil must have a resistance reading. If the reading is zero, the coil is shorted. If the reading is infinity, the coil is opened. The larger the motor, the smaller the resistance reading. After the resistance of one coil has been found, the basic electrical laws of series and parallel circuits are applied. When measuring the resistance of two coils in series, the total resistance is twice the resistance of one coil. When measuring the resistance of two coils in parallel, the total resistance is one half the resistance of one coil.

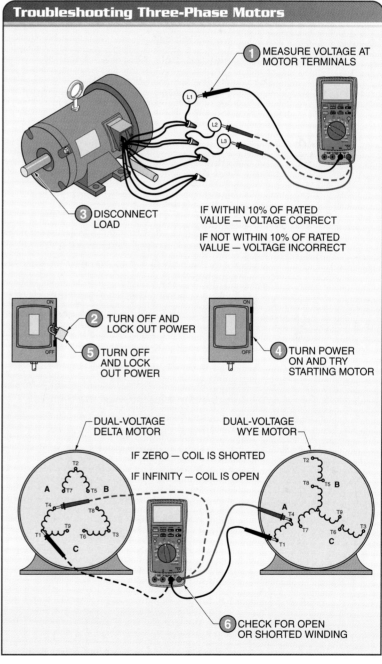

Figure 18-13. The first step in troubleshooting a 3-phase motor is checking the voltage at the motor.

Troubleshooting Single-Phase Motors

Most problems with single-phase motors involve the centrifugal switch, thermal switch, or capacitor. If the problem is in the centrifugal switch, thermal switch, or capacitor, the motor is usually serviced and repaired. However, if the motor is more than 10 years old and less than 1 HP, the motor is usually replaced. If the motor is less than ⅛ HP, it is almost always replaced.

Shaded-Pole Motors. A shaded-pole motor has very low starting torque. It is generally limited to approximately 1/20 HP and used for applications such as small fans and timing devices. Shaded-pole motors that fail are usually replaced. However, the reason for the motor failure should be found, if possible. If the motor failed due to a jammed load, etc., replacing the motor will not solve the problem. See Figure 18-14. To troubleshoot a shaded-pole motor, apply the following procedure:

1. Turn power to the motor OFF. Visually inspect the motor. Replace the motor if it is burned, the shaft is jammed, or if there is any sign of damage.
2. The only electric circuit that can be tested without taking the motor apart is the stator winding. With an ohmmeter, measure the resistance of the stator winding. Set the ohmmeter to the lowest scale for taking the reading. If the ohmmeter indicates an infinity reading, the winding is open. Replace the motor. If the ohmmeter indicates a zero reading, the winding is shorted. Replace the motor. If the ohmmeter indicates a low-resistance reading, the winding may still be good. Check the winding with a megohmmeter before replacing.

Tech Fact

When reversing a dual-voltage motor, care must be taken to ensure that the wires are attached to the correct voltages to ensure proper performance of the motor.

Split-Phase Motors. A split-phase motor has a starting and running winding. The starting winding is automatically removed by a centrifugal switch as the motor accelerates. Some split-phase motors also include a thermal switch that automatically turns the motor OFF when it overheats. Thermal switches may have a manual reset or automatic reset. Caution should be taken with any motor that has an automatic reset, as the motor can automatically restart at any time. See Figure 18-15. To troubleshoot a split-phase motor, apply the following procedure:

1. Turn power to the motor OFF. Visually inspect the motor. Replace the motor if it is burned, the shaft is jammed, or if there is any sign of damage.
2. Check to determine if the motor is controlled by a thermal switch. If the thermal switch is manual, reset the thermal switch and turn motor ON.

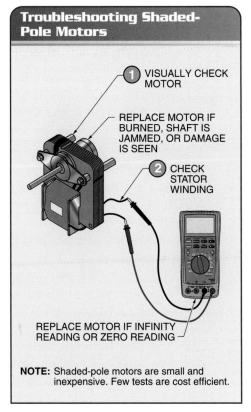

Troubleshooting Shaded-Pole Motors

① VISUALLY CHECK MOTOR

REPLACE MOTOR IF BURNED, SHAFT IS JAMMED, OR DAMAGE IS SEEN

② CHECK STATOR WINDING

REPLACE MOTOR IF INFINITY READING OR ZERO READING

NOTE: Shaded-pole motors are small and inexpensive. Few tests are cost efficient.

Figure 18-14. Shaded-pole motors that fail are usually replaced. However, the reason for the motor failure should be found, if possible.

3. If the motor does not start, use a voltmeter to check for voltage at the motor terminals. The voltage should be within 10% of the motor's listed voltage. If the voltage is not correct, troubleshoot the circuit leading to the motor. If the voltage is correct, turn power to motor OFF so the motor can be tested.

4. Turn the handle of the safety switch or combination starter OFF. Lock out and tag the starting mechanism per company policy.

5. With power OFF, connect the ohmmeter to the same motor terminals the incoming power leads were disconnected from. The ohmmeter will read the resistance of the starting and running windings. Since the windings are in parallel, their combined resistance is less than the resistance of either winding alone. If the meter reads zero, a short is present. If the meter reads infinity, an open circuit is present and the motor should be replaced.

6. Visually inspect the centrifugal switch for signs of burning or broken springs. If any obvious signs of problems are present, service or replace the switch. If not, check the switch using an ohmmeter. Manually operate the centrifugal switch. The endbell on the switch side may have to be removed. If the motor is good, the resistance on the ohmmeter will decrease. If the resistance does not change, a problem exists. Continue checking to determine the problem.

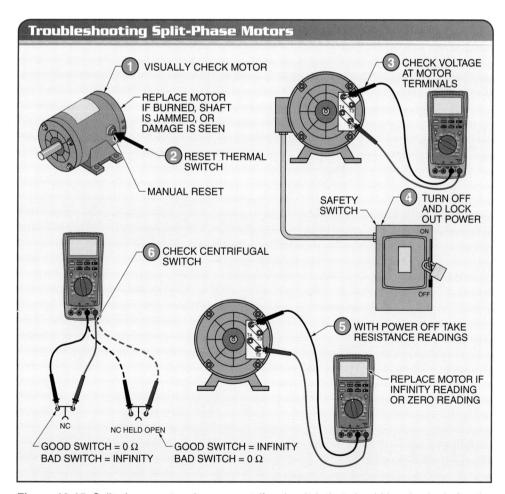

Figure 18-15. Split-phase motors have a centrifugal switch that should be checked after the coils are checked.

Capacitor Motors. A capacitor motor is a split-phase motor with the addition of one or two capacitors. Capacitors give the motor more starting and/or running torque. Troubleshooting capacitor motors is similar to troubleshooting split-phase motors. The only additional device to be considered is the capacitor. **See Figure 18-16.**

Capacitors have a limited life and are often the problem in capacitor motors. Capacitors may have a short circuit or an open circuit or may deteriorate to the point that they must be replaced. Deterioration can also change the value of a capacitor, which can cause additional problems. When a capacitor short-circuits, the winding in the motor may burn out. When a capacitor deteriorates or opens, the motor has poor starting torque. Poor starting torque may prevent the motor from starting, which will usually trip the overloads.

All capacitors are made with two conducting surfaces separated by dielectric material. Dielectric material is a medium in which an electric field is maintained with little or no outside energy supply. It is the type of material used to insulate conducting surfaces of a capacitor. Capacitor insulation is either oil or electrolytic. Oil capacitors are filled with oil and sealed in a metal container. The oil serves as the dielectric material.

More motors use electrolytic capacitors than oil capacitors. Electrolytic capacitors are formed by winding two sheets of aluminum foil separated by pieces of thin paper impregnated with an electrolyte. An electrolyte is a conducting medium in which the current flow occurs by ion migration. The electrolyte is used as the dielectric material. The aluminum foil and electrolyte are encased in a cardboard or aluminum cover. A vent hole is provided to prevent a possible explosion in the event the capacitor is shorted or overheated.

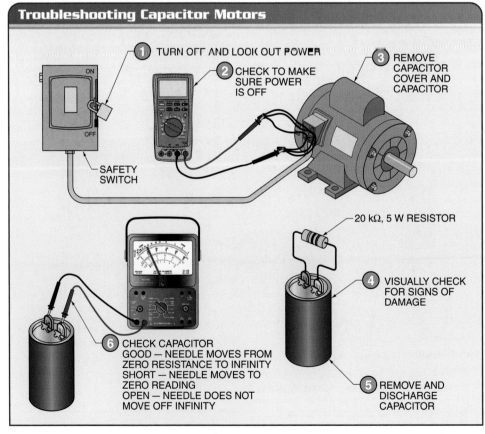

Figure 18-16. Capacitor motors are similar to split-phase motors with the addition of a capacitor.

AC capacitors are used with capacitor motors. Capacitors that are designed to be connected to AC have no polarity. To troubleshoot a capacitor motor, apply the following procedure:

1. Turn the handle of the safety switch or combination starter OFF. Lock out and tag the starting mechanism per company policy.
2. Using a voltmeter, measure the voltage at the motor terminals to make sure the power is OFF.
3. Capacitors are on the outside frame of the motor. Remove the cover of the capacitor. Caution: A good capacitor will hold a charge, even when power is removed.
4. Visually check the capacitor for leakage, cracks, or bulges. Replace the capacitor if damage is present.
5. Remove the capacitor from the circuit and discharge it. To safely discharge a capacitor, place a 20 kΩ, 5 W resistor across the terminals for five seconds.
6. After the capacitor is discharged, connect the ohmmeter leads to the capacitor terminals. The ohmmeter will indicate the general condition of the capacitor. A capacitor is either good, weak, shorted, or open.

- Good capacitor. The needle will swing to zero resistance and slowly move across the scale to infinity. When the needle reaches the halfway point, remove one of the leads and wait 30 seconds. When the lead is reconnected, the needle should swing back to the halfway point and continue to infinity. This shows the capacitor can hold a charge and is not the problem.
- Weak capacitor. When the lead is reconnected and the needle swings back to zero resistance, the capacitor cannot hold a charge and must be replaced.
- Shorted capacitor. The needle will swing to zero and not move. The capacitor is bad and must be replaced.
- Open capacitor. The needle will not move from infinity. The capacitor is bad and must be replaced.

DC MOTORS

Direct current motors are often used in applications that require very high torque. To produce the high torque, the outside power supply is connected to both the armature and field. A commutator and brushes are used to supply power to the rotating field. Because of their brushes, DC motors generally require more repair than motors that do not use brushes. The brushes should be checked first when troubleshooting DC motors.

Troubleshooting Brushes

Brushes wear faster than any other component of a DC motor. Brush wear may be mechanical or electrical. Mechanical wear is wear on the contact surface of the brush as spring pressure forces the brush against the commutator. Carbon brushes running on a commutator form a thin, low-friction film composed of copper oxide, water, and micrographite particles. The film reduces friction and extends brush life. Light loading on a motor sometimes inhibits the creation of the film and the resulting friction can quickly erode the brush.

Electrical wear resulting from contaminated film increases the resistance between the brush and the commutator surface. Contamination can increase the air gap, resulting in sparking. Sparking occurs as the current passes from the commutator to the brushes. Sparking causes heat, burning, and wear of electric parts.

Seating is the process of mating the surface of a brush to the surface of the commutator. Seating is important for good contact between the brushes and the commutator. Brushes can wear excessively and overheat if they do not fit properly against the commutator. If brushes overheat, they can cause burn marks or pitting on the commutator. The spring in the brush rigging can also become overheated and damaged.

On large machines that contain multiple sets of brushes to handle large currents, only one set of brushes should be changed at a time. This allows the new brushes to seat while the existing sets continue to run, allowing the new brushes to seat properly.

The condition of the brushes and their holders is extremely important for good motor operation. The brushes should be checked every time the motor is serviced. Most DC motors are designed so that the brushes and the commutator can be inspected without disassembling the motor. Some motors require disassembly for close inspection of the brushes and commutator. If the motor is still operable, observe the brushes as the motor is operating. The brushes should be riding on the commutator smoothly, with little or no sparking. There should be no brush noise, such as chattering. Brush sparking, chattering, or a rough commutator indicates service is required. **See Figure 18-17.** To troubleshoot brushes, apply the following procedure:

1. Turn the handle of the safety switch or combination starter OFF. Lock out and tag the starting mechanism per company policy.
2. Using a voltmeter, measure the voltage at the motor terminals to make sure the power is OFF.
3. Check the brush movement and tension. The brushes should move freely in the brush holder. The spring tension should be approximately the same on each brush. Remove the brushes.
4. Check the length of the brushes. Brushes should be replaced when they have worn down to about half of their original size. Some brushes have a small wear mark. A brush should be replaced when it wears down to the mark. If any brush does not have a wear mark, it should be replaced if it is less than half its original length. Brushes should be replaced as a set. They should never be replaced individually.
5. Check the position of the brush holder in relationship to the commutator. The brush holder should be $\frac{1}{16}''$ to $\frac{1}{8}''$ from the commutator. If the brush holder is closer, the commutator may be damaged. If the brush holder is too far away, the brush may break.
6. Check for proper brush pressure with a brush pressure tester. Too little pressure causes the brushes to arc excessively and groove the commutator. Too much pressure causes the brushes to chatter and wear faster than normal. When checking brush pressure, remove the endbell on the side in which the commutator is located. Pull back on the gauge, noting the pressure at which the piece of paper is free to move. Divide this reading by the contact area of the brush to get actual brush pressure in psi. Typical values of brush pressure are about 1.5 to 5 psi. If the original spring is in good condition, it should provide the proper pressure. If the spring is not in good condition, replace with one of the same type. Always replace brushes with brushes of the same composition. Check manufacturer's recommendations for type of brushes to use.

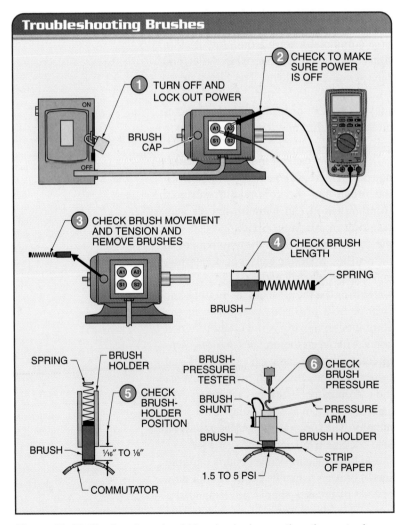

Figure 18-17. The brushes should be checked every time the motor is serviced and whenever a motor problem exists.

Setting Brush Neutral. Commutation is the short circuiting of the coil and any current induced in that coil flows through the brushes via the commutator segment. Since the brushes and the commutator segment offer no limit to the current, the current can be high and may destroy the brushes and the commutator. Brush neutral must be set by rotating the brush rigging to the correct position to aid in commutation of the armature. This sets the brush across the segments of the armature windings to be commutated (short-circuited) at the moment that they see the fewest lines of flux.

Setting brush neutral requires a centering VOM meter connected across adjacent brush holders. For a two-pole motor, the leads will be connected across the two brush riggings. For this procedure, the fields have to be flashed, or connected and then disconnected, to provide the expanding and then collapsing magnetic fields necessary for induction in the armature.

When the fields are connected to power, the meter deflects in one direction and then moves back to center. When the fields are disconnected from power, the meter deflects in the opposite direction and then moves back to center. The meter should be allowed to stop moving before disconnecting the fields. Through trial and error, the brush rigging needs to be rotated to the position where the least amount of deflection is noted when the fields are connected and disconnected. Each time the position is checked, the rigging must be tightened to make for an accurate test.

Troubleshooting Commutators

Brushes wear faster than the commutator. However, after the brushes have been changed once or twice, the commutator usually needs servicing. Any markings on the commutator, such as grooves or ruts, or discolorations other than a polished, brown color where the brushes ride, indicate a problem. **See Figure 18-18.** If the commutator is pitted or has burrs, it can be repaired with a stone or on a lathe. The brushes need to be reseated after a commutator has been repaired. To troubleshoot commutators, apply the following procedure:

1. Make a visual check of the commutator. The commutator should be smooth and concentric. A uniform dark copper oxide carbon film should be present on the surface of the commutator. This naturally occurring film acts like a lubricant by prolonging the life of the brushes and reducing wear on the commutator surface.

2. Check the mica insulation between the commutator segments. The mica insulation separates and insulates each commutator segment. It should be undercut (lowered below the surface) approximately 1/32″ to 1/16″, depending upon the size of the motor. The larger the motor, the deeper the undercut. Replace or service the commutator if the mica is raised.

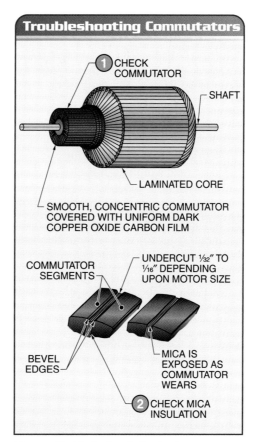

Figure 18-18. A visual check of a commutator includes inspecting the mica between the commutator segments.

Troubleshooting DC Motors

In addition to problems with brushes and commutators, other problems in DC motors include overloads, bearings, or a problem with the motor windings. Some problems can be repaired in the field, depending on the size of the motor and the extent of the problem. In many cases, the motor must be replaced and sent out for repair. In order to isolate the location of a problem, it must be determined if the problem is at the motor, upstream of the motor, or downstream from the motor. To troubleshoot DC motors, apply the following procedure:

1. Check the voltage at the motor terminals with a voltmeter or DMM. If the meter shows that the correct voltage is present, the problem is in the motor or load. If the meter shows that the correct voltage is not present, the problem is upstream of the motor, and the starter or drive, the control circuit, and the power system should be checked.

2. If the correct voltage is present at the motor, disconnect the motor from the load. If the motor starts without the load, the problem is with the load. If the motor starts without the load but vibrates excessively, the problem is with the coupling or bearings or with loose field windings. It is also possible for a motor to overheat or not be able to turn the load. The motor needs to be replaced or repaired in the field. If the motor does not start without the load, the problem is probably an open in one of the windings or an open between the brushes and the armature. All windings must be operating properly for the motor to function properly.

3. Check the field winding for an open or short. Before checking the windings, all power needs to be removed from the motor. All the coils should have continuity. Any reading of infinite resistance indicates an open circuit. All the coils should have some resistance. Any reading of 0 Ω indicates a short somewhere in the circuit.

Troubleshooting Windings. For a DC series motor, an ohmmeter is used to check the series winding at terminals S1 and S2. **See Figure 18-19.** A reading of infinite resistance between S1 and S2 indicates that there is an open in the series winding. A reading of infinite resistance between A1 and A2 indicates there is a problem with the armature winding. The armature winding circuit also includes the brushes and commutator.

For a DC shunt motor with a separately-excited shunt field, a reading can be taken directly at terminals F1 and F2. For a DC shunt motor with a self-excited shunt field, the shunt winding needs to be isolated from the armature winding by disconnecting the leads where the field and shunt windings are connected before checking at terminals F1 and F2. This is done by removing the jumper between A1 and F1 or the jumper between A2 and F2. A reading of infinite resistance between S1 and S2 indicates that there is an open in the series winding. A reading of infinite resistance between A1 and A2 indicates there is a problem with the armature winding.

For a DC compound motor, the shunt winding needs to be isolated from the armature winding by removing the series connection between the armature winding and the series winding. This is done by removing the jumper between A2 and S1. The shunt winding is tested by checking at terminals F1 and F2. The series winding is tested by checking at terminals S1 and S2. The armature winding is tested by checking at terminals A1 and A2. Any reading of infinite resistance indicates that there is an open in that winding.

> **Tech Fact**
>
> *Control transformers are used to step down the voltage for motor control circuits. Small units operating at 120/24 V are common in small commercial buildings. They are used to supply control voltages for HVAC systems. Control transformers should be checked when troubleshooting motors that do not start.*

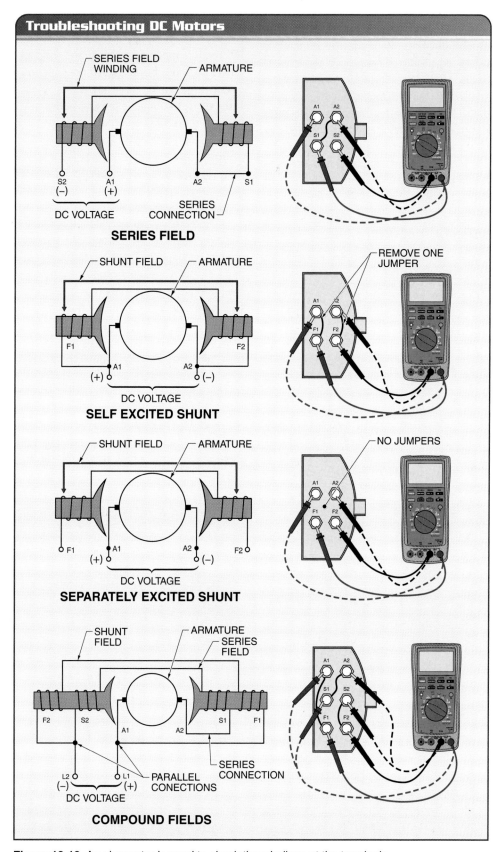

Figure 18-19. An ohmmeter is used to check the windings at the terminals.

Definition

*The **tie-down troubleshooting method** is a testing method in which one DMM probe is connected to either the L2 (neutral) or L1 (hot) side of a circuit and the other DMM probe is moved along a section of the circuit to be tested.*

MOTOR CONTROLS

Troubleshooting motor controls requires an organized, sequenced approach and requires the use of electrical test equipment, drawings and diagrams, and manufacturer specifications. Troubleshooting motor controls often involves troubleshooting contactors and motor starters, starting circuits, and circuit faults. The most common troubleshooting method is the tie-down troubleshooting method.

Tie-Down Troubleshooting Method

The *tie-down troubleshooting method* is a testing method in which one DMM probe is connected to either the L2 (neutral) or L1 (hot) side of a circuit and the other DMM probe is moved along a section of the circuit to be tested. The tie-down troubleshooting method allows a troubleshooter to work quickly on a familiar circuit that is small enough for the test probes to reach across the test points.

When using the tie-down troubleshooting method, one DMM test lead should be placed (tied down) on the L2 (neutral) conductor, and the other lead should be moved through the circuit starting with L1 (hot conductor). **See Figure 18-20.** If the correct voltage is not measured at L1 and L2, there is a power problem and the main power must be checked (for a possible fuse, circuit breaker, or main switch problem). If the proper voltage is present between L1 and L2, the DMM lead connected to L1 is moved along the circuit until the meter lead is directly at the load. If voltage is measured at the load but the load is not operating, the problem is the load on any circuit in which the load is connected directly to L2.

All loads are connected directly to L2 except when a magnetic motor starter overload contact is connected between the starter coil and L2. When a magnetic motor starter overload contact is used in a circuit, the DMM lead connected to L2 can be moved to the other side of the overload (side connected directly to the starter coil) to check if the overloads are open. **See Figure 18-21.** However, caution must be exercised when doing this because one DMM lead is still connected to L1 (hot conductor). This means the tip of the other DMM lead (the one being moved) can cause an electrical shock if touched and there is a complete path to ground through the troubleshooter's body.

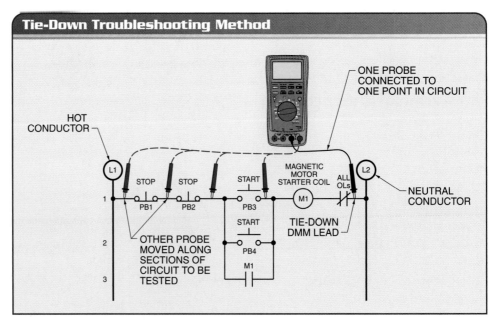

Figure 18-20. When using the tie-down troubleshooting method, one DMM test lead should be placed on the L2 (neutral) conductor, and the other lead should be moved through the circuit starting with L1 (hot conductor).

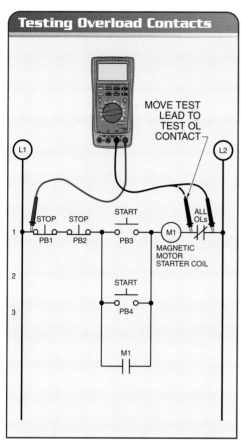

Figure 18-21. When a magnetic motor starter overload contact is used in a circuit, the DMM lead connected to L2 can be moved to the other side of the overload to check if the overloads are open.

Motor-driven pumps are often placed in banks. This allows one pump to be taken off-line for maintenance or troubleshooting.

Troubleshooting Control Circuits

Before troubleshooting an electrical circuit, an individual must understand the operation of the circuit, the sequence of events, timing or counting functions, and devices used to energize and de-energize the circuit. A line diagram shows the logic of an electrical circuit using single lines and symbols. Along with a line diagram, a DMM can be used to troubleshoot components in electrical circuits. Common electrical problems include open circuits and short circuits.

Troubleshooting Contactors and Motor Starters. Contactors or motor starters are the first devices checked when troubleshooting a circuit that does not work or has a problem. Contactors or motor starters are checked first because they are the point where the incoming power, load, and control circuit are connected. Basic voltage readings are taken at a contactor or motor starter to determine where the problem lies. The same basic procedure used to troubleshoot a motor starter works for contactors because a motor starter is a contactor with added overload protection.

The tightness of all terminals and busbar connections is checked when troubleshooting control devices. Loose connections in the power circuit of contactors and motor starters cause overheating. Overheating leads to equipment malfunction or failure. Loose connections in the control circuit cause control malfunctions. Loose connections of grounding terminals lead to electrical shock and cause electromagnetic-generated interference.

The power circuit and the control circuit are checked if the control circuit does not correctly operate a motor. The two circuits are dependent on each other, but are considered two separate circuits because they are normally at different voltage levels and always at different current levels. **See Figure 18-22.** To troubleshoot a motor starter, apply the following procedure:

1. Inspect the motor starter and overload assembly. Service or replace motor starters that show heat damage, arcing, or wear. Replace motor starters that show burning. Check the motor and driven load for signs of an overload or other problem.

2. Reset the overload relay if there is no visual indication of damage. Replace the overload relay if there is visual indication of damage.

3. Observe the motor starter for several minutes if the motor starts after resetting the overload relay. The overload relay reopens if an overload problem continues to exist.

4. Check the voltage into the starter if resetting the overload relay does not start the motor. Check circuit voltage ahead of the starter if the voltage reading is 0 V. The voltage is acceptable if the voltage reading is within 10% of the motor voltage rating. The voltage is unacceptable if the voltage reading is not within 10% of the motor voltage rating.

5. Energize the starter and check the starter contacts if the voltage into the starter is present and at the correct level. The starter contacts are good if the voltage reading is acceptable. Open the starter, turn the power OFF, and replace the contacts if there is no voltage reading.

6. Check the overload relay if voltage is coming out of the starter contacts. Turn the power OFF and replace the overload relay if the voltage reading is 0 V. The problem is downstream from the starter if the voltage reading is acceptable and the motor is not operating.

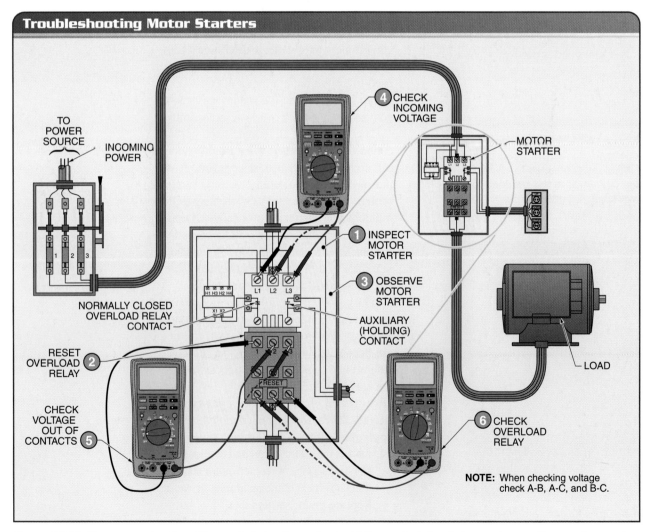

Figure 18-22. The power circuit and the control circuit are checked if the control circuit does not correctly operate a motor.

Application—Electric Motor Drive Troubleshooting

After gathering information, technicians must inspect the electric motor drive application. An inspection allows technicians to become familiar with the physical layout and operation of an application. Inspections typically yield clues as to the cause of an electric motor drive application problem. To inspect an electric motor drive application, apply the procedure:

1. Verify that all power disconnects are ON.
2. Access the fault history of the electric motor drive for information on possible causes and record the software version number.
3. Inspect the electric motor drive for physical damage and signs of overheating or fire.
4. Record the electric motor drive nameplate model number, serial number, input voltage, input current, output current, and horsepower rating.
5. Inspect the exterior of the motor and the area adjacent to the motor for debris to ensure proper ventilation to cool the motor.
6. Verify that the motor power rating corresponds to the electric motor drive power rating.
7. Verify that the motor is correctly aligned with the driven load.
8. Verify that the coupling or other connection method between the motor and driven load is not loose or broken.
9. Verify that the motor and the driven load are securely fastened in place.
10. Verify that an object is not preventing the motor or load from rotating.
11. Determine if any special equipment is required to work on the electric motor drive application.

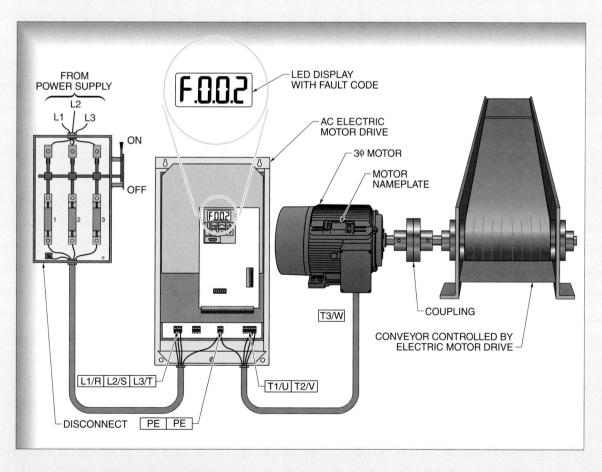

Troubleshooting Reduced-Voltage Starting Circuits. As with all motor circuits, the two main sections that must be considered when troubleshooting reduced-voltage starting circuits are the power circuit and control circuit. The power circuit connects the motor to the main power supply. In addition to including the main switching contacts and overload detection device (which can be heaters or solid-state), the power circuit also includes the power resistors (in the case of primary resistor starting) or autotransformer (in the case of autotransformer starting).

The control circuit determines when and how the motor starts. It is often powered by a step-down transformer from the power circuit. The control circuit includes the motor starter (mechanical or solid-state), overload contacts, and timing circuit. To troubleshoot the control circuit, the same troubleshooting procedure is used as when troubleshooting any other motor control circuit.

When troubleshooting the power circuit, voltage and current readings are taken. Current readings can be taken at the incoming power leads or the motor leads, since the current draw is the same at either point. The current reading during starting should be less than the current reading when starting without reduced-voltage starting. The amount of starting current varies by the starting method. With each starting method, there should be a reduction in starting current, as compared to a full-voltage start. **See Figure 18-23.** When troubleshooting a reduced-voltage power circuit, apply the following procedure:

1. Visually inspect the motor starter. Look for loose wires, damaged components, and signs of overheating (discoloration).
2. Measure the incoming voltage coming into the power circuit. The voltage should be within 10% of the voltage rating listed on the motor nameplate. If the voltage is not within 10%, the problem is upstream from the reduced-voltage power circuit.
3. Measure the voltage delivered to the motor from the reduced-voltage power circuit during starting and running. For primary resistor starting, the voltage during starting should be 10% to 50% less than the incoming measured voltage. The exact amount depends on the resistance added into the circuit. The resistance is set by using the resistor taps or adding resistors in series/parallel.

   For autotransformer starting, the voltage during starting should be 50%, 65%, or 80% less than the incoming measured voltage. The exact amount depends on which tap connection is used on the autotransformer. For part-winding starting, the voltage during starting should be equal to the incoming measured voltage.

   For wye-delta starting, the voltage during starting should equal the incoming measured voltage. For solid-state starting, the voltage during starting should be 15% to 50% less than the incoming measured voltage. The exact amount depends on the setting of the solid-state starting control switch.

   The voltage measured after the motor is started should equal the incoming voltage with each method of reduced-voltage starting. There is a problem in the power circuit or control circuit if the voltage out of the starting circuit is not correct.
4. Measure the motor current draw during starting and after the motor is running. In each method of reduced-voltage starting, the starting current should be less than the current that the motor draws when connected for full-voltage starting. The current should normally be about 40% to 80% less. After the motor is running, the current should equal the normal running current of the motor. This current value should be less than or equal to the current rating listed on the motor nameplate.

Tech Fact
AC motor starting and running torque is decreased by 20% if the motor is connected to a voltage source that is 10% less than the nameplate voltage.

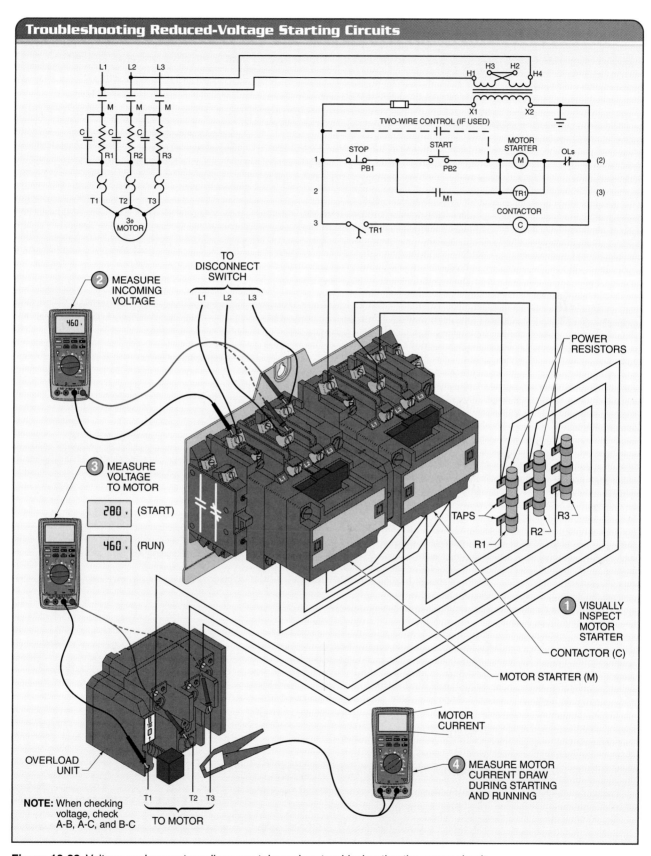

Figure 18-23. Voltage and current readings are taken when troubleshooting the power circuit.

Troubleshooting Thermal Switches. Many motors come equipped with a thermal switch to protect the motor from overheating. A thermal switch removes the motor windings from the circuit at a preset temperature. When the windings cool, the contacts close. When a thermal switch is not operating properly, a motor either does not operate or operates without thermal protection. To test a thermal switch, apply the following procedure:

1. Disconnect the motor from the power source. Remove the endbell of the motor that includes the thermal switch. Remove one of the leads connecting the thermal switch to the motor windings.

2. In a motor containing a two-terminal switch, use an ohmmeter to check the switch for continuity. **See Figure 18-24.** Set the ohmmeter on the lowest resistance scale and check across the switch contacts. If a high-resistance reading is obtained, the contacts are open and the switch is defective.

In a motor containing a three-terminal thermal switch, check for continuity across the switch contacts using an ohmmeter set on the lowest scale. If a high-resistance reading is obtained, the contacts are open and the switch is defective. Also, check for continuity across the heater element. If a high-resistance reading is obtained, the heater element is open and defective.

In a motor containing a four-terminal thermal switch, check for continuity across the switch contacts and the heater element by using an ohmmeter set on the lowest scale. If a high-resistance reading is obtained, there is an open in the contacts or the heater element. Replace the defective component.

Troubleshooting Circuit Faults

Many components in an electrical circuit can fail, such as switches, relays, and broken or damaged wires. Troubleshooting component failures requires knowledge of the proper operation of the component and a logical approach to finding the fault. Common tools include digital multimeters and continuity testers. Typical types of circuit faults are an open circuit, a short circuit, or a short to ground.

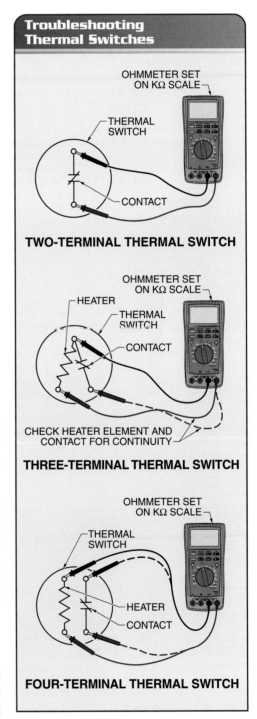

Figure 18-24. A thermal switch needs to be checked at all terminals.

Troubleshooting Open Circuits. An *open circuit* is an electrical circuit that has an incomplete path that prevents current flow. An open circuit represents a very-high-resistance path for current and is usually regarded as having infinite resistance. An open circuit in a series circuit de-energizes the entire circuit. Open circuits may be caused intentionally or unintentionally. An open circuit is caused intentionally when a switch is used to open a circuit. An open circuit may be caused unintentionally when the wiring between parts in a circuit is broken, when a component or device in a circuit malfunctions, or when a fuse blows.

A switch in the OFF position is an open circuit. Switches are tested by toggling the switch to check if the contacts open and close. A DMM set to measure voltage can be used to test a mechanical switch. A good switch indicates source voltage when open and 0 V when closed. A faulty switch indicates source voltage both when open and when closed. **See Figure 18-25.**

The proper operation of a switch must be known to determine when it is not operating properly. For example, a good solid-state switch indicates source voltage when open and a slight voltage drop when closed. This is normal due to the construction of the solid-state switch.

Definition

*An **open circuit** is an electrical circuit that has an incomplete path that prevents current flow.*

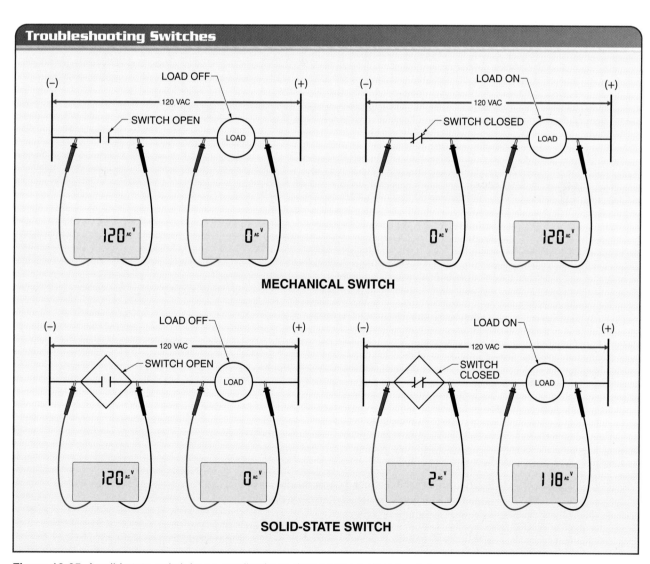

Figure 18-25. A solid-state switch has a small voltage drop across a closed switch.

Definition

A *short circuit* is a circuit in which current takes a shortcut around the normal path of current flow.

A *dead short* is a short circuit that opens the circuit as soon as the circuit is energized or when the section of the circuit containing the short is energized.

A switch may also be checked with a jumper wire. A jumper wire is placed in parallel around the switch and the circuit is energized. The jumper wire closes the circuit, energizing the load. **WARNING:** Jumper wires can cause equipment to start unexpectedly and must be removed from the circuit when no longer needed for testing.

An open circuit may occur unintentionally by a break in the wire of a circuit, a malfunctioning component, or a blown fuse. When a wire breaks, the path for current is interrupted and current flow stops. A broken wire in an individual line of a circuit de-energizes that line only. That branch of the circuit can be tested using a DMM set to measure voltage. For example, a DMM may be placed across a section of wire to determine if there is a break in that part of the wire. This test is often taken across wire connection points. The DMM indicates 0 V if the wire has no break. The DMM indicates source voltage if the wire is broken and there is no other open in the branch.

A faulty electrical component may also cause an unintentional open circuit. For example, a break in the conducting path of an electrical component, such as a burnt-out filament of a light bulb, also breaks the path for current and opens the circuit. In addition, when a fuse blows, the current flow in the circuit increases to a level that opens the conducting path inside the fuse.

Troubleshooting Short Circuits. A *short circuit* is a circuit in which current takes a shortcut around the normal path of current flow. In a short circuit, current leaves the normal current-carrying path and goes around the load and back to the power source or to ground. The low-resistance path can be due to failure of circuit components or failure in the wiring of the circuit. For example, if two pieces of wire accidentally contact each other, the wires produce a dead short across the circuit. A *dead short* is a short circuit that opens the circuit as soon as the circuit is energized or when the section of the circuit containing the short is energized. **See Figure 18-26.**

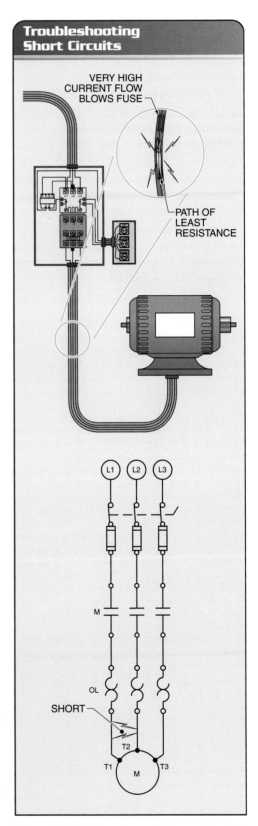

Figure 18-26. If two pieces of wire accidentally contact each other, the wires produce a dead short across the circuit.

A dead short reduces the resistance of the short-circuited part of a circuit to nearly 0 Ω. A dead short produces a surge of current in the circuit, resulting in an overload device such as a fuse being blown or a circuit breaker being tripped. In a circuit with a dead short, the fuse must be replaced or the circuit breaker reset. The circuit is inspected for the location of the short if the fuse blows or circuit breaker trips again when the circuit is energized. Signs of overheating, such as burn marks or discolored insulation, usually indicate the location of the short.

A short can be also found with a continuity tester. A continuity tester uses its own power (usually a battery) to power the circuit to determine if a short circuit exists between a wire and its housing. Once the short circuit is located, the shorted wire must be replaced. Ensure that the circuit is disconnected from its power source before testing using a continuity tester.

A DMM set to measure resistance can also be used to test for short circuits. A circuit is tested for a short circuit with all open contacts closed. In a good circuit, a DMM reads total circuit resistance when all open contacts are closed. In a circuit with a dead short, a DMM reads near 0 Ω. **See Figure 18-27.** To test each branch of a circuit, each branch is isolated by disconnecting a wire from the branch. The branch does not contain a short circuit if this produces no change in the DMM resistance reading. The branch is reconnected after the resistance reading is taken. This process is continued by isolating each branch in succession.

A branch contains a short if the DMM resistance reading jumps from 0 Ω to a high resistance when the branch is disconnected. This branch is inspected for signs of overheating and crossed, frayed, or loose wires. Further inspection is required to find the exact cause of the short. Large, complex circuits are tested one section at a time to determine which section contains the short. The individual branches of the section are then tested to find the exact location of the short. **WARNING:** The circuit must be de-energized when measuring resistance.

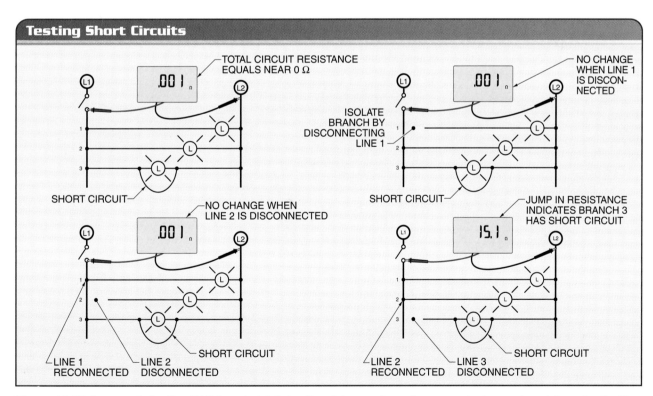

Figure 18-27. In a good circuit, a DMM reads total circuit resistance when all open contacts are closed. In a circuit with a dead short, a DMM reads near 0 Ω.

432 MOTORS

Definition

A grounded circuit is a circuit in which current leaves its normal path and travels to the frame of the motor.

Troubleshooting Grounded Circuits. A DC motor can be tested for a grounded circuit by using a test light or continuity tester. A *grounded circuit* is a circuit in which current leaves its normal path and travels to the frame of the motor. It is caused when insulation breaks down or is damaged and touches the metal frame of the motor.

A continuity tester is preferred for a quick check of a motor. A continuity tester can give results quickly when there is a problem. **See Figure 18-28.** To troubleshoot for a grounded circuit, one lead of the continuity tester is connected to the frame of the motor. The other lead is connected to one motor lead and then the other motor lead. A grounded circuit is present if the continuity tester beeps. The motor needs to be repaired or replaced.

MOTOR LEAD IDENTIFICATION

Three-phase induction motors are the most common motors used in industrial applications. Three-phase induction motors operate for many years with little or no required maintenance. It is not uncommon to find 3-phase induction motors that have been in operation for many years in certain applications. The length of time a motor is in operation may cause the markings of the external leads to become defaced. This may also happen to a new or rebuilt motor that has been in the maintenance shop for some time. To ensure proper operation, each motor lead must be re-marked before troubleshooting and reconnecting the motor to a power source.

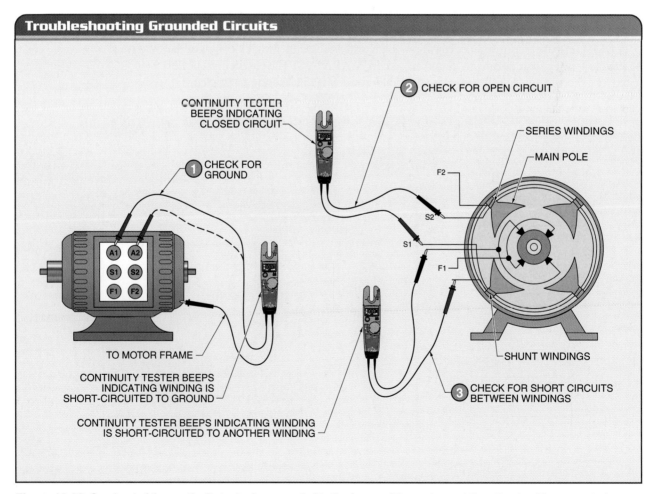

Figure 18-28. One lead of the continuity tester is connected to the frame of the motor and the other lead is connected to one motor lead and then the other motor lead.

The two most common 3-phase motors are the single-voltage, 3-phase, 3-lead motor and the dual-voltage, 3-phase, 9-lead motor. Both may be internally connected in a wye or delta configuration. The three leads of a single-voltage, 3-phase, 3-lead motor can be marked T1, T2, and T3 in any order. The motor can be connected to the rated voltage and allowed to run. Any two leads may be interchanged if the rotation is in the wrong direction. The industry standard is to interchange T1 and T3.

Determining Delta or Wye Connections

A standard dual-voltage motor has nine leads extending from it and may be internally connected as a wye or delta motor. The internal connections must be determined when re-marking the motor leads. A DMM is used to measure resistance or a continuity tester is used to determine whether a dual-voltage motor is internally connected in a wye or delta configuration.

A dual-voltage, wye-connected motor has four separate circuits, three circuits of two leads each (T1-T4, T2-T5, and T3-T6) and one circuit of three leads (T7-T8-T9). A dual-voltage, delta-connected motor has three separate circuits of three leads each (T1-T4-T9, T2-T5-T7, and T3-T6-T8). **See Figure 18-29.**

A DMM is used to determine the winding circuits (T1-T4, T2-T5, etc.) on an unmarked motor by connecting one meter lead to any motor lead and temporarily connecting the other meter lead to each remaining motor lead. *Note:* Ensure that the motor is disconnected from the power supply. A resistance reading other than infinity indicates a complete circuit.

A continuity tester may also be used to determine the winding circuits on an unmarked motor by connecting one test lead to any motor lead and temporarily connecting the other test lead to each remaining motor lead.

The continuity tester indicates a complete circuit by an audible beep. Each complete circuit can be marked by taping or pairing the leads together. All pairs of leads need to be checked with all the remaining motor leads to determine if the circuit is a two- or three-lead circuit. The motor is a wye-connected motor if three circuits of two leads and one circuit of three leads are found. The motor is a delta-connected motor if three circuits of three leads are found.

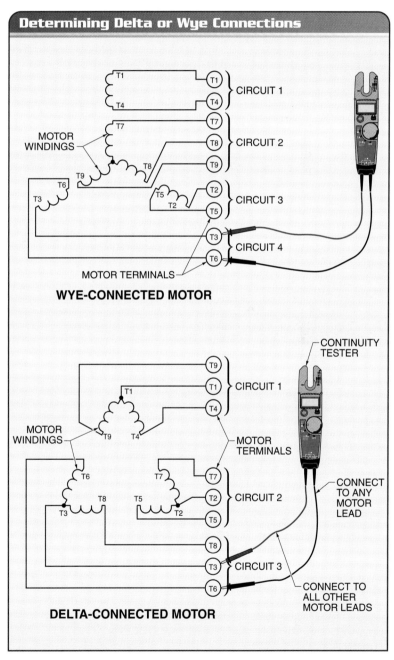

Figure 18-29. A dual-voltage, wye-connected motor has four separate circuits, three circuits of two leads each and one circuit of three leads. A dual-voltage, delta-connected motor has three separate circuits of three leads each.

Re-Marking Dual-Voltage, Wye-Connected Motors

Once a dual-voltage motor has been identified as a wye-connected motor, the individual leads can be re-marked. The power and load connectors must be removed before beginning. To re-mark a dual-voltage, wye-connected motor, apply the following procedure:

1. Determine the winding circuits using a DMM or continuity tester. **See Figure 18-30.**
2. Mark the leads of the one three-lead circuit T7, T8, and T9, in any order. Separate the other motor leads into pairs, making sure none of the wires touch.
3. Connect the motor to the correct supply voltage. Connect T7 to L1, T8 to L2, and T9 to L3. The correct supply voltage is the lowest voltage rating of the dual-voltage rating given on the motor nameplate. The low voltage is normally 220 V because the standard dual-voltage motor operates on 220/440 V. For any other voltage, all test voltages should be changed in proportion to the motor rating.
4. Turn ON the supply voltage and let the motor run. The motor should run with no apparent noise or problems. The starting voltage should be reduced through a reduced-voltage starter if the motor is too large to be started by connecting it directly to the supply voltage.
5. Measure the voltage across each of the three open circuits while the motor is running, using a DMM set on at least the 440 VAC scale. Care must be taken when measuring the high voltage of a running motor. Insulated test leads and proper PPE must be used. Connect only one test lead at a time. The voltage measured should be about 127 V or slightly less, and should be the same on all three circuits.

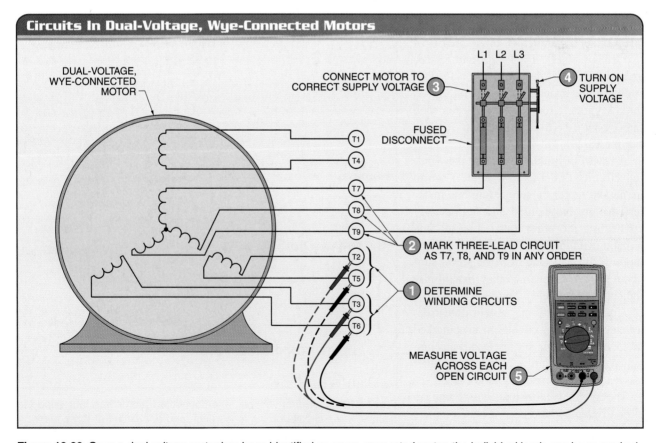

Figure 18-30. Once a dual-voltage motor has been identified as a wye-connected motor, the individual leads can be re-marked.

The voltage is read on all circuits even though the two-wire circuits are not connected to the power lines because the voltage applied to the three-lead circuit induces a voltage in the two-wire circuits. The wiring diagram for the dual-voltage, wye-connected motor can be drawn and the voltage readings added. **See Figure 18-31.**

One lead of any two-wire circuit is connected to T7 and the other lead of the circuit connected to one side of a DMM. The lead connected to T7 is temporarily marked as T4 and the lead connected to the DMM as T1. The other lead of the DMM is connected to T8 and then to T9. T1 and T4 can be marked permanently if the two voltages are the same and are approximately 335 V. Perform the same procedure on another two-wire circuit if the voltages are unequal. Mark the new terminals T1 and T4 if the new circuit gives the correct voltage (335 V). T1, T7, and T4 are found by this first test.

One lead of the two remaining unmarked two-wire circuits is connected to T8 and the other lead to one side of the DMM. The lead connected to T8 is temporarily marked as T5 and the lead connected to the DMM as T2. The other side of the DMM is connected to T7 and T9. Measurements and changes should be made until a position is found at which both voltages are the same and approximately 335 V. T2, T5, and T8 are found by this second test.

The third circuit is checked in the same way until a position is found at which both voltages are the same and approximately 335 V. T3, T6, and T9 are found by this third test.

After each motor lead is found and marked, turn OFF the motor and connect L1 to T1 and T7, L2 to T2 and T8, L3 to T3 and T9, and connect T4, T5, and T6 together. The last step is to start the motor and check the current on each power line with a clamp-on ammeter. The markings are correct and may be marked permanently if the current is approximately equal on all of the three power lines.

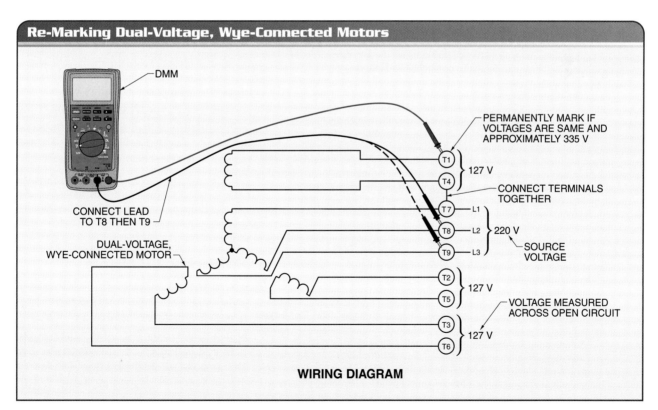

Figure 18-31. The voltage is read on all circuits even though the two-wire circuits are not connected to the power lines because the voltage applied to the three-lead circuit induces a voltage in the two-wire circuits.

Re-Marking Dual-Voltage, Delta-Connected Motors

Once a dual-voltage motor has been identified as a delta-connected motor, the individual leads can be re-marked. The power and load connectors must be removed before beginning. A dual-voltage, delta-connected motor has nine leads grouped into three separate circuits. Each circuit has three motor leads connected, which make the circuits T1-T4-T9, T2-T5-T7, and T3-T6-T8. To re-mark a dual-voltage, delta-connected motor, apply the following procedure:

1. Determine the winding circuits using a DMM or continuity tester. **See Figure 18-32.**
2. Measure the resistance of each circuit to find the center terminal. The resistance from the center terminal to the other two terminals is one-half the resistance between the other two terminals. Separate the three circuits and mark the center terminal for each circuit as T1, T2, and T3. Temporarily mark the two leads in the T1 group as T4 and T9, the two leads in the T2 group as T5 and T7, and the two leads in the T3 group as T6 and T8. Disconnect the DMM.
3. Connect the group marked T1, T4, and T9 to L1, L2, and L3 of a 220 V power supply. This should be the low-voltage rating on the nameplate of the motor. The other six leads should be left disconnected and must not touch because a voltage is induced in these leads even though these leads are not connected to power.
4. Turn the motor ON and let it run with the power applied to T1, T4, and T9.
5. Connect T4 (which is also connected to L2) to T7 and measure the voltage between T1 and T2. Set the DMM on at least a 460 VAC range. Use insulated test leads and connect one meter lead at a time. The lead markings for T4 and T9, and T7 and T5, are correct if the measured voltage is approximately 440 V. Interchange T5 with T7, or T4 with T9, if the measured voltage is approximately 380 V. Interchange both T5 with T7, and T4 with T9 if the new measured voltage is approximately 220 V. T4, T9, T7, and T5 may be permanently marked if the voltage is approximately 440 V.

To correctly identify T6 and T8, T6 and T8 are connected and the voltage measured from T1 and T3. The measured voltage should be approximately 440 V. Interchange leads T6 and T8 if the voltage does not equal 440 V. T6 and T8 may be permanently marked if the voltage is approximately 440 V.

Turn the motor OFF and reconnect the motor to a second set of motor leads. Connect L1 to T2, L2 to T5, and L3 to T7. The motor is restarted and the direction of rotation observed. The motor should rotate in the same direction as with the previous connection. Turn the motor OFF and reconnect the motor to the third set of motor leads (L1 to T3, L2 to T6, and L3 to T8) after the motor has run and the direction is determined.

Restart the motor and observe the direction of rotation. The motor should rotate in the same direction as the first two connections. Start over carefully, re-marking each lead if the motor does not rotate in the same direction for any set of leads.

Turn the motor OFF and reconnect the motor for the low-voltage connection. Connect L1 to T1-T6-T7, L2 to T2-T4-T8, and L3 to T3-T5-T9. The motor is restarted and current readings taken on L1, L2, and L3 with a clamp-on ammeter. The markings are correct if the motor current is approximately equal on each line.

Re-Marking DC Motors

The three basic types of DC motors are the series, shunt, and compound types. **See Figure 18-33.** All three types may have the same armature and frame but differ in the way the field coil and armature are connected. For DC motors, terminal markings A1 and A2 always indicate the armature leads. Terminal markings S1 and S2 always indicate the series field leads. Terminal markings F1 and F2 always indicate the shunt field leads.

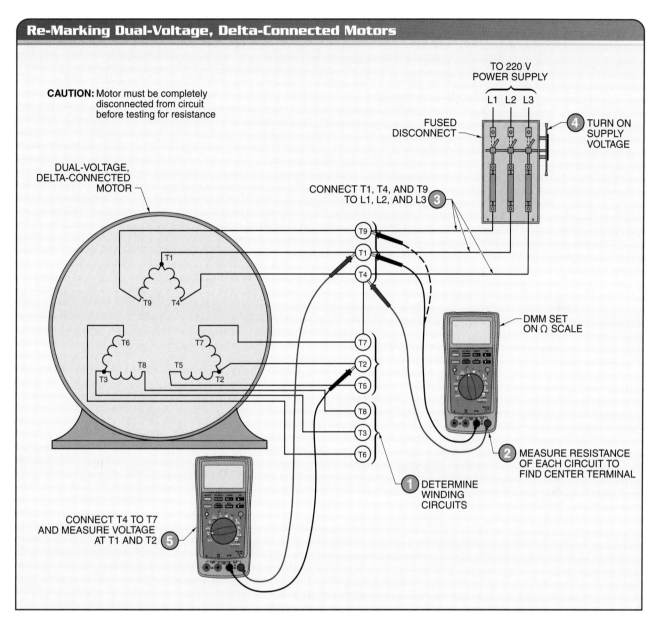

Figure 18-32. A dual-voltage, delta-connected motor has nine leads grouped into three separate circuits.

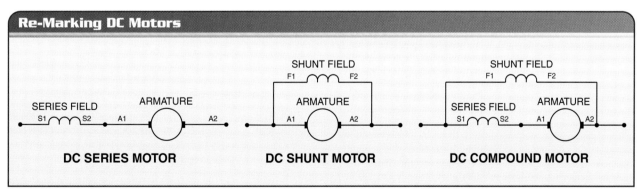

Figure 18-33. For DC motors, terminal markings A1 and A2 always indicate the armature leads. Terminal markings S1 and S2 always indicate the series field leads. Terminal markings F1 and F2 always indicate the shunt field leads.

DC motor terminals can be re-marked using a DMM by measuring the resistance of each pair of wires. A pair of wires must have a resistance reading or they are not a pair. The field reading can be compared to the armature reading because each DC motor must have an armature. The series field normally has a reading lower than that of the armature. The shunt field has a reading considerably larger than that of the armature. The armature can be easily identified by rotating the shaft of the motor when taking the readings. The armature varies the DMM reading as it makes and breaks different windings. One final check can be made by lifting one of the brushes or placing a piece of paper under the brush. The DMM moves to the infinity reading across the armature.

From this information, a motor is either a DC series or DC shunt motor if it has two pairs of leads (four wires) coming out. A coil is the series field if the reading of the coil is less than the armature coil resistance. A coil is the shunt field if the reading is considerably larger than the armature resistance.

TEST TOOLS

There are many tools that electricians use in their daily work. Voltmeters, ammeters, and ohmmeters are commonly used, either alone or combined into a digital multimeter (DMM). Two of the most common tools used in troubleshooting motors are phase sequence testers and megohmmeters (Meggers®).

Phase Sequence Testers

The phase sequence of power lines can be verified using a phase sequence tester. In addition, the rotation of a motor can be tested before the motor is connected to an electrical circuit or mechanical system.

Power Source Phase Sequence. To verify the phase sequence of the three 3-phase power lines, apply the following procedures:
1. Connect the three test leads from the phase sequence tester to the three power lines being tested. The test leads on a phase sequence tester are typically color-coded and include alligator clips. Connect phase A test lead to what should be power line phase A (L1/R), phase B test lead to B (L2/S), and phase C test lead to C (L3/T). **See Figure 18-34.**
2. Verify that all three 3-phase indicator lights are ON. If a light is not ON, use a voltmeter to test why the phase is not powered (for example, a fuse may be blown). When all phase indicator lights are ON, there is no open phase.
3. Check the phase sequence lamps. When the phases are in the correct order (the test leads are connected to the system A to A, B to B, and C to C), the phase tester "ABC" light will be ON. If the lines are not in the right order, the phase tester "BAC" light will be ON. Interchange the test leads until the "ABC" light is ON. Mark each phase line for proper identification.
4. Reconnect the phase sequence tester to test any additional parts of a system. Reconnect the tester because the phase sequence will change if the power lines are not correctly connected before and after a device.
5. Remove the phase sequence tester from the circuit.

Motor Rotation Phase Sequence. To test the direction of motor rotation of a 3-phase motor, apply the following procedures:
1. When the motor is not disconnected from the power lines, verify that the circuit power to the motor is OFF using a voltmeter. Some motor rotation test instruments have a button that is pressed to test that all power in the system is OFF. When a system is powered, a warning light will turn ON. **See Figure 18-35.**
2. Connect the test leads of the motor rotation tester to the motor input terminals T1 (U), T2 (V), and T3 (W).
3. Rotate the motor shaft clockwise by hand. When the clockwise light is lit, the motor will run in the clockwise direction when T1 is connected to L1, T2 to L2, and T3 to L3. When the counterclockwise light is lit, the motor will run in the counterclockwise direction when connected T1 to L1, T2 to L2, and T3 to L3.

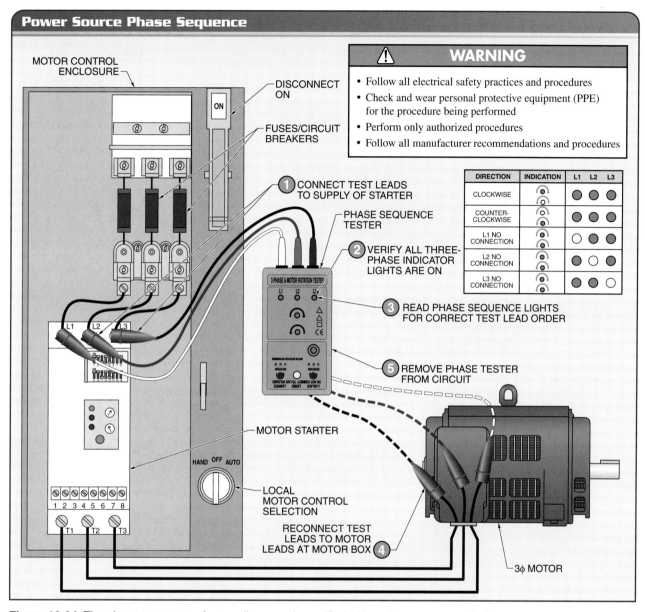

Figure 18-34. The phase sequence of power lines can be verified with a phase sequence tester.

4. When the motor must run in the opposite direction, interchange two power (or motor) lines. Interchanging 1 and 3 is the industrial standard.

5. Remove the motor rotation tester from the motor.

Megohmmeters

A megohmmeter is used to measure insulation deterioration on various wires by measuring high resistance values during high-voltage test conditions. Megohmmeter test voltages range from 50 V to 5000 V. A megohmmeter detects insulation failure or potential failure of insulation caused by excessive moisture, dirt, heat, cold, corrosive substances, vibration, and aging. Insulation in good working order has a high resistance. Insulation in poor working order has a low resistance. The ideal megohmmeter measurement is infinite resistance between a conductor or motor winding and ground. Infinite resistance is depicted by an infinity symbol (∞).

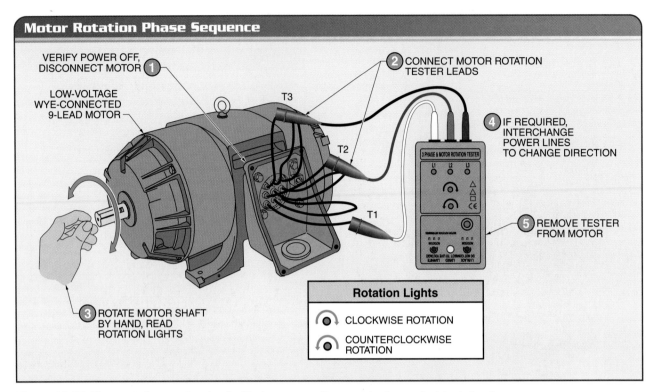

Figure 18-35. Motor rotation can be checked before connecting a motor to an electrical circuit.

Definition

*An **insulation spot test** is a short-term test that verifies the integrity of insulation on electrical devices.*

Megohmmeter measurements typically follow the general rule of thumb of 1 MΩ of resistance for every 1000 V of insulation rating. When using a megohmmeter, the test voltage is typically rounded up to the nearest 1000 V. For example, wire used in 480 VAC or 240 VAC distribution systems has a rating of 600 V. For testing purposes, consider the wire to have 1000 V insulation. The stator winding insulation for inverter duty motors has a rating of approximately 1500 V. For testing purposes, consider the stator winding to have 2000 V rated insulation.

Several megohmmeter readings must be taken over time, because the resistance of good insulation varies with time. Megohmmeter readings are typically taken when an electrical device such as a motor is installed and at regular intervals thereafter. An electrical device is in need of service when a megohmmeter measurement is below the minimum acceptable value.

Insulation Spot Testing. An *insulation spot test* is a short-term test that verifies the integrity of insulation on electrical devices. An insulation spot test is taken when a motor is placed in service. It should be repeated every 6 months thereafter. An insulation spot test should also be taken after a motor has had maintenance that could affect the windings or been rewound. **See Figure 18-36.**

During an insulation spot test, the test leads of the meter are connected across the conductor and insulation being tested. The test leads are also connected across any other conductor that may come in contact with the insulation being tested. The test voltage is applied for 60 sec to allow for the most accurate measurements.

Interpretation of the measured resistance values requires knowledge of previous resistance measurements, so recordkeeping is important. Megohmmeter manufacturers include charts for recording and plotting resistance measurements over time. To perform an insulation spot test using a megohmmeter, apply the following procedures:

Insulation Spot Testing

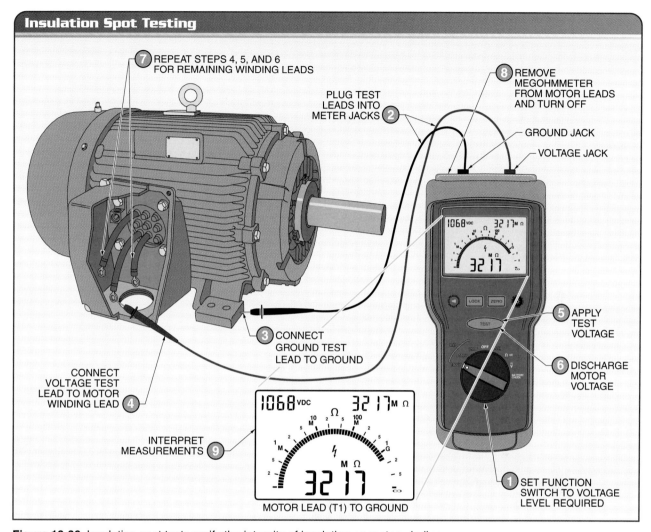

Figure 18-36. Insulation spot tests verify the integrity of insulation on motor windings.

1. Set the function switch of the megohmmeter to the proper test voltage level. The test voltage is typically set higher than the voltage rating of the insulation being tested to stress the insulation. The 1000 V setting is typically used for motors and conductors operating at 480 VAC. When a megohmmeter does not have a 1000 V setting, use the voltage setting closest to but not greater than 1000 V.
2. Plug the test leads of the megohmmeter into the proper meter jacks.
3. Connect the black test lead of the megohmmeter to a grounded surface.
4. Connect the red test lead of the megohmmeter to one of the motor winding leads or an individual conductor.
5. Apply the test voltage for 60 sec. Record the megohmmeter reading. Record the lowest reading on an insulation spot-test graph when all readings are above the minimum acceptable reading. The lowest reading is used because a motor or a set of feeder conductors is only as good as the weakest point.
6. Discharge the circuit being tested.
7. Repeat steps 4, 5, and 6 for the remaining motor winding leads or individual conductors.
8. Remove the megohmmeter from the motor leads and turn OFF the meter to prevent battery drain.
9. Interpret the measurements taken.

Megohmmeter readings must be interpreted. **See Figure 18-37.** A motor installed outdoors and tested two days in a row can have two different readings depending on the weather (foggy conditions one day would result in low MΩ and sunny conditions the next day would result in high MΩ). In general, megohmmeter readings are the most useful when taken semiannually over a period of years. A sudden drop in a resistance measurement of a motor, such as 100 MΩ to 2 MΩ over a six-month period, is an indication of a problem, even when the measurement is above the accepted value. A large difference between resistance measurements of motor leads (L1 = 20 MΩ, L2 = 21 MΩ, and L3 = 1 MΩ) also indicates a problem.

The cause of low resistance readings must be determined. The cause can be moisture, dirt, or damaged insulation. Typically, low resistance readings require the motor or conductors to be repaired or replaced. The repaired or replaced items must be tested with a megohmmeter before being placed into service.

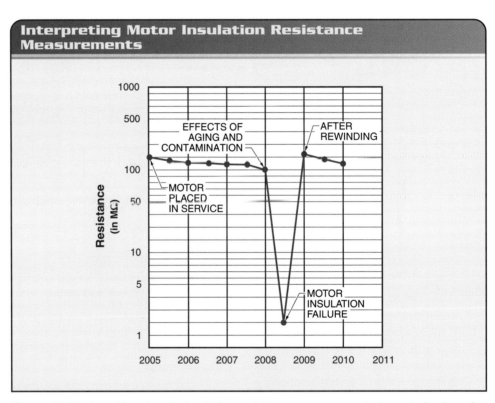

Figure 18-37. A sudden drop in insulation resistance measurements is an indication of a problem.

Application—Troubleshooting Guide

Troubleshooting guides for motors state a problem, its possible cause(s), and corrective action(s) that may be taken. These easy-to-reference guides, while general in nature, may be used to quickly determine potential problems and possible courses of action.

Troubleshooting Guide For Three-Phase Motors

Problem	Possible Cause	Corrective Action
Motor will not start	Wrong motor connections	Most 3ϕ motors are dual voltage. Check for proper motor connections.
	Blown fuse or open CB	Test the OCPD. If voltage is present at the input but not the output of the OCPD, the fuse is blown or the CB is open. Check the rating of the OCPD. It should be at least 125% of the motor's FLC.
	Motor overload on starter tripped	Allow overloads to cool. Reset overloads. If reset overloads do not start the motor, test the starter.
	Low or no voltage applied to motor	Check the voltage at the motor terminals. The voltage must be present and within 10% of the motor nameplate voltage. If voltage is present at the motor but the motor is not operating, remove the motor from the load the motor is driving. Reapply power to the motor. If the motor runs, the problem is with the load. If the motor does not run, the problem is with the motor. Replace or service the motor.
	Open control circuit between incoming power and motor	Check for cleanliness, tightness, and breaks. Use a voltmeter to test the circuit starting with the incoming power and moving to the motor terminals. Voltage generally stops at the problem area.
Fuse, CB, or overloads retrip after service	Power not applied to all three lines	Measure voltage at each power line. Correct any power supply problems.
	Blown fuse or open CB	Test the OCPD. If voltage is present at the input but not the output of the OCPD, the fuse is blown or the CB is open. Check the rating of the OCPD. It should be at least 125% of the motor's FLC.
	Motor overload on starter tripped	Allow overloads to cool. Reset overloads. If reset overloads do not start the motor, test the starter.
	Low or no voltage applied to motor	Check the voltage at the motor terminals. The voltage must be present and within 10% of the motor nameplate voltage. If voltage is present at the motor but the motor is not operating, remove the motor from the load the motor is driving. Reapply power to the motor. If the motor runs, the problem is with the load. If the motor does not run, the problem is with the motor. Replace or service the motor.
	Open control circuit between incoming power and motor	Check for cleanliness, tightness, and breaks. Use a voltmeter to test the circuit starting with the incoming power and moving to the motor terminals. Voltage generally stops at the problem area.
	Motor shaft does not turn	Disconnect the motor from the load. If the motor shaft still does not turn, the bearings are frozen. Replace or service the motor.
Motor overheats	Motor is single phasing	Check each of the 3ϕ power lines for correct voltage.
	Improper ventilation	Clean all ventilation openings. Vacuum or blow dirt out of motor with low-pressure, dry, compressed air.
	Motor is overloaded	Check the load for binding. Check shaft straightness. Measure motor current under operating conditions. If the current is above the listed current rating, remove the motor. Remeasure the current under no-load conditions. If the current is excessive under load but not when unloaded, check the load. If the motor draws excessive current when disconnected, replace or service the motor.

Application—Troubleshooting Motors

Troubleshooting Guide for Direct Current Motors

Problem	Possible Cause	Corrective Action
Motor will not start	Blown fuse or open CB	Test the OCPD. If voltage is present at the input, but not the output of the OCPD, the fuse is blown or the CB is open. Check the rating of the OCPD. It should be at least 125% of the motor's FLC.
	Motor overload on starter tripped	Allow overloads to cool. Reset overloads. If reset overloads do not start the motor, test the starter.
	No brush contact	Check brushes. Replace, if worn.
	Open control circuit between incoming power and motor	Check for cleanliness, tightness, and breaks. Use a voltmeter to test the circuit starting with the incoming power and moving to the motor terminals. Voltage generally stops at the problem area.
Fuse, CB, or overloads retrip after service	Excessive load	If the motor is loaded to excess or is jammed, the circuit OCPD will open. Disconnect the load from the motor. If the motor now runs properly, check the load. If the motor does not run and the fuse or CB opens, the problem is with the motor or control circuit. Remove the motor from the control circuit and connect it directly to the power source. If the motor runs properly, the problem is in the control circuit. Check the control circuit. If the motor opens the fuse or CB again, the problem is in the motor. Replace or service the motor.
	Motor shaft does not turn	Disconnect the motor from the load. If the motor shaft still does not turn, the bearings are frozen. Replace or service the motor.
Brushes chip or break	Brush material is too weak or the wrong type for motor's duty rating	Replace with better grade or type of brush. Consult manufacturer if problem continues.
	Brush face is overheating and losing brush bonding material	Check for an overload on the motor. Reduce the load as required. Adjust brush holder arms.
	Brush holder is too far from commutator	Too much space between the brush holder and the surface of the commutator allows the brush end to chip or break. Set correct space between brush holder and commutator.
	Brush tension is incorrect	Adjust brush tension so the brush rides freely on the commutator.
Brushes spark	Worn brushes	Replace worn brushes. Service the motor if rapid brush wear, excessive sparking, chipping, breaking, or chattering is present.
	Commutator is concentric	Grind commutator and undercut mica. Replace commutator if necessary.
	Excessive vibration	Balance armature. Check brushes. They should be riding freely.
Rapid brush wear	Wrong brush material, type, or grade	Replace with brushes recommended by manufacturer.
	Incorrect brush tension	Adjust brush tension so the brush rides freely on the commutator.
Motor overheats	Improper ventilation	Clean all ventilation openings. Vacuum or blow dirt out of motor with low-pressure, dry, compressed air.
	Motor is overloaded	Check the load for binding. Check shaft straightness. Measure motor current under operating conditions. If the current is above the listed current rating, remove the motor. Remeasure the current under no-load conditions. If the current is excessive under load but not when unloaded, check the load. If the motor draws excessive current when disconnected, replace or service the motor.

Troubleshooting Guide For Split-phase Motors . . .

Problem	Possible Cause	Corrective Action
Motor will not start	Thermal cutout switch is open	Reset the thermal switch. **Caution:** Resetting the thermal switch may automatically start the motor.
	Blown fuse or open CB	Test the OCPD. If voltage is present at the input but not the output of the OCPD, the fuse is blown or the CB is open. Check the rating of the OCPD. It should be at least 125% of the motor's FLC.
	Motor overload on starter tripped.	Allow overloads to cool. Reset overloads. If reset overloads do not start the motor, test the starter.
	Low or no voltage applied to motor	Check the voltage at the motor terminals. The voltage must be present and within 10% of the motor nameplate voltage. If voltage is present at the motor but the motor is not operating, remove the motor from the load the motor is driving. Reapply power to the motor. If the motor runs, the problem is with the load. If the motor does not run, the problem is with the motor. Replace or service the motor.
	Open control circuit between incoming power and motor	Check for cleanliness, tightness, and breaks. Use a voltmeter to test the circuit starting with the incoming power and moving to the motor terminals. Voltage generally stops at the problem area.
	Starting winding not receiving power	Check the centrifugal switch to make sure it connects to the starting winding when the motor is OFF.
Fuse, CB, or overloads retrip after service	Blown fuse or open CB	Test the OCPD. If voltage is present at the input but not the output of the OCPD, the fuse is blown or the CB is open. Check the rating of the OCPD. It should be at least 125% of the motor's FLC.
	Motor overload on starter tripped	Allow overloads to cool. Reset overloads. If reset overloads do not start the motor, test the starter.
	Low or no voltage applied to motor	Check the voltage at the motor terminals. The voltage must be present and within 10% of the motor nameplate voltage. If voltage is present at the motor but the motor is not operating, remove the motor from the load the motor is driving. Reapply power to the motor. If the motor runs, the problem is with the load. If the motor does not run, the problem is with the motor. Replace or service the motor.
	Open control circuit between incoming power and motor	Check for cleanliness, tightness, and breaks. Use a voltmeter to test the circuit starting with the incoming power and moving to the motor terminals. Voltage generally stops at the problem area.
	Motor shaft does not turn	Disconnect the motor from the load. If the motor shaft still does not turn, the bearings are frozen. Replace or service the motor.
Motor produces electric shock	Broken or disconnected ground strap	Connect or replace ground strap. Test for proper ground.
	Hot power lead at motor connecting terminals is touching motor frame	Disconnect the motor. Open the motor terminal box and check for poor connections, damaged insulation, or leads touching the frame. Service and test motor for ground.
	Motor winding shorted to frame	Remove, service, and test motor.
Motor overheats	Starting windings are not being removed from circuit as motor accelerates	When the motor is turned OFF, a distinct click should be heard as the centrifugal switch closes.
	Improper ventilation	Clean all ventilation openings. Vacuum or blow dirt out of motor with low-pressure, dry, compressed air.
	Motor is overloaded	Check the load for binding. Check shaft straightness. Measure motor current under operating conditions. If the current is above the listed current rating, remove the motor. Remeasure the current under no-load conditions. If the current is excessive under load but not when unloaded, check the load. If the motor draws excessive current when disconnected, replace or service the motor.

Application—Troubleshooting Motors

...Troubleshooting Guide For Split-phase Motors

Problem	Possible Cause	Corrective Action
Excessive noise	Dry or worn bearings	Dry or worn bearings cause noise. The bearings may be dry due to dirty oil, oil not reaching the shaft, or motor overheating. Oil the bearings as recommended. If noise remains, replace the bearings or the motor.
	Dirty bearings	Clean or replace bearings.
	Excessive end play	Check end play by trying to move the motor shaft in and out. Add end-play washers as required.
	Unbalanced motor or load	An unbalanced motor or load causes vibration, which causes noise. Realign the motor and load. Check for excessive end play or loose parts. If the shaft is bent, replace the rotor or motor.
	Dry or worn bearings	Dry or worn bearings cause noise. The bearings may be dry due to dirty oil, oil not reaching the shaft, or motor overheating. Oil the bearings as recommended. If noise remains, replace the bearings or the motor.
	Excessive grease	Ball bearings that have excessive grease may cause the bearings to overheat. Overheated bearings cause noise. Remove any excess grease.

Troubleshooting Guide for Shaded-Pole Motors

Problem	Possible Cause	Corrective Action
Motor will not start	Blown fuse or open CB	Test the OCPD. If voltage is present at the input, but not the output of the OCPD, the fuse is blown or the CB is open. Check the rating of the OCPD. It should be at least 125% of the motor's FLC.
	Motor overload on starter tripped	Allow overloads to cool. Reset overloads. If reset overloads do not start the motor, test the starter.
	Low or no voltage applied to motor	Check the voltage at the motor terminals. The voltage must be present and within 10% of the motor nameplate voltage. If voltage is present at the motor but the motor is not operating, remove the motor from the load the motor is driving. Reapply power to the motor. If the motor runs, the problem is with the load. If the motor does not run, the problem is with the motor. Replace or service the motor.
	Open control circuit between incoming power and motor	Check for cleanliness, tightness, and breaks. Use a voltmeter to test the circuit starting with the incoming power and moving to the motor terminals. Voltage generally stops at the problem area.
Fuse, CB, or overloads retrip after service	Excessive load	If the motor is loaded to excess or jammed, the circuit OCPD will open. Disconnect the load from the motor. If the motor now runs properly, check the load. If the motor does not run and the fuse or CB opens, the problem is with the motor or control circuit. Remove the motor from the control circuit and connect it directly to the power source. If the motor runs properly, the problem is in the control circuit. Check the control circuit. If the motor opens the fuse or CB again, the problem is in the motor. Replace or service the motor.
Excessive noise	Unbalanced motor or load	An unbalanced motor or load causes vibration, which causes noise. Realign the motor and load. Check for excessive end play or loose parts. If the shaft is bent, replace the rotor or motor.
	Dry or worn bearings	Dry or worn bearings cause noise. The bearings may be dry due to dirty oil, oil not reaching the shaft, or motor overheating. Oil the bearings as recommended. If noise remains, replace the bearings or the motor.
	Excessive grease	Ball bearings that have excessive grease may cause the bearings to overheat. Overheated bearings cause noise. Remove any excess grease.

Troubleshooting Guide for Contactors and Motor Starters

Problem	Possible Cause	Corrective Action
Humming noise	Magnetic pole faces misaligned	Realign. Replace magnet assembly if realignment is not possible.
	Too low voltage at coil	Measure voltage at coil. Check voltage rating of coil. Correct any voltage that is 10% less than coil rating.
	Pole face obstructed by foreign object, dirt, or rust	Remove any foreign object and clean as necessary. Never file pole faces.
Loud buzz noise	Shading coil broken	Replace coil assembly.
Controller fails to drop out	Voltage to coil not being removed	Measure voltage at coil. Trace voltage from coil to supply. Search for shorted switch or contact if voltage is present.
	Worn or rusted parts causing binding	Replace worn parts. Clean rusted parts.
	Contact poles sticking	Check for burning or sticky substance on contacts. Replace burned contacts. Clean dirty contacts.
	Mechanical interlock binding	Check to ensure interlocking mechanism is free to move when power is OFF. Replace faulty interlock.
Controller fails to pull in	No coil voltage	Measure voltage at coil terminals. Trace voltage loss from coil to supply voltage if voltage is not present.
	Too low voltage	Measure voltage at coil terminals. Correct voltage level if voltage is less than 10% of rated coil voltage. Check for a voltage drop as large loads are energized.
	Coil open	Measure voltage at coil. Remove coil if voltage is present and correct but coil does not pull in. Measure coil resistance for open circuit. Replace if open.
	Coil shorted	Shorted coil may show signs of burning. The fuse or breakers should trip if coil is shorted. Disconnect one side of coil and reset if tripped. Remove coil and check resistance for short if protection device does not trip. A shorted coil has zero or very low resistance. Replace shorted coil. Replace any coil that is burned.
	Mechanical obstruction	Remove any obstructions.
Contacts badly burned or welded	Too high inrush current	Measure inrush current. Check load for problem of higher-than-rated load current. Change to larger controller if load current is correct but excessive for controller.
	Too fast load cycling	Change to larger controller if load cycles ON and OFF repeatedly.
	Too large overcurrent protection device	Size overcurrent protection to load and controller.
	Short circuit	Check fuses or circuit breakers. Clear any short circuit.
	Insufficient contact pressure	Check to ensure contacts are making good connection.
Nuisance tripping	Incorrect overload size	Check size of overload against rated load current. Adjust size as permissible per NEC®.
	Lack of temperature compensation	Check setting of overload if controller and load are at different ambient temperatures.
	Loose connections	Check for loose terminal connection.

Summary

- Motor failure commonly occurs due to overheating, improper power supply, improper application, improper maintenance, and motor defects.

- The life of the insulation is shortened as the temperature in a motor increases beyond the temperature rating of the insulation.

- Common types of power problems are phase unbalance, voltage unbalance, single phasing, and surge voltages.

- Common causes of application failures include overloads, overcycling, and exposure to moisture.

- Common causes of motor failure from improper maintenance include improper belt tension and misalignment and vibration.

- Three-phase motors have fewer components that may malfunction than other motor types and usually operate for many years without any problems.

- Most problems with single-phase motors involve the centrifugal switch, thermal switch, or capacitor.

- Because of their brushes, DC motors generally require more repair than motors that do not use brushes. The brushes should be checked every time the motor is serviced.

- A common problem in DC motors is an open or short in the series or shunt windings.

- Troubleshooting motor controls often involves troubleshooting contactors and motor starters, starting circuits, and circuit faults.

- When troubleshooting a control circuit, the contactor or motor starter is checked first because they are the point where the incoming power, load, and control circuit are connected.

- Typical types of circuit faults are an open circuit, a short circuit, or a short to ground.

- A dual-voltage, wye-connected motor has four separate circuits, three circuits of two leads each (T1-T4, T2-T5, and T3-T6) and one circuit of three leads (T7-T8-T9).

- A dual-voltage, delta-connected motor has three separate circuits of three leads each (T1-T4-T9, T2-T5-T7, and T3-T6-T8).

- For all DC motors, terminal markings A1 and A2 always indicate the armature leads. Terminal markings S1 and S2 always indicate the series field leads. Terminal markings F1 and F2 always indicate the shunt field leads.

- A coil is the series field when the reading of the coil is less than the armature coil resistance.

- A coil is the shunt field when the reading is considerably larger than the armature resistance.

Glossary

Phase unbalance is the unbalance that occurs when lines are out of phase.

Voltage unbalance is the unbalance that occurs when the voltages at the motor terminals are not equal.

Single phasing is the operation of a motor designed to operate on three phases operating on only two phases because one phase is lost.

A **voltage surge** is any higher-than-normal voltage that temporarily exists on one or more of the power lines.

An **overload** is the application of too much load to a motor.

Overcycling is the process of turning a motor ON and OFF more often than the motor design allows.

The **tie-down troubleshooting method** is a testing method in which one DMM probe is connected to either the L2 (neutral) or L1 (hot) side of a circuit and the other DMM probe is moved along a section of the circuit to be tested.

An **open circuit** is an electrical circuit that has an incomplete path that prevents current flow.

A **short circuit** is a circuit in which current takes a shortcut around the normal path of current flow.

A **dead short** is a short circuit that opens the circuit as soon as the circuit is energized or when the section of the circuit containing the short is energized.

A **grounded circuit** is a circuit in which the current leaves its normal path and travels to the frame of the motor.

An **insulation spot test** is a short-term test that verifies the integrity of insulation on electrical devices.

Review

1. Explain why overheating causes motor failure.

2. Explain why an improper power supply causes motor failure.

3. Calculate the voltage unbalance of a system with readings of L1 to L2 = 468 V, L1 to L3 = 474 V, L2 to L3 = 458 V.

4. Explain why an overload or overcycling causes motor failure.

5. Summarize the steps to follow in troubleshooting a 3-phase motor.

6. Summarize the steps to follow in troubleshooting a shaded-pole motor.

7. Summarize the steps to follow in troubleshooting a split-phase motor.

8. Summarize the steps to follow in troubleshooting a capacitor motor.

9. Summarize the steps to follow in troubleshooting a DC motor.

10. Summarize the steps to follow in troubleshooting a contactor or motor starter.

11. Summarize the steps to follow in troubleshooting a reduced-voltage starting circuit.

12. Summarize the steps to follow in troubleshooting a thermal switch.

13. Explain how to identify a motor with unmarked leads as wye or delta connected.

Refer to the CD-ROM for Quick Quiz® questions related to chapter content.

MOTORS

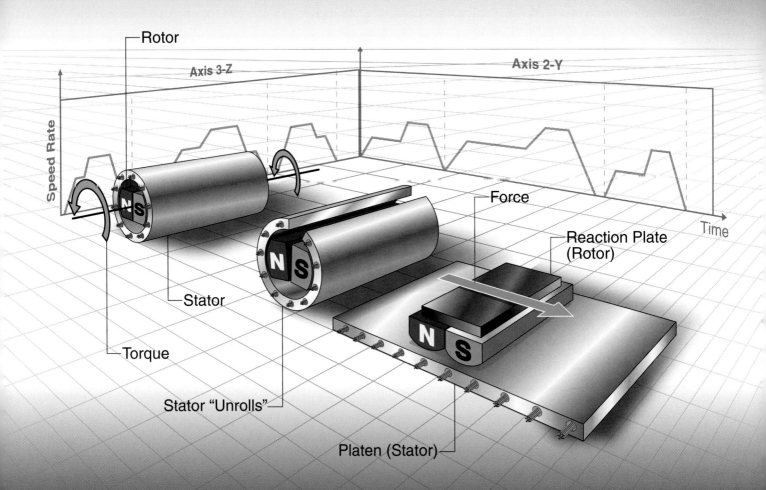

Special-Application Motors

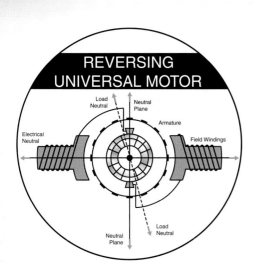

Motion Control Motors 452
Universal Motors 456
Linear Induction Motors 458
Rotary Phase Converters 459
Supplemental Topic—Stepper Motor Resolution 460
Supplemental Topic—Servomotor Encoder Resolution 462
Application—Stepper Motor Shaft Position 464
Summary 464
Glossary 465
Review 465

OBJECTIVES

- Describe the operation of stepper motors and servomotors.
- Explain how universal motors are similar to a DC series motor.
- Explain how a linear induction motor is similar to a standard AC induction motor.
- Describe the operation of a rotary phase converter.

There are several types of motors that are becoming more important because of their unique characteristics. Stepper motors and servomotors are becoming more common in motion control applications. Universal motors can operate on single-phase AC or on DC power. Linear induction motors operate similarly to a standard induction motor, but with a linear, rather than rotary, movement. Rotary phase converters are used to create a 3-phase source from a single-phase source.

Definition

A **stepper motor** is a motor that uses discrete voltage and current pulses to control the movement of a load.

Trajectory control is the predictable path of speed changes a stepper motor undergoes as it moves a load from its starting position to its desired end position.

MOTION CONTROL MOTORS

There are many applications where motors are used to control the motion or position of an object. Stepper motor and servomotor are general terms for two types of motors used in motion control. These types of motors use many different technologies to provide position control. The motors are typically reversible and can be safely stalled in any position.

Stepper Motors

A *stepper motor* is a motor that uses discrete voltage and current pulses to control the movement of a load. Stepper motors translate incoming voltage and current pulses, through a stepper translator, into mechanical motion. Stepper motor controllers are used to program the required stepper motor motions. A stepper motor can be accelerated, decelerated, or maintained by controlling the pulse rate output from a PLC stepper module or other controller. Under controlled conditions, a stepper motor's motion follows the number of input pulses. This ability to respond to a fixed input enables the system to operate in an open-loop mode, leading to cost savings in the total system. However, in some applications, closed-loop operation is used.

Trajectory control is the predictable path of speed changes a stepper motor undergoes as it moves a load from its starting position to its desired end position. A stepper motor drive varies the number of pulses to control the position of the load, varies the frequency of the pulses to control the speed of the load, and varies the rate of change of the frequency of the pulses to control the acceleration and deceleration of the load.

Stepper motors are almost always DC motors. They are used in applications that require precise control of the position of the motor shaft. Typical applications include pen positioning, rotary and indexing table control, robotic positioning, machine tools, laser positioning, and printer control. **See Figure 19-1.**

The torque output of stepper motors typically ranges from 0.5 oz-in. (0.0026 lb-ft) for small motors to 5000 oz-in. (26 lb-ft) for larger motors. Stepper motors are good for applications with a constant load. They should not be used with varying loads. Stepper motors provide good positional accuracy both at rest and while in motion. However, to hold the shaft in position, power must be maintained to the stator winding. This power causes heat in the windings and must be considered in applications requiring a load to be held in position. Using a higher-rated motor and providing ventilation usually takes care of any heat problems.

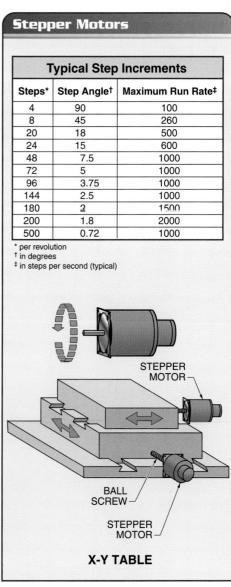

Stepper Motors

Typical Step Increments

Steps*	Step Angle†	Maximum Run Rate‡
4	90	100
8	45	260
20	18	500
24	15	600
48	7.5	1000
72	5	1000
96	3.75	1000
144	2.5	1000
180	2	1500
200	1.8	2000
500	0.72	1000

* per revolution
† in degrees
‡ in steps per second (typical)

X-Y TABLE

Figure 19-1. A stepper motor divides shaft rotation into discrete steps.

Although stepper motors provide precise positioning, they cannot operate large loads because their torque output is low compared to other motors. Stepper motors are prone to losing steps at higher speed. Motor size is relatively large for the amount of torque output. Most stepper motors used in industrial applications have 200 to 400 steps per revolution.

Stepper Motor Operation. Stepper motors operate on the principle that like magnetic poles repel each other and unlike magnetic poles attract each other. Stepper motors rely on electronic commutation to provide motion control. In motion control applications, commutation is accomplished through electronic control of the drive voltages and currents in the stator windings. As the commutation changes the voltage and current in the stator windings, the stator poles change and rotor moves from one discrete position to another to align the unlike rotor stator poles. When the commutation holds the voltage and current at the same values, a stepper motor holds its position with a known holding torque. In a common design, stepper motors have 50 teeth on the inside of the stator and the outside of the rotor. However, the number of teeth on the rotor and stator do not have to be the same. **See Figure 19-2.**

The shaft of a stepper motor rotates at fixed angle when it receives an electric pulse, or step. Each input step produces shaft rotation through the stepper motor's rated step angle. For example, a typical stepper motor rotates 1.8° for every pulse. If 50 pulses are applied to a 1.8° stepper motor, the shaft rotates exactly 90° (50 × 1.8 = 90) and then stops. If 200 pulses are applied to the same motor, the shaft rotates exactly 360°, or one full revolution, and then stops. The steps/revolution parameter gives the resolution of the motor position control. A stepper motor with a resolution of 200 steps/revolution rotates the shaft 1.8° per step. If the 200 steps were applied over a period of one minute, the motor speed would be 1 rpm. If the 200 steps were applied over a period of one second, the motor speed would be 1 revolution per second, or 60 rpm.

Advanced stepper motor drives provide better position control through microstepping. *Microstepping* is stepper motor control that divides the standard stepper control pulse intervals into smaller intervals and sends modified current to the motor coils to increase the resolution of a stepper motor system. This provides more precise position control and smoother movement because of the smaller steps.

Definition

Microstepping is stepper motor control that divides the standard stepper control pulse intervals into smaller intervals and sends modified current to the motor coils to increase the resolution of a stepper motor system.

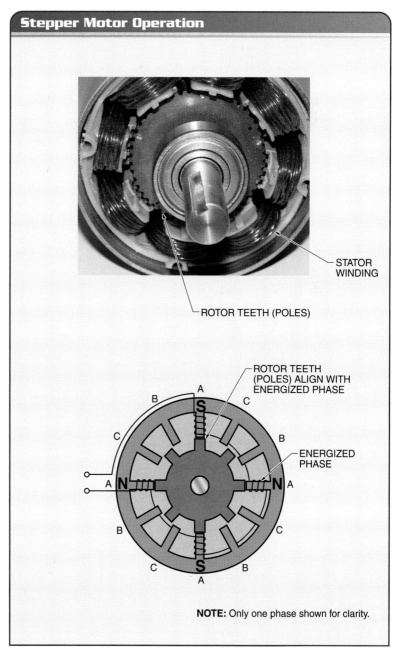

Figure 19-2. The rotor teeth (poles) align with the energized stator poles.

Definition

A *servomotor* is a motor that uses feedback signals to provide position and speed control.

Servomotors

A *servomotor* is a motor that uses feedback signals to provide position and speed control. The shortened term, servo, is often used to describe a servomotor. A typical servomotor drive uses a closed-loop control system with feedback from an encoder mounted on the motor or load. **See Figure 19-3.** The feedback provides information on the exact location of the shaft or load. The control system adjusts the signal to the servomotor drive to place the shaft or load where it should be. The servomotor drive controls the voltage and current that flows through the servomotor armature and windings.

Applications that once employed clutch-gear systems or other mechanical arrangements to perform motion control now often use servomotor controls. The advantages of servomotor control are shorter positioning time, higher accuracy, better reliability, and improved repeatability in the coordination of axis motion. Typical applications of servomotor positioning include grinders, metal-forming machines, transfer lines, and the precise control of valves in continuous process applications.

The two primary types of servomotors are brush servomotors and brushless servomotors. Brush servomotors are fairly simple in design and cost effective in many applications. A brush servomotor can be reversed by changing the direction of current flow through the armature. Brushless servomotors are more expensive than brush servomotors, but typically have lower maintenance costs because there are no brushes to wear.

Acceleration and Deceleration

For both stepper motors and servomotors, motion control includes both the acceleration and deceleration of the motor. The acceleration part of the move is the interval required to achieve the continuous speed of the motor. The continuous speed is the frequency of pulses (in pulses/sec, or Hz) sent to the motor when the motor is at full speed. This frequency typically varies from 1 kHz to 20 kHz. Conversely, the deceleration part of the move is the interval for the speed to decrease to 0 Hz. Acceleration and deceleration ramps are specified as the rate of change of the speed.

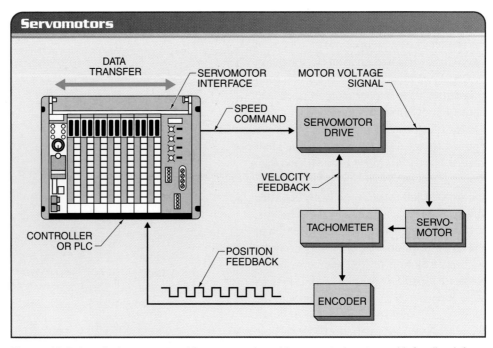

Figure 19-3. A typical servomotor drive uses a closed-loop control system with feedback from an encoder mounted on the motor or load.

The two modes of motion control are single-step profile mode and continuous profile mode. In single-step profile mode, a PLC processor sends individual move sequences to the motor. These sequences include the acceleration and deceleration rates of the move, along with the final or continuous speed. Once this move sequence is terminated, the processor may start another one by transferring the next move's profile information and commands. The processor can store several single-step mode profiles and send them to the module through the PLC program control. **See Figure 19-4.**

In continuous profile mode, the motion profile is cycled through various accelerations, decelerations, and continuous speed rates to form a blended motion profile. Rather than requiring additional commands for motion speed changes, an interface in continuous mode receives the whole move profile in a single block of instructions. The interface then performs the step motor control duty until the motion is completed and the processor sends the next profile. As in the single-step mode, the processor can store several continuous-mode profiles in its memory and send them to the interface during the program execution.

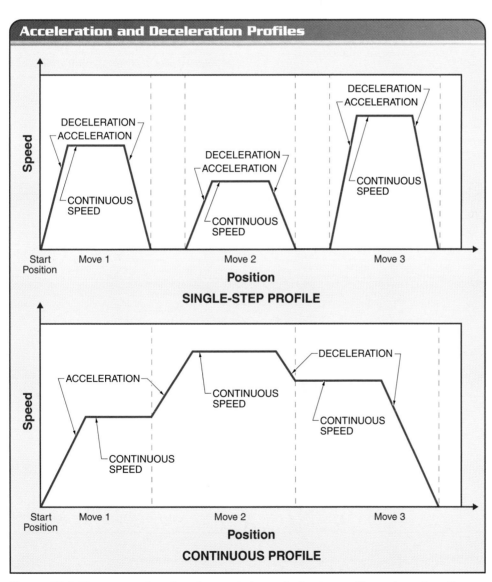

Figure 19-4. The two modes of motion control are single-step profile mode and continuous profile mode.

Definition

A *universal motor* is a motor that can be operated on either single-phase AC or DC power.

Each controller used to control a motor controls an axis, since the motion generated causes a movement about the X-, Y-, or Z-axis. Depending on the PLC manufacturer, more than one axis may be controlled using several PLC module interfaces. When multiple-axis motions are required, the axes can be controlled either independently or synchronously. **See Figure 19-5.** When controlled independently, each axis is independent of the other, executing its own single-step or continuous profile mode. The beginning and end time of each axis motion may be different. When controlled synchronously, the beginning and end of the motion commands for each axis occur at the same time. A profile of one of the axes may start later or end before the other axes, but the move that follows will not occur until all axes have started and ended their motions.

UNIVERSAL MOTORS

A *universal motor* is a motor that can be operated on either single-phase AC or DC power. The motor characteristics are approximately the same on AC as on DC, provided the AC voltage does not exceed 60 Hz. A universal motor is electrically the same as a DC series motor. Current flows from the supply, through the field, through the armature windings, and back to the supply. The main parts of a universal motor are the field windings, which are stationary, and the armature, which rotates. **See Figure 19-6.**

Tech Fact

Universal motors are typically available in sizes less than 1 HP. They are used frequently in portable tools, such as drills, saws, and routers, and in small household appliances.

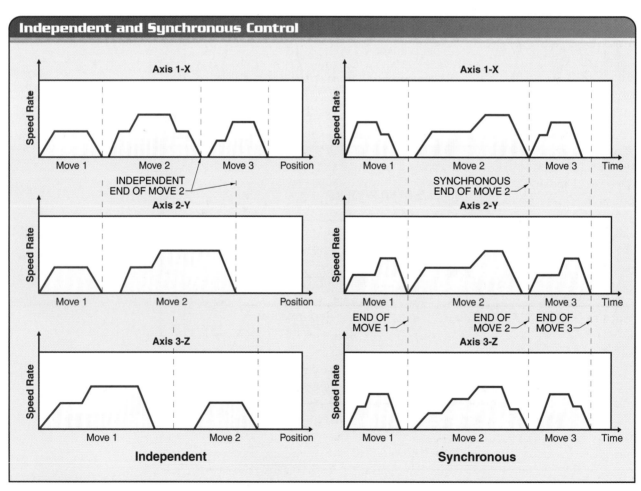

Figure 19-5. When multiple-axis motions are required, the axes can be controlled either independently or synchronously.

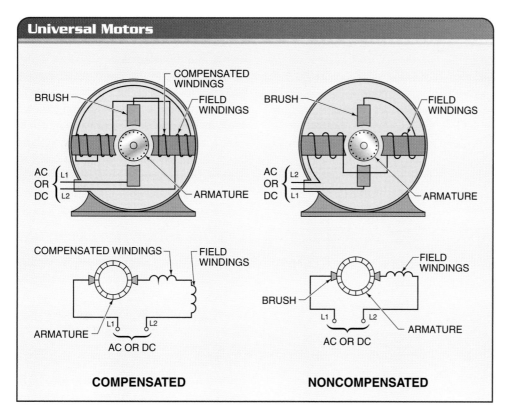

Figure 19-6. The main parts of a universal motor are the field windings, which are stationary, and the armature, which rotates. Compensated windings may also be present.

Definition

A compensated universal motor is a universal motor with extra windings added to the field poles to reduce sparking.

A noncompensated universal motor is a universal motor without extra windings added to the field poles.

Universal motors are the most common motors found in residences. The main advantages of universal motors are high torque, high speed, and small size when compared to other AC motors. They provide high torque in a minimum of space. Universal motors are commonly used in such applications as kitchen appliances, vacuum cleaners, bench tools, and portable hand tools, such as saws, drills, routers, and grinders.

Universal motors operate on the same principles as DC series motors. The ends of the armature coils are connected to segments of a commutator. Soft carbon brushes are mounted in fixed, insulated holders that allow them to slide across the commutator segments as the armature rotates.

When a universal motor is connected to an AC voltage supply, the current flows through the armature and the field windings. The field produces an AC flux that reacts with the current flowing through the armature to produce torque. Because the armature current and the flux reverse simultaneously as the AC current changes direction, the torque always acts in the same direction. Unlike an AC motor, no revolving field is produced in a universal motor.

A *compensated universal motor* is a universal motor with extra windings added to the field poles to reduce sparking. These extra windings are connected in series with the armature windings. Compensated universal motors are usually ¼ HP and larger and are the most common type of universal motor.

A *noncompensated universal motor* is a universal motor without extra windings added to the field poles. Noncompensated universal motors are simpler in construction and are less expensive than compensated universal motors. They are commonly used for lower-power output and higher-speed applications. Operating a noncompensated universal motor on AC poses some problems, such as inefficiency due to hysteresis and eddy-current losses.

Definition

*A **linear induction motor** is an AC motor that uses induction to create linear movement.*

Reversing Universal Motors

Universal motors may be reversing or nonreversing. **See Figure 19-7.** A nonreversing universal motor has two power leads and a ground wire coming out of the motor. The ground wire is used for grounding the frame of the motor. Grounding is required on portable tools due to the danger of electrocution. The standard direction of rotation for a nonreversing motor is counterclockwise facing the end opposite the shaft extension.

A reversing motor also has two power leads and a ground wire coming out of the motor. Additionally, it has three or more leads available at the reversing switch. A universal motor may be reversed by using one coil for the forward direction and the other for the reverse direction. A universal motor may also be reversed by reversing the direction of current in the field with a double-pole, double-throw (DPDT) switch. The same field is used for both directions.

LINEAR INDUCTION MOTORS

A *linear induction motor* is an AC motor that uses induction to create linear movement. It consists of a platen and a reaction plate. The platen is the fixed, flat bed of a linear induction motor and the reaction plate is the moving part. A linear induction motor can be visualized as a standard induction motor with the stator unrolled into a platen. **See Figure 19-8.** For example, a stator with an inside diameter of 4′ would unroll into a linear motor of 12.6′ (4 × π = 12.6). In actual practice, a stator cannot be unrolled, but it is a useful way to visualize a linear induction motor.

A reaction plate is the equivalent of the rotor in an induction motor. It is typically constructed of a conductive sheet of copper or aluminum. A bearing assembly is needed to maintain the air gap between the stator and reaction plate. Another design for linear induction motors uses two flat platens with the reaction plate between them. Horizontal travel is limited only by the length of the linear motor. Long conveyors can be constructed of multiple linear motors.

The advantages of linear motors over other types of motors include high repeatability resolution for consistent part placement; no backlash as is seen with gears; fast acceleration to shorten cycle times; and low maintenance costs because there are no contacting parts.

Linear-induction motors are used for motion control of reciprocal motion machines, people movers, and monorail trains. Monorails and reciprocal motion motors are usually two sided, while people movers are more likely to be one-sided motors.

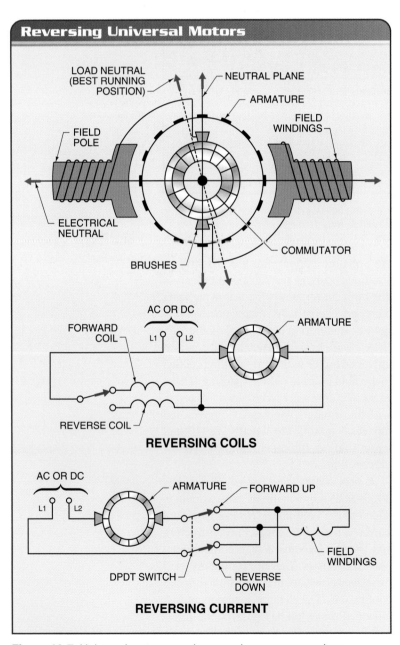

Figure 19-7. Universal motors may be reversing or nonreversing.

Chapter 19—Special-Application Motors **459**

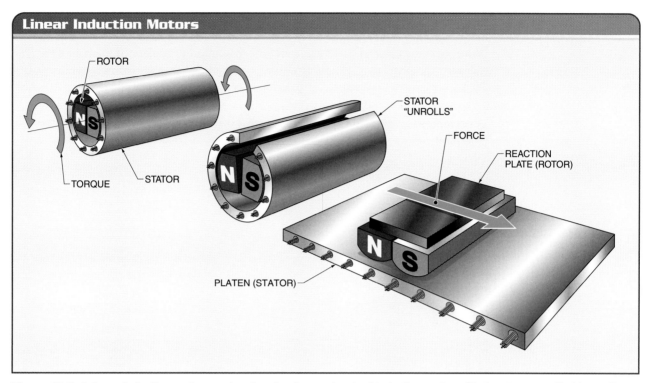

Figure 19-8. A linear induction motor can be visualized as a standard induction motor with the stator unrolled into a linear stator, or platen.

ROTARY PHASE CONVERTERS

A *rotary phase converter* is a 3-phase power source consisting of a single-phase power source, a standard 3-phase induction motor, and a standard 3-phase disconnect used to supply power to a 3-phase load. **See Figure 19-9.** If 3-phase motors are needed in areas where only single-phase power is available, a single-phase to 3-phase converter is required. There are several methods of providing 3-phase power from a single-phase source. The selection of the method depends on cost, expandability, and application. Rotary phase converters are a common method used to provide 3-phase power in locations where only single-phase power is available.

A rotary phase converter uses a 3-phase motor powered by a single-phase source. The 3-phase induction motor is sometimes called an idler motor because no load is connected to the shaft. It is only used to induce the 3rd phase. When single-phase power is applied to the windings of a 3-phase motor, the 3rd phase is induced in the windings. Three-phase power can be taken from the rotary phase converter and used to operate any type of 3-phase load.

When single-phase power is applied to a 3-phase motor, the motor does not produce the rotating field that is necessary for rotation in the three-phase motor. Some method must be used to develop a rotating field to start the motor in the rotary phase converter. A common method used to develop a rotating field is to use start capacitors wired across each of the phases. Manufacturers design rotary phase converters with the correct capacitance for the rating of the rotary phase converter.

The horsepower rating of the rotary phase converter must be at least as large as the largest motor that will be started by the rotary phase converter. If the motor is to be started under load, the rotary phase converter should be about 1.5 times the motor horsepower. Even though a rotary phase converter can start only one motor of a size comparable to the rotary phase converter,

Definition

A *rotary phase converter* is a 3-phase power source consisting of a single-phase power source, a standard 3-phase induction motor, and a standard 3-phase disconnect used to supply power to a 3-phase load.

the converter can provide power to operate more than one motor. For example, a rotary phase converter rated at 5 HP can start an unloaded motor of up to 5 HP. However, the rotary phase converter can provide power to operate motors up to about 15 HP total, depending on the loads. The motors may need to be derated to about 80%. The manufacturer of the rotary converter should be consulted.

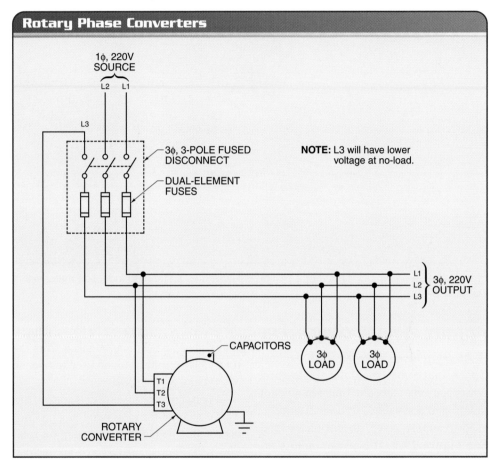

Figure 19-9. A rotary phase converter consists of a single-phase power source, a standard 3-phase induction motor, and a standard 3-phase disconnect used to supply power to a 3-phase load.

Supplemental Topic

Stepper Motor Resolution

A stepper controller generates a pulse train that indicates distance, rate, and direction commands to the motor. The number of pulses sent to the stepper, which translates into linear or rotational units of travel, defines position displacement. Therefore, the number of pulses sent to the motor from the module determines the motor's final position. The motion can be rotational or linear, such as the forward or backward movement of a linear slide using leadscrews, which translate rotational movements from a servomotor into linear displacement. The actual location also depends on the resolution of the stepper and the application, given as the number of threads per inch of travel in the leadscrew. The resolution may also be given as inches of travel per thread. A leadscrew with a pitch of 4 threads per inch has a resolution of 0.25 inches per thread or 0.25 inches per turn.

The information needed to determine stepper resolution is the leadscrew pitch and the stepper frequency. The distance moved for each step (resolution) is calculated as follows:

$$d = \frac{1}{pitch \times f}$$

where

d = distance moved for each step, in in.

$pitch$ = leadscrew pitch, in threads per in.

f = stepper frequency, in steps/revolution

For example, a typical linear slide using a stepper motor makes one revolution per 200 steps with a pitch of 4 threads per inch on the leadscrew. The resolution is calculated as follows:

$$d = \frac{1}{pitch \times f}$$

$$d = \frac{1}{4 \times 200}$$

$$d = \frac{1}{800}$$

$$d = \mathbf{0.00125''}$$

Since the leadscrew has a pitch of 4 threads per inch, it takes 4 revolutions to travel 1″ and it takes 800 steps to travel 1″. To calculate the step angle, the number of degrees in one revolution (360°) is divided by the number of steps required to turn the motor one full revolution. Therefore, the step angle is 1.8°/step (360 ÷ 200 = 1.8). This gives a resolution of 1.8°/step or 1/200 of one revolution per step.

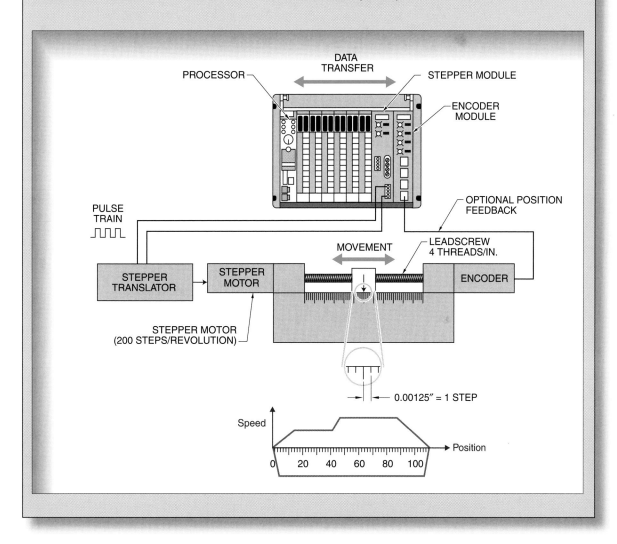

Supplemental Topic

Servomotor Encoder Resolution

The input voltage to the servomotor is used to control the speed. A PLC servomotor module sends the drive controller a ±10 VDC signal, which defines the forward and reverse speeds of the servomotor. These modules are generally used when axis motion control, either linear or rotational, is required. A common linear motion example is a leadscrew assembly, which translates rotational movements from a servomotor into linear displacement.

Servomotor position controllers operate in a closed-loop system, requiring feedback information in the form of speed or position. Servomotor controllers may receive speed feedback in the form of a tachometer input, or position feedback in the form of an encoder input, or both. The feedback signal provides the module with information about the actual speed of the servomotor and the position of the axis. This information is then compared with the desired speed and the desired position of the axis and the module output is adjusted to correct the error.

PLCs that have position control capabilities typically require two PLC modules. One module is used to implement the servomotor control task and the other is used to receive feedback and close the loop. Some manufacturers offer complete servomotor control for one axis in a single module. Servomotor control can occur in either single-step or continuous positioning mode. Depending on the manufacturer, multiaxis control can also be synchronized in either single-step or continuous mode.

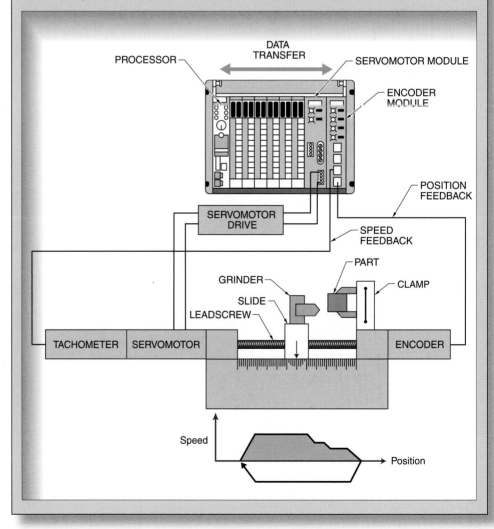

The PLC processor sends all of the move and position information, including acceleration, deceleration, and the final and feed velocities, to the servomotor module. When the module is operating, the processor monitors its status without interfering with the module's complex, rapid calculations. The processor updates the module with a new move for an axis when the previous move has been completed and the module is ready for a new profile. The acceleration and deceleration parameters are given in inches per minute per second (ipm/s) at a specific resolution.

When servomotor controllers are used for position control, the feedback resolution provided by the system is a key issue. For example, if an interface uses a leadscrew (a rotational-to-linear motion translator) for axis displacement and an encoder to provide a feedback signal to the servomotor module, the user must know the leadscrew pitch, the number of encoder pulses per revolution, and the multiplier value in the encoder section of the controller. Some controllers allow the user to select a multiplier, thus providing better feedback resolution without changing the encoder.

Each servomotor controller has a predefined resolution, which typically varies from 0.001″ to 0.0001″. A tradeoff exists between axis speed and feedback resolution because resolution decreases as the speed increases. Typical axis positioning speeds range from 500 to 1000 inches per minute (ipm) and encoder feedback input frequencies range up to 250 kHz. Many servomotor modules include an encoder multiplier that increases the resolution. The feedback resolution is calculated as follows:

$$d = \frac{1}{pitch \times f \times m}$$

where

d = distance moved for each step, in in.

pitch = leadscrew pitch, in threads per in.

f = encoder frequency, in pulses/revolution

m = encoder multiplier

For example, a PLC system uses a servomotor controller to perform a one-axis positioning of a metal part. This part will be machined at a defined profile, which is stored in processor memory. A leadscrew with a pitch of 8 threads per inch, which allows travel of ⅛″ (0.125″) per revolution, moves the part along an X-axis. An encoder with 250 pulses per revolution supplies position feedback information. The encoder is connected to an encoder feedback terminal in the servomotor controller that can provide a software programmable multiplier of x1, x2, and x4 increments per pulse, depending on the encoder design. In this case, a multiplier of x2 is chosen. The feedback resolution from the encoder is calculated as follows:

$$d = \frac{1}{pitch \times f \times m}$$

$$d = \frac{1}{8 \times 250 \times 2}$$

$$d = \frac{1}{4000}$$

$$d = \mathbf{0.00025 \text{ in. per pulse}}$$

The feedback resolution from the encoder is 0.00025 in. per pulse, or 4000 pulses per in. (1 ÷ 0.00025 = 4000). Therefore, an encoder monitoring a part moving 1″ generates 4000 pulses. If the part needs to move 2.5″, the encoder will generate 10,000 pulses (2.5 × 4000 = 10,000). The controller counts the pulses to position the part.

Application—Stepper Motor Shaft Position

A stepper motor rotor can be made from permanent magnets or electromagnets. The stator windings are electromagnets. When the control switch is closed, the stator windings are magnetized by the current flowing through the coil. When windings are energized, a stator pole is created and the motor generates torque to align the rotor poles with the stator pole.

If Switch 1 is opened and Switch 2 is closed, the rotor moves clockwise a fixed interval to align the nearest rotor pole with the new stator pole energized by Switch 2. If the switches are opened and closed again in the same order, the rotor continues to move in the same direction.

Changing the order in which the switches are opened and closed reverses the direction of the motor. A stepper motor can rotate in either direction, based on the order in which the control switches are activated. Although control switches can be used to energize the stator windings, solid-state switches or PLCs are used for most applications.

Switch Table for Clockwise Rotation

Rotation (Degrees)	SW1	SW2	SW3	SW4
0	X			
30		X		
60			X	
90				X
120	X			
150		X		
180			X	
210				X
240	X			
270		X		
300			X	
330				X
360	X			

STEPPING MOTOR OPERATION

Summary...

- Stepper motors are almost always DC motors that translate incoming voltage and current pulses into mechanical motion.

- A stepper motor drive varies the number of pulses to control the position of the load, varies the frequency of the pulses to control the speed of the load, and varies the rate of change of the frequency of the pulses to control the acceleration and deceleration of the load.

- Stepper motors operate on the principle that like magnetic poles repel each other and unlike magnetic poles attract each other. As the commutation changes the voltage and current in the stator windings, the poles change and motor moves from one discrete position to another to align the unlike poles.

- A typical servomotor drive uses a closed-loop control system with feedback from an encoder mounted on the motor or load.

- The two modes of motion control are single-step profile mode and continuous profile mode. In single-step mode, a PLC processor sends individual move sequences to the interface. In continuous profile mode, the motion profile is cycled through various accelerations, decelerations, and continuous speed rates to form a blended motion profile.

Summary

- Universal motors operate on the same principles as DC series motors. With AC operation, the armature current and the flux reverse simultaneously as the AC current changes direction so the torque always acts in the same direction.

- A rotary phase converter uses a 3-phase motor powered by a single-phase source. The 3-phase induction motor is sometimes called an idler motor because no load is connected to the shaft. It is only used to induce the third phase.

Glossary

A **stepper motor** is a motor that uses discrete voltage and current pulses to control the movement of a load.

Trajectory control is the predictable path of speed changes a stepper motor undergoes as it moves a load from its starting position to its desired end position.

Microstepping is stepper motor control that divides the standard stepper control pulse intervals into smaller intervals and sends modified current to the motor coils to increase the resolution of a stepper motor system.

A **servomotor** is a motor that uses feedback signals to provide position and speed control.

A **universal motor** is a motor that can be operated on either single-phase AC or DC power.

A **compensated universal motor** is a universal motor with extra windings added to the field poles to reduce sparking.

A **noncompensated universal motor** is a universal motor without extra windings added to the field poles.

A **linear induction motor** is an AC motor that uses induction to create linear movement.

A **rotary phase converter** is a 3-phase power source consisting of a single-phase power source, a standard 3-phase induction motor, and a standard 3-phase disconnect used to supply power to a 3-phase load.

Review

1. Explain how a stepper motor uses voltage and current pulses to control movement.

2. Demonstrate how to calculate the distance moved for each step of a stepper motor with a resolution of 200 steps per revolution and a leadscrew with a pitch of 6 threads per inch.

3. Explain how a servomotor uses feedback to control movement.

4. Demonstrate how to calculate the distance moved for each pulse of a servomotor encoder that has a resolution of 100 pulses per revolution, a leadscrew with a pitch of 4 threads per inch, and a software programmable multiplier of 1.

5. Describe the difference between single-step profile mode and continuous profile mode acceleration and deceleration.

6. Explain how a universal motor is electrically the same as a DC series motor.

7. Explain how a linear induction motor is similar to a standard induction motor.

8. Describe how a rotary phase converter is used to convert single-phase power to 3-phase power.

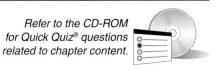

Refer to the CD-ROM for Quick Quiz® questions related to chapter content.

Answer Key

Chapter 1 – Magnetism and Induction

1. A magnet is a substance that produces a magnetic field and attracts magnetic materials. A permanent magnet is a magnet that can hold its magnetism for a long period. A temporary magnet is a magnet that retains only trace amounts of magnetism after a magnetizing force has been removed. An electromagnet is a temporary magnet produced when electricity passes through a conductor, such as a coil, that concentrates the magnetic field. This is important because a temporary magnet can be turned off while de-energized, such as a coil in a magnetic starter. In addition, the magnetic strength can be easily controlled by regulating the amount of current.

3. Electron current flow is a description of current flow as the flow of electrons from the negative terminal to the positive terminal of a power source. In a battery or other power source, electrons move to the negative terminal, creating a separation of charge, with the negative terminal having a negative charge and the positive terminal having a positive charge. Conventional current flow is a description of current flow as the flow of positive charges from the positive terminal to the negative terminal of a power source. In a battery or other power source, conventional current flow states that positive charges move to the positive terminal, creating a separation of charge and a potential difference.

5. The magnetic field surrounding a coil interacts with another wire or coil. The interaction between the magnetic field around the first coil and the second coil induces a current in the second coil. The induced current in the second coil creates its own magnetic field that opposes the initial magnetic field.

Chapter 2 – Motor Nameplates

1. Electrical ratings describe the electrical requirements for the operation of a motor.

3. Operating ratings describe how a motor is designed and how that design relates to where the motor is to be used.

5. Mechanical-design codes describe the different types of motor design features. Following the guidelines on mechanical-design codes when installing or servicing a motor results in improved efficiency and lower operating cost.

7. Definite-purpose motors are designed for applications such as washdown-rated motors, submersible pumps, hazardous locations, instantly reversible motors, extra-high-torque motors, farm-duty or agricultural motors, irrigation, and many other applications.

9. An open motor enclosure is a motor enclosure with openings to all passage of air to cool the windings. Open motor enclosures include general, drip-proof, splashproof, guarded, semiguarded, and drip-proof fully guarded. A totally enclosed motor enclosure is a motor enclosure that prevents air from entering the motor. Totally enclosed motor enclosures include fan-cooled, nonventilated, pipe-ventilated, water-cooled, explosionproof, dust-ignition-proof, and waterproof. Totally enclosed motor enclosures can be used in moist, dirty locations where a fan would clog; can be installed in the airstream of a driven blower or fan; or can be used with water-cooling in enclosed areas or for high power at low speeds.

Chapter 3 – Motor Protection

1. A short circuit, a ground fault, and an overload are all types of overcurrents. A short circuit is an excessive current that leaves the normal current-carrying path by going around the load and back to the power source or ground. A ground fault is a type of short circuit consisting of an unintentional connection between an ungrounded conductor and any grounded raceway, box, enclosure, or fitting. An overload is an excessive current that is confined to the normal current-carrying conductors and is caused by a load that exceeds the full-load torque rating of the motor.

3. An Edison-base fuse is a plug fuse that incorporates a screw configuration that is interchangeable with fuses of other amperage ratings. A type S fuse is a plug fuse that incorporates a screw and adapter configuration that is not interchangeable with fuses of another amperage rating. Since Edison-base fuses have an interchangeable screw configuration, it is possible to overfuse a circuit by using the wrong size fuse. Type S fuses are noninterchangeable with fuses of a lower amperage rating, which protects the circuit from the possibility of overfusing.

5. Common types of CBs include inverse-time CBs, adjustable-trip CBs, non-adjustable-trip CBs, and instantaneous-trip CBs.

7. An OCPD must clear short circuits and ground faults, but must not open the circuit because of the normal momentary inrush current of a motor starting. OCPDs are installed in the combination starter, safety switch, or fuse panel. After a motor has started and has accelerated to its rated speed, the motor draws enough current from the power lines to remain running. The overload relays in a motor starter are time-delay devices that allow temporary overloads without disconnecting the motor. If an overload is present for longer than the preset time, the overload relays trip and disconnect the motor from the circuit.

9. An electromagnetic overload relay is a relay that operates on the principle that as the level of current in a circuit increases so does the strength of the magnetic field produced by that current flow. When the level of current through a current coil in the circuit reaches the preset value, the increased magnetic field acts as a solenoid and opens a set of contacts. An electronic overload relay is a device that has built-in circuitry to sense changes in current and temperature. An electronic overload monitors the current in the load directly by measuring the current in the power lines leading to the load.

Chapter 4 – Three-Phase Motors

1. The iron sheets are electrically separated from each other by an insulating coating. The separation reduces the cross-sectional

area of the core and shortens the conductive path for damaging eddy currents. The laminations break the potential conductive path in the core into smaller sections and reduce the loss.

3. A rotor consists of a core, windings, and the shaft. The rotor core consists of many thin iron sheets laminated together. The rotor windings vary, depending on the type of motor. The shaft provides support for the rotor and transfers power from the motor to the load.

5. The synchronous speed of a 2-pole motor is calculated as follows:
$$\Omega_S = \frac{7200}{P}$$
$$\Omega_S = \frac{7200}{2}$$
$$\Omega_S = 3600 \text{ rpm}$$

7. In a wye-connected, 3-phase motor, one end of each of the three phases is internally connected to the other phases. The remaining end of each phase is then brought out externally to form T1, T2, and T3. In a delta-connected, 3-phase motor, each phase is wired end-to-end to form a completely closed circuit. At each point where the phases are connected, leads are brought out externally to form T1, T2, and T3.

9. Power factor is the ratio of true power to apparent power. For a 3-phase, 5 HP, 240 V motor that has 94% efficiency and draws a full-load current of 12 A, the power factor is calculated as follows:
$$pf = \frac{hp \times 746}{V \times i \times \sqrt{3} \times \varepsilon}$$
$$pf = \frac{5 \times 746}{240 \times 12 \times \sqrt{3} \times 0.94}$$
$$pf = 0.80$$

Chapter 5 – Induction Motors

1. The stator consists of a core and windings.

3. The rotor consists of a core and windings, mounted on a shaft.

5. With Motor Design A, the bars are placed near the surface of the rotor and have low reactance. The low reactance allows for large current to flow through the bars and a large torque to be developed. With Motor Design B, the bars are narrow and are placed deep in the iron, which increases the reactance and lowers the current, while maintaining normal starting torque. With Motor Design C, the rotor has two conductor bars in each slot, with one above the other. The top bar is a high-resistance conductor that carries most of the current during starting. With Motor Design D, high-resistance bars are placed deep in the iron, creating low starting current.

Chapter 6 – Wound-Rotor Motors

1. A wound-rotor motor can be distinguished from a squirrel-cage induction motor by the presence of the coils of wire in the winding slots instead of the solid conductor bars, by the presence of the three slip rings on the shaft, and by the presence of an external resistance bank.

3. Once the motor starts, the speed of the rotor increases and the induced frequency decreases, decreasing the induction and induced current in the rotor. Resistance is removed from the rotor circuit to increase the current and torque to allow the motor to continue to accelerate. This can continue in steps, with the number of steps depending on the motor design. For each step, the motor accelerates, the current decreases, and more resistance is removed, until the motor is at full-load speed.

5. A wound-rotor motor starter contains low-wattage resistors and should only be used during startup. The resistors must be removed from the rotor circuit after the motor is up to speed. A wound-rotor motor regulator contains high-wattage resistors and is used for starting and speed control. The resistors can remain in the rotor circuit continuously to allow for speed control.

Chapter 7 – Synchronous Motors

1. The rotor of a synchronous motor has induction windings as well as synchronous windings. The rotor starts and accelerates to near-synchronous speed, similarly to a standard induction motor. When the DC power is applied to the rotor field, the discharge resistor is switched out of the circuit. The purpose of applying the DC to the field winding is to create a magnetic field (electromagnet) that synchronizes with the stator field.

3. Most often, the DC is applied to the rotor through slip rings mounted on the shaft. Other methods of applying a DC exciting current to the rotor windings include DC shaft-mounted exciters, motor-generator (M-G) sets, static exciters, and brushless exciters. In addition, some rotors are manufactured with permanent magnets to eliminate the need for excitation.

5. The exciter circuit includes relays or a control module that switches the motor from inductive operation to synchronous operation. The reactive voltage in the reactor is reduced as the frequency decreases. At the optimum point, the frequency is low enough that the PFFR AC coil cannot force the DC flux through the armature. The armature opens, closing the PFFR NC contact in series with the M contact in the exciter circuit. This energizes relay F, and the contactor closes the two NO contacts, connecting the exciter generator output to the rotor.

7. Reluctance torque is torque developed by the salient rotor poles that occurs before the poles are excited by the external DC power. Pull-in torque is the maximum torque required to accelerate a synchronous motor into synchronization at the rated voltage and frequency. Pull-out torque is the torque produced by a motor overload that pulls the rotor out of synchronization. Torque angle is the angle between the rotor and stator fields as a synchronous motor is running under load.

Chapter 8 – Single-Phase Motors

1. The start winding is placed 90 mechanical degrees from the run windings. The start winding is made from smaller wire, giving it a higher resistance, and fewer turns, giving it a lower reactance, than the run winding. In most motor designs, the start winding must be removed from the circuit after the motor reaches about 60% to 80% of full-load speed.

3. The shading coil receives its power as a result of transformer action by the motion of the flux in the iron. The induced current is about 90° behind the main pole.

5. A capacitor motor introduces capacitance into an AC circuit to create a phase shift between the start and the run windings. It creates a greater phase shift than a start winding alone. The capacitor is added to provide a higher starting torque at lower starting current than is delivered by the split-phase motor.

7. The most common application of split-phase motors is for use in easy-to-start applications such as fans, business machines, machine tools, and centrifugal pumps.

Chapter 9 – AC Alternators

1. A stator consists of a core and windings. The stator is enclosed within a housing.

3. Centrifugal force is larger for high-speed rotation and large diameters and is smaller for low-speed rotation and small diameters. Salient poles and spider rings are used at slow speeds. Salient poles are not used at high speeds. In addition, the rotor diameter is minimized at high speeds.

5. A voltage regulator senses the voltage output of an alternator. If the voltage is not at the setpoint, the regulator adjusts the strength of the alternator magnetic field to vary the strength of the poles. This changes the interaction between the stator and the rotor and changes the voltage output.

7. When the synchroscope needle is stationary at the twelve o'clock position, the frequencies and phases are matched and the alternator can be brought on-line. In actual practice, the frequency of the alternator should be slightly faster than the frequency on the power line.

9. A controller measures the parameters of the incoming source and the busbar and adjusts the alternator to match the busbar. An automatic synchronizer energizes a relay and closes a contact when the incoming source and busbar match.

Chapter 10 – DC Motors and Generators

1. DC motors and generators consist of field windings in the field frame, an armature, and a commutator and brushes on the shaft. Both machines depend on induction, where there is relative motion between a magnetic field and a conductor.

3. When power is applied to a DC motor, both the armature windings and the field windings generate magnetic fields that become distorted when they interact. When there is no current in the armature, the magnetic field between the field poles is undistorted and the neutral plane is at right angles to the field flux. When there is current in the armature, the magnetic field between the field poles is distorted and the neutral plane is at an angle to the original position.

5. A DC series motor has high starting torque and low running torque. A DC shunt motor has moderate starting torque with steady torque as the motor speeds up. A DC compound motor has moderate starting torque that drops off as the motor speeds up. A DC permanent-magnet motor has high starting torque that drops off as the motor speeds up.

Chapter 11 – Starting

1. In an open-circuit transition, a motor is temporarily disconnected from the voltage source when switching from a reduced starting voltage level to a running voltage level, before reaching full motor speed. In a closed-circuit transition, a motor remains connected to the voltage source when switching from a reduced starting voltage level to a running voltage level, before reaching full motor speed.

3. Full-voltage starting is the simplest and least expensive method of starting a motor, but has high starting current and high starting torque.

5. Autotransformer starting provides the highest torque per ampere of line current, with the motor current greater than line current during starting, but is relatively expensive, especially in smaller motors, and has a low power factor.

7. Wye-delta starting has low starting current and works well with high-inertia, long-starting loads, but requires a special motor.

9. A typical autotransformer may have a turns ratio of 1:0.8. Autotransformer starting can use the turns ratio advantage to provide more current on the load side of the transformer than on the line side.

Chapter 12 – Braking

1. Braking is used when it is necessary to stop a motor more quickly than coasting allows. Hazard braking may be required to protect an operator even if braking is not part of the normal stopping method.

3. Friction braking has the advantage of being relatively inexpensive with simple maintenance. Few expensive electrical components are needed. Friction braking has the disadvantage of requiring frequent maintenance to inspect and replace the brake shoes.

5. Plugging is more expensive than friction braking but less expensive than other braking methods. It allows for rapid stopping and can be used as an emergency stop. Plugging can only be used with motors that can be reversed at full speed. Plugging draws high current and generates considerable heat, so a motor with a high service factor should be used.

7. Electric braking provides a quick and smooth braking action on all types of loads. Maintenance is minimal because there are no parts that wear from physical contact. However, electric braking is relatively expensive because of the electronic components required to switch the DC field. Electric braking requires a power source at all times. Electric braking cannot stop a motor in the event of a power failure, and it cannot be used for holding a load.

9. Because of the heat generated during dynamic braking, the number of allowed stops per hour may be limited. Dynamic braking is relatively expensive because of the electronic components required to reconnect the motor as a generator. The braking action is strongest as the leads are reversed and decreases as the motor slows, because the generator action decreases as the motor slows. Therefore, dynamic braking cannot brake a motor to a complete stop and cannot be used for holding a load.

Chapter 13 – Multispeed Motors

1. A series connection of two windings has one pole. The current flows through both coils in the same direction so both coils have the same pole. The poles merge since they are adjacent. The same windings connected in parallel have two poles. The current flows through the coils in opposite directions so the coils have opposite poles. The poles are separate since they are opposite. A motor with four poles when wired in series has eight poles when wired in parallel.

3. In this type of control circuit, pressing the low-speed pushbutton energizes the low-speed starter coil and starts the motor in low speed. The NC interlock contact from the low-speed starter prevents the high-speed starter coil from being energized. The motor can also be started by pressing the high-speed pushbutton. This energizes the high-speed starter coil and starts the motor in high speed. In this case, the NC contact from the high-speed starter prevents the low-speed starter coil from being energized. No matter how the motor is started, the stop pushbutton must be pressed before changing from low to high speed or from high to low speed.

5. The compelling circuit logic compels the operator to first start the motor at low speed before changing to high speed. This arrangement prevents the motor and driven machinery from starting at high speed. The motor and driven machinery are allowed to

accelerate to low speed before accelerating to high speed. The circuit also compels the operator to press the stop pushbutton before changing speed from high to low.

Chapter 14 – Adjustable-Speed Drives

1. The converter section of an adjustable-speed drive is used to convert the AC source power into DC power that can be inverted back to variable-frequency voltage in the inverter section. Converter sections of adjustable-speed drives are single-phase full-wave rectifiers, single-phase bridge rectifiers, or three-phase full-wave rectifiers.

3. The inverter section of an adjustable-speed drive controls the voltage level, voltage frequency, and amount of current that a motor receives. Common inverter designs for AC motors use pulse-width modulation to vary the frequency of the voltage applied to the motor. Common inverter designs for DC motors use SCRs or chopper circuits to vary the voltage applied to the motor.

5. Conductors between an adjustable-speed drive and motor have line-to-line (phase-to-phase) capacitance and line-to-ground (phase-to-ground) capacitance. Longer conductors produce higher capacitance that causes high-voltage spikes in the voltage to a motor. Voltage spikes are a problem because spikes stress motor insulation.

7. A closed-loop vector drive uses shaft-mounted sensors to determine the rotor position and speed of a motor and send the information back to the adjustable-speed drive. Feedback from sensors, like encoders and tachometers, allows an adjustable-speed drive to automatically make adjustments to better meet the motor requirements. An open-loop vector drive, or sensorless vector drive, has no feedback from the motor. An open-loop vector drive uses an internal model of the motor and load to control the speed.

Chapter 15 – Bearings

1. A radial load is a load applied perpendicular to the rotating shaft, straight through the ball toward the center of the shaft. A rotating shaft resting horizontally on, or being supported by, a bearing surface at each end has a radial load due to the weight of the shaft itself. An axial load is a load applied parallel to the rotating shaft. A rotating vertical shaft has an axial load due to the weight of the shaft itself.

3. Press fit is a bearing installation where the bore of the inner rotating ring is smaller than the diameter of the shaft and considerable force must be used to press the bearing onto the shaft. Push fit is a bearing installation where the diameter of the outer fixed ring is smaller than the diameter of the bearing housing and the ring can be pushed in by hand.

5. Devices used for lubricating bearings include grease fittings, pressure cups, oil cups, and oil wicks.

7. Brinell damage is bearing damage where applied force exceeds the yield strength of the surface and presses the balls into the surface to cause indentations. It results in indentations in the surface that are spaced the same as the distance between the balls. Fluting is the elongated and rounded grooves or tracks left by the etching of each roller on the rings of an improperly grounded roller bearing when current passes through the bearing. Fluting marks do not occur at the same distance apart as the ball bearings, as happens with Brinell damage.

Chapter 16 – Drive Systems and Clutches

1. Belts are attached to a motor shaft and a load shaft. As the motor shaft turns, friction between the belt and a pulley provide the torque to needed to turn the other shaft. Pulleys are used to change the speed of a driven load relative to the motor speed. A pulley system depends on friction between the V-belt and the sheave.

3. A timing belt provides a drive system that has no slippage or creep, does not stretch, needs no lubricant, requires low belt tension, and has very little backlash. Timing belts can provide more efficiency than V-belts by combining the advantages of a flat belt drive system and a positive synchronous drive system.

5. A gear drive is a synchronous mechanical drive system that uses the meshing of two or more gears to transfer motion from one shaft to another. Gear drives provide positive contact between gears and the teeth prevent significant slippage. The gear ratio determines how fast the driven gear rotates in relation to the drive gear.

7. A friction clutch uses the force of friction between two or more rotating disks, drums, or cones to engage or disengage the two shafts. Engaging the clutch brings the clutch components together and starts power transfer from the motor to the load. Disengaging the clutch separates the clutch components and stops power transfer from the motor to the load.

Chapter 17 – Motor Alignment

1. Considerations for good alignment include preparation of foundations and base plates; piping strain; anchoring; machinery adjustments during alignment; soft foot, and thermal expansion.

3. A jackscrew is inserted through a block attached to a machine base plate. Jackscrews are used to move a machine horizontally. Shims are used as spacers between machine feet and a base plate. Shim stock is used to make shims. Shims are used to move a machine vertically.

5. In the shaft-deflection method, the shaft is checked for deflection when anchor bolts are tightened or loosened. The anchor bolt for one foot at a time is loosened while keeping the bolts for the other three feet tightened. If the shaft moves more than 0.002″, critical distortion has occurred and correction is necessary.

7. Total indicator readings are found by subtracting one reading from the other reading and taking the absolute value of the difference.

9. $S = g \times \dfrac{D}{d}$

where
S = shim thickness (in in.)
g = measured gap at coupling (in in.)
d = distance between gap measurements (in in.)
D = distance between adjustment and pivot points (in in.)
Given that $g = 0.030$, $D = 10$, and $d = 4$, the shim thickness is calculated as follows:

$S = 0.030 \times \dfrac{10}{4}$
$S = 0.030 \times 2.5$
$S = 0.075″$

Chapter 18 – Troubleshooting Motors

1. As the temperature in a motor increases beyond the temperature rating of the insulation, the life of the insulation is shortened. The higher the temperature, the sooner the insulation will fail. When motor insulation is damaged, the windings short and the motor is no longer functional.

3. Calculate the average voltage, Va.

$$V_a = \frac{468 + 474 + 458}{3}$$

$V_a = 467$ V

Calculate the maximum voltage deviation, V_d.

$V_d = 468 - 467 = 1$
$V_d = 474 - 467 = 7$
$V_d = 458 - 467 = -9$

V_d is the largest deviation from the average. In this case, $V_d = 9$ V. Calculate the voltage unbalance, V_u.

$$V_u = \frac{V_d}{V_a} \times 100$$

$$V_u = \frac{9}{467} \times 100$$

$V_u = 0.0193 \times 100$

$V_u = 1.93\%$

5.
   1. Measure the voltage at the motor terminals. If the voltage is present and at the correct level on all three phases, the motor must be checked. If the voltage is not present on all three phases, the incoming power supply must be checked.
   2. If voltage is present but the motor is not operating, turn the handle of the safety switch or combination starter OFF.
   3. Disconnect the motor from the load.
   4. Turn power ON to try restarting the motor. If the motor starts, check the load.
   5. If the motor does not start, turn it OFF and lock out the power.
   6. Check for open or shorted windings.
   7. Visually inspect the motor. Reset the thermal switch (if present) and try to start the motor. Test voltage at the motor terminals. Turn OFF and lock out the motor. Measure the resistance of the windings to check for an open or short. Inspect and operate the centrifugal switch.

9. The first step is to check the brushes and commutator. Check the brush movement, tension, and brush length. Check for brush location and pressure. Check the commutator film and mica insulation. If the brushes or commutator are not a problem, the next step is to check the rest of the motor, the source, and the load.
   1. Check the voltage at the motor terminals. If the meter shows that the correct voltage is present, the problem is in the motor or load. If the meter shows that the correct voltage is not present, the problem is upstream of the motor and the starter or drive, the control circuit, and the power system should be checked.
   2. If the motor starts without the load, the problem is with the load. If the motor starts without the load but vibrates excessively, the problem is with the coupling or bearings or with loose field windings. If the motor does not start without the load, the problem is probably an open in one of the windings or an open between the brushes and the armature.
   3. Check the field winding for an open or short. Before checking the windings, all power needs to be removed from the motor.

11.
   1. Visually inspect the motor starter.
   2. Measure the incoming voltage coming into the power circuit.
   3. Measure the voltage delivered to the motor from the reduced-voltage power circuit during starting and running.
   4. Measure the motor current draw during starting and after the motor is running.

13. A dual-voltage, wye-connected motor has four separate circuits, three circuits of two leads each (T1-T4, T2-T5, and T3-T6) and one circuit of three leads (T7-T8-T9). A dual-voltage, delta-connected motor has three separate circuits of three leads each (T1-T4-T9, T2-T5-T7, and T3-T6-T8). A continuity tester may also be used to determine the winding circuits on an unmarked motor by connecting one test lead to any motor lead and temporarily connecting the other test lead to each remaining motor lead. All pairs of leads need to be checked with all the remaining motor leads to determine if the circuit is a two- or three-lead circuit. The motor is a wye-connected motor if three circuits of two leads and one circuit of three leads are found. The motor is a delta-connected motor if three circuits of three leads are found.

Chapter 19 – Special-Application Motors

1. Stepper motors operate on the principle that like magnetic poles repel each other and unlike magnetic poles attract each other. In motion control applications, commutation is accomplished through electronic control of the drive voltages and currents in the stator windings. As the commutation changes the voltage and current in the stator windings, the stator poles change and rotor moves from one discrete position to another to align the unlike rotor stator poles.

3. The feedback provides information on the exact location of the shaft or load. The control system adjusts the signal to the servomotor drive to place the shaft or load where it should be. The servomotor drive controls the voltage and current that flows through the servomotor armature and windings.

5. In single-step profile mode, a PLC processor sends individual move sequences to the motor. These sequences include the acceleration and deceleration rates of the move, along with the final or continuous speed. In continuous profile mode, the motion profile is cycled through various accelerations, decelerations, and continuous speed rates to form a blended motion profile. Rather than requiring additional commands for motion speed changes, an interface in continuous mode receives the whole move profile in a single block of instructions.

7. A linear induction motor can be visualized as a standard induction motor with the stator unrolled into a linear stator, or platen. A reaction plate is the equivalent of the rotor in an induction motor. It is typically constructed of a conductive sheet of copper or aluminum.

Appendix

Multispeed Motor Connections	472-473
3φ, 230 V Motors and Circuits—240 V System	474
3φ, 460 V Motors and Circuits—480 V System	475
Motor Frame Dimensions	476-477
Motor Frame Table	478
Motor Frame Letters	478
IEC — Reference Chart	479
AC Motor Characteristics	480
DC and Universal Motor Characteristics	480
Coupling Selections	481
V-Belts	481
Typical Motor Power Factors	481
Common Service Factors	481
Full-Load Currents—DC Motors	482
Full-Load Currents—1φ, AC Motors	482
Full-Load Currents—3φ, AC Induction Motors	482
Typical Motor Efficiencies	482

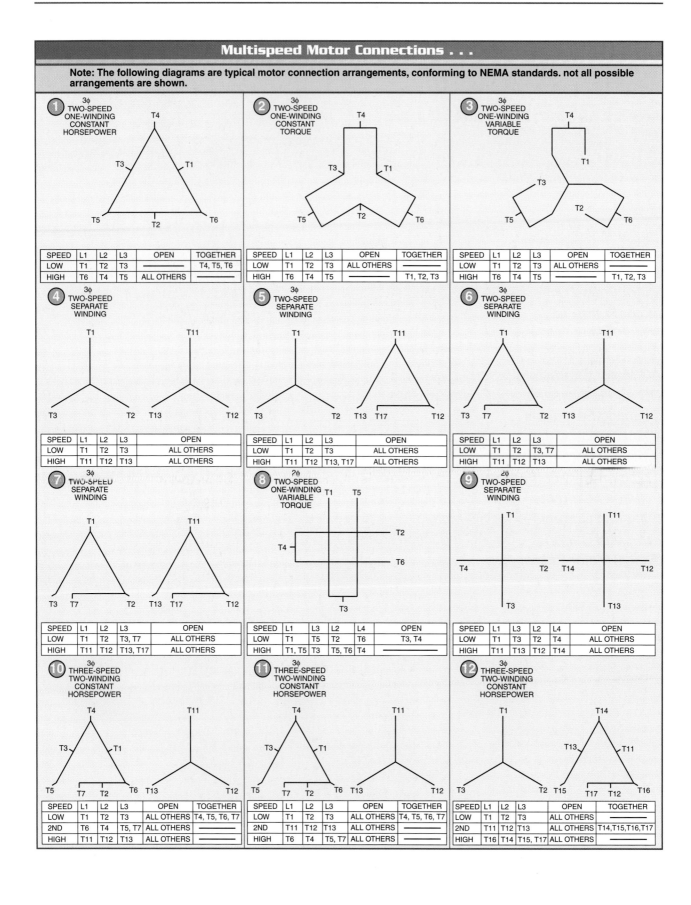

...Multispeed Motor Connections

Note: The following diagrams are typical motor connection arrangements, conforming to NEMA standards. not all possible arrangements are shown.

13 — 3φ THREE-SPEED TWO-WINDING CONSTANT TORQUE

SPEED	L1	L2	L3	OPEN	TOGETHER
LOW	T1	T2	T3, T7	ALL OTHERS	—
2ND	T6	T4	T5	ALL OTHERS	—
HIGH	T11	T12	T13	ALL OTHERS	T1, T2, T3, T7

14 — 3φ THREE-SPEED TWO-WINDING CONSTANT TORQUE

SPEED	L1	L2	L3	OPEN	TOGETHER
LOW	T1	T2	T3, T7	ALL OTHERS	—
2ND	T11	T12	T13	ALL OTHERS	—
HIGH	T6	T4	T5	ALL OTHERS	T1, T2, T3, T7

15 — 3φ THREE-SPEED TWO-WINDING CONSTANT TORQUE

SPEED	L1	L2	L3	OPEN	TOGETHER
LOW	T1	T2	T3	ALL OTHERS	—
2ND	T11	T12	T13, T17	ALL OTHERS	—
HIGH	T16	T14	T15	ALL OTHERS	T11, T12, T13, T17

16 — 3φ THREE-SPEED TWO-WINDING VARIABLE TORQUE

SPEED	L1	L2	L3	OPEN	TOGETHER
LOW	T1	T2	T3	ALL OTHERS	—
2ND	T6	T4	T5	ALL OTHERS	T1, T2, T3
HIGH	T11	T12	T13	ALL OTHERS	—

17 — 3φ THREE-SPEED TWO-WINDING VARIABLE TORQUE

SPEED	L1	L2	L3	OPEN	TOGETHER
LOW	T1	T2	T3	ALL OTHERS	—
2ND	T11	T12	T13	ALL OTHERS	—
HIGH	T6	T4	T5	ALL OTHERS	T1, T2, T3

18 — 3φ THREE-SPEED TWO-WINDING VARIABLE TORQUE

SPEED	L1	L2	L3	OPEN	TOGETHER
LOW	T1	T2	T3	ALL OTHERS	—
2ND	T11	T12	T13	ALL OTHERS	—
HIGH	T16	T14	T15	ALL OTHERS	T11, T12, T13

19 — 3φ FOUR-SPEED TWO-WINDING CONSTANT HORSEPOWER

SPEED	L1	L2	L3	OPEN	TOGETHER
LOW	T1	T2	T3	ALL OTHERS	T4, T5, T6, T7
2ND	T6	T4	T5, T7	ALL OTHERS	—
3RD	T11	T12	T13	ALL OTHERS	T14, T15, T16, T17
HIGH	T16	T14	T15, T17	ALL OTHERS	—

20 — 3φ FOUR-SPEED TWO-WINDING CONSTANT HORSEPOWER

SPEED	L1	L2	L3	OPEN	TOGETHER
LOW	T1	T2	T3	ALL OTHERS	T4, T5, T6, T7
2ND	T11	T12	T13	ALL OTHERS	T14, T15, T16, T17
3RD	T6	T4	T5, T7	ALL OTHERS	—
HIGH	T16	T14	T15, T17	ALL OTHERS	—

21 — 3φ FOUR-SPEED TWO-WINDING CONSTANT TORQUE

SPEED	L1	L2	L3	OPEN	TOGETHER
LOW	T1	T2	T3, T7	ALL OTHERS	—
2ND	T6	T4	T5	ALL OTHERS	T1, T2, T3, T7
3RD	T11	T12	T13, T17	ALL OTHERS	—
HIGH	T16	T14	T15	ALL OTHERS	T11, T12, T13, T17

22 — 3φ FOUR-SPEED TWO-WINDING CONSTANT TORQUE

SPEED	L1	L2	L3	OPEN	TOGETHER
LOW	T1	T2	T3, T7	ALL OTHERS	—
2ND	T11	T12	T13, T17	ALL OTHERS	—
3RD	T6	T4	T5	ALL OTHERS	T1, T2, T3, T7
HIGH	T16	T14	T15	ALL OTHERS	T11, T12, T13, T17

23 — 3φ FOUR-SPEED TWO-WINDING VARIABLE TORQUE

SPEED	L1	L2	L3	OPEN	TOGETHER
LOW	T1	T2	T3	ALL OTHERS	—
2ND	T6	T4	T5	ALL OTHERS	T1, T2, T3
3RD	T11	T12	T13	ALL OTHERS	—
HIGH	T16	T14	T15	ALL OTHERS	T11, T12, T13

24 — 3φ FOUR-SPEED TWO-WINDING VARIABLE TORQUE

SPEED	L1	L2	L3	OPEN	TOGETHER
LOW	T1	T2	T3	ALL OTHERS	—
2ND	T11	T12	T13	ALL OTHERS	—
3RD	T6	T4	T5	ALL OTHERS	T1, T2, T3
HIGH	T16	T14	T15	ALL OTHERS	T11, T12, T13

3ɸ, 230 V Motors and Circuits — 240 V System

Size of Motor		Motor Overload Protection Low-peak or Fusetron®		Switch 115% minimum or HP rated or fuse holder size	Minimum size of starter	Controller termination temperature rating				Minimum size of copper wire and trade conduit	
		Motor less than 40° C or greater than 1.15 SF (Max fuse 125%)	All other motors (Max fuse 115%)			60° C		75° C		Wire size (AWG or kcmil)	Conduit (inches)
HP	AMP					TW	THW	TW	THW		
½	2	2½	2¼	30	00	•	•	•	•	14	½
¾	2.8	3½	3²⁄₁₀	30	00	•	•	•	•	14	½
1	3.6	4½	4	30	00	•	•	•	•	14	½
1½	5.2	6¼	5⁶⁄₁₀	30	00	•	•	•	•	14	½
2	6.8	8	7½	30	0	•	•	•	•	14	½
3	9.6	12	10	30	0	•	•	•	•	14	½
5	15.2	17½	17½	30	1	•	•	•	•	14	½
7½	22	25	25	30	1					10	½
10	28	35	30	60	2	•	•	•		8	¾
									•	10	½
15	42	50	45	60	2	•	•	•		6	1
									•	6	¾
20	54	60	60	100	3	•	•			4	1
25	68	80	75	100	3	•	•	•		3	1¼
								•		3	1
									•	4	1
30	80	100	90	100	3	•	•	•		1	1¼
									•	3	1¼
40	104	125	110	200	4	•	•	•		2/0	1½
									•	1	1¼
50	130	150	150	200	4	•	•	•		3/0	2
									•	2/0	1½
75	192	225	200	400	5	•	•	•		300	2½
									•	250	2½
100	248	300	250	400	5	•	•	•		500	3
									•	350	2½
150	360	450	400	600	6	•	•	•		300-2/ɸ*	2-2½*
									•	4/0-2/ɸ*	2-2*

*two sets of muliple conductors and two runs of conduit required

3φ, 460 V Motors and Circuits — 480 V System											
Size of Motor		Motor Overload Protection Low-peak or Fusetron®		Switch 115% minimum or HP rated or fuse holder size	Minimum size of starter	Controller termination temperature rating				Minimum size of copper wire and trade conduit	
		Motor less than 40° C or greater than 1.15 SF (Max fuse 125%)	All other motors (Max fuse 115%)			60° C		75° C		Wire size (AWG or kcmil)	Conduit (inches)
HP	AMP					TW	THW	TW	THW		
½	1	1¼	1⅛	30	00	•	•	•	•	14	½
¾	1.4	1⁶⁄₁₀	1⁶⁄₁₀	30	00	•	•	•	•	14	½
1	1.8	2¼	2	30	00	•	•	•	•	14	½
1½	2.6	3²⁄₁₀	2⁶⁄₁₀	30	00	•	•	•	•	14	½
2	3.4	4	3½	30	00	•	•	•	•	14	½
3	4.8	5⁶⁄₁₀	5	30	0	•	•	•	•	14	½
5	7.6	9	8	30	0	•	•	•	•	14	½
7½	11	12	12	30	1	•	•	•	•	14	½
10	14	17½	15	30	1	•	•	•	•	14	½
15	21	25	20	30	2	•	•	•	•	10	½
20	27	30	30	60	2	•	•	•		8	¾
									•	10	½
25	34	40	35	60	2	•	•	•		6	1
									•	8	¾
30	40	50	45	60	3	•	•	•		6	1
									•	8	¾
40	52	60	60	100	3	•	•	•		4	1
									•	6	1
50	65	80	70	100	3	•	•	•		3	1¼
									•	4	1
60	77	90	80	100	4	•	•	•		1	1¼
									•	3	1¼
75	96	110	110	200	4	•	•	•		1/0	1½
									•	1	1¼
100	124	150	125	200	4	•	•	•		3/0	2
									•	2/0	1½
125	156	175	175	200	5	•	•	•		4/0	2
									•	3/0	2
150	180	225	200	400	5	•	•	•		300	2½
									•	4/0	2
200	240	300	250	400	5	•	•	•		500	3
									•	350	2½
250	302	350	325	400	6	•	•	•		4/0-2/φ*	2-2*
									•	3/0-2/φ*	2-2*
300	361	450	400	600	6	•	•	•		300-2/φ*	2-1½*
									•	4/0-2/φ*	2-2*

*two sets of multiple conductors and two runs of conduit required

Motor Frame Dimensions . . .

Frame No.	Shaft U	Shaft V	Key W	Key T	Key L	Dimensions – Inches A	B	D	E	F	BA
48	1/2	1 1/2*	flat	3/64	—	5 5/8*	3 1/2*	3	2 1/8	1 3/8	2 1/2
56	5/8	1 7/8*	3/16	3/16	1 3/8	6 1/2*	4 1/4*	3 1/2	2 7/16	1 1/2	2 3/4
143T	7/8	2	3/16	3/16	1 3/8	7	6	3 1/2	2 3/4	2	2 1/4
145T	7/8	2	3/16	3/16	1 3/8	7	6	3 1/2	2 3/4	2 1/2	2 1/4
182	7/8	2	3/16	3/16	1 3/8	9	6 1/2	4 1/2	3 3/4	2 1/4	2 3/4
182T	1 1/8	2 1/2	1/4	1/4	1 3/4	9	6 1/2	4 1/2	3 3/4	2 1/4	2 3/4
184	7/8	2	3/16	3/16	1 3/8	9	7 1/2	4 1/2	3 3/4	2 3/4	2 3/4
184T	1 1/8	2 1/2	1/4	1/4	1 3/4	9	7 1/2	4 1/2	3 3/4	2 3/4	2 3/4
203	3/4	2	3/16	3/16	1 3/8	10	7 1/2	5	4	2 3/4	3 1/8
204	3/4	2	3/16	3/16	1 3/8	10	8 1/2	5	4	3 1/4	3 1/8
213	1 1/8	2 3/4	1/4	1/4	2	10 1/2	7 1/2	5 1/4	4 1/4	2 3/4	3 1/2
213T	1 3/8	3 1/8	5/16	5/16	2 3/8	10 1/2	7 1/2	5 1/4	4 1/4	2 3/4	3 1/2
215	1 1/8	2 3/4	1/4	1/4	2	10 1/2	9	5 1/4	4 1/4	3 1/2	3 1/2
215T	1 3/8	3 1/8	5/16	5/16	2 3/8	10 1/2	9	5 1/4	4 1/4	3 1/2	3 1/2
224	1	2 3/4	1/4	1/4	2	11	8 3/4	5 1/2	4 1/2	3 3/8	3 1/2
225	1	2 3/4	1/4	1/4	2	11	9 1/2	5 1/2	4 1/2	3 3/4	3 1/2
254	1 1/8	3 1/8	1/4	1/4	2 3/8	12 1/2	10 3/4	6 1/4	5	4 1/8	4 1/4
254U	1 3/8	3 1/2	5/16	5/16	2 3/4	12 1/2	10 3/4	6 1/4	5	4 1/8	4 1/4
254T	1 5/8	3 3/4	3/8	3/8	2 7/8	12 1/2	10 3/4	6 1/4	5	4 1/8	4 1/4
256U	1 3/8	3 1/2	5/16	5/16	2 3/4	12 1/2	12 1/2	6 1/4	5	5	4 1/4
256T	1 5/8	3 3/4	3/8	3/8	2 7/8	12 1/2	12 1/2	6 1/4	5	5	4 1/4
284	1 1/4	3 1/2	1/4	1/4	2 3/4	14	12 1/2	7	5 1/2	4 3/4	4 3/4
284U	1 5/8	4 5/8	3/8	3/8	3 3/4	14	12 1/2	7	5 1/2	4 3/4	4 3/4
284T	1 7/8	4 3/8	1/2	1/2	3 1/4	14	12 1/2	7	5 1/2	4 3/4	4 3/4
284TS	1 5/8	3	3/8	3/8	1 7/8	14	12 1/2	7	5 1/2	4 3/4	4 3/4
286U	1 5/8	4 5/8	3/8	3/8	3 3/4	14	14	7	5 1/2	5 1/2	4 3/4
286T	1 7/8	4 3/8	1/2	1/2	3 1/4	14	14	7	5 1/2	5 1/2	4 3/4
286TS	1 5/8	3	3/8	3/8	1 7/8	14	14	7	5 1/2	5 1/2	4 3/4
324	1 5/8	4 5/8	3/8	3/8	3 3/4	16	14	8	6 1/4	5 1/4	5 1/4
324U	1 7/8	5 3/8	1/2	1/2	4 1/4	16	14	8	6 1/4	5 1/4	5 1/4
324S	1 5/8	3	3/8	3/8	1 7/8	16	14	8	6 1/4	5 1/4	5 1/4
324T	2 1/8	5	1/2	1/2	3 7/8	16	14	8	6 1/4	5 1/4	5 1/4
324TS	1 7/8	3 1/2	1/2	1/2	2	16	14	8	6 1/4	5 1/4	5 1/4
326	1 5/8	4 5/8	3/8	3/8	3 3/4	16	15 1/2	8	6 1/4	6	5 1/4
326U	1 7/8	5 3/8	1/2	1/2	4 1/4	16	15 1/2	8	6 1/4	6	5 1/4
326S	1 5/8	3	3/8	3/8	1 7/8	16	15 1/2	8	6 1/4	6	5 1/4
326T	2 1/8	5	1/2	1/2	3 7/8	16	15 1/2	8	6 1/4	6	5 1/4
326TS	1 7/8	3 1/2	1/2	1/2	2	16	15 1/2	8	6 1/4	6	5 1/4
364	1 7/8	5 3/8	1/2	1/2	4 1/4	18	15 1/4	9	7	5 5/8	5 7/8

* Not NEMA standard dimensions

... Motor Frame Dimensions

Frame No.	Shaft U	Shaft V	Key W	Key T	Key L	A	B	D	E	F	BA
364S	1⅝	3	⅜	⅜	1⅞	18	15¼	9	7	5⅝	5⅞
364U	2⅛	6⅛	½	½	5	18	15¼	9	7	5⅝	5⅞
364US	1⅞	3½	½	½	2	18	15¼	9	7	5⅝	5⅞
364T	2⅜	5⅝	⅝	⅝	4¼	18	15¼	9	7	5⅝	5⅞
364TS	1⅞	3½	½	½	2	18	15¼	9	7	5⅝	5⅞
365	1⅞	5⅜	½	½	4¼	18	16¼	9	7	6⅛	5⅞
365S	1⅝	3	⅜	⅜	1⅞	18	16¼	9	7	6⅛	5⅞
365U	2⅛	6⅛	½	½	5	18	16¼	9	7	6⅛	5⅞
365US	1⅞	3½	½	½	2	18	16¼	9	7	6⅛	5⅞
365T	2⅜	5⅝	⅝	⅝	4¼	18	16¼	9	7	6⅛	5⅞
365TS	1⅞	3½	½	½	2	18	16¼	9	7	6⅛	5⅞
404	2⅛	6⅛	½	½	5	20	16¼	10	8	6⅛	6⅝
404S	1⅞	3½	½	½	2	20	16¼	10	8	6⅛	6⅝
404U	2⅜	6⅞	⅝	⅝	5½	20	16¼	10	8	6⅛	6⅝
404US	2⅛	4	½	½	2¾	20	16¼	10	8	6⅛	6⅝
404T	2⅞	7	¾	¾	5⅝	20	16¼	10	8	6⅛	6⅝
404TS	2⅛	4	½	½	2¾	20	16¼	10	8	6⅛	6⅝
405	2⅛	6⅛	½	½	5	20	17¾	10	8	6⅞	6⅝
405S	1⅞	3½	½	½	2	20	17¾	10	8	6⅞	6⅝
405U	2⅜	6⅞	⅝	⅝	5½	20	17¾	10	8	6⅞	6⅝
405US	2⅛	4	½	½	2¾	20	17¾	10	8	6⅞	6⅝
405T	2⅞	7	¾	¾	5⅝	20	17¾	10	8	6⅞	6⅝
405TS	2⅛	4	½	½	2¾	20	17¾	10	8	6⅞	6⅝
444	2⅜	6⅞	⅝	⅝	5½	22	18½	11	9	7¼	7½
444S	2⅛	4	½	½	2¾	22	18½	11	9	7¼	7½
444U	2⅞	8⅜	¾	¾	7	22	18½	11	9	7¼	7½
444US	2⅛	4	½	½	2¾	22	18½	11	9	7¼	7½
444T	3⅜	8¼	⅞	⅞	6⅞	22	18½	11	9	7¼	7½
444TS	2⅜	4½	⅝	⅝	3	22	18½	11	9	7¼	7½
445	2⅜	6⅞	⅝	⅝	5½	22	20½	11	9	8¼	7½
445S	2⅛	4	½	½	2¾	22	20½	11	9	8¼	7½
445U	2⅞	8⅜	¾	¾	7	22	20½	11	9	8¼	7½
445US	2⅛	4	½	½	2¾	22	20½	11	9	8¼	7½

* Not NEMA standard dimensions

Motor Frame Table

Frame No. Series	Third/Fourth Digit of Frame No.							
	D	1	2	3	4	5	6	7
140	3.50	3.00	3.50	4.00	4.50	5.00	5.50	6.25
160	4.00	3.50	4.00	4.50	5.00	5.50	6.25	7.00
180	4.50	4.00	4.50	5.00	5.50	6.25	7.00	8.00
200	5.00	4.50	5.00	5.50	6.50	7.00	8.00	9.00
210	5.25	4.50	5.00	5.50	6.25	7.00	8.00	9.00
220	5.50	5.00	5.50	6.25	6.75	7.50	9.00	10.00
250	6.25	5.50	6.25	7.00	8.25	9.00	10.00	11.00
280	7.00	6.25	7.00	8.00	9.50	10.00	11.00	12.50
320	8.00	7.00	8.00	9.00	10.50	11.00	12.00	14.00
360	9.00	8.00	9.00	10.00	11.25	12.25	14.00	16.00
400	10.00	9.00	10.00	11.00	12.25	13.75	16.00	18.00
440	11.00	10.00	11.00	12.50	14.50	16.50	18.00	20.00
500	12.50	11.00	12.50	14.00	16.00	18.00	20.00	22.00
580	14.50	12.50	14.00	16.00	18.00	20.00	22.00	25.00
680	17.00	16.00	18.00	20.00	22.00	25.00	28.00	32.00

Frame No. Series	Third/Fourth Digit of Frame No.							
	D	8	9	10	11	12	13	14	15
140	3.50	7.00	8.00	9.00	10.00	11.00	12.50	14.00	16.00
160	4.00	8.00	9.00	10.00	11.00	12.50	14.00	16.00	18.00
180	4.50	9.00	10.00	11.00	12.50	14.00	16.00	18.00	20.00
200	5.00	10.00	11.00	—	—	—	—	—	—
210	5.25	10.00	11.00	12.50	14.00	10.00	18.00	20.00	22.00
220	5.50	11.00	12.50	—	—	—	—	—	—
250	6.25	12.50	14.00	16.00	18.00	20.00	22.00	25.00	28.00
280	7.00	14.00	16.00	18.00	20.00	22.00	25.00	28.00	32.00
320	8.00	16.00	18.00	20.00	22.00	25.00	28.00	32.00	36.00
360	9.00	18.00	20.00	22.00	25.00	28.00	32.00	36.00	40.00
400	10.00	20.00	22.00	25.00	28.00	32.00	36.00	40.00	45.00
440	11.00	22.00	25.00	28.00	32.00	36.00	40.00	45.00	50.00
500	12.50	25.00	28.00	32.00	36.00	40.00	45.00	50.00	56.00
580	14.50	28.00	32.00	36.00	40.00	45.00	50.00	56.00	63.00
680	17.00	36.00	40.00	45.00	50.00	56.00	63.00	71.00	80.00

Motor Frame Letters

Letter	Designation
G	Gasoline pump motor
K	Sump pump motor
M and N	Oil burner motor
S	Standard short shaft for direct connection
T	Standard dimensions established
U	Previously used as frame designation for which standard dimensions are established
Y	Special mounting dimensions required from manufacturer
Z	Standard mounting dimensions except shaft extension

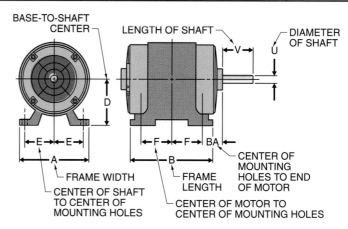

IEC—Reference Chart

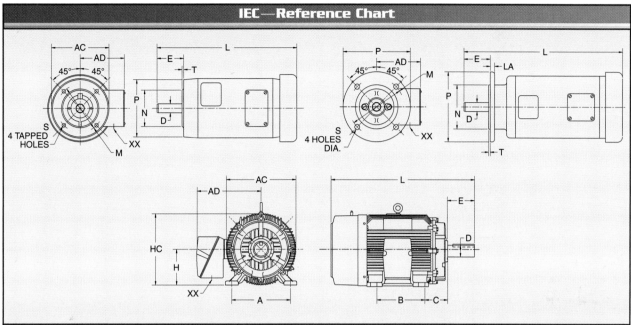

IEC Frame	Type	Foot Mounting* A	B	C	H	Shaft* D	E	B5 Flange* LA	M	N	P	S	T	B14 Face* M	N	P	S	T	General* L	AC	AD	HC	XX
63	300	3.397	3.150	1.570	2.480	.433	.906	.313	4.528	3.740	5.512	.354	.118	2.953	2.362	3.540	M5	.098	†	4.690	4 / 4.567‡	4.760 / 5.375‡	.500 / .880‡
71	300 / 400	4.409	3.543	1.770	2.800	.551	1.181	.313	5.118	4.331	6.299	.393	.138	3.347	2.756	4.130	M6	.098	†	4.690 / 5.690‡	4	5.140 / 5.880‡	.690 / .844‡
80	400 / 500	4.921	3.937	1.969	3.150	.748	1.575	.500	6.496	5.118	7.874	.430	.138	3.937	3.150	4.724	M6	.118	†	5.690 / 6.614‡	4.510 / 5.120	6 / 6.380‡	.880 / .844‡
90	S / L	5.511	3.937 / 4.921	2.205	3.543	.945	1.969	.500	6.496	5.118	7.874	.472	.138	4.530	3.740	5.512	M8	.118	†	6.614 / 5.687‡	5.120 / 4.250‡	6.810 / 6.531‡	.880 / .844‡
100	S / L	6.300	4.409 / 5.512	2.480	3.937	1.102	2.362	.562	8.465	7.087	9.840	.560	.160	5.108	4.331	6.299	M8	1.38	†	7.875	5.875 / 6.060‡	7.906 / 9.440‡	1.062
112	S / M	7.480	4.488 / 5.512	2.760	4.409	1.102	2.362	.562	8.465	7.087	9.840	.560	.160	5.108	4.331	6.299	M8	.138	†	7.875	5.875	8.437	1.062
132	S / M	8.504	5.512 / 7.008	3.504	5.197	1.496	3.150	.562	10.433	9.055	11.811	.560	.160	6.496	5.118	7.874	M8	.138	†	9.562	7.375	10.062	1.062
160	M / L	10	8.268 / 10	4.252	6.299	1.654	4.331	.787	11.811	9.842	13.780	.748	.200	8.465	7.087	9.840	M12	.160	†	12.940	9.510	12.940	1.375
180	M / L	10.984	9.488 / 10.984	4.764	7.087	1.890	4.331		11.811	9.842	13.780	.748	.200						†	15.560	13.120	14.640	2.008
200	L / M	12.520	10.512 / 12.008	5.236	7.874	2.165	4.331		13.780	11.811	15.748	.748							†	17.375	14.125	16.375	2.500
225	S / M	14.016	11.260 / 12.244	5.866	8.858	2.362	5.512		15.748	13.780	17.716	.748							†	19.488	15.079	19.016	2.500
250	S / M	15.984	12.244 / 13.740	6.614	9.843	2.756	5.512												†	20.472	17.992	20.197	2.500
280	S / M	17.992	14.488 / 16.496	7.485	11.025	3.150	6.693												†	24.252	19.567	22.874	2.500
315	S / M	20	16 / 18	8.500	12.400	3.346	6.693												†	29.900	26.880	28.840	4
355	S / L	24	19.690 / 24.800	10	13.980	3.346	6.693												†	29.900	26.880	28.320	4

* In in.
† Contact manufacturer for "L" dimensions
‡ DC motor

AC Motor Characteristics

Motor Type 1φ	Typical Voltage	Starting Ability (Torque)	Size (HP)	Speed Range (rpm)	Cost*	Typical Uses
Shaded-pole	115 V, 230 V	Very low 50% to 100% of full load	Fractional ½ HP to ⅓ HP	Fixed 900, 1200, 1800, 3600	Very low 75% to 85%	Light-duty applications such as small fans, hair dryers, blowers, and computers
Split-phase	115 V, 230 V	Low 75% to 200% of full load	Fractional ⅓ HP or less	Fixed 900, 1200, 1800, 3600	Low 85% to 95%	Low-torque applications such as pumps, blowers, fans, and machine tools
Capacitor-start	115 V, 230 V	High 200% to 350% of full load	Fractional to 3 HP	Fixed 900, 1200, 1800	Low 90% to 110%	Hard-to-start loads such as refrigerators, air compressors, and power tools
Capacitor-run	115 V, 230 V	Very low 50% to 100% of full load	Fractional to 5 HP	Fixed 900, 1200, 1800	Low 90% to 110%	Applications that require a high running torque such as pumps and conveyors
Capacitor-start-and-run	115 V, 230 V	Very high 350% to 450% of full load	Fractional to 10 HP	Fixed 900, 1200, 1800	Low 100% to 115%	Applications that require both a high starting and running torque such as loaded conveyors
3φ Induction	230 V, 460 V	Low 100% to 175% of full load	Fractional to over 500 HP	Fixed 900, 1200, 3600	Low 100%	Most industrial applications
Wound rotor	230 V, 460 V	High 200% to 300% of full load	½ HP to 200 HP	Varies by changing resistance in rotor	Very high 250% to 350%	Applications that require high torque at different speeds such as cranes and elevators
Synchronous	230 V, 460 V	Very low 40% to 100% of full load	Fractional to 250 HP	Exact constant speed	High 200% to 250%	Applications that require very slow speeds and correct power factors

* based on standard 3φ induction motor

DC and Universal Motor Characteristics

Motor Type	Typical Voltage	Starting Ability (Torque)	Size (HP)	Speed Range (rpm)	Cost*	Typical Uses
DC Series	12 V, 90 V, 120 V, 180 V	Very high 400% to 450% of full load	Fractional to 100 HP	Varies 0 to full speed	High 175% to 225%	Applications that require very high torque such as hoists and bridges
Shunt	12 V, 90 V, 120 V, 180 V	Low 125% to 250% of full load	Fractional to 100 HP	Fixed or adjustable below full speed	High 175% to 225%	Applications that require better speed control than a series motor such as woodworking machines
Compound	12 V, 90 V, 120 V, 180 V	High 300% to 400% of full load	Fractional to 100 HP	Fixed or adjustable	High 175% to 225%	Applications that require high torque and speed control such as printing presses, conveyors, and hoists
Permanent-magnet	12 V, 24 V, 36 V, 120 V	Low 100% to 200% of full load	Fractional	Varies from 0 to full speed	High 150% to 200%	Applications that require small DC-operated equipment such as automobile power windows, seats, and sun roofs
Stepping	5 V, 12 V, 24 V	Very low** .5 to 5000 oz/in.	Size rating is given as holding torque and number of steps	Rated in number of steps per sec (maximum)	Varies based on number of steps and rated torque	Applications that require low torque and precise control such as indexing tables and printers
AC/DC Universal	115 VAC, 230 VAC, 12 VDC, 24 VDC, 36 VDC, 120 VDC	High 300% to 400% of full load	Fractional	Varies 0 to full speed	High 175% to 225%	Most portable tools such as drills, routers, mixers, and vacuum cleaners

* based on standard 3φ induction motor
** torque is rated as holding torque

Coupling Selections

Coupling Number	Rated Torque (lb-in)	Maximum Shock Torque (lb-in)
10-101-A	16	45
10-102-A	36	100
10-103-A	80	220
10-104-A	132	360
10-105-A	176	480
10-106-A	240	660
10-107-A	325	900
10-108-A	525	1450
10-109-A	875	2450
10-110-A	1250	3500
10-111-A	1800	5040
10-112-A	2200	6160

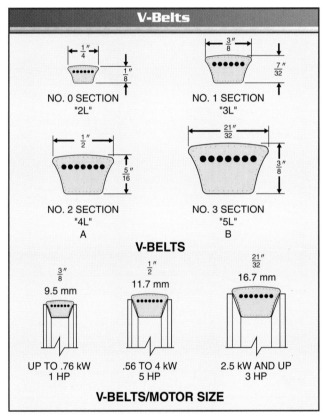

V-BELTS

V-BELTS/MOTOR SIZE

Typical Motor Power Factors

HP	Speed (rpm)	Power Factor at ½ Load	¾ Load	Full Load
0–5	1800	.72	.82	.84
5.01–20	1800	.74	.84	.86
20.1–100	1800	.79	.86	.89
100.1–300	1800	.81	.88	.91

Common Service Factors

Equipment	Service Factors
Blowers	
Centrifugal	1.00
Vane	1.25
Compressors	
Centrifugal	1.25
Vane	1.50
Conveyors	
Uniformly loaded or fed	1.50
Heavy-duty	2.00
Elevators	
Bucket	2.00
Freight	2.25
Extruders	
Plastic	2.00
Metal	2.50
Fans	
Light-duty	1.00
Centrifugal	1.50
Machine tools	
Bending roll	2.00
Punch press	2.25
Tapping machine	3.00
Mixers	
Concrete	2.00
Drum	2.25
Paper mills	
De-barking machines	3.00
Beater and pulper	2.00
Bleacher	1.00
Dryers	2.00
Log haul	2.00
Printing presses	1.50
Pumps	
Centrifugal—general	1.00
Centrifugal—sewage	2.00
Reciprocating	2.00
Rotary	1.50
Textile	
Batchers	1.50
Dryers	1.50
Looms	1.75
Spinners	1.50
Woodworking machines	1.00

Full-Load Currents – DC Motors

Motor rating (HP)	Current (A)	
	120 V	240 V
1/4	3.1	1.6
1/3	4.1	2.0
1/2	5.4	2.7
3/4	7.6	3.8
1	9.5	4.7
1 1/2	13.2	6.6
2	17	8.5
3	25	12.2
5	40	20
7 1/2	48	29
10	76	38

Full-Load Currents – 1φ, AC Motors

Motor rating (HP)	Current (A)	
	115 V	230 V
1/6	4.4	2.2
1/4	5.8	2.9
1/3	7.2	3.6
1/2	9.8	4.9
3/4	13.8	6.9
1	16	8
1 1/2	20	10
2	24	12
3	34	17
5	56	28
7 1/2	80	40

Full-Load Currents – 3φ, AC Induction Motors

Motor rating (HP)	Current (A)			
	208 V	230 V	460 V	575 V
1/4	1.11	.96	.48	.38
1/3	1.34	1.18	.59	.47
1/2	2.2	2.0	1.0	.8
3/4	3.1	2.8	1.4	1.1
1	4.0	3.6	1.8	1.4
1 1/2	5.7	5.2	2.6	2.1
2	7.5	6.8	3.4	2.7
3	10.6	9.6	4.8	3.9
5	16.7	15.2	7.6	6.1
7 1/2	24.0	22.0	11.0	9.0
10	31.0	28.0	14.0	11.0
15	46.0	42.0	21.0	17.0
20	59	54	27	22
25	75	68	34	27
30	88	80	40	32
40	114	104	52	41
50	143	130	65	52
60	169	154	77	62
75	211	192	96	77
100	273	248	124	99
125	343	312	156	125
150	396	360	180	144
200	—	480	240	192
250	—	602	301	242
300	—	—	362	288
350	—	—	413	337
400	—	—	477	382
500	—	—	590	472

Typical Motor Efficiencies

HP	Standard Motor (%)	Energy-Effiecient Motor (%)	HP	Standard Motor (%)	Energy-Efficient Motor (%)
1	76.5	84.0	30	88.1	93.1
1.5	78.5	85.5	40	89.3	93.6
2	79.9	86.5	50	90.4	93.7
3	80.8	88.5	75	90.8	95.0
5	83.1	88.6	100	91.6	95.4
7.5	83.8	90.2	125	91.8	95.8
10	85.0	90.3	150	92.3	96.0
15	86.5	91.7	200	93.3	96.1
20	87.5	92.4	250	93.6	96.2
25	88.0	93.0	300	93.8	96.5

Glossary

A

accelerating circuit logic: A control function that permits the operator to select a motor speed so that the control circuit automatically accelerates the motor to that speed.

adhesive wear or galling: A bonding, shearing, and tearing away of material from two contacting, sliding metals.

adjustable-trip circuit breaker (ATCB): A circuit breaker (CB) whose trip setting can be changed by adjusting the current setpoint, trip-time characteristics, or both, within a particular range.

alignment: The process where the centerlines of two machine shafts are placed within specified tolerances.

alternator: A synchronous machine that produces alternating current (AC).

ambient temperature rating: The maximum allowable temperature of the air surrounding an object.

amortisseur windings or damper windings: Squirrel-cage conducting bars placed in slots on the pole faces and connected at the ends.

anchoring: Any means of fastening a mechanism securely to a base or foundation.

angular misalignment: A condition where one shaft is at an angle to the other shaft.

angular soft foot: A condition that exists when one machine foot is bent and not on the same plane as the other feet.

antifriction bearing or rolling-contact bearing: A bearing that contains rolling elements that provide a low-friction support surface for rotating or sliding surfaces.

apparent power: The power, in VA or kVA, that is the vector sum of true power and reactive power.

armature: The rotating part of a DC motor, consisting of the laminated core with slots for the coils, the main shaft, and the commutator and brushes.

armature reaction: The distortion of the magnetic fields that happens when a current-carrying wire is placed within a fixed magnetic field.

autotransformer starting: A method of reduced-voltage starting that uses a tapped three-phase autotransformer to provide reduced voltage for starting.

axial load or thrust load: A load applied parallel to the rotating shaft.

B

babbitt metal: An alloy of soft metals such as copper, tin, and lead, and a hardening material such as antimony.

backlash: The amount of movement, or play, between meshing gear teeth.

ball bearing: A rolling-contact bearing that permits free motion between a moving part and a fixed part by means of balls confined between inner and outer rings.

base plate: A rigid steel support for firmly coupling and aligning two or more rotating devices.

bearing: A machine component used to reduce friction and maintain clearance between stationary and moving parts.

bearing loss: Any energy lost from friction between the motor shaft, the bearing, and the bearing support.

bevel gear: A gear with straight tapered teeth used in applications where shaft axes intersect.

bimetallic-strip overload relay: A relay consisting of two joined pieces of dissimilar metals with different expansion rates constructed into a strip in such a way that the strip bends when heated and opens a set of contacts.

bolt bound: The condition where the horizontal movement of a machine is restricted because the machine anchor bolts contact the sides of the machine anchor holes.

breakdown torque: The maximum torque a motor can provide without an abrupt reduction in motor speed (stalling).

Brinell damage: Bearing damage where applied force exceeds the yield strength of the surface and presses the balls into the surface to cause indentations.

brush: A sliding contact that rides against a rotating component to provide a connection to a stationary circuit.

brushless exciter: A rectifier assembly mounted on the main rotor shaft along with the exciter generator.

brush neutral: The position of the brushes where commutation can occur with minimal induced voltage in the armature coils.

brush rigging: The entire assembly of the brush, brush holder, insulators, and any wiring included in the assembly.

C

capacitor: A device that stores an electric charge.

capacitor motor: A single-phase motor with a capacitor connected in series with the start windings to produce phase displacement in the start winding.

carrier frequency: The frequency that controls the number of times the solid-state switches in the inverter section of a PWM adjustable-speed drive turn ON and OFF.

cartridge fuse: A snap-in cylindrical fuse constructed of a fusible metallic link or links that is designed to open at predetermined current levels to protect circuit conductors and equipment.

centrifugal switch: A switch that opens to disconnect the start winding when the rotor reaches a certain preset speed and reconnects the start winding when the speed falls below a preset value.

chain drive: A synchronous mechanical drive system that uses a chain to transfer torque from one sprocket to another.

chopper armature-voltage control: A method of using a high-speed chopper circuit to control the voltage applied to the armature of a DC motor.

circuit breaker (CB): An overcurrent protection device with a mechanical mechanism that automatically opens a circuit when an overload condition or short-circuit occurs.

closed-loop vector drive: A vector drive that uses shaft-mounted sensors to determine the rotor position and speed of a motor and send the information back to the adjustable-speed drive.

clutch: A coupling between a motor and a load that connects or disconnects the motor shaft to a drive shaft while the motor is running.

commutation: The process where the armature current is periodically reversed in order to keep the motor torque in the same direction during the entire armature rotation.

commutator: A ring made of insulated segments that keep the armature windings in the correct polarity to interact with the main fields.

compelling circuit logic: A control function that requires the operator to start and operate a motor in a predetermined order.

compensated universal motor: A universal motor with extra windings added to the field poles to reduce sparking.

compensating windings or pole-face windings: Field windings placed in slots on the main poles.

compound motor: A DC motor with the field connected in both series and shunt with the armature.

compound-wound generator: A generator that includes series and shunt field windings.

conducting ring or shorting ring: A metal ring used to electrically connect the bars of a squirrel-cage rotor at the end of the cage frame.

Conrad bearing: A single-row ball bearing that has races that are deeper than normal.

consequent-pole motor: A motor with stator windings that can be connected in two or more different ways so that the number of stator poles can be changed.

conventional current flow: A description of current flow as the flow of positive charges from the positive terminal to the negative terminal of a power source.

core loss: The total energy loss in the stator and rotor cores due to circulating currents and to the magnetic field escaping from the core.

countervoltage or counter EMF (CEMF): A voltage induced in the windings that is opposite in polarity to that of the power supply.

coupling: A device that connects the ends of rotating shafts.

current rating: 1. The amount of current a motor draws when delivering full rated power output. **2.** The continuous amount of current that can be safely carried by an overcurrent protection device without blowing or tripping.

cylindrical roller bearing: A roller bearing having cylinder-shaped rollers.

D

damper windings or amortisseur windings: Squirrel-cage conducting bars placed in slots on the pole faces and connected at the ends.

DC cumulative-compounded motor or DC overcompounded motor: A motor where current flows in the same direction in the series and shunt coils and the flux surrounding the coils adds.

DC differential-compounded motor or DC undercompounded motor: A motor where the current flows in the opposite direction in the series and shunt coils and the resulting net flux is the difference between the two fluxes.

DC field-current control: A method of controlling the voltage applied to a shunt field.

DC injection braking or electric braking: A method of braking in which a DC voltage is applied to the stator windings of a motor after the AC voltage is removed.

DC permanent-magnet motor: A motor that uses magnets, not a coil of wire, for the field windings.

dead short: A short circuit that opens the circuit as soon as the circuit is energized or when the section of the circuit containing the short is energized.

decelerating circuit logic: A control function that permits the operator to select a low motor speed so that the control circuit automatically decelerates the motor to that speed.

design A motor: A seldom-used integral-horsepower 3-phase induction motor design made for full-voltage starting, with normal values of breakdown torque and locked-rotor torque.

design B motor: An integral-horsepower 3-phase induction motor design made for full-voltage starting, with normal values of breakdown torque and locked-rotor torque not exceeding a specified value.

design C motor: An integral-horsepower 3-phase induction motor design made for full-voltage starting and high locked-rotor torque, with locked-rotor current not exceeding a specified value.

design D motor: An integral-horsepower 3-phase induction motor design made for full-voltage starting and locked-rotor torque of at least 275% of full-load torque, with locked-rotor current not exceeding a specified value.

design L motor: An integral-horsepower single-phase motor design made for full-voltage starting and locked-rotor torque higher than for design M, N, and O motors.

design M motor: An integral-horsepower single-phase motor design made for full-voltage starting and locked-rotor current not exceeding specified values.

design N motor: A fractional-horsepower single-phase motor design made for full-voltage starting and locked-rotor current not exceeding specified values.

design O motor: A fractional-horsepower single-phase motor designed for full-voltage starting and locked-rotor current not exceeding specified values, which are higher than for design N motors.

diamagnetic material: A material that can be very weakly magnetized in the opposite direction as the applied magnetic field.

direct current generator or DC generator: A power source that supplies DC when the armature is rotated.

direct current motor or DC motor: A machine that uses DC connected to the field windings and armature to produce shaft rotation.

discharge resistor: A resistor used to discharge any AC potential that builds up in the DC field winding of a rotor in a synchronous motor.

dowel effect: A condition that exists when the bolt hole of a machine is so large that the bolt head forces the washer into the hole opening on an angle.

drum switch: A rotating control device used to switch resistors in or out of a wound-rotor circuit.

dual-voltage motor: A motor that operates at more than one voltage level.

duty cycle or operating time rating: The amount of time a motor can be operated without being turned OFF to allow for cooling.

dynamic braking: A method of motor braking where the braking energy is dissipated as heat in a resistor as a motor is reconnected to act as a generator immediately after it is turned OFF.

E

eddy current: An undesired current circulating in the stator and rotor core caused by magnetic induction.

eddy-current clutch: Uses a magnetic field to couple a motor to a load.

Edison-base fuse: A plug fuse that incorporates a screw configuration that is interchangeable with fuses of other amperage ratings.

electrical pitting: Bearing damage in the form of pits formed on the balls or race caused by electrical discharge through the bearing.

electric braking or DC injection braking: A method of braking in which a DC voltage is applied to the stator windings of a motor after the AC voltage is removed.

electromagnet: A temporary magnet produced when electricity passes through a conductor, such as a coil, that concentrates the magnetic field.

electromagnetic overload relay: A relay that operates on the principle that as the level of current in a circuit increases so does the strength of the magnetic field produced by that current flow.

electromagnetism: The temporary magnetic field produced when electricity passes through a conductor.

electron current flow: A description of current flow as the flow of electrons from the negative terminal to the positive terminal of a power source.

electronic overload relay: A device that has built-in circuitry to sense changes in current and temperature.

enclosure type: The type of protection given to a motor to shield the motor from the outside environment as well as to protect individuals from the electrical and rotating parts of the motor.

eutectic alloy: An alloy that has a constant melting-point temperature due to the combination of the given components.

exciter generator: An assembly consisting of a small three-phase alternator used to supply current to an alternator rotor.

F

false Brinell damage: Bearing damage caused by vibration or other forces that move one ring relative to another and cause axial elliptical indentations at ball positions when the bearing is not rotating.

ferrite: A ferromagnetic nonconducting ceramic alloy with a high permeability.

ferromagnetic material: A material that is easily magnetized and has high permeability.

field frame: The stationary part in a DC motor or generator.

field poles: Metal pieces mounted to the field frame that are used as field windings.

field windings: Magnets or stationary windings used to produce the magnetic field in an alternator or motor.

flat belt: A belt that has a rectangular cross section and relies on friction for proper operation.

flexible coupling: A coupling with a resilient center that flexes under temporary torque or misalignment due to thermal expansion.

flexible drive: A system in which a resilient flexible belt is used to drive one or more shafts.

fluting: The elongated and rounded grooves or tracks left by the etching of each roller on the rings of an improperly grounded roller bearing when current passes through the bearing.

flux density: The amount of concentration of magnetic flux through a specific area.

flux-linkage loss: The loss of flux in the air gap because the air gap has increased reluctance compared to the cores.

foundation: An underlying base or support.

fractional-horsepower (FHP) V-belt: A V-belt designed for light-duty applications.

frame size: A number designating standard dimensions of a motor housing, shaft, and mounting holes.

frequency rating: The power line frequency at which a motor is designed to operate.

fretting corrosion: The rusty appearance that results when two metals in contact are vibrated, rubbing loose minute metal particles that become oxidized.

friction bearing: A bearing consisting of a stationary bearing surface, such as machined metal or pressed-in bushings, that provides a low-friction support surface for rotating or sliding surfaces.

friction clutch: Uses the force of friction between two or more rotating disks, drums, or cones to engage or disengage the two shafts.

full-load current (FLC): The amount of current drawn when the motor is connected to the maximum load the motor is designed to drive.

full-load torque: The torque required to produce the rated power at full speed of the motor.

full-voltage starting: A method of starting a motor with the full line voltage placed across the terminals.

fundamental frequency: The frequency of the voltage used to control motor speed.

fuse: An overcurrent protection device with a fusible link that melts and opens the circuit when an overcurrent occurs.

G

galling or adhesive wear: A bonding, shearing, and tearing away of material from two contacting, sliding metals.

gearbox: A sealed container that has an input shaft and an output shaft and houses at least one set of mating gears.

gear drive: A synchronous mechanical drive system that uses the meshing of two or more gears to transfer motion from one shaft to another.

gear ratio: The ratio between the diameter of the drive gear and the driven gear.

generator: A machine that converts mechanical energy into electrical energy by means of electromagnetic induction.

grease fitting: A hollow tubular fitting used to direct grease to bearing components.

ground fault: An unintentional connection between an ungrounded conductor and any grounded raceway, box, enclosure, or fitting.

grounded circuit: A circuit in which the current leaves its normal path and travels to the frame of the motor.

H

helical gear: A gear with teeth that are not parallel to the shaft axis.

herringbone gear: A gear with two rows of helical teeth.

high-inertia load: A load that has a relatively large amount of momentum.

hysteresis: A loss due to the power consumed to realign the magnetic domains in the iron twice every electrical cycle.

I

induced soft foot: A condition that exists when external forces are applied to the machine.

inductance: The property of a device or circuit that causes it to store energy in an electromagnetic field.

induction motor: An electric motor that uses the principles of mutual induction to develop current and torque in the rotor.

inductive reactance: The opposition to the flow of AC in a circuit due to inductance.

inrush current, or starting current, or locked-rotor current (LRC): The amount of current a motor draws on startup or when the rotor is locked.

instantaneous-trip circuit breaker (ITB): A circuit breaker (CB) with no delay between the fault- or overload-sensing element and the tripping action of the device.

insulation class: A code letter signifying the maximum operating temperature of the insulation used in a motor.

insulation spot test: A short-term test that verifies the integrity of insulation on electrical devices.

interpoles: Auxiliary poles placed between the main field poles of the motor.

interrupting rating: The maximum amount of current that can be safely applied to an OCPD while still maintaining its physical integrity when reacting to fault currents.

inverse-time circuit breaker (ITCB): A circuit breaker (CB) with an intentional delay between the time when the fault or overload is sensed and the time when the CB operates.

inverter drive or scalar drive: A standard adjustable-speed drive that uses pulse-width modulation to control the speed and torque.

inverter duty motor: An electric motor specifically designed to work with AC adjustable-speed drives.

J

jackscrew: A screw inserted through a block attached to a machine base plate that is used to move a machine horizontally.

L

laser rim-and-face alignment: A method of coupling alignment in which laser devices are placed opposite each other to measure alignment.

left-hand coil rule: An explanation of the direction of a magnetic field around a coil relative to the direction of the current in the coil, where a left hand is used to illustrate the relationship.

left-hand conductor rule: An explanation of the direction of a magnetic field around a conductor relative to the direction of the current in the conductor, where a left hand is used to illustrate the relationship.

left-hand generator rule: An explanation of the relationship between the direction of motion of the conductor within a magnetic field in a generator, the direction of the magnetic field existing around the conductor, and the direction of induced current in a conductor.

linear induction motor: An AC motor that uses induction to create linear movement.

locked in step: The lack of rotation when the stator's field and the rotor's field are parallel to one another.

locked rotor: A condition when a motor is loaded so heavily that the motor shaft cannot turn.

locked-rotor current (LRC), or starting current, or inrush current: The amount of current a motor draws on startup or when the rotor is locked.

locked-rotor indicating code letter: A designation for the range of locked-rotor current draw per motor horsepower.

locked-rotor torque: The torque a motor produces when the rotor is stationary and full power is applied to the motor.

loss-of-excitation relay: A relay used to protect a synchronous motor from damage caused by loss of excitation in the DC winding.

M

magnetic field: A force produced by a magnet that exerts a force on moving electric charges or on other magnets.

magnetic flux (field flux): The invisible lines of force that make up the total quantity of a magnetic field.

magnetic motor starter: An electrically operated switch that includes motor overload protection.

magnetism: A force caused by a magnet that acts at a distance on other magnets.

magnet: A substance that produces a magnetic field.

manual contactor: A control device that uses pushbuttons to energize or de-energize the load connected to it.

manual starter: A contactor with an added overload protection device.

mechanical drive: A combination of mechanical components that transfer torque from one location to another.

melting-alloy overload relay (heater): An overload relay that uses a heater coil to produce the heat to melt a eutectic alloy.

microstepping: Stepper motor control that divides the standard stepper control pulse intervals into smaller intervals and sends modified current to the motor coils to increase the resolution of a stepper motor system.

misalignment: The condition where the centerlines of two machine shafts are not aligned within specified tolerances.

misalignment wear: Bearing damage that occurs when the two bearing rings are not aligned with one another and the rolling-contact points cause eccentric wear.

motor bearing: A machine component used to reduce friction and maintain clearance between stationary and moving parts.

motor control circuit: The circuit of a motor control apparatus or system that carries electric signals directing the performance of the controller, but does not carry the main power current.

motor efficiency: 1. A measure of the effectiveness with which a motor converts electrical energy to mechanical energy. **2.** The ratio of useful work performed by a motor to the energy used by the motor to produce the work.

motor-generator (M-G) set: A motor and a generator with shafts connected and used to convert one form of power to another form.

motor shaft: A cylindrical bar used to carry the revolving rotor and to transfer power from the motor to the load.

mutual induction: The ability of an inductor (coil) in one circuit to induce a voltage in another circuit or conductor.

N

nameplate: A metal tag permanently attached to an electric motor frame that gives the required electrical ratings, operating ratings, and mechanical-design codes of the motor.

narrow V-belt: A V-belt having a smaller cross section and a higher profile than a standard belt that is designed for heavy-duty applications.

needle bearing: An rolling-contact roller-type bearing with long rollers of small diameter.

NEMA design letter: A code letter representing a National Electrical Manufacturers Association (NEMA) motor classification for the torque and current curves of a motor.

neutral plane: A line through the armature cross section that is perpendicular to the maximum amount of magnetic flux.

non-adjustable-trip circuit breaker (NATCB): A circuit breaker (CB) whose settings for the amperage-trip setpoint or the time-trip setpoint cannot be changed.

noncompensated universal motor: A universal motor without extra windings added to the field poles.

nonparallel misalignment: Misalignment where two pulleys or shafts are not parallel.

non-time-delay fuse (NTDF): A single-element fuse that can detect an overcurrent and open the circuit almost instantly.

O

offset misalignment: A condition where two shafts are parallel but are not on the same axis.

oil cup: An oil reservoir located on a bearing housing to provide lubrication to a bearing.

open circuit: An electrical circuit that has an incomplete path that prevents current flow.

open-loop vector drive or sensorless vector drive: A vector drive that has no feedback from the motor.

open motor enclosure: A motor enclosure with openings to allow passage of air to cool the windings.

operating time rating or duty cycle: The amount of time a motor can be operated without being turned OFF to allow for cooling.

out-of-step relay (OSR): An overload relay that is used to protect a synchronous motor from damage from induced currents caused by the rotor falling out of step with the rotating stator field.

overcurrent: Any current over the normal current level.

overcurrent protection device (OCPD): A set of fuses or circuit breakers added to provide overcurrent protection of a switched circuit.

overcycling: The process of turning a motor ON and OFF more often than the motor design allows.

overload: 1. An excessive current that is confined to the normal current-carrying conductors and is caused by a load that exceeds the full-load torque rating of the motor. **2.** The application of too much load to a motor.

overload class rating: An indicator of the maximum length of time it takes for the overload relay to trip at 600% overload.

overload rating: The load above the normal load that can be carried for a specified period.

P

parallel soft foot: A condition that exists when one or two machine feet are higher than the others and parallel to the base plate.

paramagnetic material: A material that can be weakly magnetized in the same direction as the applied magnetic field.

part-winding starting: A method of reduced-voltage starting that applies voltage to only part of the motor coil windings for starting and then applies power to the remaining coil windings for normal running.

permanent magnet: A magnet that can hold its magnetism for a long period.

permeability: A measure of the ability of a material to conduct magnetic flux.

permissible temperature rise: The difference between the ambient temperature and the nameplate ambient rating of a motor.

phase: The power phase (1φ, 3φ, or DC) that a motor requires for operation.

phase unbalance: The unbalance that occurs when lines are out of phase.

pigtail: An extended, flexible connection or a braided copper conductor.

plug fuse: A screw-in fuse that uses a metallic strip that melts when a predetermined amount of current flows through it.

plugging: A method of motor braking in which the motor connections are reversed so that the motor develops a countertorque that acts as a braking force.

polarized field frequency relay (PFFR): A relay used to apply current to the DC field windings of a synchronous motor and to remove the discharge resistor from the starting circuit.

pole-face windings or compensating windings: Field windings placed in slots on the main poles.

power factor: The ratio of true power, in W or kW, to apparent power, in VA or kVA, in a circuit.

power factor correction capacitor: A capacitor used to improve a facility's power factor by improving voltage levels, increasing system capacity, and reducing line losses.

power rating: The amount of power a motor can deliver to a load.

press fit: A bearing installation where the bore of the inner rotating ring is smaller than the diameter of the shaft and considerable force must be used to press the bearing onto the shaft.

pressure cup: A pressurized grease reservoir that provides constant lubrication to a bearing.

primary-resistor starting: A method of reduced-voltage starting that places resistors in series in the motor power circuit to reduce the voltage to the motor.

prime mover: The power source used to create the relative motion between the coil and the magnetic field.

pull-in torque: The maximum torque required to accelerate a synchronous motor into synchronization at the rated voltage and frequency.

pull-out torque: The torque produced by a motor overload that pulls the rotor out of synchronization.

pull-up torque: The accelerating torque required to bring a load up to the correct speed.

pulse-width modulation (PWM): A method of controlling the voltage sent to a motor by controlling the amount of time a transistor is ON and conducting current.

push fit: A bearing installation where the diameter of the outer fixed ring is smaller than the diameter of the bearing housing and the ring can be pushed in by hand.

R

race: The bearing surface of a rolling-contact bearing that supports the rolling elements.

radial load: A load applied perpendicular to the rotating shaft, straight through the ball toward the center of the shaft.

reactive power: The power, in VAR or kVAR, stored and released by the magnetic field around inductors and capacitors.

regenerative braking: A method of dynamic braking that reuses the braking energy to the AC source instead of dissipating the energy as heat.

reluctance torque: Torque developed by the salient rotor poles before the poles are excited by external DC power.

resistance loss (copper loss or I^2R loss): The energy loss in a motor due to current flowing through conductors and coils that have resistance.

retentivity: A measure of the ability of a magnet to retain magnetism after the magnetizing force has been removed.

reverse dial alignment: A method of coupling alignment that uses two dial indicators to take readings of opposing sides of coupling rims, giving two sets of shaft runout readings.

revolving-field alternator: An alternator where a magnetic field is created in the rotor, which turns within the fixed stator windings, and AC power is supplied through the stator windings.

revolving-rotor alternator: An alternator where a fixed magnetic field is created in the stator, with the rotor turning within the stator, and AC power is supplied through the rotor slip rings and brushes.

right-hand motor rule: An explanation of the relationship between the direction of the applied current in a conductor, the direction of the magnetic field around the conductor, and the direction of the induced motion of the conductor within a motor.

rigid coupling: A device that joins two precisely aligned shafts within a common frame.

rim-and-face alignment: A method of coupling alignment in which the offset and angular misalignment of two shafts is determined by measuring at the rim and face of a coupling.

roller bearing: A rolling-contact bearing that has parallel or tapered steel rollers confined between inner and outer rings.

roller chain: A chain that contains roller, pin, and connecting master links.

rolling-contact bearing or antifriction bearing: A bearing that contains rolling elements that provide a low-friction support surface for rotating or sliding surfaces.

rotary phase converter: A 3-phase power source consisting of a single-phase power source, a standard 3-phase induction motor, and a standard 3-phase disconnect used to supply power to a 3-phase load.

rotor: **1.** The rotating, moving part of a motor, consisting of a core and windings, which convert the rotating magnetic field of the stator into the torque that rotates the shaft. **2.** The rotating moving part of an alternator or generator, consisting of a core and windings, that converts torque to magnetic energy.

rotor frequency: The rate at which the stator magnetic field passes the poles in the rotor.

runout: An out-of-round variation from a true circle.

S

salient pole (projecting pole): A pole that extends away from the core toward the stator or extends away from the stator toward the rotor.

saturation: 1. The condition where a magnetic core has substantially all the magnetic domains aligned with the field and any increases in current no longer result in a stronger electromagnet. **2.** The loss of magnetic lines of flux out of the core when the core cannot carry any more lines of flux with an increase in current.

scalar drive or inverter drive: A standard adjustable-speed drive that uses pulse-width modulation to control the speed and torque.

Schrage motor: A 3-phase AC motor with a rotor fed by a commutator and a set of brushes connected to an external circuit.

SCR armature-voltage control: A method of using SCR bridge rectifiers to control the voltage applied to the armature of a DC motor.

self-excited shunt field: A shunt field connected to the same power supply as the armature.

self-induction: The ability of an inductor in a circuit to generate inductive reactance, which opposes change in the circuit.

sensorless vector drive or open-loop vector drive: A vector drive that has no feedback from the motor.

separately excited shunt field: A shunt field connected to a different power supply than the armature.

series motor: A DC motor that has the field winding connected in series with the armature.

series-wound generator: A generator that has its field windings connected in series with the armature and the external circuit (load).

service factor rating: A multiplier that represents the amount of load, beyond the rated load, that can be placed on a motor without causing damage.

servomotor: A motor that uses feedback signals to provide position and speed control.

shaded pole: A short-circuited winding, consisting of a single turn of copper wire, that acts on only a portion of the stator windings.

shaded-pole motor: An AC motor that uses a shaded stator pole for starting.

shading coil: A single turn of copper wire wrapped around part of the salient pole of a shaded-pole motor.

sheave: A grooved wheel used to hold a V-belt.

short circuit: 1. An excessive current that leaves the normal current-carrying path by going around the load and back to the power source. **2.** A circuit in which current takes a shortcut around the normal path of current flow.

shorting ring or conducting ring: A metal ring used to electrically connect the bars of a squirrel-cage rotor at the end of the cage frame.

shunt motor: A DC motor that has the field wiring connected in parallel with the armature.

shunt-wound generator: A generator that has its field windings connected as a shunt in parallel with the armature and the external circuit.

silent chain: A synchronous chain that consists of a series of links joined together with bushings and pins.

silicon-controlled rectifier (SCR): A solid-state device used to switch a set of resistors in or out of a wound-rotor circuit.

single phasing: The operation of a motor designed to operate on three phases operating on only two phases because one phase is lost.

single-voltage motor: A motor that operates at only one voltage level.

slip: The difference between the synchronous speed and rated speed of a motor.

slip ring: A metallic ring mounted on a motor shaft and electrically insulated from the shaft.

soft foot: A condition that occurs when one or more feet of a machine do not make complete contact with its base plate.

soft starter: A device that provides a gradual voltage increase (ramp up) during AC motor starting.

solid-state starter: A motor starter that uses a solid-state device, such as an insulated gate bipolar transistor (IGBT) or silicon-controlled rectifier (SCR), to control motor voltage, current, torque, and speed during acceleration.

sound loss: Energy lost in a motor when the motor makes noise.

spalling: General wear of rolling contacts where metal pieces flake away the surfaces in contact, leaving a roughened surface.

speed rating: The approximate speed at which the rotor of a motor rotates when delivering rated power to a load.

split-phase motor: A single-phase AC motor that includes a run winding and a resistive start winding that creates a phase-shift for starting.

springing soft foot: A condition that exists when a dial indicator at the shaft shows soft foot but feeler gauges show no gaps.

spur gear: A gear with straight teeth cut parallel to the shaft axis.

squirrel-cage rotor: An induction motor rotor consisting of conductors made from solid bars assembled into a cage frame resembling a squirrel cage.

stabilizing field winding: A small series field winding placed over the top of a shunt field winding that improves stability of the fields while running with reduced current in the field circuit.

standard V-belt: A V-belt designed for moderate-duty applications.

starting current, or locked-rotor current (LRC), or inrush current: The amount of current a motor draws on startup or when the rotor is locked.

stator: 1. The fixed, unmoving part of a motor, consisting of a core and windings, or coils, that converts electrical energy to the energy of a magnetic field. **2.** The fixed unmoving part of a generator, consisting of a core and windings, that converts the energy of a magnetic field to electrical energy.

stepper motor: A motor that uses discrete voltage and current pulses to control the movement of a load.

straightedge alignment: A method of coupling alignment in which a straightedge is used to align couplings and a taper gauge is used to measure the gap between the coupling halves.

surface reaction: Damage to bearing surfaces caused by chemical or electrochemical reactions between the lubricant and the metal of the bearing.

synchronous condenser: A synchronous motor operated at no load in order to provide power factor correction.

synchronous motor: A motor that rotates at exactly the same speed as the rotating magnetic field of the stator.

synchronous speed: The theoretical speed of a motor based on line frequency and the number of poles of the motor.

synchronous torque: The torque required to keep the rotor turning at synchronous speed and represents the torque available to drive the load.

synchroscope: A device that indicates whether two AC sources to be connected in parallel are in the correct phase relationship.

T

tapered roller bearing: A roller bearing having tapered rollers.

temperature damage: Bearing damage caused by high temperatures in the bearing.

temperature rise: The difference between the winding temperature of a running motor and the ambient temperature.

temporary magnet: A magnet that retains only trace amounts of magnetism after a magnetizing force has been removed.

thermal expansion: The dimensional change in a substance due to a change in temperature.

thermal overload relay: An overload relay that uses the resistive heating to open a set of contacts.

three-pole rheostat: A switch with tapped resistors used to switch a set of resistors in or out of a wound-rotor circuit.

thrust damage: Bearing damage due to axial force.

thrust load or axial load: A load applied parallel to the rotating shaft.

tie-down troubleshooting method: A testing method in which one DMM probe is connected to either the L2 (neutral) or L1 (hot) side of a circuit and the other DMM probe is moved along a section of the circuit to be tested.

time-delay fuse (TDF): A dual-element fuse that can detect and remove a short circuit almost instantly, but allow small overloads to exist for a short period.

timing belt: A flat belt containing gear teeth that are used for synchronous drive systems.

torque: A turning or twisting force that causes an object to rotate.

torque angle: The angle between the rotor and stator fields as a synchronous motor is running under load.

totally enclosed motor enclosure: A motor enclosure that prevents air from entering the motor.

trajectory control: The predictable path of speed changes a stepper motor undergoes as it moves a load from its starting position to its desired end position.

true power: The power, in W or kW, drawn by a motor that produces useful work.

type S fuse: A plug fuse that incorporates a screw-and-adapter configuration that is not interchangeable with fuses of another amperage rating.

U

universal motor: A motor that can be operated on either single-phase AC or DC power.

usage rating: A description of the expected or allowed application of a motor.

V

V-belt: A flexible drive belt that has a cross-section in the shape of a V.

vector drive: A variable-speed drive that uses a microprocessor and PWM, usually with feedback, to calculate the precise vector between voltage and frequency that is needed to provide better control of speed and torque.

voltage rating: 1. The voltage level that a motor can use. **2.** The maximum amount of voltage that can safely be applied to an overcurrent protection device and determines the ability of a fuse to suppress the internal arcing after the fusible link melts.

voltage regulation: The ability of a source to vary the output voltage in order to maintain system voltage as the load varies.

voltage surge: Any higher-than-normal voltage that temporarily exists on one or more of the power lines.

voltage unbalance: The unbalance that occurs when the voltages at the motor terminals are not equal.

volts-per-hertz ratio (V/Hz): The ratio of voltage to frequency in a motor.

W

windage loss: Energy lost by blowing air past a motor to remove heat.

worm: A screw thread that rotates the worm gear.

worm gear: A spur gear that is driven by a worm.

wound-rotor motor: An induction motor with the squirrel-cage conductor bars replaced with coils of wire, and with added slip rings, brushes, and a resistor circuit.

wound-rotor motor regulator: A device containing high-wattage resistors that is designed to control the speed of a wound-rotor motor and operate in variable-speed mode for as long as needed.

wound-rotor motor starter: A device containing low-wattage resistors that is designed to provide rotor circuit resistance during startup and remove that resistance when the motor is up to speed.

Index

A

AC alternators, *190,* 190–200
  operating principles, 192–198
  output frequency, 194
  output voltage, 193–194
  overload rating, 198
  paralleling, 198–200
    automatic synchronization, 200, *201*
    lights-out methods, 199, *200*
    manual synchronization, 199
  prime movers, 198
  revolving-field, 192
  revolving-rotor, 193
  rotor, 191, *192*
    speed, 194
    torque, *191*
  single-phase, 195, *196*
  stator, *190,* 190–191
  three-phase, 197
accelerating circuit logic, 287, *288*
AC drives, 306–310
  closed-loop vector, 307–308
  current-source-inverter (CSI), 308–310
  inverter, 306, *307*
  variable-voltage-inverter (VVI), 308–310
  vector, 307
AC motors
  dynamic braking, 273
  lead length, 304–306, *305*
  speed control, *299,* 299–310
  troubleshooting, 412–417
    single-phase, 414–417
    three-phase, 412–413, *413*
adhesive wear, 340
adjustable-speed drives, *296,* 296–313
  AC, *299,* 306–310
  applied frequency, 301–304
    pulse-width modulation (PWM), 302–303

applied voltage, 306
  carrier frequency, *303,* 303–304
    power derating curve, *304*
  converter section, 296–297
  DC bus section, 298
  frequencies, *303*
  fundamental frequency, 303
  installation, 299
  inverter duty motors, 300–301
  inverter section, 298
  names, 300
  three-phase full-wave rectifiers, *297*
adjustable-trip circuit breaker (ATCB), 63
alignment, 372–395, *376*
  adjustments, *376,* 376–378
    jackscrews, 376, *377*
  anchoring, 374–375, *375*
    bolt bound, 374
    dowel effect, 375
  angular movements, 387–389, *388*
  dial indicators, *383,* 383–384
    rod sag, 384
  horizontal movements, 387
  methods, 384–395, *385*
    accuracy, 385–386
    electronic reverse dial alignment, 395
    laser rim-and-face alignment, *393,* 393–394
    reverse dial alignment, *394,* 394–395
    rim-and-face alignment, 392–394, *393*
  motor placement, 373–382
    base plates, *373*
    foundations, 373
  piping strain, 373, *374*
  runout, 378
  shim stock, 376–378, *377*
  soft foot, 378–381, *379*
    angular, 378–379
    induced, 379
    measuring, *380,* 380–381

    parallel, 378
    springing, 379
  straightedge, *389,* 389–392, *390*
  thermal expansion, 381–382, *382*
  tolerance, 386
alternating current, 9–10, *10*
alternators, *190,* 190–200
  operating principles, 192–198
  output frequency, 194
  output voltage, 193–194
  overload rating, 198
  paralleling, 198–200
    automatic synchronization, 200, *201*
    lights-out methods, 199, *200*
    manual synchronization, 199
  prime movers, 198
  revolving-field, 192
  revolving-rotor, 193
  rotor, 191, *192*
    speed, 194
    torque, *191*
  single-phase, 195, *196*
  stator, *190,* 190–191
  three-phase, 197
ambient temperature rating, 34, *35*
ambient temperature derating, *36*
permissible temperature rise, 34, *35*
amortisseur windings, 142
anchoring, 374–375, *375*
  bolt bound, 374
  dowel effect, 375
angular misalignment, 372, *392*
angular soft foot, 378–379, *379*
antifriction bearings, 324–327
apparent power, 103
armature, *210,* 210–211
  commutation, 214–215, *215*
  commutator, 210
  neutral plane, 213
  reaction, 212
  torque, *213,* 213–214

armature reaction, 212
at-each-foot method, 381
autotransformer starting, *244*, 244–246
   circuit, 244–246, *245*
axial loads, 324

B

babbitt metals, 328
backlash, 358
ball bearings, 325–326, *326*
base plates, 373
bearing loss, *105*, 106
bearings, 324–341
   adhesive wear, 340
   Brinell damage, 339
   contamination, 336
   electrical pitting, 341
   failure analysis, 338–341
   false Brinell damage, 339
   fluting, 341
   fretting corrosion, 340
   galling, 340
   heating and cooling, 330–331, *331*
   installation, 329–332
     parts preparation, 329–330
   lubrication, 332–336, *333*
   machine run-in, 336–337
   misalignment wear, 340
   mounting, 330–332, *331*
   operation, 332–337
   removal, *337*, 337–341
   selection, 329
   spalling, 338
   surface reaction, 338
   temperature damage, 339, *340*
   thrust damage, 340, *341*
belt drives, *350*, 350–355
   centrifugal tension, 350, *351*
   flat belts, 353
   speed, *355*
   tension, 411
   timing belts, 353
   V-belts, 351–352, *352*
     fractional-horsepower (FHP) V-belts, 351
     narrow V-belts, 351
     pulleys, *353*, 353–354
     standard V-belts, 351
     wedging action, *352*
bevel gear, 358, *359*
bimetallic-strip overload relay, 70

bolt bound, 374, *375*
braking, 264–274
   comparison of methods, 274
   dynamic braking, 272–274
     AC motor, 273
     DC motor, 272, *273*
     electric motor drive, 273, *274*
     regenerative braking, 272
   electric braking, *269*, 269–271
     DC electric braking circuits, 270, *271*
     limitations, 271
     operation, 270
   friction braking, 264–265
     brake shoes, 265
     friction brakes, *264*
     limitation, 265
     solenoid operation, 264, *265*
   plugging, 265–268, *266*
     continuous plugging, 266, *267*
     for emergency stops, 266–267, *267*
     limitations, 269
     plugging switch operation, 266
     using timing relays, 267–269, *268*
Brinell damage, 339
brushes, 128, 210–211, *211*
   brush neutral, 213
   brush rigging, 210
   pigtail, 210
   troubleshooting, 417–419, *418*
brushless exciter, 147–148, *148*, 192
brush neutral, 213
brush rigging, 128, 210

C

capacitor, 176, *178*
capacitor motors, 176–181
   capacitor-run, 179–181, *180*
     speed control, *180*, 180–181
     three-speed, 181
   capacitor-start, 178–179, *179*
   capacitor start-and-run, *181*
   construction, 176–177
   operating principles, 178
   phase angle between windings, *178*
   troubleshooting, *416*, 416–417
carrier frequency, *303*, 303–304
   power derating curve, *304*
cartridge fuse, 59
centrifugal switch, 175
centrifugal tension, 350, *351*
chain drives, 355–357

roller chain, *355*, 355–356
silent chain, *356*, 356–357
sprockets, *356*
chopper armature-voltage control, 313
circuit breaker (CB), *57*, *61*, 61–63
   adjustable-trip circuit breaker (ATCB), 63
   instantaneous-trip circuit breaker (ITB), 63
   inverse-time circuit breaker (ITCB), 63
   non-adjustable-trip circuit breaker (NATCB), 63
   operation, *62*
circuit logic
   accelerating, 287, *288*
   compelling, 286, *287*
   decelerating, 287–288, *288*
closed-circuit transitions, 239, *240*
closed-loop vector drives, 307–308
clutches, 360–362
   eddy-current clutches, 361–362, *362*
   friction clutches, 361
commutation, 214–215, *215*
   in DC generators, *226*
commutators, 210
   troubleshooting, 419
compelling circuit logic, 286, *287*
compensated universal motor, 457
compensating windings, 210
compound-wound generator, 228
compound motors, *221*, 221–223
   DC differential-compounded motor, 222
   DC undercompounded motor, 222
   torque, 222
conducting ring, 116
Conrad bearing, 326
consequent-pole motors, *282*, 282–284
   circuits, 285–288
     logic, 285–288
     two-speed, *286*, *287*
   connections, *283*
contactors
   troubleshooting, 423–424
control circuits
   troubleshooting, 423–428
conventional current flow, 11
core loss, *105*, 105–106
counter-electromotive force (CEMF), 12, 212
countervoltage, 212
couplings, 360
   flexible, 360
   rigid, 360

current
  alternating, 9–10, *10*
  direct, 9
  eddy, 106
  flow, 9–11, *11*
    conventional, 11
    direction, 10–11
    electron, 10–11
  full-load current (FLC), 67
  induced, *119*
  locked-rotor, 25
    indicating code letter, 25
  overcurrent, 54
  rating, *24,* 24–26
  readings, *411*
current rating, *24,* 24–26
  of overcurrent protection device, 64
current-source-inverter (CSI) drives, 308–310
current transformer, 72
cylindrical roller bearings, 326

D

damage, 412
damper windings, 142
DC cumulative-compounded motor, 222
DC differential-compounded motor, 222
DC drives, 311–313, *312*
  DC field-current control, 311–313, *312*
DC exciter generator, *145,* 145–148
  shaft-mounted, 146
DC generators, 208, 224–228, *225*
  commutation, *226*
  compound-wound, 228
  construction, 225
  field frame, 209
  operating principles, 225
  series-wound, *227*
  shunt-wound, 227, *228*
  types, 227–228
  voltage regulation, 225
DC motors, *208,* 208–223
  armature, *210,* 210–211
  compound motors, *221,* 221–223
    cumulative-compounded, 222
    overcompounded, 222
  dynamic braking, 272, *273*
  operating principles, 211–215, *212*
  permanent-magnet motor, *223,* 223–224
  re-marking, 437

reversing, 215–217, *216*
series motors, *217,* 217–218
shunt motors, *219,* 219–220
speed control, *310,* 310–313
troubleshooting, 417–420, *421*
  brushes, 417–419, *418*
  commutators, 419
  types, 217–224
DC overcompounded motor, 222
DC permanent-magnet motor, *223,* 223–224
  torque, 224
dead shorts, 430
decelerating circuit logic, 287–288, *288*
design A motor, 40
design B motor, 40, *41*
design C motor, 40, *41*
design D motor, 40, *42*
design L motor, 40
design M motor, 41
design N motor, 41
design O motor, 41
dial indicators, *383,* 383–384
  rod sag, 384
diamagnetic materials, 7
direct current, 9
discharge resistor, 151, *152*
  torque, 158–159, *159*
dowel effect, 375
drives
  AC, 306–310
    closed-loop vector, 307–308
    current-source-inverter (CSI), 308–310
    inverter, 306, *307*
    open-loop vector, 308
    variable-voltage-inverter (VVI), 308–310
    vector, 307
  belt, *350,* 350–355
    centrifugal tension, 350, *351*
    flat belts, 353
    speed, *355*
    tension, 411
    timing belts, 353
    V-belts, 351–352, *352*
  DC, 311–313, *312*
    field-current control, 311–313, *312*
  flexible, *350,* 350–355
  mechanical, 355–360
    chain, 355–357
    gear drives, 357–360
drum switch, 132

dual-voltage motor, 96–97
  delta-connected, three-phase, *97*
  wye-connected, three-phase, *96*
duty cycle, 31, *32*
dynamic braking, 272–274
  AC motor, 273
  DC motor, 272, *273*
  electric motor drive, 273, *274*
  regenerative braking, 272

E

eddy current, 106
eddy-current clutches, 361–362, *362*
Edison-base fuse, *58,* 59
electrical degrees, 87
electrical pitting, 341
electric braking, *269,* 269–271
  DC electric braking circuits, 270, *271*
  limitations, 271
  operation, 270
electric motor drives
  dynamic braking, 273, *274*
electromagnetic overload relay, *71,* 71–72
electromagnetism, 7–9, *8*
  inductance, 11
  mutual induction, 13
  saturation, 8–9
  self-induction, 12
electromagnets, 7–8
  saturation, 8–9
electron current flow, 10–11, *11*
electronic overload relay, *72,* 72–73
electronic reverse dial alignment, 395
enclosure type, 41–45
  open motor enclosure, 42, *43*
  totally enclosed motor enclosure, 42, *44*
energy losses, 104–106, *105*
  bearing loss, *105,* 106
  core loss, *105,* 105–106
  flux-linkage loss, 106
  resistance loss, 105
  sound loss, *105,* 106
  windage loss, *105,* 106
eutectic alloy, 68–69, *69*
exciter
  brushless, 147–148, *148*
exciter generator, 192, *193*
  DC, 145–148
    shaft-mounted, *146*

F

false Brinell damage, 339
feeler gauge, *389*
ferromagnetic materials, 7
field flux. *See* magnetic flux
field frame, 209
   interpoles, 209
field poles, 209
field windings, 193, 209
flat belts, 353
flexible couplings, 360
flexible drives, *350,* 350–355
fluting, 341
flux-linkage loss, 106
foundations, 373
fractional-horsepower motors
   frames, 36, *38*
fractional-horsepower (FHP) V-belts, 351
frame size, 35–41, *37*
   fractional-horsepower motors, 36, *38*
   integral-horsepower motors, 36–38, *39*
frequency rating, 25–26, *26*
fretting corrosion, 340
friction bearings, 327–329, *328*
   journal, 327
   materials, *328*
friction braking, 264–265
   brake shoes, 265
   friction brakes, *264*
   limitation, 265
   solenoid operation, 264, *265*
friction clutches, 361
full-load current (FLC), 67
full-voltage starting, 238
fundamental frequency, 303
fuse, *57,* 58–63
   cartridge fuse, 59
   non-time-delay fuse (NTDF), 60
   plug fuse, *58,* 58–59
   time-delay fuse (TDF), *60,* 60–63

G

galling, 340
gearbox, *359,* 359–360
gear drives, 357–360
   backlash, 358
   couplings, 360
   gear ratio, 357
gear ratio, 357

gears, *359*
   bevel, 358, *359*
   helical, 358, *359*
   herringbone, 358
   spur, 358, *359*
   worm, 358, *359*
generators, 208, 224–228, *225*
   compound-wound, 228
   construction, 225
   field frame, 209
   operating principles, 225
   series-wound, 227
   shunt-wound, 227, *228*
   types, 227–228
   voltage regulation, 225
grease fitting, 332, *333*
grounded circuits
   troubleshooting, 432
ground fault, 55

H

heater, *68,* 68–70, *69, 70*
helical gear, 358, *359*
herringbone gear, 358, *359*
high-inertia load, 101
horsepower, *100*
hysteresis, 106

I

impedance reduced-voltage starting, 246
induced current, *119*
induced soft foot, 379
inductance, 11
induction
   mutual, *13*
   self-, 12
induction motor, *114,* 114–118
   operating principles, 118, *119*
   rotor, 115–117
   slip, 118
   stator, 115
   torque, 118
inductive reactance, 175
inrush current, 25
instantaneous-trip circuit breaker (ITB), 63

insulated gate bipolar transistors (IGBTs), *302,* 302–303
insulation class, 34–35, *36*
insulation spot testing, 440–442, *441*
integral-horsepower motors, 36–38
   frames, *39*
interpoles, 209
interrupting rating
   of overcurrent protection device, 64
inverse-time circuit breaker (ITCB), 63
inverter drives, 306, *307*
   motor torque characteristics, *308*

J

jackscrews, 376, *377*
journal, 327

L

laser rim-and-face alignment, *393,* 393–394
lights-out methods, 199, *200*
linear induction motors, 458, *459*
line frequency, 98
locked in step, 169–170
locked rotor, 238
   current (LRC), 25
   indicating code letter, 25
loss-of-excitation relay, 150, *151*

M

magnetic flux, 4–5, *5*
   flux density, 4–5
   permeability, 6–7, *7*
magnetic motor starter, 239
magnetic polarity, 5
magnetism, 4–9
   diamagnetic materials, 7
   electromagnetism, 7–9, *8*
   ferromagnetic material, 7
   magnetic field, 4
      direction, 13–14
      left-hand coil rule, 13–14, *14*

Index **497**

left-hand conductor rule, 13
left-hand generator rule, 14
motor action, 15
right-hand motor rule, 14, *15*
magnetic flux, 4–5, *5*
magnets, 4–5
molecular theory of, 6
paramagnetic materials, 7
permeability, 6–7, *7*
polarity, 5
retentivity, 4
magnets, *4*, 4–5
permanent, 4
temporary, 4
maintenance
improper, 411–412
manual contactors, 238, *239*
manual starters, 238, *239*
mechanical-design codes, 35–45
enclosure type, 41–45
frame size, 35–41, *37*
motor bearings, 45
NEMA design letter, 38–41, *40*
mechanical drives, 355–360
chain drives, 355–357
roller chain, *355*, 355–356
silent chain, *356*, 356–357
sprockets, *356*
gear drives, 357–360
backlash, 358
couplings, 360
gear ratio, 357
megohmmeters, 439–442
insulation spot testing, 440–442, *441*
interpreting readings, *442*
melting-alloy overload relay, *68*, 68–70, *69*, *70*
microstepping, 453
misalignment, 372
angular, 372
effects, *372*
offset, 372
misalignment wear, 340
molecular theory of magnetism, 6
motion control motors, 452–453
acceleration and deceleration, 454–456, *455*
independent and synchronous control, *456*
motor bearings, 45
motor controls
troubleshooting, 422–432

tie-down troubleshooting method, *422*, *423*
motor efficiency, 31–33, *33*, 104
motor leads
identification, 432–438
determining wye or delta connection, 433
re-marking DC motors, 436–438, *437*
re-marking dual-voltage, delta-connected motors, 436, *437*
re-marking dual-voltage, wye-connected motors, *434*, 434–435, *435*
motor-generator (M-G) set, 146, *147*
motors
AC
troubleshooting, 412–417, *413*
capacitor, 176–181
capacitor-run, 179–181, *180*, *181*
capacitor-start, 178–179, *179*
capacitor start-and-run, 181
construction, 176–177
operating principles, 178
phase angle between windings, *178*
consequent-pole, *282*, 282–284
circuits, 285–288
connections, *283*
damage, 412
DC, *208*, 208–223
armature, *210*, 210–211
compound motors, *221*, 221–223
field frame, 209
operating principles, 211–215, *212*
permanent-magnet motor, *223*, 223–224
re-marking, 437
reversing, 215–217, *216*
series motors, *217*, 217–218
shunt motors, *219*, 219–220
speed control, *310*, 310–313
troubleshooting, 417–420, *421*
types, 217–224
efficiency, 104
failure, *404*, 404–412
improper application, 410
improper maintenance, 411–412
improper power supply, 406–408
overheating, 405
overloading, 410, *411*
induction, *114*, 114–118
operating principles, 118
inverter duty, 300–301
lead length, 304–306, *305*

linear induction, 458, *459*
matching to load, 100–101, *101*
motion control, 452–453
acceleration and deceleration, 454–456, *455*
independent and synchronous control, 456
multispeed, 282–288
placement of, 373–382
Schrage, 284–285, *285*
servomotors, 454
acceleration and deceleration, 454–456, *455*
independent and synchronous control, 456
shaded-pole, 172–174, *173*
construction, 172
operating principles, 172–174
rotating magnetic field, *174*
single-phase, *168*
construction of, 168, *169*
dual-voltage, 170, *171*
locked in step, 169–170
operating principles, 169–174
poles, *170*
reversing, 171, *172*
stator windings, *169*
split-phase, *175*, 175–176
operating principles, 176
speed control, 176, *177*
starting, 238–255
method comparison, 254–255, *255*, *256*
stepper, *452*, 452–454
acceleration and deceleration, 454–456, *455*
independent and synchronous control, 456
microstepping, 453
operation, 453
synchronous, *140*, 140–165
DC exciter generator, *145*, 145–148
motor-generator (M-G) set, 146, *147*
operating principles, 148–159
power factor, 160–165
starting, 152–155, *153*, *154*
synchronous condenser, 161
torque characteristics, *308*
types, 101–102
constant-horsepower, 101, *102*
constant-torque, 101, *102*
variable-torque, 102

universal, 456–458, *457*
  compensated, 457
  noncompensated, 457
  reversing, 458
wound-rotor, *126*, 126–133
  operating principles, 130–133
  regulator, 132–133
  slip rings, 128
  speed control, 131
  starter, 131–132, *132*
  starting, 130–131
  torque, 130–131, *131*
wye-connected, 248
wye-delta, 251
motor shaft, 86
motor starters
  troubleshooting, 423–424, *424*
mutiple-speed motors, 101
mutual induction, 13

N

nameplate, 22
  ambient temperature rating, 34, *35*
  current rating, *24*, 24–26
  duty cycle, 31, *32*
  frequency rating, 25–26, *26*
  insulation class, 34–35, *36*
  motor efficiency, 31–33, *33*
  operating ratings, 27–35
  phase rating, 26–27, *27*
  power rating, 27–29, *28*
  service factor rating, 30
  speed rating, 30–31, *31*
  usage rating, *29*, 29–30
  voltage rating, 22–24, *23*
  wiring diagram, *89*
narrow V-belt, 351
needle bearings, 327
NEMA design letter, 38–41, *40*
  design A motor, 40
  design B motor, 40, *41*
  design C motor, 40, *41*
  design D motor, 40, *42*
  design L motor, 40
  design M motor, 41
  design N motor, 41
  design O motor, 41
neutral plane, 213
non-adjustable-trip circuit breaker (NATCB), 63

non-time-delay fuse (NTDF), 60
noncompensated universal motor, 457
nonparallel misalignment, *392*

O

offset misalignment, 372, *392*
oil cup, *333*, 334
oil seal, *334*, 334–336, *335*
oil wick, *333*
open circuits
  troubleshooting, 429–430
    switches, *429*
open-circuit transitions, 239, *240*
open-loop vector drives, 308
open motor enclosure, 42, *43*
operating ratings, 27–35
operating time rating, 31, *32*
out-of-step relay (OSR), 150–151, *151*
overcurrent, 54, *57*
overcurrent protection device (OCPD), *56*, 56–64
  circuit breaker (CB), *61*, 61–63
    operation, *62*
  current rating, 64
  fuse, 58–61
  interrupting rating, 64
  voltage rating, 64
overcycling, 410, *411*
overheating, 405
  improper ventilation, *406*
overload, 55
  protection, *67*, 67–73
overloading, *404*, 410, *411*
overload relays, 54, 67
  bimetallic-strip overload relay, 70
  electromagnetic overload relay, *71*, 71–72
  electronic overload relay, *72*, 72–73
  eutectic alloy, 68–69, *69*
  melting-alloy overload relay, *68*, 68–70, *69*, *70*
  overload class ratings, 68
  thermal overload relay, 68–70

P

parallel soft foot, 378, *379*
paramagnetic materials, 7

part-winding starting, 247–248
  circuit, 247
  wye-connected motors, 248
permanent magnets, 4
permissible temperature rise, 34, *35*
phase rating, 26–27, *27*
phase sequence testers, 438–439, *439*, *440*
phase unbalance, 406, *407*
pigtail, 128, 210
piping strain, 373, *374*
pitting, 341
plug fuse, *58*, 58–59
  Edison-base fuse, *58*, 59
  type S fuse, *58*, 59
plugging, 265–268, *266*
  continuous plugging, 266, *267*
  for emergency stops, 266–267, *267*
  limitations, 269
  plugging switch, *266*
    operation, 266
  using timing relays, 267–269, *268*
polarized field frequency relay (PFFR), *149*, 149–150
  operation, *150*
pole-face windings, 210
power, 103–104
  apparent, 103
  rating, 27–29, *28*
  reactive, 103
  true, 103
power factor, 104, *160*, 160–165
  correction, 104, 161, *162*
power factor correction capacitor, 104, *105*
power rating, 27–29, *28*
pressure cup, 332, *333*
press fit, 330, *331*
primary-resistor starting, 240–242, *241*
  circuit, 242
prime movers, 198
projecting pole, 86
pulleys, 353–354, *354*
  belt speed, 355
  misalignment, *392*
pull-in torque, 156–157, *157*
pull-out torque, *157*, 157–158
pulse-width modulation (PWM), *302*, 302–303
push fit, 330

R

race, 325
rack and pinion gear, *359*
radial loads, 324
reactance
  inductive, 175
reactive power, 103
reduced-voltage starting, 239–253, *240*
  autotransformer starting, *244*, 244–246
    circuit, 244–246, *245*
  closed-circuit transitions, 239, *240*
  DC, 242–243
    circuit, 243
  impedance reduced-voltage starting, 246
  open-circuit transitions, 239, *240*
  part-winding starting, 247–248
    circuit, 247
    wye-connected motors, 248
  primary-resistor starting, 240–242
  solid-state starting, 251–253, *252*
    circuits, 251–253, *252*
  wye-delta starting, *249*, 249–251
    circuit, 250
reduced-voltage starting circuits
  troubleshooting, 426, *427*
regenerative braking, 272
regulators
  for wound-rotor motors, 132–133
    drum switch, 132
    silicon-controlled rectifier (SCR), 133
    three-pole rheostat, *133*
relays, 149–151
  loss-of-excitation relay, 150, *151*
  out-of-step relay (OSR), 150–151, *151*
  polarized field frequency relay (PFFR), *149*, 149–150
  operation, *150*
reluctance torque, 155, *156*
resistance loss, 105
resistors, 128–130, *129*
  discharge resistor, 151, *152*
retentivity, 4
reverse dial alignment, *394*, 394–395
  electronic reverse dial alignment, 395
rigid couplings, 360
rim-and-face alignment, 392–394, *393*
  laser rim-and-face alignment, *393*, 393–394
rod sag, 384
roller bearings, 325, 326, *327*
roller chain, *355*, 355–356
rolling-contact bearings, 324–327
  race, 325
rotary phase converters, 459–460, *460*
rotor, 168
  construction, 85–86, *86*
  frequency, 88
  of induction motor, 115–117
  NEMA design code letters, 117
  resistors, 128–130, *129*
  salient pole, 86
  squirrel-cage, 115–116, *116*
  starting, *153*
  synchronization, *143*
  of synchronous motor, *142*, 142–145
    amortisseur windings, 142
    high-speed, 143, *144*
    low-speed, 144, *145*
  of wound-rotor motor, *127*, 127–130
  wye-wound, *127*
rotor frequency, 88
runout, 378

S

salient pole, 86
saturation, 106
Schrage motors, 284–285, *285*
SCR armature-voltage control, 313
self-excited shunt field, 219
self-induction, 12
sensorless vector drives, 308
separately excited shunt field, 219
series-wound generator, 227
series motors, *217*, 217–218
  torque, 218
service factor rating, 30
servomotors, 454
  acceleration and deceleration, 454–456, *455*
  independent and synchronous control, 456
shaded-pole motors, 172–174, *173*
  construction, 172
  operating principles, 172–174
  rotating magnetic field, *174*
shading coil, 172
shaft-deflection method, 380
sheaves, 353
shim stock, 376–378, *377*
short circuits, 54, 55
  dead shorts, 430
  troubleshooting, *430*, 430–431
  testing, *431*
shorting ring, 116
shunt motors, *219*, 219–220
  self-excited shunt field, 219
  separately excited shunt field, 219
  stabilizing field winding, 220
  torque, 220
shunt-wound generator, 227, *228*
silent chain, *356*, 356–357
silicon-controlled rectifier (SCR), 133
single phasing, 408, *409*
single-phase motors, *168*
  construction of, 168, *169*
  dual-voltage, 170, *171*
  locked in step, 169–170
  operating principles, 169–174
  poles, *170*
  reversing, 171, *172*
  stator windings, *169*
  troubleshooting, 414–417
    capacitor motors, *416*, 416–417
    shaded-pole, *414*
    split-phase, 414–415, *415*
single-voltage motor, 89–94
  delta-connected, *93*
  wye-connected, *90*
slip, 30, *31*
  in induction motors, 118
slip ring, 128
soft foot, 378–381, *379*
  angular, 378–379
  induced, 379
  measuring, *380*, 380–381
    at-each-foot method, 381
    shaft-deflection method, 380
  parallel, 378
  springing, 379
soft starter, 253
soft starting, 253
  circuit, *254*
solid-state starter, 251, *252*
solid-state starting, 251–253, *252*
  circuits, 251–253, *252*
  soft starter, 253
  soft starting, 253
    circuit, *254*
sound loss, *105*, 106
spalling, 338
speed rating, 30–31, *31*

slip, 30, *31*
split-phase motors, *175,* 175–176
   operating principles, 176
   speed control, 176, *177*
springing soft foot, 379
sprockets, 356
spur gear, 358, *359*
squirrel-cage rotor, 115–116, *116*
stabilizing field winding, 220
standard V-belt, 351
starting, 238–255
   full-voltage, 238
   method comparison, 254–255, *255, 256*
   primary-resistor, 240–242, *241*
     circuit, 242
   reduced-voltage, 239–253
     DC, 242–243, *243*
   small-motor, *238*
starting current, 25
stator, *82, 85,* 168
   construction, 82–85
   core, *83*
   of induction motor, 115
   rotating field, 88–89
     in delta-connected motors, *95*
     in wye-connected motors, *92*
   of synchronous motor, 141
   windings, *83, 84*
   of single-phase motor, *169*
   of wound-rotor motor, 126
stepper motors, *452,* 452–454
   acceleration and deceleration, 454–456, *455*
   independent and synchronous control, 456
   microstepping, 453
   operation, 453
   trajectory control, 452
straightedge alignment, *389,* 389–392, *390*
surface reaction, 338
switches
   centrifugal, 175
   troubleshooting, *429*
synchronous condenser, 161
synchronous motor, *140,* 140–165
   accessories, 149–151
   operating principles, 148–159
   power factor, 160–165
     correction, 161, *162*
   relays, 149–151
   rotor, *142,* 142–145

   amortisseur windings, 142
   high-speed, 143, *144*
   low-speed, 144, *145*
   rotor synchronization, *143*
   starting, 152–155
     circuit operation, 153–155, *154*
     rotor starting, *153*
   stator, 141
   synchronous condenser, 161
   torque, 155–158
synchronous speed, 87–88
synchronous torque, 157–158
synchroscope, 199

T

tapered roller bearings, 326
taper gauge, *389*
temperature
   internal motor temperature protection, 71
temperature rise, 34
   permissible, 34
temporary magnets, 4
test tools, 438–442
   megohmmeters, 439–442
     insulation spot testing, *441*
     interpreting readings, *442*
   phase sequence testers, 438–439, *439, 440*
thermal expansion, 381–382, *382*
thermal overload relay, 68–70
thermal switches
   troubleshooting, 428
three-phase motors
   construction, 82–86
   housing, 85
   motor shaft, 86
   multiple-speed, 101
   operating principles, 87–98
   reversing direction, 98, *99*
   rotor construction, 85–86, *86*
   stator construction, *82,* 82–85, *83, 85*
   troubleshooting, 412–413, *413*
   twelve-lead, 97, *98*
three-phase power, 89
   rotary phase converters, 459–460, *460*
three-pole rheostat, 133
thrust damage, 340, *341*
thrust loads, 324
tie-down troubleshooting method, 422, *423*
time-delay fuse (TDF), *60,* 60–63

timing belts, 353
torque, 99–100, 155–158, 191, *213,* 213–214
   angle, 158
   breakdown, 100
     in compound motors, 222
     in DC permanent-magnet motor, 224
   discharge resistor, 158–159, *159*
   full-load, 100
     in induction motors, 118
   locked-rotor, 99–100, *100*
   pull-in torque, 156–157, *157*
   pull-out torque, *157,* 157–158
   pull-up, 100
   reluctance torque, 155, *156*
     in series motors, 218
     in shunt motors, 220
   synchronous torque, 157–158
   of wound-rotor motor, 130–131, *131*
torque angle, 158
totally enclosed motor enclosure, 42, *44*
trajectory control, 452
troubleshooting
   AC motors, 412–417
   circuit faults, 428–432
   contactors, 423–424
   control circuits, 423–428
   DC motors, 417–420, *421*
     brushes, 417–419, *418*
     commutators, 419
   grounded circuits, 432
   insulation spot testing, 440–442, *441*
     interpreting readings, *442*
   motor controls, 422–432
     tie-down troubleshooting method, 422, *423*
   motor lead identification, 432–438
     determining wye or delta connection, 433
     re-marking DC motors, 436–438, *437*
     re-marking dual-voltage, delta-connected motors, 436, *437*
     re-marking dual-voltage, wye-connected motors, *434,* 434–435, *435*
   motor starters, 423–424, *424*
   open circuits, 429–430
   switches, *429*
   reduced-voltage starting circuits, 426, *427*
   short circuits, *430,* 430–431
     testing, *431*

single-phase motors, 414–417
   capacitor motors, *416,* 416–417
   shaded-pole, 414
   split-phase, 414–415, *415*
  with test tools, 438–442
   megohmmeters, 439–442, *441, 442*
   phase sequence testers, 438–439, *439, 440*
   thermal switches, 428
  three-phase motors, 412–413, *413*
true power, 103
twelve-lead, three-phase motors, 97, *98*
type S fuse, *58,* 59

U

universal motors, 456–458, *457*
  compensated, 457
  noncompensated, 457
  reversing, 458
usage rating, *29,* 29–30

V

V-belts, 351–352, *352*
  fractional-horsepower (FHP) V-belts, 351
  narrow V-belts, 351
  pulleys, *353,* 353–354
   belt speed, *355*
  standard V-belt, 351
  wedging action, *352*
variable-voltage-inverter (VVI) drives, 308–310
vector drives, 307
  motor torque characteristics, *308*
ventilation
  improper, *406*
voltage
  changing in adjustable-speed drive, *296*
  control
   chopper armature-voltage control, 313
   SCR armature-voltage control, *313*
  control of applied, 306
  full-load voltage drop, *297*
  and motor lead length, *305*
  rating, 22–24, *23*
  unbalance, 23
voltage rating, 22–24, *23*
  of overcurrent protection device, 64
voltage regulation, 225
voltage surge, 408, *409*
voltage unbalance, 23, 406, *407*
volts-per-hertz ratio (V/Hz), *301,* 304

W

waveforms
  sawtooth, 311, *312*
windage loss, *105,* 106
windings, 82
  construction, *83, 84*
worm gear, 358, *359*
wound-rotor motor, *126,* 126–133
  operating principles, 130–133
  regulator, 132–133
  rotor, *127,* 127–130
   wye-wound, *127*
  slip rings, 128
  speed control, 131
  starter, 131–132, *132*
  starting, 130–131
  stator, 126
  torque, 130–131, *131*
wound-rotor motor regulator, 132–133
wound-rotor motor starter, 131–132, *132*
wye-delta starting, *249,* 249–251
  circuit, 250

USING THE MOTORS CD-ROM

Before removing the CD-ROM from the protective sleeve, please note that the book cannot be returned for refund or credit if the CD-ROM sleeve seal is broken.

System Requirements

To use this Windows®-compatible CD-ROM, your computer must meet the following minimum system requirements:
- Microsoft® Windows Vista™, Windows XP®, Windows 2000®, or Windows NT® operating system
- Intel® Pentium® III (or equivalent) processor
- 256 MB of available RAM
- 90 MB of available hard-disk space
- 800 × 600 monitor resolution
- CD-ROM drive
- Sound output capability and speakers
- Microsoft® Internet Explorer 5.5, Firefox 1.0, or Netscape® 7.1 web browser and Internet connection required for Internet links

Opening Files

Insert the CD-ROM into the computer CD-ROM drive. Within a few seconds, the home screen will be displayed allowing access to all features of the CD-ROM. Information about the usage of the CD-ROM can be accessed by clicking on USING THIS CD-ROM. The Quick Quizzes®, Illustrated Glossary, Flash Cards, Motor Animations, and ATPeResources.com can be accessed by clicking on the appropriate button on the home screen. Clicking on the American Tech web site button (www.go2atp.com) accesses information on related educational products. Unauthorized reproduction of the material on this CD-ROM is strictly prohibited.

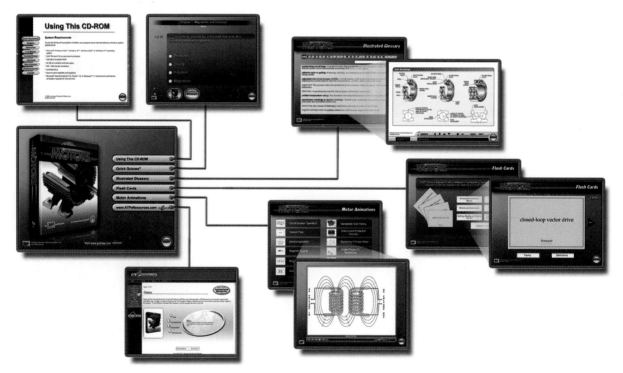

Intel and Pentium are registered trademarks of Intel Corporation or its subsidiaries in the United States and other countries. Microsoft, Windows Vista, Windows XP, Windows 2000, Windows NT, and Internet Explorer are either registered trademarks or trademarks of Microsoft Corporation in the United States and/or other countries. Adobe, Acrobat, and Reader are either registered trademarks or trademarks of Adobe Systems Incorporated in the United States and/or other countries. Netscape is a registered trademark of Netscape Communications Corporation in the United States and other countries. Quick Quiz and Quick Quizzes are registered trademarks of American Technical Publishers, Inc. All other trademarks are the properties of their respective owners.